MECHANICAL BEHAVIOR OF MATERIALS

MECHANICAL BEHAVIOR OF MATERIALS

KEITH BOWMAN
Materials Engineering Department
Purdue University

JOHN WILEY & SONS, INC.

ACQUISITIONS EDITOR	Joseph Hayton
EDITORIAL ASSISTANT	Mary Moran
MARKETING MANAGER	Katherine Hepburn
PRODUCTION EDITOR	Patricia McFadden
SENIOR DESIGNER	Karin Gerdes Kinchebe
PRODUCTION MANAGEMENT SERVICES	Argosy Publishing

This book was set in Times by Argosy Publishing Services and printed and bound by Hamilton Printing Company. The cover was printed by Lehigh Press.

This book is printed on acid-free paper. 

ISBN 0-471-24198-9
WIE ISBN 0-471-45231-9

Printed in the United States of America

10 9 8 7 6 5 4 3 2 1

PREFACE

GOALS AND FEATURES

The idea for this book began with efforts at Purdue to teach a materials engineering curriculum focused on common themes that occur across all types of materials. Writing the text came about due to frustration with existing texts that seemed to be trying to serve as both handbooks and teaching texts. Many of these texts had grown quite large, offered little integration between the chapters, and had little or no historical perspective. The present text attempts to remedy those shortcomings. Concurrently, a strong interest in applying the quantitative tools undergraduate students have learned (and debunking a common assertion that there is not much math in materials science) led me to write some problems that utilize math software tools.

I have always been surprised at the inability of engineering graduate students and practicing engineers (of all disciplines) to handle three-dimensional deformation problems, even though they are responsible for modeling production processes or component performance using sophisticated software tools. Much of the material in Chapters 2 and 3 was developed to improve these skills. Another goal in writing the text was to try and improve on the figures available to represent important materials phenomena. Other texts have relied on just a few sources for figures representing dislocations and fracture. I hope that some of the perspective drawings and gray-scale representations in this text help improve the situation.

Audience

The text came about from notes used in the Purdue Mechanical Behavior course MSE 382, a required course that occurs in the 6th semester of our materials engineering curriculum, and a senior elective for other engineers. It is also appropriate as introductory material for graduate-level mechanical behavior classes on fracture, high-temperature deformation mechanisms, and deformation processing. We often teach these classes on television and enrollment can include engineers from any discipline.

Organization and Use

The book can be utilized in several ways. Educators using the text can go straight through the first seven chapters as we do in the one-semester junior-level course at Purdue. Then Chapters 8–10 can be used as supplements if there is time. If you would like to avoid the challenge of fourth-rank tensors, skipping the latter sections of Chapter 2 can be done without any critical losses to later content. If you are teaching a senior/graduate course that is geared toward deformation processing Chapters 1–5, and Chapter 10 can provide the course foundation. If you are teaching a senior/graduate level course on deformation mechanisms, Chapters 1–6 and Section 8.1 should fulfill most requirements for such a course. I was extremely fortunate in getting the permission to reprint a good number of the deformation mechanism maps from an out-of-print book by Frost and Ashby. If you are teaching a course on fracture and want to avoid much discussion of microscale deformation mechanisms, Chapters 1–3, 7, and Section 8.2 on fracture mechanism maps can provide most of the necessary materials.

ACKNOWLEDGMENTS

This book would not have been possible without the family, students, colleagues, and mentors that have brought me to this point. My first thanks go to Ron Gibala, who through a period of many years saw things in me that I hadn't yet seen myself. His example has been a critical one that I have tried to apply wisely. The other students and faculty at Case Western Reserve University (CWRU), the Materials Science and Engineering department, the faculty in Western Reserve College, the volunteers at the campus radio station, WRUW, and the folks in Housing all made some indelible marks that I can't seem to lose and would never want to. The professionalism and enthusiasm of the materials faculty at CWRU, especially Terry Mitchell, Arthur Heuer, Alfred Cooper, Bob Hehemann, Jack Wallace, and Lynn Ebert, had a profound influence on what you see here. These folks helped prepare three CWRU students for a journey northward to the University of Michigan. The other two of those students, Ron Noebe and Jim Fekete, were both major influences on all of the research and teaching I have participated in and I use what I learned from them to help my students. The trip to Ann Arbor probably wouldn't have been worth taking if J. Wayne Jones, T. Y. Tien, and William Hosford had not been there and if Bob Pehlke, Larry Van Vlack, and Bill Leslie had not shown so much skepticism in what would come of it all.

In the field of ceramics, I first have to thank I-Wei Chen for helping guide my first efforts and providing important support along the way. Others, including Arthur Heuer (again), Fred Lange, Manfred Rühle, Paul Becher, Raj Bordia, Gary Messing, and Kathy Faber have showed that the best lead by example. A special group from the National Institute of Standards and Technology (NIST), Carol Handwerker, Shelley Wiederhorn, Ed Fuller, Mark Vaudin, and John Blendell have all been valuable contributors to my understanding and enjoyment of ceramic materials.

In the field of texture and anisotropy, I would remain disoriented without the inspired energies of Hans Bunge and Rudy Wenk. They both helped get my career into the right "space." Others in the texture community that have strongly influenced the presentation on directional properties and orientation include Heinz Mecking, John Jonas, Werner Skrotzki, Stuart Wright, and Günter Gottstein.

Major portions of the text were tried out and written at the University of Technology, Darmstadt, while I was on two sabbaticals at Jürgen Rödel's institute, Nichtmetallische Anorganische Werkstoffe, in Darmstadt, Germany. Both came about with the financial help of the best academic foundation in the world, the Von Humboldt Foundation. The care and feeding of this visiting professor by Jürgen, his students, his staff (viz. Emil, Roswitha, and Marion) and family (Daniel, Karen, and Ling) has been an important part of my life and I hope they all remain part of it.

The text was finished while on sabbatical at the University of New South Wales (Sydney, Australia) in the School of Materials Science and Engineering. Mark Hoffman and his wife Carmen graciously provided the right environment for the completion of the text, and the excellent library at the University of New South Wales proved invaluable.

Purdue University undergraduate and graduate students have been key to the writing for the text and many have contributed to the descriptions and explanations. In particular, many Purdue graduate students have been important in shaping many of the ideas and presentations here. Of these many students, special thanks go to Robert Moon, Keith Kruger, Ryan Roeder, and Farnjeng Lee for their contributions to the figures and problems in this book.

The folks at Wiley have been especially supportive and I appreciate their help. The initial confidence of the late Cliff Robichaud started the whole thing, and encouragement from Wayne Anderson and Charlie Dresser sustained me through to the end. Nancy Kotary from Argosy has been a joy in nailing down the manuscript details. I also need to thank all the other publishers and authors who gave permissions for use of figures and data or helped in locating particular figures. In particular I would like to thank Naoki Kondo, Alain Kounga, Robert Moon, Michael F. Ashby, and David Green.

David Johnson receives special recognition for his serving as "guinea pig" teacher from the text in MSE 382. His willingness to do so, along with his input and support in the several classes we have taught together were critical to getting this project completed. He and the other Purdue MSE faculty (Bob Spitzer, Rod Trice, Eric Kvam, Alex King, and Matt Krane) have also been very supportive. The person who has dealt with it the most, Cheryl Waller, typed and retyped major portions of the document. She also astounds me with her patience. Cheryl, Donna Bystrom, and Vicki Cline have made every day I work at Purdue a bit easier.

David Gaskell, Jerry Liedl, C. T. Sun, and Mysore Dayananda have all been there from the start of my Purdue career. Each has had a major influence on how I think about research and education. Elliott Slamovich and Kevin Trumble are incredible colleagues and friends. Elliott, Kevin, and their families have been the best part of living by the Wabash. Other victims of hearing about the textbook too much include my friends Gwendolyn, Denise, Judi, Diana, Joan, Darlene, Sheena, Diane, Mike, and Brenda. I am sure they are relieved!

CONTENTS

CHAPTER 1

INTRODUCTION

THE MECHANICAL responses and failure processes described in this book employ the principles of physics and mathematics learned through the second year of university for science and engineering students. Any additional background in materials science and mechanics will also help. Materials concepts for crystalline and noncrystalline materials spanning the mechanical processes of elastic distortion, permanent or time-dependent deformation, and failure are enriched with a background in chemistry and the microstructures of materials. A distinct effort is made to consider these elements of mechanical behavior without narrowly focusing on metals, ceramics, polymers, and composites, but on the similarities and differences in mechanical response within and between the material classes.

The goal of this chapter is to "level the playing field" so that we all start with a basic foundation and begin to consider the distinctions among elasticity, plasticity, and rupture of materials. Many of these concepts may have been presented previously to the student, so the goal here is to present a brief but reasonably complete review. Many of the concepts are presented in the introductory materials science and mechanics of materials textbooks listed at the end of this chapter.

The reader should remember that mechanical properties are only some of the considerations required to design with materials. The design process is not simply a process of materials selection but an optimization process wherein the materials selection and design are best made in concert. The alternative is a process of rigid design with materials selection (or substitution) as an afterthought. An understanding of mechanisms for mechanical behavior is essential to applications of new materials and new designs using established materials.

Nearly all the mechanisms described in this book were established within the twentieth century. For this reason, the publication of this book at the beginning of the twenty-first century affords an opportunity to identify the people who proposed important concepts, and when they did so. Short biographical or historical comments are presented throughout the book to identify many of these individuals. At the beginning of the twentieth century, the connections of crystal or molecular structure with elasticity, plasticity, and fracture were debated using ideas that now may seem absurd. The passage of time often allows us to believe we are smarter than our ancestors.

Recent intense interest in the tragedy of the Titanic, owing in part to the discovery of the wreckage and in part to a popular movie, has driven abundant analyses of the fracture and, in particular, the "inferiority" of the construction materials and the design. Nearly every time I ask, "Why did the Titanic sink?" some student enthusiastically responds with a description of sulfur in the steel or a similar technical answer. My response is that it is unwise always to blame the material. After all, no iceberg, no problem.

1.1 STRAIN

The last definition of strain in most dictionaries is typically something similar to: "a deformation of a material body through the action of applied forces." In real components, the strains can vary with position as a result of gradients of the forces or stresses acting on the material. The strains can be elastic or plastic, and they may vary with time. Types of strains include linear strains, shear strains, and volumetric or *dilatant* strains. To allow for this, we must be able to define strain as occurring at a point. This description of strain enables us to establish equivalency between distortions in components or structures that are size-independent. We also need to describe strain of three-dimensional objects in three dimensions. The next several sections develop these concepts for defining strain (see also Kelly, Groves, and Kidd, 2000).

1.1.1 One-Dimensional Strain

Consider a linear object such as the worm shown in Fig. 1.1. Assign an arbitrary origin 0 and a point M—the distance between M and 0 is x. When the worm stretches uniformly, point M moves to M'. Then the distance between the two points is $x + u$, where x represents position and u represents displacement. In this instance, the displacement results in a tensile strain. Unlike displacement, which can be given in distance units of inches or millimeters, strain is dimensionless. The same tensile strain in worms of different sizes results in different displacements. If the displacement of the worm is reversed, the displacement on this scale is negative, resulting in a compressive strain.

Deformation of $\overline{MN}$, where N is another point located near the head of the worm and away from M, gives an engineering strain

$$e_{\text{eng}} = \frac{\Delta u}{\Delta x} = \frac{\overline{M'N'} - \overline{MN}}{\overline{MN}}$$

The most familiar treatment of e_{eng} is in the context of a tensile test with

$$e_{\text{eng}} = \frac{\Delta l}{l} = \frac{\text{Change in length}}{\text{Original length}} \tag{1.1}$$

Strain is conventionally reported as a fraction (e.g., 0.05) or in percent (e.g., 5%).

The engineering strain should be used only if the deformation strains are small in magnitude (e.g., $e_{\text{eng}} < 0.10$) or only limited precision is required. Calculations of strains for

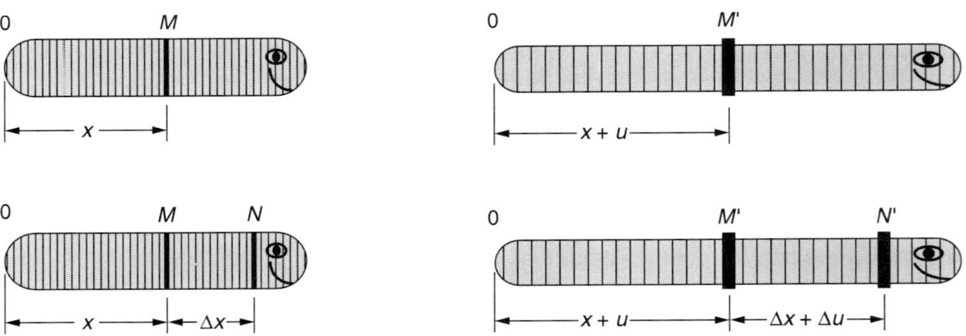

FIG. 1.1 Stretching of a worm, demonstrating linear strain.

multiple-step forming processes or for operations with precise requirements on final dimensions require a measure of the instantaneous or *true* strain. This requires consideration of strain at a point. Then, what is the strain at point M on the worm? The true strain is

$$e_{\text{true}} = \frac{du}{dx} = \lim_{x \to 0} \frac{\Delta u}{\Delta x}$$

In terms of a conventional tensile test,

$$e_{\text{true}} = \int_{l}^{l_f} \frac{dl}{l} = \ln \frac{l_f}{l} = \ln \frac{\text{Final length}}{\text{Original length}} = \ln\left(1 + e_{\text{eng}}\right) \tag{1.2}$$

For small or elastic strains, $e_{\text{eng}} \approx e_{\text{true}}$, so we will often use e_{eng} for elastic strains, but for large plastic strains, e_{true} is better. We can assign reference coordinates to the strain for the worm by calling the linear strain e_{x} or e_{xx}. The term e_{x} is shorthand that is often used for strain e_{xx}. The term e_{xx} represents the displacement in the x-direction of an element initially parallel to x. This strain is often also designated as e_1 or e_{11}, wherein the subscript represents the first axis of an x, y, z (or 1, 2, 3) right-handed coordinate system.

1.1.2 Multiaxial Strain

The worm in Fig. 1.1 also undergoes a change in diameter or diametral strain in addition to the strain changing its length. Truly one-dimensional strains occur under very complex loading conditions. A systematic approach for representing strains in three dimensions must be employed to describe shape change. We can get most of the way there by considering two-dimensional or *plane strain* conditions. Plane strain is the condition wherein a body undergoes no strain in one direction. Deformation then occurs in the plane normal to this direction.

Fig. 1.2 shows permanent deformation of an initially rectangular sheet of material. In addition to orthogonal tensile or compressive strains, strains that generate shear distortion become possible. After deformation, M and N move to M' and N', and displacement in the x-direction is again u and displacement in the y-direction is again v. The full set of coordinates is then

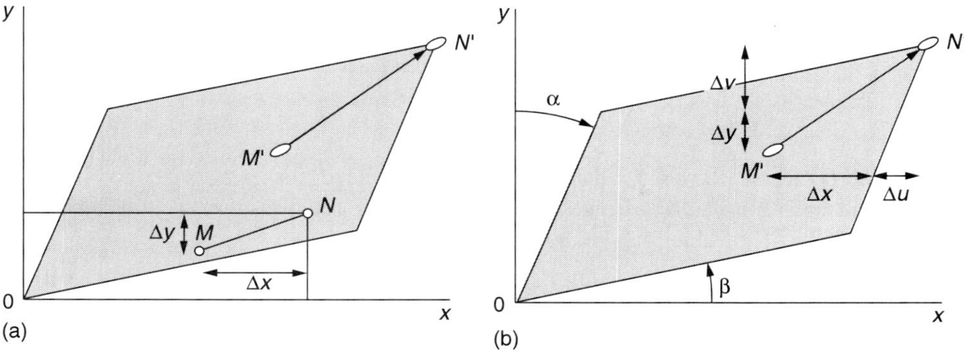

FIG. 1.2 (a) The deformation of the shaded region changes the relative positions of points M and N to the prime positions. After strain, the region and the circles surrounding the points become distorted, as shown by the shaded region. (b) The respective x- and y- displacements are defined by Δu and Δv. The angles corresponding to the shear deformation are defined as α and β.

$$M \rightarrow (x, y) \qquad M' \rightarrow (x+u, x+v)$$
$$N \rightarrow (x+\Delta x, y+\Delta y) \qquad N' \rightarrow (x+u+\Delta x+\Delta u, y+v+\Delta y+\Delta v)$$

These expressions become too complicated to express unless we define them in terms of strains. To express displacements in strain, we assume that the displacements are small. Otherwise, the specific dimensions that are lost in converting to strains will be necessary for a reasonable description of the shape change. The four parameters that separately describe the types of shape changes in Fig. 1.2 are the strains

$$e_{11} \text{ or } e_{\text{xx}} = \frac{\Delta u}{\Delta x} \qquad e_{12} \text{ or } e_{\text{xy}} = \frac{\Delta u}{\Delta y}$$
$$e_{21} \text{ or } e_{\text{yx}} = \frac{\Delta v}{\Delta x} \qquad e_{22} \text{ or } e_{\text{yy}} = \frac{\Delta v}{\Delta y}$$

If limits are taken, these strains can be represented by a single term describing multiaxial strain at a point as

$$e_{\text{ij}} = \frac{du_{\text{i}}}{dx_{\text{j}}} \tag{1.3}$$

or

$$du_{\text{i}} = e_{\text{ij}} dx_{\text{j}}$$

The subscript i represents the direction of displacement, and j is related to the direction of the original length that was distorted by the deformation. The total displacements can be written in terms of the original position and these strains by

$$u = e_{11}x + e_{12}y$$
$$v = e_{21}x + e_{22}y$$

We can then separate the components of the strain. The strains e_{11} and e_{22} are the tensile or compressive strains in the respective x- and y-directions. The shear strains represent distortions that include a combined shape change and rotation. The shear strains can be represented as angles of rotation by the tangent of the angular change. For small angular changes,

$$\begin{aligned} \tan\alpha &\approx \frac{\Delta u}{\Delta y} = e_{12} \approx \alpha \\ \tan\beta &\approx \frac{\Delta v}{\Delta x} = e_{21} \approx \beta \end{aligned} \tag{1.4}$$

where α and β are expressed in radians. Unfortunately, the tensor e_{ij} measures not only distortions and rotations resulting from deformation, but also rotations that take place without any deformation. The values can be defined as a matrix or *tensor*

$$e_{\text{ij}} = \begin{vmatrix} e_{11} & e_{12} \\ e_{21} & e_{22} \end{vmatrix}$$

We can separate this tensor e_{ij} into

$$e_{\text{ij}} = \varepsilon_{\text{ij}} + \omega_{\text{ij}}$$

where

$$\varepsilon_{ij} = \frac{1}{2}\left(e_{ij} + e_{ji}\right)$$

and

$$\omega_{ij} = \frac{1}{2}\left(e_{ij} - e_{ji}\right)$$

Consider that

$$\varepsilon_{ij} = \varepsilon_{ji} = \frac{1}{2}\left(e_{ij} + e_{ji}\right) = \frac{1}{2}\left(e_{ji} + e_{ij}\right)$$

so that ε_{ij} is a symmetric matrix producing deformation or shape change. The tensor ω_{ij} is antisymmetric, producing rotation of a rigid body. This tensor representation of strains can be applied if the strains are small. The separate components of ε and ω are given in matrix form as

$$e_{ij} = \begin{vmatrix} e_{11} & \frac{1}{2}\left(e_{12} + e_{21}\right) \\ \frac{1}{2}\left(e_{21} + e_{12}\right) & e_{22} \end{vmatrix} + \begin{vmatrix} 0 & \frac{1}{2}\left(e_{12} - e_{21}\right) \\ \frac{1}{2}\left(e_{21} - e_{12}\right) & 0 \end{vmatrix} \tag{1.5}$$

The simple shear strain e_{ij}, with $i \neq j$, is also often represented by the Greek letter γ. This strain includes both the shape change and rotation associated with motion of dislocations in solids or the gradients of deformation in fluids subjected to shear stress. The separate components of strain and rotation are defined in Fig. 1.3.

To add the third dimension to a definition of strain, the process follows naturally from the definition given for plane strain. The three displacements are defined relative to the initial position as

$$u = e_{11}x + e_{12}y + e_{13}z$$
$$v = e_{21}x + e_{22}y + e_{23}z$$
$$w = e_{31}x + e_{32}y + e_{33}z$$

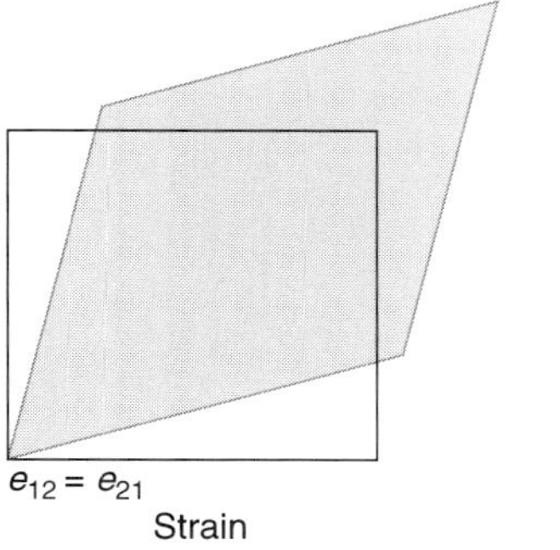

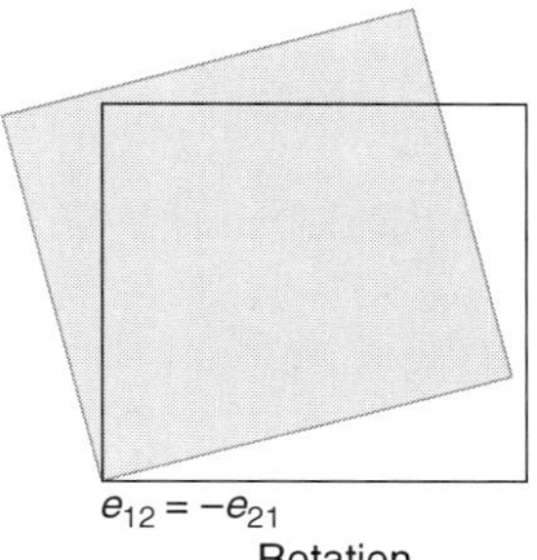

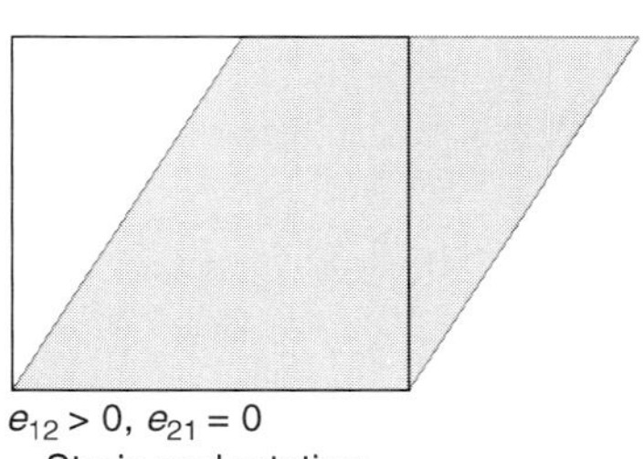

FIG. 1.3 Examples of strain, rotation, and combined strain and rotation.

and then

$$e_{ij} = \begin{vmatrix} e_{11} & e_{12} & e_{13} \\ e_{21} & e_{22} & e_{23} \\ e_{31} & e_{32} & e_{33} \end{vmatrix}$$

A three-dimensional tensor for shape change can be defined as

$$\varepsilon_{ij} = \frac{1}{2}\left(\frac{du_i}{dx_j} + \frac{du_j}{dx_i}\right) \quad (i, j = 1, 2, 3) \tag{1.6}$$

with

$$\varepsilon_{13} = \frac{1}{2}(e_{13} + e_{31})$$

and

$$\omega_{13} = \frac{1}{2}(e_{13} - e_{31})$$

The strains ε_{31} and ω_{31} have equivalent definitions. A three-dimensional tensor for shape change is then

$$\varepsilon_{ij} = \begin{vmatrix} e_{11} & \frac{1}{2}(e_{12} + e_{21}) & \frac{1}{2}(e_{13} + e_{31}) \\ \frac{1}{2}(e_{21} + e_{12}) & e_{22} & \frac{1}{2}(e_{23} + e_{32}) \\ \frac{1}{2}(e_{31} + e_{13}) & \frac{1}{2}(e_{32} + e_{23}) & e_{33} \end{vmatrix} \tag{1.7}$$

EXAMPLE 1.1 *Strain and Displacement*

Consider a deck of playing cards that following removal from the package has been arranged so that the stack of cards is "sheared" at an angle of 30° (see Fig. 1.4). We will define the direction of shear as x, and the plane parallel to the cards is normal to the z-direction. Because a deck of 54 cards (including jokers) is about 15 mm tall, we can find how far the top card is displaced from the horizontal position of the bottom card. The simple shear, including rotation, can be defined as

$$\gamma = e_{13} \approx \tan 30° = 0.58$$

Then the top card, which is at Δz = 15 mm above the bottom card, is displaced in the x-direction by

$$\Delta u = 0.58(15 \text{ mm}) = 8.7 \text{ mm}$$

The shape strain for the stack of cards is

$$\varepsilon = \begin{vmatrix} 0 & 0 & \frac{1}{2}(0.58 + 0) \\ 0 & 0 & 0 \\ \frac{1}{2}(0 + 0.58) & 0 & 0 \end{vmatrix} = \begin{vmatrix} 0 & 0 & 0.29 \\ 0 & 0 & 0 \\ 0.29 & 0 & 0 \end{vmatrix}$$

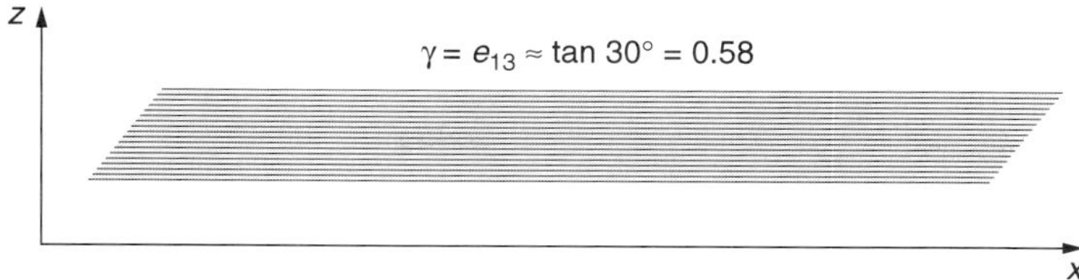

FIG. 1.4 "Sheared" stack of cards.

The pure rotation (often called rigid body rotation) undergone by the stack of cards is

$$\omega = \begin{vmatrix} 0 & 0 & \frac{1}{2}(0.58-0) \\ 0 & 0 & 0 \\ \frac{1}{2}(0-0.58) & 0 & 0 \end{vmatrix} = \begin{vmatrix} 0 & 0 & 0.29 \\ 0 & 0 & 0 \\ -0.29 & 0 & 0 \end{vmatrix}$$

The sum $\varepsilon + \omega$ returns that $e_{13} = 0.58$ and all other $e_{ij} = 0$. The cards are, of course, separate objects that can readily be returned to their original state. Because strains are differential quantities, large strains are most accurately calculated from summing a series of small displacements that produce the desired total strain. If we perform the same deformation in increments of 5° and sum them, the total strain is $e_{13} \approx 0.52$. ■

1.2 STRESS

Just as strain provides a measure of displacements that scale to the dimensions of the deformed object, stress provides a measure of force scaled to the area over which the force operates. An additional important distinction between force and stress is that stress is manifested by a balance of forces that does not result in acceleration. Stresses that result in extension, or tensile stresses, are defined as positive, and compressive stresses are negative in sign.

1.2.1 Stress Definitions

The simplest definition of stress is a measure of force F applied over a part with an area A_o before deformation, as shown in Fig. 1.5*a*. This gives the definition of the *engineering stress* as

$$\sigma_{\text{eng}} = \frac{F}{A_o} \tag{1.8}$$

for F applied normal to a cross section of magnitude A_o before application of the force. The engineering stress is often employed for elastic stresses or stresses for components deformed to small plastic strains. At large strains, the change in cross-sectional area significantly alters the actual stresses.

The *true stress* is

$$\sigma_{\text{true}} = \frac{F}{A} \tag{1.9}$$

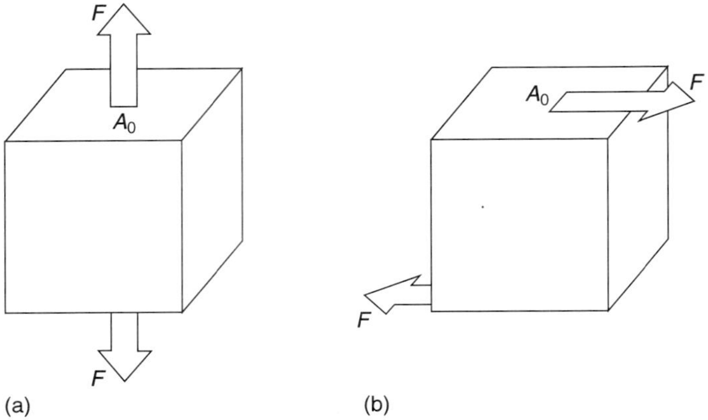

FIG. 1.5 (a) A normal tensile stress and (b) a state of shear stress (without the equal and opposite force couple required to prevent rotational acceleration).

where A is the instantaneous area. The true stress can also be given in terms of a limit for small elements of area. Often an assumption of constant volume can be made so that

$$A_i l_i = A_f l_f$$

where A_i is the initial area, l_i is the initial length, A_f is the final area, and l_f is the final length. The true stress can then be defined in terms of the engineering stress and engineering strain as

$$\sigma_{\text{true}} = \sigma_{\text{eng}}\left(1 + e_{\text{eng}}\right)$$

1.2.2 Multiaxial Stresses

A shear stress is easily represented by a pair of force couples, of the type shown in Fig. 1.5*b*. If all forces are equal, there should be no net acceleration. The shear stress is defined by a force applied within the planar faces of area A.

The *shear stress* is

$$\tau = \frac{F}{A} \tag{1.10}$$

where F is a force couple applied across A (remember that an additional couple with an opposite sense is required to prevent rotational acceleration).

Most mechanical testing strategies are designed to make the relation between applied stress and the resulting strains as simple as possible. Few load-bearing components undergo only simple uniaxial stresses. Often the stress state is at least biaxial and varies with position in the part. For a defined, infinitesimal location within a component, the stress state can be defined using the two-dimensional or plane stress element shown in Fig. 1.6. Although most stress states are three-dimensional, many of the examples examined in elementary mechanics are given in terms of plane stress for simplicity. Stress states, including shear in three dimensions, can be expressed using nine terms just as we have seen for strain. Fortunately, we will demonstrate in Chapter 2 that even the most complex stress states can be simplified to three normal tensile or compressive stresses if the proper orientation frame is chosen.

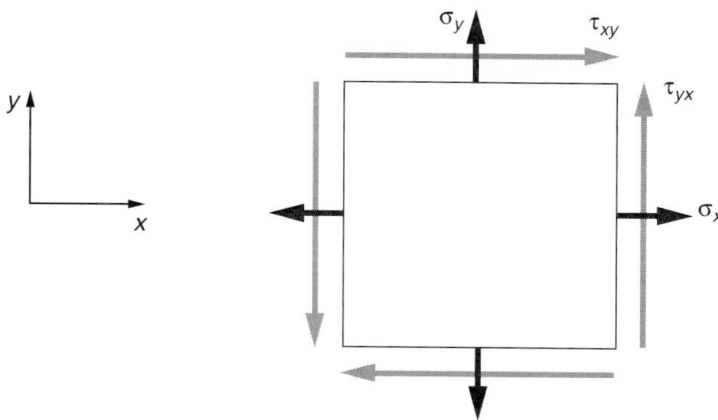

FIG. 1.6 Two-dimensional element showing combination of normal and shear stresses for a material loaded in plane stress where $\sigma_x = \sigma_{11}$ and $\sigma_y = \sigma_{22}$.

For *plane stress,* a stress state wherein no force is present normal to the plane, the stresses can be defined as a tensor using

$$F_1 = \sigma_{11}A_1 + \sigma_{12}A_2$$
$$F_2 = \sigma_{21}A_1 + \sigma_{22}A_2$$

where F_i is the vector of force components and σ_{ij} is the stress components. The A_i values are vector expressions for area with a magnitude defined by the size of the area and the direction defined by its normal. This expression can be written in a shortened form as

$$F_i = \sigma_{ij}A_j \tag{1.11}$$

where

$$\sigma = \begin{vmatrix} \sigma_{11} & \sigma_{12} \\ \sigma_{21} & \sigma_{22} \end{vmatrix}$$

A complete discussion of three-dimensional stress states is given in Chapter 2.

EXAMPLE 1.2 *Pure Shear*

Consider the forces $F = 10N$ applied (Fig. 1.7*a*) to the sides of a cube with faces of area $A = 10\ \text{mm}^2$. The stresses are then

$$\sigma = \begin{vmatrix} 0 & \dfrac{10N}{10\ \text{mm}^2} \\ \dfrac{10N}{10\ \text{mm}^2} & 0 \end{vmatrix} = \begin{vmatrix} 0 & 1 \\ 1 & 0 \end{vmatrix} \text{MPa}$$

Unlike strain, which can include symmetric and asymmetric tensor components, the shear components for stress must be symmetric. If not, the loaded component will undergo rotational acceleration. If the same stresses are applied to the same size cube rotated by 45°, the stresses in the new coordinate system can be determined by resolving the force components on the rotated cube shown in Fig. 1.7*b*. The stresses at the new orientation are

$$\sigma' = \begin{vmatrix} 1 & 0 \\ 0 & -1 \end{vmatrix} \text{MPa}$$

We will demonstrate transformations from rotated coordinate systems in Chapter 2. ■

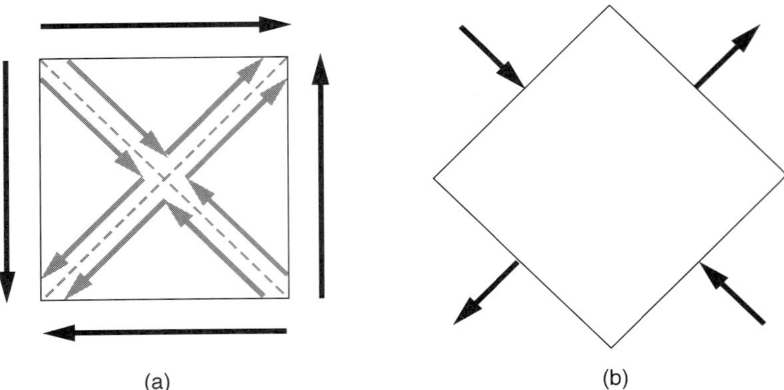

FIG. 1.7 Applied shear stresses in black can be resolved into normal components applied across the corners of these elements. For an element rotated 45°, these stresses have the same magnitude, but are now normal stresses.

1.3 MECHANICAL TESTING

Assessment of mechanical properties is part of materials research, development of materials for application, quality control, and forensic analysis of failed components or structures. Simplicity and reproducibility are important considerations in the design of any test. Additionally, it is important to have standards for testing that make it possible to compare results from different sources. Organizations that maintain these properties include the American Society for Testing and Materials (ASTM), the German Standards Bureau (DIN), the Japanese Industrial Standard (JIS), and similar groups from other countries and international organizations. Professional societies, manufacturers, and governmental agencies also establish guidelines for mechanical performance and testing. Sample sizes and designs, measurement equipment, and testing parameters are part of these standards. The standards are established by demonstrating that the guidelines produce essentially the same results in round-robin studies. Round-robin studies certify that the results found using the guidelines on the same batch of materials submitted to different testing laboratories are comparable.

1.3.1 The Tension Test

Tension tests are probably the most common form of mechanical test performed on ductile materials to evaluate the resistance to deformation and failure of materials. These tests are performed using one of several types of mechanical testing machines similar to that shown in Fig. 1.8*a*. All modern mechanical testing machines apply force through a force transducer, called a load cell, and gripping mechanisms. Most tensile specimens have a reduced central section that is uniform in cross section over a region called the gage length, as shown in Fig. 1.8*b*. Strain is either inferred from known displacements of the machine or determined by directly attaching displacement transducers called extensometers or strain gages to the specimen gage section. The gage section geometry is often standardized as shown in Fig. 1.9. Over some temperature ranges, many materials are sensitive to the rate of deformation. Consequently, the strain rate $de/dt = \dot{e}$ (sec^{-1}) is an important parameter in many mechanical tests.

Data recorded in the tension test are the specimen dimensions, including the gage length and cross-sectional area, the force measured by the load cell, and the displacement

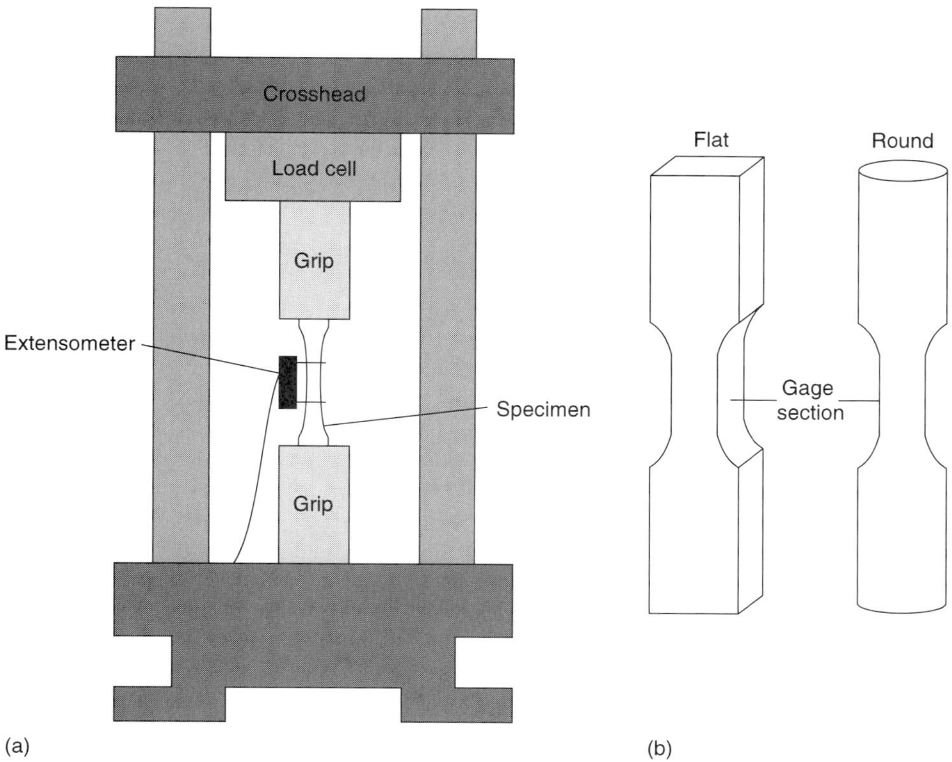

FIG. 1.8 (a) Schematic of screw-driven mechanical testing machine and (b) examples of two different tensile specimen geometries.

or strain from the beginning to the end of the test. The appearance and dimensions of the fracture surface are often documented following the tension test.

Compression tests are conducted on specimens with square, rectangular, or round cross sections compressed between hard platens. Friction effects affect the accuracy of compression tests; however, compression tests can allow deformation to large strains without fracture.

1.3.2 Bending

Any imposed bending moment during tensile testing (e.g., from misalignment) and surface finish can affect the stresses at which materials fracture. These effects are magnified in

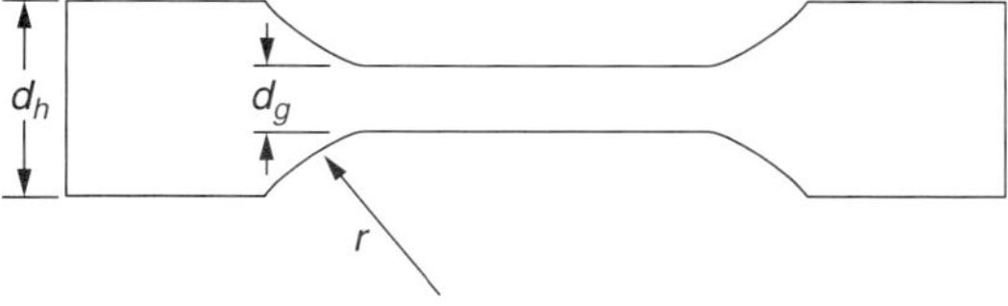

FIG. 1.9 Round tensile test specimen geometries for ASTM and other standards specify the radius of curvature r for reduction from the head diameter d_h to the gage diameter d_g. The ratios of r and d_h to d_g can be used to specify dimensions for a specimen of any size. Dimensions of standard specimens with square cross sections have similar relations.

brittle ceramics, glasses, and polymers. Additionally, the added expense of making tensile specimens, as well as gripping problems, often result in the use of bending as a mode for mechanical testing.

Three- and Four-Point Bending Two bending geometries are common for low-temperature and high-temperature strength and fracture toughness measurements. Three- and four-point loading geometries are shown in Fig. 1.10. In strength tests, the maximum tensile stress in the bend specimen at failure is often termed the *modulus of rupture*. For elastic loading, the expressions for the maximum stress are given as

$$\sigma_{\max} = \frac{Plh}{8I} \tag{1.12}$$

for three-point bending and

$$\sigma_{\max} = \frac{Plh}{4I} \tag{1.13}$$

for four-point bending. In both equations, P is the applied force, l is the spacing between the two outer loading points in three-point bending, h is the specimen height, and I is the moment of inertia. In four-point bending, the inner loading points are designated by l. The moment of inertia for bend specimens with uniform rectangular cross sections is given by

$$I = \frac{wh^3}{12} \tag{1.14}$$

where w is the sample width. The stress varies linearly across the specimen height for specimens with rectangular cross sections. For a circular cross section, the diameter d is used for the moment of inertia

$$I = \frac{\pi d^4}{64} \tag{1.15}$$

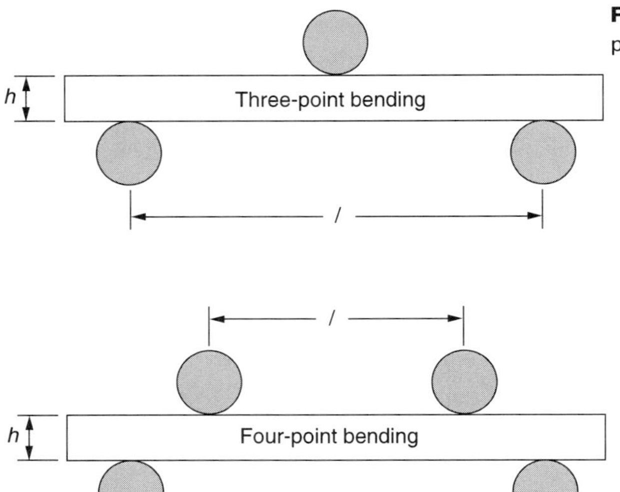

FIG. 1.10 Geometry for three-point and four-point bend tests.

The moment of inertia allows collection of all terms describing the cross section. Mechanics texts or engineering handbooks provide moments of inertia for other cross-sectional shapes. Problems at the end of the chapter also include other expressions for moment of inertia.

As shown in Fig. 1.11, for three-point bending the tensile stress is greatest at the specimen surface opposite to the central loading point because the greatest bending moment occurs there. At this point, the sample has the smallest radius of curvature. For a rectangular cross section, the stress varies linearly through the thickness. In four-point bending, the portion of the specimen between the inner loading points has a uniform radius of curvature or constant bending moment. Consequently, the entire portion of the specimen between these loading points has the same stress distribution. Because bending tests load the sample through a range of stresses with only a small volume at the maximum stress, comparisons of fracture results from modulus of rupture tests on specimens tested in three-point bending or four-point bending or tension tests should be made with care. Also, because the expressions for bending given here apply only to elastic loading, they are not applicable to plastic deformation.

1.3.3 Hardness Tests

Hardness tests measuring the penetration resistance of materials are quick and efficient ways to estimate strength, and thus they are frequently employed for screening and quality control considerations. Other hardness tests, such as the scratch (Moh's) hardness test, are qualitative evaluations. Conventional hardness tests involve recording the depth or area of an indentation made into the surface of a material by a hard spherical, conical, or pyramidal indenter. Most of the tests are conducted to obtain an empirical number allowing relative comparisons of materials. Only comparisons between similar materials have much meaning, because the combination of indentation and load is typically specific to the class of materials. The four most common hardness tests performed on metals and ceramics are Brinell, Rockwell, Vickers, and Knoop tests.

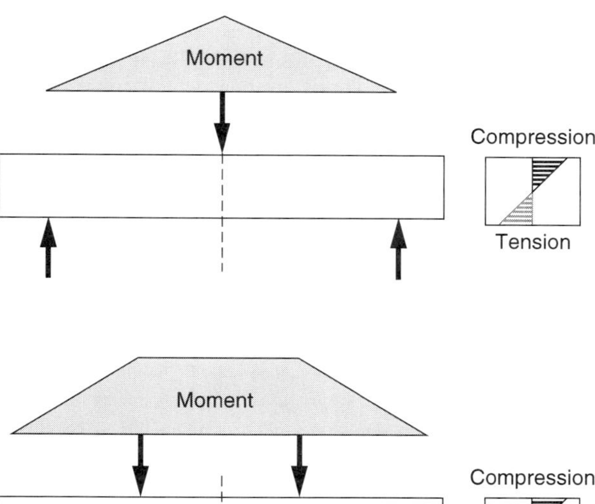

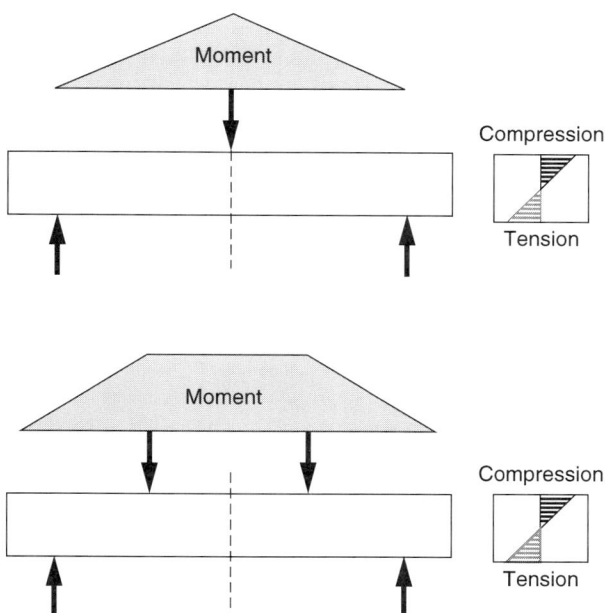

FIG. 1.11 Moment diagrams and stress distributions along the beam length at the centers of rectangular specimens in three- and four-point bending.

The Brinell test, in use since about 1900, indents surfaces with a 10-mm-diameter ball at a load P of 3000 kg for hard metals and 500 kg for soft metals for 30 seconds. The diameter of the hardness indentation is measured with a low-power microscope to give the Brinell hardness number as

$$\text{BHN} = \frac{P}{(\pi D/2)\left(D - \sqrt{D^2 - d^2}\right)} = \frac{P}{\pi D t} \tag{1.16}$$

where P is load in kilograms, D is ball diameter in millimeters, d is indentation diameter in millimeters, and t is impression depth in millimeters. Although it might make more sense to give the BHN in units of stress, the standard is in kilograms per square millimeter.

The Rockwell hardness tests are a series of measurements commonly employed in industry for consistency, speed, and resolution. The depth of the indentation, as shown in Fig. 1.12, is the parameter that distinguishes the Rockwell hardness. Through a combination of varying loads, depths, and indenter shapes, the Rockwell hardness numbers are determined as shown in Table 1.1.

The two microhardness tests shown in Table 1.1, Vickers and Knoop, employ sharp diamond indenters to make small indentations that are measured using a microscope. This allows measurement of spatial variations in hardness in polished specimens, including the hardnesses of phases or the cross sections of hardened surfaces.

The Vickers test uses a square pyramid diamond indenter to measure the hardness. The microhardness number obtained in this test is the Vickers Hardness Number (VHN) or Diamond Pyramid Hardness (DPH). The VHN is calculated from measurements of the diagonals L across the indentation and the applied load P in kilograms by

$$\text{VHN} = \frac{1.72P}{L^2} \tag{1.17}$$

The Vickers test is also employed to estimate fracture toughness from cracks that form in brittle materials. Indentation toughness measurements are discussed in Chapter 7.

Determination of hardness on very small size scales can be accomplished using a microhardness test at low loads. Gradients in hardness near carburized or decarburized surfaces and the hardnesses of specific phases are often measured using a Knoop indenter.

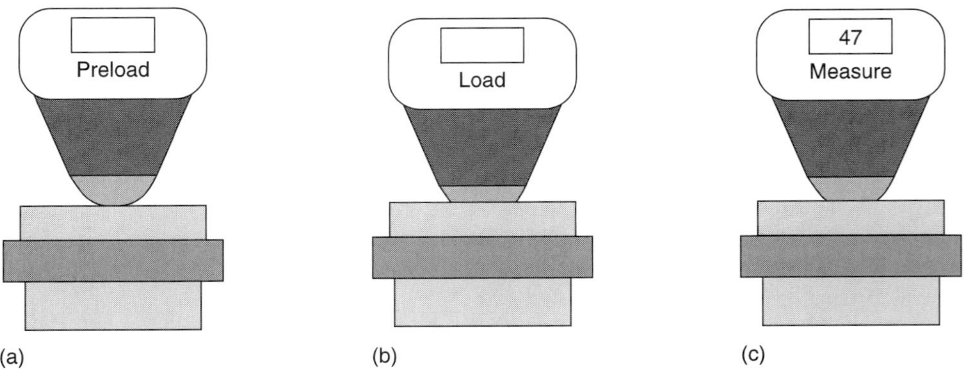

FIG. 1.12 The Rockwell hardness test employs an initial load (a) to locate the specimen at a small applied load. The specified load (b) is applied for a fixed time. At the end of this time, the specimen is unloaded (c) to the initial load for measurement of the new ball position. The final depth t is recorded in microns.

TABLE 1.1 Parameters for the Most Common Rockwell Hardness Tests and the Corresponding Indenter Shapes*

Test	Indenter	Side view	Top view	Load
Vickers	Diamond pyramid	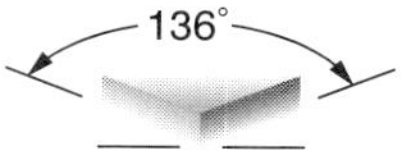 136°	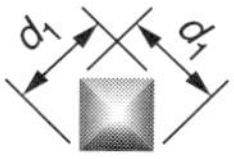 d_1, d_1	P
Knoop microhardness	Diamond pyramid	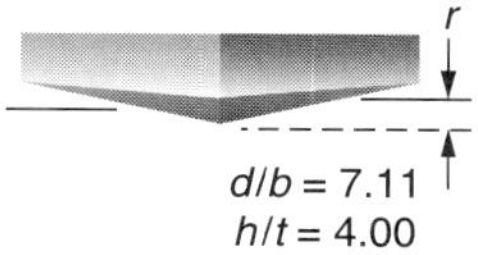 r; $d/b = 7.11$; $h/t = 4.00$	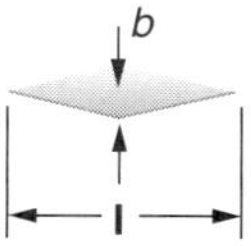 b	P
Rockwell				
A, C, D	Diamond cone	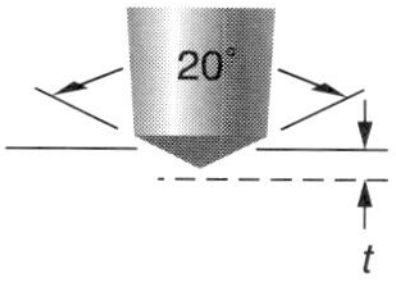 20°; t		60 kg, 130 kg, 100 kg
B, F, G	$\frac{1}{16}$-in.–diameter steel sphere	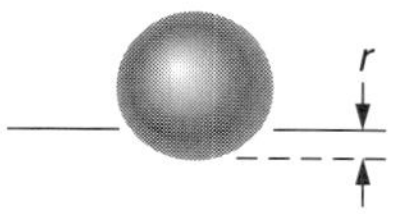 r	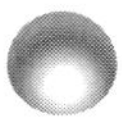	100 kg, 60 kg, 150 kg
E, H	$\frac{1}{4}$-in.–diameter steel sphere			100 kg, 60 kg

Source: Adapted from Callister, 1996, Wiley, used with permission.

*The A, C, and D tests are calculated using $100–500t$ = Rockwell number, with t being depth of penetration in microns. The B, F, G, E, and H tests are calculated using $130–500t$ = Rockwell number, with t being depth of penetration in microns.

The Knoop indenter is a pyramidal diamond indenter that produces an indentation with long and short diagonals in a ratio of 7:1. The Knoop Hardness Number (KHN) is given as

$$\mathrm{KHN} = \frac{CP}{L^2} \tag{1.18}$$

where P is load in kilograms, L is long diagonal length in millimeters, and C is a constant for a specific indenter, typically 14.2.

1.3.4 Fracture Toughness

The resistance of a material to the propagation of a crack under a stress applied normal to the crack plane is defined as the fracture toughness K_{Ic}. The theoretical value of the fracture toughness of a brittle material is defined as

$$\mathrm{K_{Ic}} = \sqrt{EG} \tag{1.19}$$

where E is Young's modulus and G is called the critical strain energy release rate. For a brittle material, $G = 2\gamma$, where γ is the surface energy. For ductile materials, $G > 2\gamma$. Direct determination of solid surface energies is very difficult. Thus, mechanical tests are conducted on samples with known preexisting flaws. The specimen and loading geometries

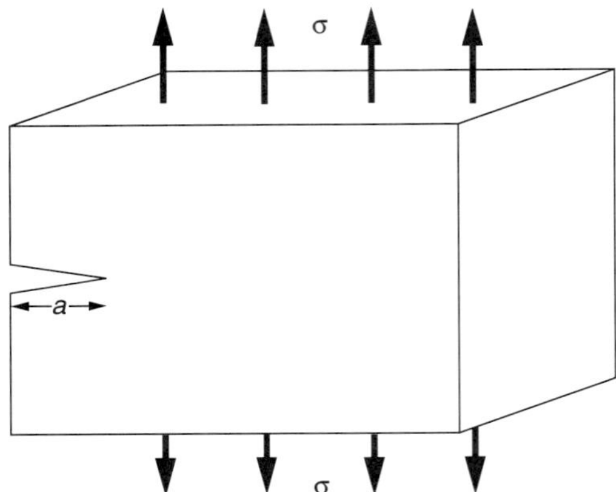

FIG. 1.13 Geometry for plane strain fracture toughness test.

are very important in fracture toughness tests. The calculation for obtaining K_{Ic} for a specimen with the geometry shown in Fig. 1.13 uses the equation

$$K_{\mathrm{Ic}} = \beta\sigma\sqrt{\pi a} \tag{1.20}$$

where β is a correction factor corresponding to geometry, σ is the applied stress, and a is the crack length. For the specimen in Fig. 1.13, the correction factor is $\beta = 1.1$. The basis for these relationships and the geometric factors for different specimen shapes are given in Chapter 7.

1.3.5 Creep and Stress Relaxation

Mechanical testing at elevated temperatures, called creep testing, consists of time-dependent deformation at a fixed load or stress. The specimen is usually heated in a furnace and a load is applied by means of a set of weights. Creep tests can be conducted in tension, compression, and bending. Strain is documented versus time, as shown in Fig. 1.14*a*. The strain rates at different temperatures or stress levels are then compared to predict lifetimes of parts.

The complement to creep testing is stress relaxation testing. Stress relaxation tests are conducted by loading a specimen to a fixed strain in a mechanical testing machine. The decay in load with time is termed stress relaxation. Stress relaxation is shown in Fig. 1.14*b*. Both creep and stress relaxation are discussed further in Chapter 6.

1.3.6 Cyclic Deformation and Fatigue

Many failures occur in moving parts resulting in distinctive fractures indicative of fatigue (see Fig. 1.15). These parts can be subjected to repeated tension, compression, or bending. The loading may also consist of alternating tension and compression or bending in alternating directions. Many distinct mechanical tests have been developed to simulate cyclic loading that leads to failure in fatigue. Tests in specimens that have intentional flaws are performed to evaluate crack propagation in fatigue. Tests conducted on smooth specimens can be used to evaluate crack initiation or just the number of cycles to failure.

Failure of components by fatigue in aircraft structures almost never occurs despite the fact that microscopic flaws and cracks may form early in the duty cycle. This excellent safety record persists because research has been conducted to understand the allowable sizes for these flaws without catastrophic failure, and we have a good understanding of crack propagation rates. Cyclic loading and failure by fatigue are discussed in Chapter 9.

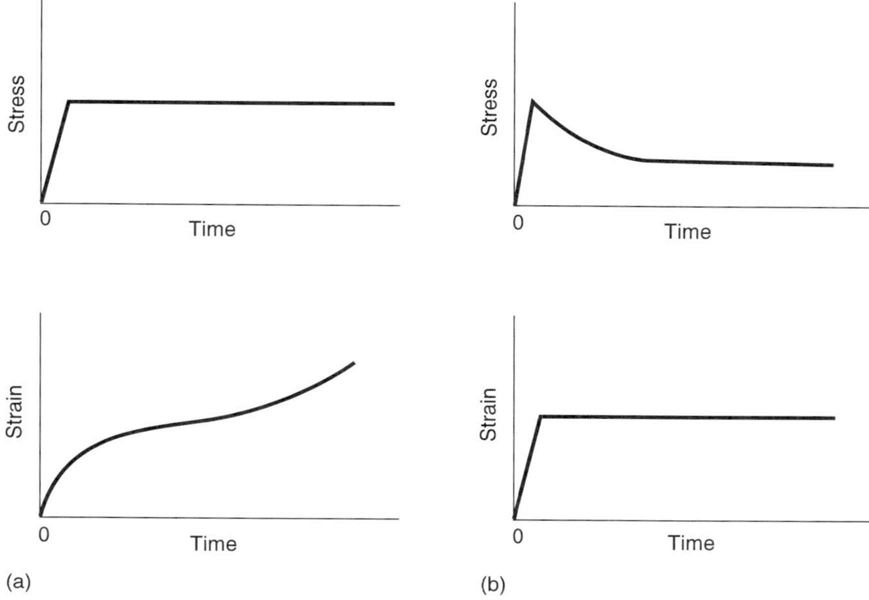

FIG. 1.14 (a) Hypothetical creep test showing that at constant load the specimen undergoes creep. (b) Hypothetical stress relaxation data showing that when held at a fixed strain the specimen undergoes a load that decreases with time.

FIG. 1.15 Fatigue failure in service of a 1050 steel shaft subjected to rotating bending (ASM Handbook, Vol. 12: Fractography, ASM International, Materials Park, OH 44073-0002, Fig. 218, page 269, with permission of ASM International®). The crack propagated from the surface to the interior and the final failure occurred in the dark region on the left. By this picture, final failure occurred when only about 25% of the surface was attached.

1.4 MECHANICAL RESPONSES TO DEFORMATION

Mechanical properties of most materials are dependent on the microstructure inherited from processing, the test temperature, and often the loading rate. For that reason, relying only on property tables published in handbooks or product literature without some consideration of microstructure can be risky. Elastic properties of most metals and ceramics are not as greatly influenced by processing and microstructural changes as are the elastic

TABLE 1.2 Approximate Elastic Properties of Various Materials (and Dilute Alloys) at Room Temperature

Material	Young's modulus, E (GPa)	Shear modulus, μ (GPa)	Poisson's ratio, ν
Aluminum and aluminum alloys	69–72	24–26	0.35
Copper and copper alloys	125–135	47–50	0.34
Irons and steels	205–215	80–84	0.29
Stainless steels	190–200	75–78	0.33
Titanium and titanium alloys	115	42–44	0.32
Aluminum oxide	380–390	155–165	0.25
Silicon carbide	440–460	195–200	0.14
Glass	70–90	28–32	0.27
Polyethylene (PE)	0.2–2	*	0.4
Polymethylmethacrylate (PMMA)	2–3	*	0.4
Polystyrene (PS)	2–4	*	0.35
Bone**	5–30	3–8	0.25–0.5
Tendon**	0.8–1.5	—	—

*These values are typically about one-third of Young's modulus.

**Bone and tendon are not only strongly anisotropic, but the stiffness also varies with type, position, and moisture content.

properties of polymers and composites. Table 1.2 provides a brief set of room-temperature elastic properties. Table 1.3 provides a brief set of room-temperature plastic deformation properties. The ranges given for these data signify the approximate upper and lower bounds expected for conventional processing. Although product literature is often reasonably accurate (with a slight optimistic leaning), any subsequent heating or deformation of the material can change the properties.

TABLE 1.3 Typical Deformation Properties of Various Materials at Room Temperature

Material	Yield stress (MPa)	Elongation (%)	Reduction in area (%)	Strain hardening exponent, n
Aluminum alloys	80–600	10–20	10–30	0.1–0.25
Copper alloys	100–800	5–29	10–60	0.5
Iron alloys and steels	200–1800	10–50	15–60	0.1–0.2
Stainless steels	250–600	20–50	10–40	0.4–0.6
Titanium alloys	200–1200	5–15	20–30	*
Aluminum oxide	5000	≈ 0	≈ 0	*
Silicon carbide	10,000	≈ 0	≈ 0	*
Glass	3000–4000	≈ 0	≈ 0	*
Polyethylenes (PE)	5–50	50–1000	—	**
Polymethylmethacrylate (PMMA)	20–50	3–6	—	**
Polystyrene (PS)	30–80	1–3	—	**

*These materials fail without plastic deformation in tension. Given values are for compression.

**The stress-strain behavior does not normally fit Eq. 1.25.

1.4.1 Elasticity

For isotropic materials, the elastic constants can be defined very simply. Elastic properties of materials are based on an assumption of small strains and linear stress-strain relationships. Although the elastic properties are considered to have no time dependence, time-dependent values are used to describe noncrystalline polymers and glasses. The familiar elastic properties defined in strength of materials are as follows.

Young's Modulus Young's modulus is expressed as

$$E = \frac{\sigma}{e} \tag{1.21}$$

where σ is an applied tensile or compressive stress and e is an elastic tensile or compressive strain in the direction of applied stress. This relationship holds as long as this is the only stress applied to the material. If the loading is more complex, additional elastic constants are required. The units for Young's modulus are typically pascals, N/m^2, with most materials exhibiting values of 1 to 400×10^9 Pa or 1 to 400 GPa (gigapascals) at room temperature. In the English system, the unit is pounds per square inch (psi). One easy way to remember the conversion *and* know the approximate value of Young's modulus for steel at room temperature is to memorize the relationship $E = 210$ GPa $\approx 30 \times 10^6$ psi. The relative values of Young's modulus for various materials can be found from the stress-strain behavior shown in Fig. 1.16.

Shear Modulus The shear modulus[1] is expressed as

$$\mu = \frac{\tau}{\gamma} \tag{1.22}$$

where τ is an applied shear stress and γ is the resulting elastic simple shear. From the definition of simple shear, $\gamma_{12} = e_{12} = 2\varepsilon_{12}$. The resistance to elastic torsion (as undergone by an axle or driveshaft) is expressed as

$$\mu = \frac{ML}{I\theta}$$

BIOGRAPHY

THOMAS YOUNG (1773–1829)

Young was a doctor of medicine whose most famous discovery involved light interference. Young was encouraged to quit as a lecturer at the Royal College because students rarely understood his lectures. His writing was so difficult that not until 100 years later did Lord Rayleigh provide a complete assessment of his works. His word description of what we now call Young's modulus is much more complex than Eq. 1.21 (see Timoshenko, 1953).

[1]The letter G is also used to represent shear modulus. In order to avoid confusion with the use of G to also represent the strain energy release rate, μ is employed throughout this text as the shear modulus.

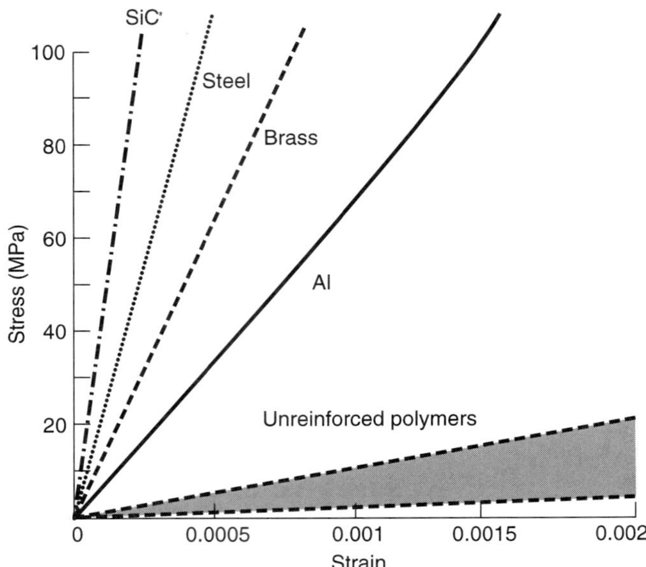

FIG. 1.16 Linear stress-strain behavior at small strains for various materials in tensile loading at room temperature. The slope of the linear behavior is called Young's modulus. Polymers have much lower elastic constants and span a broad range that depends on the sizes of molecules and their arrangement.

where M is the applied moment, L is the length, I is the moment of inertia, and θ is the twist angle. The shear modulus is often called the rigidity modulus. As shown in Table 1.2, shear moduli are typically about one-third of Young's moduli in isotropic materials.

Poisson's Ratio When a material is strained elastically in one axial direction, strains occur transverse to that direction. The transverse strains on orthogonal axes are of the opposite sign and usually of smaller magnitude than the axial strain, as defined by Poisson as

$$\nu = -\frac{e_{\text{transverse}}}{e_{\text{axial}}} \tag{1.23}$$

where e_{axial} is the strain in the direction of applied stress and $e_{\text{transverse}}$ is the strain in a direction orthogonal to the applied stress. Poisson's ratio is really a proportionality between elastic constants, and is therefore dimensionless. Typical values of Poisson's ratio are 0.2 to 0.4.

Hooke's Law For multiaxial stress states given in only normal or *principal* stresses (i.e., all shear stresses equal zero),

$$\sigma = \begin{vmatrix} \sigma_x & 0 & 0 \\ 0 & \sigma_y & 0 \\ 0 & 0 & \sigma_z \end{vmatrix}$$

Young's modulus and Poisson's ratio can be used to find the resulting elastic strains. Hooke's law relative to normal stress on three orthogonal axes yields

$$\begin{aligned} e_x &= \frac{1}{E}\left[\sigma_x - \nu\left(\sigma_y + \sigma_z\right)\right] \\ e_y &= \frac{1}{E}\left[\sigma_y - \nu\left(\sigma_x + \sigma_z\right)\right] \\ e_z &= \frac{1}{E}\left[\sigma_z - \nu\left(\sigma_y + \sigma_x\right)\right] \end{aligned} \tag{1.24}$$

This expression should be applied only to elastic deformation and for materials wherein the properties are uniform in all directions, or *isotropic*. Similar expressions that can be applied to plastic deformation are explored in Chapter 3.

1.4.2 Plasticity in Uniaxial Loading

Plastic deformation is the permanent shape change remaining after the removal of all significant external loading. A material that has undergone a uniform permanent shape change has undergone both elastic and plastic deformation while the shape change was occurring. However, as the applied stresses are removed the elastic strains return to zero (or nearly so). Consider Fig. 1.17, which shows a section of a stress-strain curve for a ductile metal.

The properties related to plastic deformation determined from a tension test include the following.

Young's Modulus The slope of the initial linear portion of the curve and the unloading between A and B in Fig. 1.17 is called the Young's modulus (if plastic deformation has not altered the arrangement of crystals or molecules to affect elastic properties).

Yield Strength The stress at which the onset of plastic deformation occurs is normally considered to be the yield strength Y (or σ_{ys}). Conventionally, a particular deviation, or offset from the linearity of the tensile stress-strain curve with the most common offset, is given as 0.2% (or 0.002).

Flow Stress, σ_{flow} Stress given as a function of plastic strain, $\sigma_{flow} = f(e_{plastic})$, is called the flow stress. The flow stress must be met in order to continue plastic deformation.

Plastic Strain The plastic strain is the deformation strain remaining after all external loading has been removed. As shown in Fig. 1.17, for a stress of magnitude A the

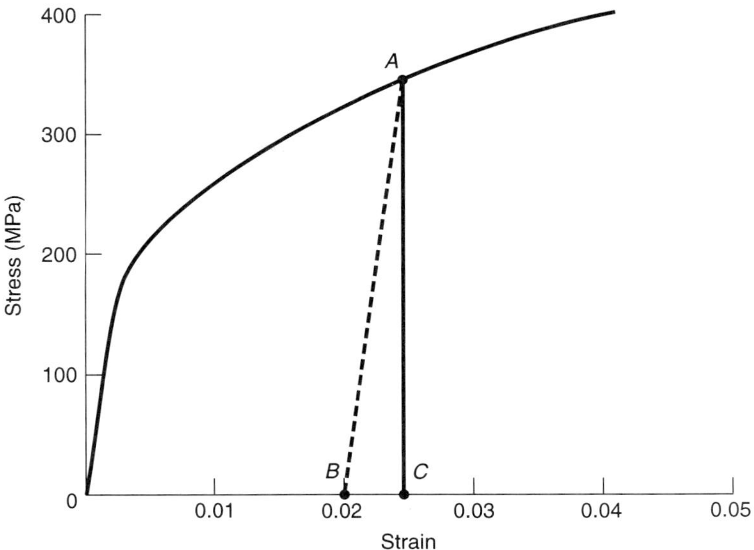

FIG. 1.17 Hypothetical room-temperature stress-strain curve from a tension test of a ductile metal.

total strain is given by the value at C. The plastic strain is the value at B after the elastic strain recovered during unloading from A is subtracted from C.

Elongation The elongation is the plastic strain to failure referenced to an initial length along the tensile bar. Necking interferes with uniform elongation, but elongation and the next term, reduction in area, provide a reasonable measure of ductility. Elongation is usually expressed as a percentage.

Ultimate Tensile Strength (UTS) The UTS (or σ_f) is the maximum stress found in an engineering stress-strain curve for a material that undergoes necking (see Fig. 1.18 for an example). Necking is the local nonuniform deformation that precedes failure in ductile metals. Necking is the result of an instability in deformation, which is why there is a drop in the force necessary to continue extension of a tensile specimen. Examples of necking are shown in Chapter 3.

Reduction in Area The reduction in area provides a semiquantitative measure of the fracture surface cross section. It is usually expressed as a percentage. The reduction in area includes the uniform plastic strain prior to necking and the reduction that occurs during the necking process.

Strain Hardening Strain hardening is the increase in the deformation resistance of the material with increasing strain. One common expression given for strain hardening relates the true stress and true plastic strain:

$$\sigma_{\text{true}} = K\left(e_{\text{true}}\right)^{n} \tag{1.25}$$

where n is the strain hardening exponent and K is a fitting parameter with units of stress. For a tension test, the necking strain is approximately equal to n. The value given here is also the flow stress for materials that undergo strain hardening. This expression applies only to the plastic portion of the stress-strain curve and does not model the elastic properties. For ductile materials such as the brass in Fig. 1.18, the elastic portion is insignificant relative to the large plastic strains at necking.

The responses of different materials to tension testing cover a wide range of behaviors, and the values defined do not provide a complete picture of the deformation behaviors of many materials. Only the complete tensile curves, as shown in Fig. 1.19, give a full description of the tensile behaviors. The tensile behaviors of materials with the same chemistry can be strongly influenced by processing history. The influence of processing history on brass and aluminum alloys is shown in Figs. 1.20 and 1.21. Fig. 1.21 also shows the effect of strain rate in a strengthened aluminum alloy.

The rate dependence of deformation in many polymers is substantial at room temperature. Figure 1.22 shows the rate dependence for deformation of PET at room temperature. PET undergoes highly localized deformation that rapidly proceeds to failure at both strain rates. The tensile curve for polyurethane shown in Fig. 1.23 exhibits the extended necking that occurs in polymers that undergo cold drawing. Changes in strain rate during the test are shown after large plastic strains.

1.4.3 Failure and Strength

The maximum engineering stress sustained by a material in tensile deformation (ultimate tensile strength, or UTS) follows plastic deformation in ductile materials. As shown by

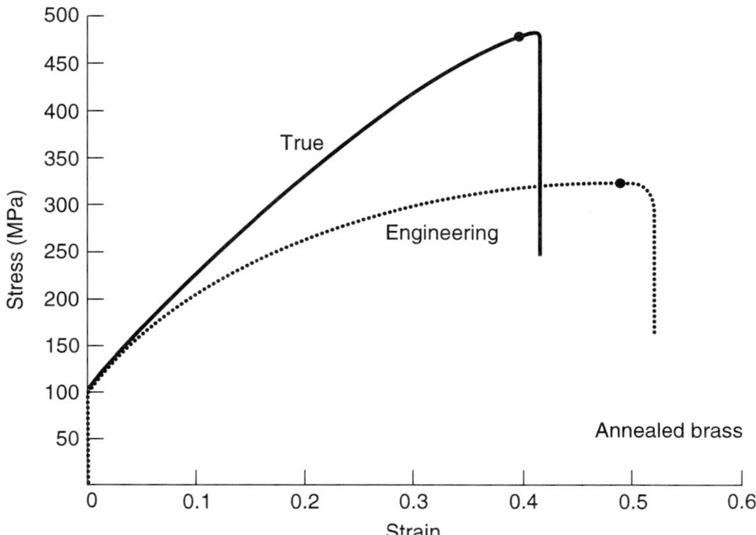

FIG. 1.18 Room-temperature stress-strain curves for annealed brass plotted as engineering and true values. The brass is a cartridge brass consisting of an alloy of about 70 percent copper by weight and the balance zinc that has been annealed at 550°C for one hour after being obtained in the "half-hard" condition. The nominal strain rate is $3 \times 10^{-3}\ s^{-1}$. On this curve the strain axis is sufficiently large that the high Young's modulus of the material results in a very steep slope that appears to be almost parallel to the stress axis. The small circles indicate the start of localized deformation or necking. For the engineering stress-strain curve, this point is the maximum in the curve.

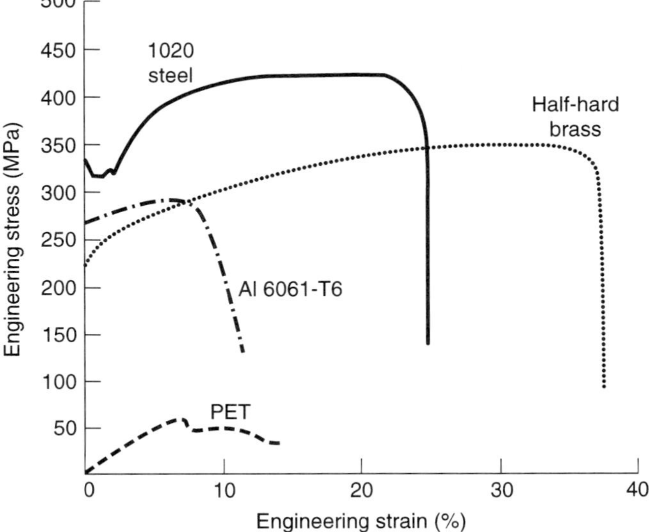

FIG. 1.19 Engineering stress-strain behavior in tension at room temperature for several ductile materials. 1020 steel is a low-carbon steel containing 0.20 percent carbon by weight. The half-hard brass is the same material as the annealed brass in Fig. 1.18, but has undergone strain hardening from cold work. Al 6061-T6 is an aluminum alloy with about 0.3 percent copper, 1.0 percent magnesium, and 0.6 percent silicon (by weight) that undergoes a special heat treatment to attain maximum strength by growing a distribution of fine second-phase precipitates within the crystals. The PET curve is for an amorphous polyethylene terephthalate.

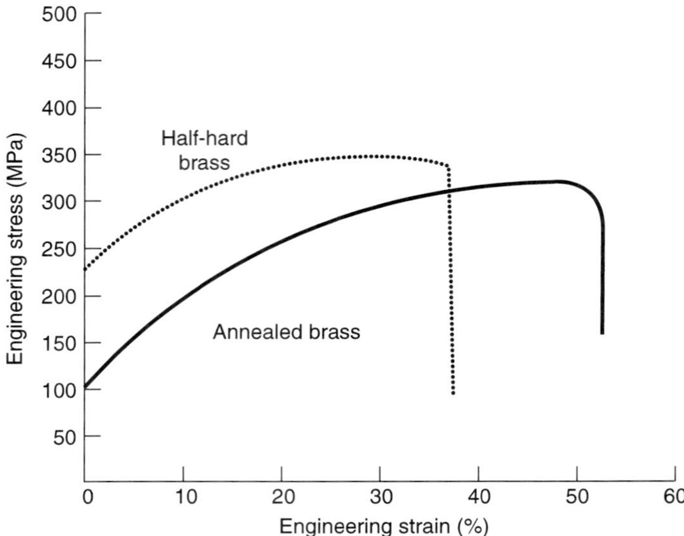

FIG. 1.20 The engineering stress-strain curves for annealed and half-hard brass from Fig. 1.18 and 1.19 plotted together.

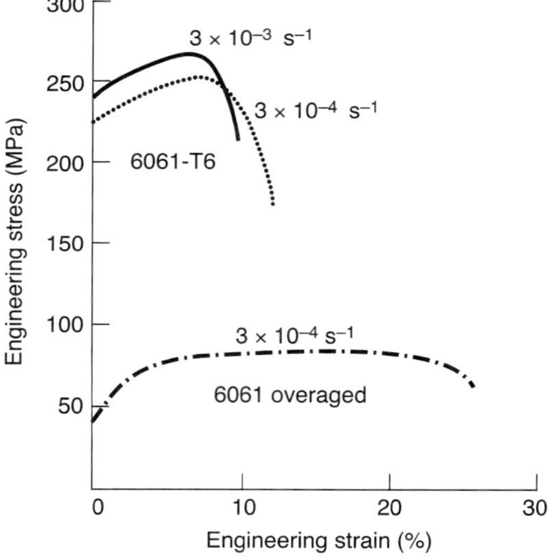

FIG. 1.21 The engineering stress-strain curve for the 6061-T6 material plotted in Fig. 1.19 at two different strain rates compared with the same material overaged by heat treatment at 450°C for 2 hours.

comparing the curves in Fig. 1.20 and 1.21, the behavior of the material after necking begins (at the UTS on the engineering stress–engineering strain curve) can be very different. The fracture stress for brittle materials is preceded by little or no plastic deformation. For these materials the fracture stress is often controlled by flaws that initiate failure.

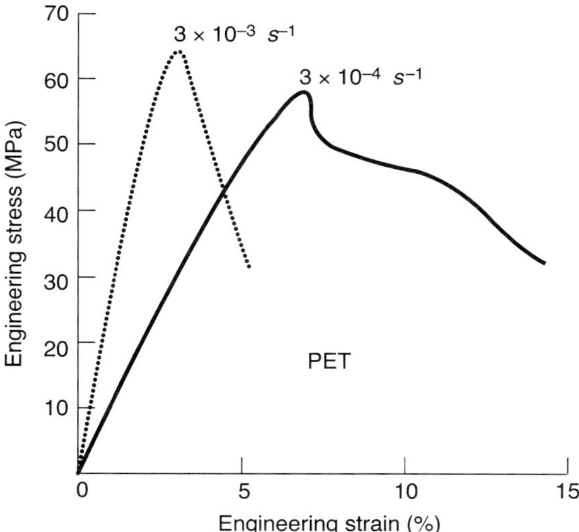

FIG. 1.22 PET tested at two different strain rates showing an effect on the apparent elastic behavior, necking, and ductility.

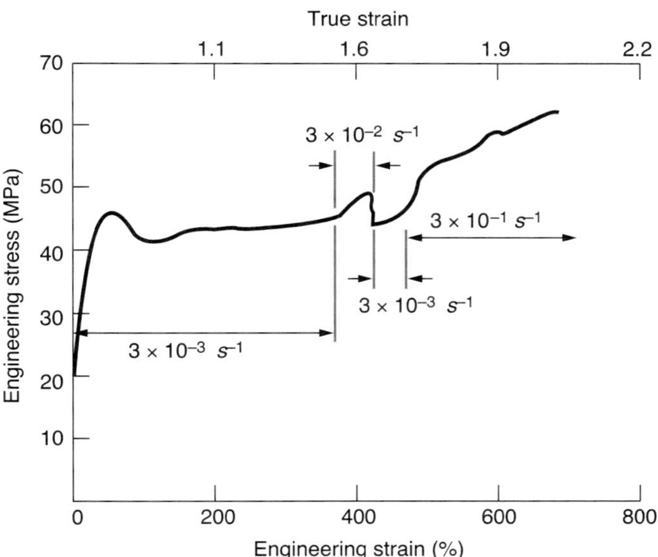

FIG. 1.23 Polyurethane strained at various rates, showing an increase in stress level for higher strain rates.

Because the critical flaw size for failure is quite small in brittle materials, the population of flaws in a material affects the likelihood of failure. For this reason, failure stresses are highly variable and often dependent on the size of the specimen. Table 1.4 gives a comparison of UTS and K_{Ic}. K_{Ic} was defined in Section 1.3.4 and is the primary topic of Chapter 7.

TABLE 1.4 Fracture Behaviors of Various Materials at Room Temperature*

Material	Ultimate tensile strength (UTS) (MPa)	Fracture toughness, K_{Ic} (MPa · $m^{1/2}$)
Aluminum alloys	100–800	20–60
Copper alloys	200–1000	140–120
Iron alloys and steels	250–2000	10–200
Stainless steels	300–700	50–200
Titanium alloys	300–1300	40–120
Aluminum oxide	200–600	3–5
Silicon carbide	250–600	2–5
Glass	70–200	0.5–1.5
Polyethylenes (PE)	20–100	2–6
Polymethylmethacrylate (PMMA)	40–120	1–3
Polystyrene (PS)	40–100	1–2
Bone	60–150	2–10
Tendon	45–50	—

*Data in this table are often meaningful only for K_{Ic} values less than 10 MPa/$m^{1/2}$. For values greater than 10 MPa/$m^{1/2}$, the component thicknesses necessary to fulfill plane strain conditions are not practical for many applications. This will be discussed further in Chapter 7.

1.5 HOW BONDING INFLUENCES MECHANICAL PROPERTIES

The nature of the bonds in materials determines the responses of the materials to applied stress as much as the melting temperature and the crystalline or molecular arrangement. The three categories of strong bonds—ionic, covalent, and metallic—comprising the bonds between atoms and ions within crystals and molecules and the secondary bonds between crystals and molecules determine the mechanical responses of materials (see Rohrer, 2001).

1.5.1 Linear Elasticity

The net result of attractive and repulsive energies, $E_T = E_{\text{attractive}} + E_{\text{repulsive}}$, for a particular bond is an asymmetric energy relationship for the spacing r between atom or ion centers shown in Fig. 1.24*a*. As the influence of atomic vibration increases at higher temperatures, the average spacing $\bar{r}$ increases, representing the thermal expansion shown in Fig. 1.24*b*. The relationship of the force F between atoms on ions with spacing is found by taking the slope dE_T/dr of the energy relationships.

Elastic properties are only defined over small strains so that the slope of the F versus r curve defining the effective spring constant for small stretches of the bonds can be treated as linear. Then, if $F = kr$, Young's modulus, E, must scale with K.

The shapes of the energy and force curves in Fig. 1.24 determine the stiffness of the bonds. They also give the corresponding strengths of the bonds. The effects of thermal vibration have an impact on the elastic stiffness and also the thermal expansion coefficient. Young's modulus decreases with increasing temperature and the thermal expansion coefficient decreases with increasing temperature. Data for Young's modulus versus temperature are given in Fig. 1.25. Figure 1.26 shows that the elastic stiffness of the bonds indicated by the room-temperature Young's modulus scales with the melting temperature. Detailed plots

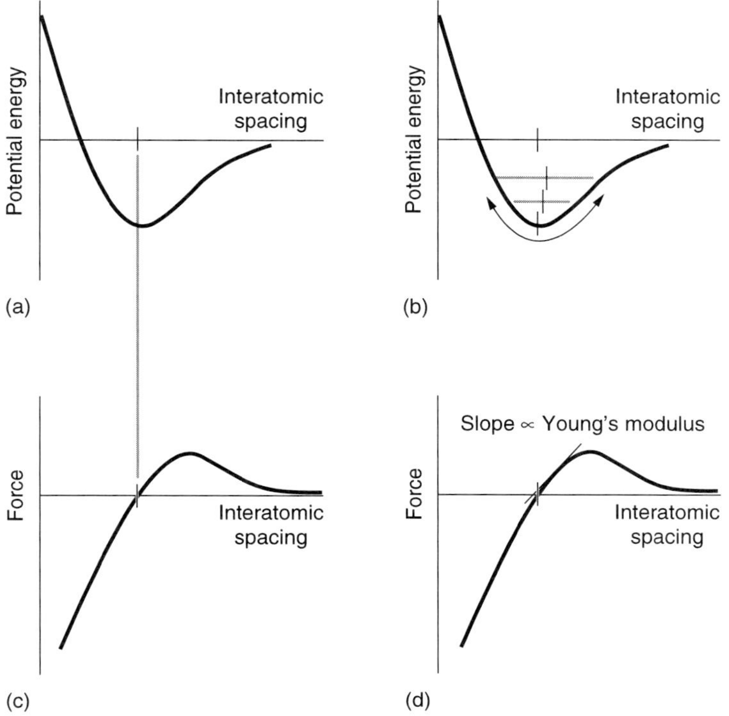

FIG. 1.24 (a) Interatomic potential energy versus spacing. (b) The effect of thermal vibration on interatomic spacing. (c) The force versus spacing derived by finding $F = dE/dr$, where r is interatomic spacing. (d) The slope of the force versus spacing relationship, approximately linear for small strains, is proportional to Young's modulus.

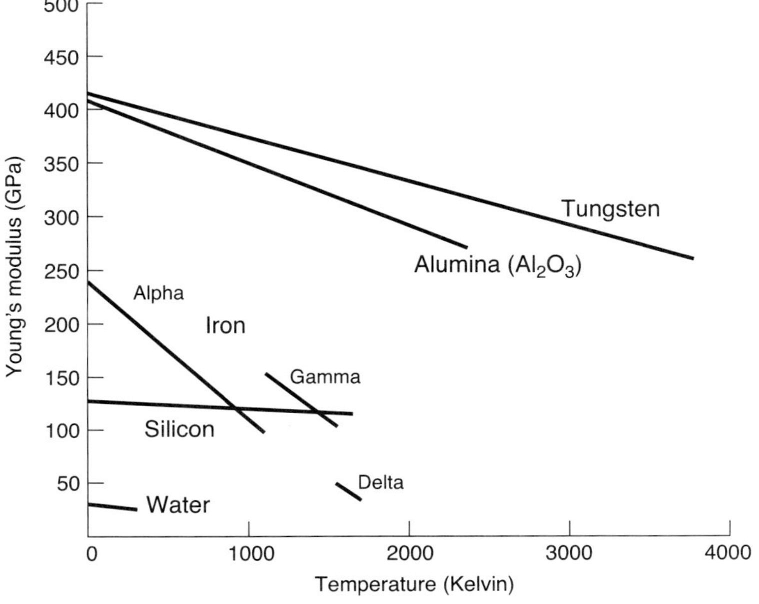

FIG. 1.25 Young's modulus versus temperature for several materials.

Pure iron undergoes a phase change at the interruptions shown in the Young's modulus versus temperature behavior in Fig. 1.25. Young's modulus of the close-packed (face-centered cubic, or FCC) gamma iron structure is greater than that of the alpha (body-centered cubic or BCC) iron structure. These phase changes have effects on the mechanical properties as well. In Chapter 8 we will show how the deformation and fracture mechanisms are completely different just above and just below these transformation temperatures.

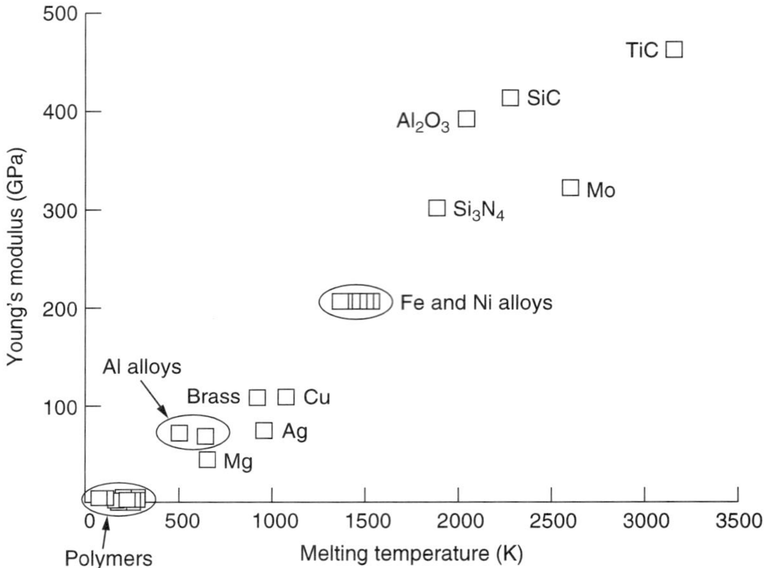

FIG. 1.26 Young's modulus at room temperature versus melting temperature for several materials.

of materials with similar chemistries show trends that follow quite well the systematic patterns determined by the periodic chart positions. Another indication of the strengths of bonds is given by thermal expansion coefficients. Figure 1.27 shows the relationship between the thermal expansion coefficient near room temperature and the room-temperature Young's modulus for a variety of materials.

Not all properties show clear trends when plotted against one another. On the other hand, different material classes span different property ranges. Figure 1.28 shows no clear relationship between room-temperature density and elastic constants. Rather, the message from Fig. 1.27 is that the specific stiffness (the ratio of Young's modulus to density) of ceramics is much higher than those of most metals and polymers.

1.5.2 Time-Dependent Transport Properties

Plastic deformation of materials becomes strongly time- or rate-dependent at temperatures greater than one-third to one-half of the absolute melting temperature. These transport properties provide a primary mechanism for deformation at very high temperatures and also interact synergistically with other mechanisms at lower temperatures. Time-dependent deformation properties can also be influenced by changes in chemical distribution, phase

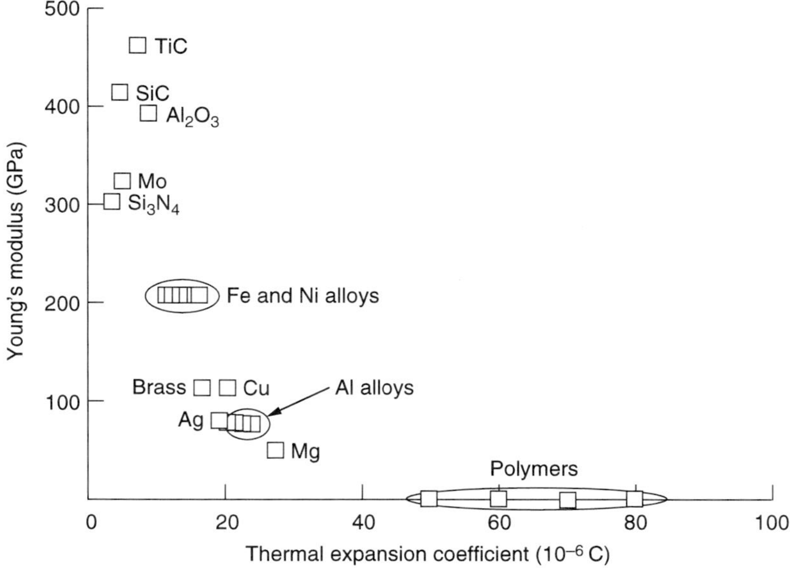

FIG. 1.27 Young's modulus versus thermal expansion coefficient for several materials.

changes, or degradation mechanisms such as oxidation or corrosion. The evaluation of creep stress relaxation in crystalline materials is often connected with transport of material by diffusion. Self-diffusion, the net motion of atomic species that depends on transport of vacancies or interstitials, contributes strongly to most creep mechanisms (see Chapter 6). Figure 1.29 shows that the activation barrier to diffusion of vacancies through a crystal does depend on melting temperature. This is another strong indication of how fundamental mechanisms for rearrangement in crystals are tied to the strength of bonding.

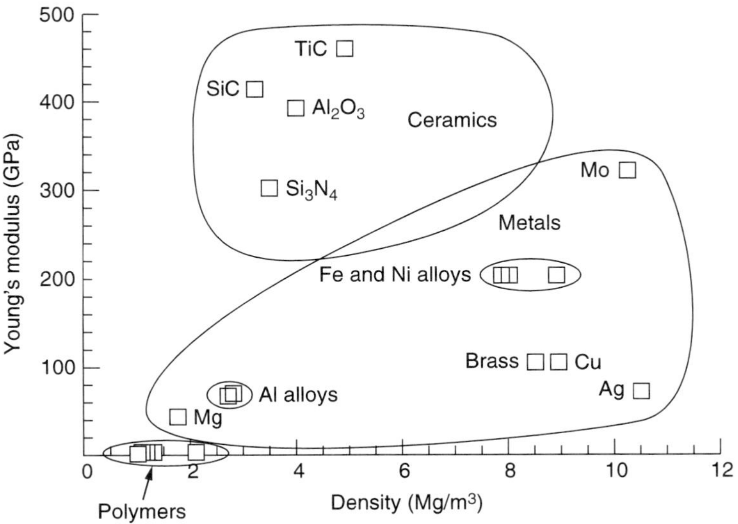

FIG. 1.28 Young's modulus versus density for several materials.

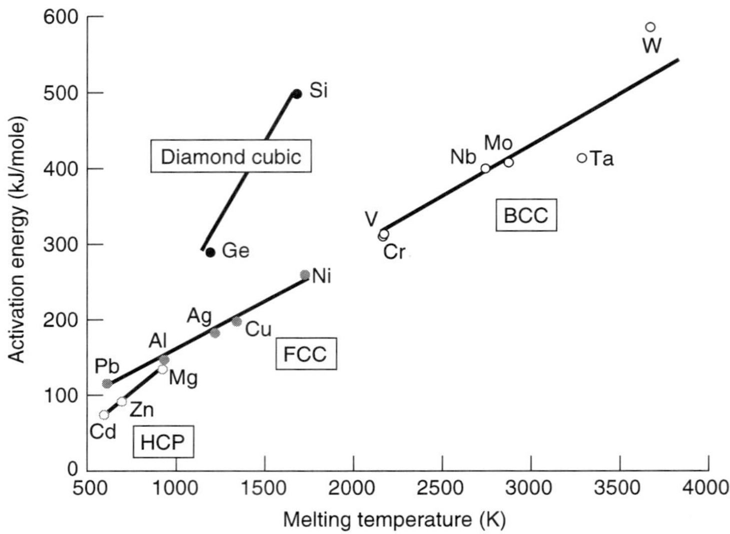

FIG. 1.29 Activation energy for lattice diffusion versus melting temperature for various metals.

1.6 REFERENCES

Introductory Materials Texts

M. F. Ashby and D. R. H. Jones, *Engineering Materials 1,* Pergamon Press, 1980.

M. F. Ashby and D. R. H. Jones, *Engineering Materials 2,* Pergamon Press, 1986.

W. D. Callister, Jr., *Materials Science and Engineering: An Introduction,* 4th Ed., Wiley, 1996.

R. A. Flinn and P. K. Trojan, *Engineering Materials and Their Applications,* 4th Ed., Houghton Mifflin, 1990.

A. Kelly, G. W. Groves, and P. Kidd, *Crystallography and Crystal Defects,* Rev. Ed., Wiley, 2000.

G. S. Rohrer, *Structure and Bonding of Crystalline Materials*, Cambridge, 2001.

J. F. Schackelford, *Introduction to Materials Science for Engineers,* 3rd. Ed., Macmillan, 1992.

L. H. Van Vlack, *Elements of Materials Science and Engineering,* 6th Ed., Addison-Wesley, 1989.

Introductory Mechanics and Strength of Materials Texts

F. B. Beer and E. R. Johnston, Jr., *Mechanics of Materials,* McGraw-Hill, 1993.

W. F. Riley and L. Zachary, *Introduction to Mechanics of Materials,* Wiley, 1989.

Other Books on Mechanical Behavior of Materials

T. H. Courtney, *Mechanical Behavior of Materials,* McGraw-Hill, 1990.

G. E. Dieter, *Mechanical Metallurgy,* 3rd. Ed., McGraw-Hill, 1986.

N. E. Dowling, *Mechanical Behavior of Materials: Engineering Methods for Deformation, Fracture, and Fatigue,* 2nd Ed., Prentice-Hall, 1998.

D. K. Felbeck and A. G. Atkins, *Strength and Fracture of Engineering Solids,* Prentice-Hall, 1996.

D. J. Green, *An Introduction to the Mechanical Properties of Ceramics,* Cambridge, 1998.

R. W. Hertzberg, *Deformation and Fracture Mechanics of Engineering Materials,* 4th Ed., Wiley, 1996.

F. A. McClintock and A. S. Argon, Editors, *Mechanical Behavior of Materials,* Addison-Wesley, 1966.

N. G. McCrum, C. P. Buckley, and C. B. Bucknall, *Principles of Polymer Engineering,* Oxford, 1988.

M. A. Meyers and K. K. Chawla, *Mechanical Behavior of Materials,* Prentice-Hall, 1999.

S. P. Timoshenko, *History of Strength of Materials,* Dover Publications Reprint, New York, with original McGraw-Hill Copyright 1953.

J. B. Wachtman, *Mechanical Properties of Ceramics,* Wiley, 1996.

1.7 PROBLEMS

A.1.1 How do the points in the worm in Fig. 1.1 change with strain? (Think of the mathematical definition of a point.)

A.1.2 Take a string that is originally 1 m in length and stretch it in three stages.

(a) Stretch it by 0.1 m.

(b) Stretch it by an additional 0.2 m.

(c) Stretch it by an additional 0.2 m.

For each stage, calculate and compare the engineering and true strains.

A.1.3 Plot true strain versus engineering strain for stretching of a string from 1 m to 1.5 m. Plot enough points to yield a smooth curve.

A.1.4 Measure the displacements of M and N on the worm in Fig. 1.1 in millimeters. From these measurements, calculate the tensile strain in engineering strain and true strain.

A.1.5 Measure the displacements of M and N in Fig. 1.2 in millimeters. From these measurements, estimate all four strains e_{ij}.

A.1.6 Measure the displacements of the strained square in Fig. 1.3 in millimeters. From these measurements, estimate all four strains e_{ij}.

A.1.7 Show how SI units of stress can also be expressed as energy per unit volume.

A.1.8 The Brinell hardness of a new alloy is measured using a load of 3000 kg. The indentation diameter is 2 mm. What is the BHN? What is the indentation depth?

A.1.9 Calculate Young's modulus and the 0.2% offset yield stress for the stress-strain curve in Fig. 1.17.

A.1.10 Give common SI units for the following mechanical properties.

(a) Stress __________ or __________/m^2

(b) Fracture toughness __________ or __________/m$^{3/2}$

(c) Creep rate __________

(d) Yield strength __________

(e) Young's modulus __________

(f) Poisson's ratio __________

B.1.1 Assume that a rectangle plotted in the x-z plane has dimensions of 2 cm × 5 cm. Apply the following values of strain to the rectangle.

$$\begin{vmatrix} 0.1 & 0 & 0.1 \\ 0 & 0 & 0 \\ 0.3 & 0 & -0.1 \end{vmatrix}$$

Give the separate components of shape-change strain and rotational strain.

B.1.2 Write the strain tensor for $du_i = 0.01$, $du_2 = 0.02$, $du_3 = 0.01$, $dx_1 = dx_2 = dx_3 = 1$. Describe the shape change. Describe the rotation.

B.1.3 Deform a cube by stretching it 20 percent in the x-direction and reducing it 20 percent in the y-direction. Does it maintain the same volume?

B.1.4 Assume Fig. 1.17 is a true stress–true strain plot. Replot this data and overlay it with the engineering stress-strain curve for the same data (similar to Fig. 1.18). Why are the curves so similar?

B.1.5 (a) Using Eq. 1.25, calculate the values of K and n that best fit the data for brass in Fig. 1.18.

(b) Are your values similar to that predicted in the description of Eq. 1.25?

(c) Estimate the strain hardening rate $d\sigma/d\varepsilon$ for this curve at the point of necking indicated on the true stress–true strain curve.

B.1.6 Use Hooke's law for multiaxial stresses to calculate the stresses required to produce the strains described in Problem B.1.3. (Assume the material is steel with a Young's modulus ≈ 210 GPa and a Poisson's ratio ≈ 0.2). Is the magnitude of the stresses too large for just elastic distortion?

B.1.7 Your nearby engineering library should have ASTM standards for mechanical testing. The standards include geometries and dimensions of typical samples. Find the correct ASTM standards for

(a) Tensile testing of metals or polymers

(b) Bend testing of ceramics

(c) Fracture testing of metals

(d) Fracture testing of ceramics.

B.1.8 Estimate the apparent linear Young's modulus for the PET samples shown in Fig. 1.22. By finding the slope up to an engineering stress of 40MPa, predict the modulus for a sample tested at a strain rate of $3 \times 10^{-2}\ s^{-1}$.

B.1.9 Assume that a beam of square cross section is bent elastically in four-point bending. Describe the changes that would occur in the cross section between the inner loading points, and show how you would calculate the shape.

C.1.1 (a) A wooden plank with a rectangular cross section will be used to span a small stream on a hiking trail. The plank will be expected to carry one person at a time. The plank will be supported only on each end. The width of the crossing is 3 m and the plank will lie about 3 m above the stream. The maximum width of stock available is 0.6 m. The fracture stress of the wood along the growth direction is approximately 8 MPa. Give your design for a safe crossing.

(b) The trail engineer has suggested that the possibility of two people crossing simultaneously should be considered. Show how the safe design of the crossing would change for the possibility of two people crossing together. Is the crossing safer if the two people are closer together or farther apart?

(c) The trail engineer has suggested that a round log would make a more interesting span. Limited to a maximum diameter of 0.6 m, will the log be more or less safe than your design for part (a)?

C.1.2 Two materials are proposed for the same application. The most critical performance criterion is the change in thickness on loading under a biaxial stress. The proposed part is a thin plate measuring 4 mm by 4 mm by 0.1 mm. The part will undergo a biaxial tensile load of 2 MPa. For the properties given below, rank the materials from best to worst.

$$M1 - E = 200 \text{ GPa}, \nu = 0.3$$

$$M2 - E = 180 \text{ GPa}, \nu = 0.25$$

$$M3 - E = 220 \text{ GPa}, \nu = 0.32$$

$$M4 - E = 250 \text{ GPa}, \nu = 0.35$$

C.1.3 (a) Estimate the yield stress for the half-hard and annealed brasses in Fig. 1.20.

(b) Calculate the true stress–true strain curves for the half-hard and annealed brasses in Fig. 1.20.

(c) Fit the data from parts (a) and (b) to Eq. 1.25. Which material has the higher strain hardening rate?

(d) For the annealed brass, fit only the data after the flow stress has reached the yield stress of the half-hard brass. Why is the deformation so similar to that of the half-hard brass?

C.1.4 (a) Estimate the yield stress for all three materials in Fig. 1.21.

(b) Calculate the true stress–true strain curves for the three materials in Fig. 1.21.

(c) Fit the data for the two materials in the T6 condition to Eq. 1.25. Which material has the higher strain hardening rate? How does this compare with the point at which necking takes place? Estimate the engineering strain at which necking starts and calculate the corresponding true strain. Is this consistent with your expectations?

(d) For the annealed brass, fit the data up to a true strain of 0.1 to Eq. 1.25. Compare the value of n with those of the two other aluminum samples.

C.1.5 The plank bridge in Problem C.1.1(a) has collapsed as a result of a fire. The use of a steel tube has been proposed. A steel tube is available with a 1-meter outer diameter and a 1-cm wall thickness. The moment of inertia for a tube is

$$I = \frac{\pi}{64}\left(d_o^4 - d_i^4\right)$$

where d_o is the outer diameter and d_i is the inner diameter. Will these tubes safely support expected loads? Explain your answer.

CHAPTER 2

TENSORS AND ELASTICITY

STRESS, STRAIN, AND ELASTICITY are most conveniently represented by tensors. Tensors and the related matrices provide an opportunity to utilize the mathematics taken through the second year of university in a very powerful approach for understanding elasticity and deformation of materials. The geometric relations given in this chapter also provide handy tools for solution of many problems related to crystals and composites. The necessity for tensors may seem obvious when one considers the properties of fiber composites, but cubic metals can show strong variations in elastic and plastic properties with the direction of applied stress. Both the student and the instructor may find that tensors are not easy to learn or to teach. If we restrict ourselves to the solutions given for uniaxial conditions or rely on equations in books for directional properties, we can solve only idealized problems. The goal of this chapter is to demonstrate the enabling capabilities derived from an understanding that tensorial relations are the foundation for physical and mechanical properties.

A common misconception is that cubic materials are isotropic—with all properties the same in all directions. As shown later in this chapter, many properties of cubic single crystals are isotropic, but elasticity and some similar properties called electrostriction and magnetostriction are anisotropic even in cubic materials. One very important application of single crystals is in the hotter stages of a turbine engine. The turbine blades located in this part of the engine typically consist of nickel-base single crystals grown with a specific orientation designed to resist long-term deformation. These single crystals consist of a disordered phase reinforced with cuboidal grains of an ordered compound called an intermetallic. The anisotropy of the cubic grains is demonstrated by a variation in elastic constants with direction in the material. The magnitude of this elastic anisotropy in a single crystal turbine blade can be surprisingly large, as shown by the values of Young's modulus given for specific crystallographic directions in Fig. 2.1.

Other applications of cubic single crystals with strong elastic anisotropies include silicon and gallium arsenide in microelectronics and alkali halides used for their optical properties. Figure 2.2 defines a standard cubic coordinate system with specific directions labeled. The values of Young's modulus in these low-index crystal directions are given in Table 2.1 for several cubic metals. Tensor representations enable calculation of any elastic constant in an arbitrary direction from the information given in this table. Other materials show anisotropy as a result of preferred orientation or texture of the crystals or molecules that comprise the material. Oriented materials can have strong anisotropies, as shown in Table 2.2. The magnitude of the anisotropy for many of these materials is dependent on the degree of preferred orientation. The strain and thermal history of deformation processing along with the mechanisms of deformation influence the degree of orientation and thereby the degree of anisotropy.

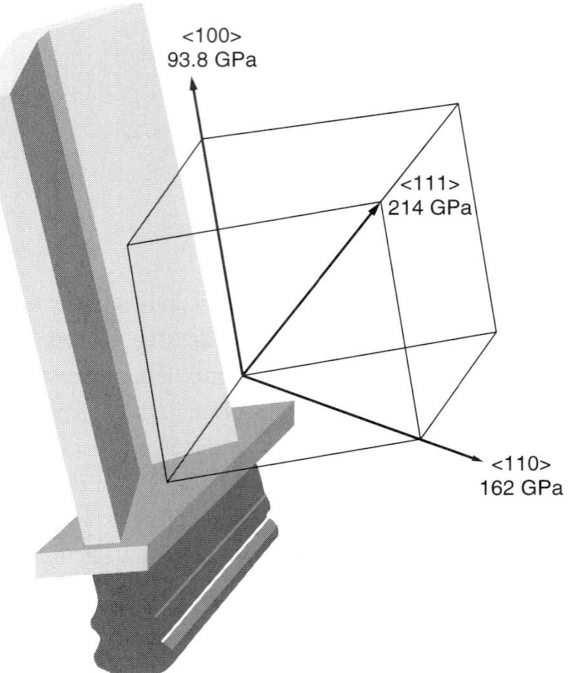

FIG. 2.1 A schematic depiction of a single crystal turbine blade with a <100> direction along its length and the magnitude of reported Young's modulus anisotropy for families of crystal directions (data are from promotional literature, Howmet, Whitehall, MI).

When single crystal turbine blades are grown, it is difficult to control the tranverse orientation of the blade such that the elastic behavior of blades in the same assembly may be different. The behavior of the large assemblies containing blades is more predictable when the extra effort is made to match or even control the transverse orientation.

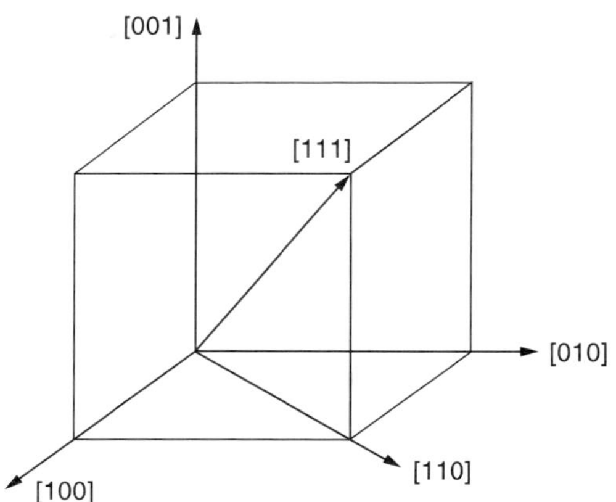

FIG. 2.2 Cubic unit cell showing crystallographic directions.

TABLE 2.1 Young's Modulus at 25°C (GPa)

Direction	Fe	Mo	W	Al	Cu
•<100>	125	360	390	64	67
•<110>	210	305	390	72	130
•<111>	272	292	390	76	190

TABLE 2.2 Young's Modulus (GPa) Versus Direction in Several Oriented Materials

Material	Rolling or axisymmetric direction	45°	Transverse or radial direction
Cold-rolled iron*	230	205	273
Cold-rolled copper*	139	109	140
Rolled and recrystallized copper*	70	123	67
Hot-forged silicon nitride	301	—	319
Kevlar	124	—	5
Wood	9–16	—	0.6–1
Drawn copper wire*	46	—	103
Bone**	17–30	—	11–18

* McClintock and Argon, (1966).
**Van Buskirk, Cowin, and Ward, (1981).

As already shown in Chapter 1, three-dimensional stress and strain states can be expressed as tensors if the increments of strain are small (i.e., <0.15). We will establish ground rules for specifying tensors and reporting changes in the defined orientation of the reference coordinates to provide a complete description of anisotropic properties.

2.1 WHAT IS A TENSOR?

A tensor is a specific type of matrix representation that can relate the directionality of either a material property in the case of property tensors (e.g., conductivity, elasticity) or a state of a particular material in the case of condition tensors (e.g., stress, strain). Tensors are a convenient form for representing the properties of single crystals, polycrystalline materials with preferentially oriented crystals, or directional composites if these materials are anisotropic (see Nye, 1957). Measurement and prediction of properties are provided in a concise manner by tensor representation. Specific types of tensors are classified by *rank* as follows.

- A tensor of zero rank is a scalar quantity, such as temperature or density, that does not have a directional dependence.
- A tensor of first rank is a vector quantity, such as force, electric field, or flux of atoms. $\mathbf{T} = T_i = (T_1, T_2, T_3)$, where T_i are the components given versus the axes of reference.
- A tensor of second rank is a tensor that relates two vector quantities, such as those in Table 2.3, with the arrangement

$$\mathbf{T} = T_{ij} = \begin{vmatrix} T_{11} & T_{12} & T_{13} \\ T_{21} & T_{22} & T_{23} \\ T_{31} & T_{32} & T_{33} \end{vmatrix} \tag{2.1}$$

- A tensor of third rank is a tensor that relates a vector and a second-rank tensor. *Example:* converse piezoelectricity, which relates an electric field to strain.
- A tensor of fourth rank is a tensor that relates two second-rank tensors. *Example:* Elasticity relates the two second-rank tensors of stress and strain. Two types of elasticity tensors, the stiffness tensor and the compliance tensor, are commonly used. Electrostriction relates strain to the square of the electric field, a vector quantity.

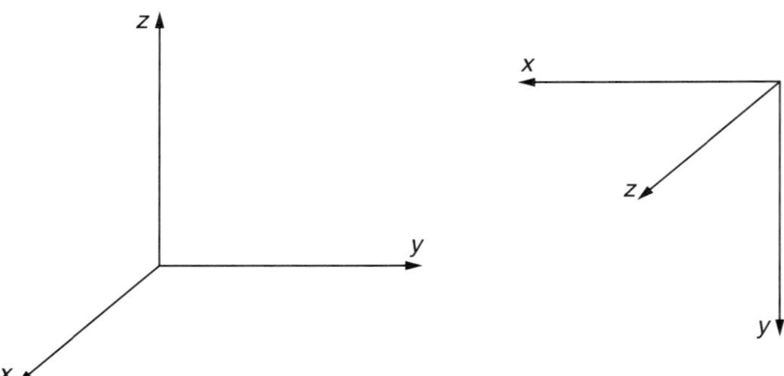

FIG. 2.3 A perspective view of coordinate axes *x*, *y*, and *z*, which obey a right-hand rule.

TABLE 2.3 Second-Rank Tensors

Vector	=	Second-rank tensor	×	Vector	Type
Current density		Electrical conductivity		Electric Field	Property
Flux of atoms		Diffusivity		Concentration gradient	Property
Displacement element		Strain (e_{ij})		Linear element	Condition
Force (or tractions)		Stress (σ_{ij})		Normal to area	Condition

2.2 TRANSFORMATION OF TENSORS

2.2.1 Defining Coordinate Systems

The key to understanding property or condition tensors is to recognize that tensors can be specified with reference to some coordinate system that is usually defined in three-dimensional space by orthogonal axes that obey a right-hand rule (Fig. 2.3). There are many ways to define the axes and their relationship with subscripts relating the tensors to a coordinate system. Common designations include (x, y, z), (1, 2, 3), and (x_1, x_2, x_3).

These axes are usually specified with reference to physical characteristics of the object. The axes must be defined for any set of tensors to have physical meaning. Often we wish to define a value of the tensor at an orientation other than the most obvious axes described by the object's shape. In this case, we need to understand how rotation of the axes transforms tensors. When defined with reference to a set of principal axes, the two vectors **M** and **N** can be related by a second-rank tensor T_{ij} that consists of only the diagonal components given in Eq. 2.1, T_{11}, T_{22}, and T_{33}, such that

$$M_1 = T_{11}N_1, \qquad M_2 = T_{22}N_2, \qquad \text{and} \qquad M_3 = T_{33}N_3$$

If we choose new axes $\{x', y', z'\}$ from an original set of axes $\{x, y, z\}$, the relationship

$$M_i = T_{ij}N_j \tag{2.2}$$

becomes

$$M_i' = T_{ij}'N_j' \tag{2.3}$$

To define the relationship for expressing the second rank tensor T_{ij}' we must perform the following steps:

1. **M′** in terms of **M** 2. **M** in terms of **N** 3. **N** in terms of **N′**

Step 1 is given by

$$M_i' = a_{ik} M_k \tag{2.4}$$

which is inclusive of the explicit expressions

$$M_1' = a_{11}M_1 + a_{12}M_2 + a_{13}M_3$$
$$M_2' = a_{21}M_1 + a_{22}M_2 + a_{23}M_3$$
$$M_3' = a_{31}M_1 + a_{32}M_2 + a_{33}M_3$$

where **a** is a rotation matrix used to relate the two coordinate systems. The definition of **a** is given in Table 2.4, where a_{32} is the cosine of the angle between the new axis z' and the old axis y. Also, a_{11} is the cosine of the angle between x' and x. (See next section for more detail on the rotation matrix).

For Step 2, we use

$$M_k = T_{km} N_m \tag{2.5}$$

and Step 3 is similar to Step 1:

$$N_m = a_{jm} N_j' \tag{2.6}$$

Then, making some substitutions and rewriting Eq. 2.3 as

$$T_{ij}' = \frac{M_i'}{N_j'}$$

yields

$$T_{ij}' = a_{ik} a_{jm} T_{km} \tag{2.7}$$

For tensors used in this book, standard *implied summation* is used such that the left-hand side (LHS) of an equation represents only a single term if both subscripts are defined. For example,

$$T_{23}' = \sum_{k}^{3} \sum_{m}^{3} a_{2k} a_{3m} T_{km}$$

represents the components of the old tensor before rotation, T_{kl}, factored with their corresponding rotation matrix components. Alphabet subscripts on the right-hand side of the equation include all three terms. For example, T_{23}' given with k expanded is

$$T_{23}' = \sum_{m}^{3} \left(a_{21} a_{3m} T_{1m} + a_{22} a_{3m} T_{2m} + a_{23} a_{3m} T_{3m} \right)$$

and summed over m is

$$T_{23}' = a_{21}a_{31}T_{11} + a_{21}a_{32}T_{12} + a_{21}a_{33}T_{13} + a_{22}a_{31}T_{21} + a_{22}a_{32}T_{22} + a_{22}a_{33}T_{23} + a_{23}a_{31}T_{31} + a_{23}a_{32}T_{32} + a_{23}a_{33}T_{33}$$

TABLE 2.4 Rotation Matrix

		Old		
		x	*y*	*z*
	x'	a_{11}	a_{12}	a_{13}
New	*y'*	a_{21}	a_{22}	a_{23}
	z'	a_{31}	a_{32}	a_{33}

Matrix multiplication provides a convenient format for performing tensor rotations. For a second-rank tensor we can write

$$\mathbf{T}' = \mathbf{a} \cdot \mathbf{T} \cdot \mathbf{a}^{\mathrm{T}} \tag{2.8}$$

where $\mathbf{T}$ is a matrix expression of the original tensor, $\mathbf{T}'$ is a matrix expression of the tensor after rotation, $\mathbf{a}$ is the rotation matrix, and $\mathbf{a}^{\mathrm{T}}$ is the transpose of $\mathbf{a}$. This matrix operation provides exactly the same expressions as those given in Eq. 2.7. Also, the diagonal of the second-rank tensor is invariant to rotations.

2.2.2 The Rotation Matrix and Euler Angles

Several schemes can be used to produce a rotation matrix. For situations in which the directions are specified by indices referenced to a crystal, the rotation matrix can be readily obtained from the reference direction cosines defined by the direction indices. Expressing the direction cosines for three orthogonal axes employs nine terms, which can be expressed in terms of three sequential rotations called Euler angles. The three Euler angles are given in Fig. 2.4 as three counterclockwise rotations: (1) A rotation about z, defined as $\phi 1$; (2) a rotation about the new x, which is x', defined as Φ; and (3) a rotation about the second z-position, which is z', defined as $\phi 2$.

From this sequence of rotations, the rotation matrix $\mathbf{a}$ is given by matrix multiplication of the rotation matrices of each individual rotation, so that $\mathbf{a} = a_{\{\phi 1, \Phi, \phi 2\}} = a_{ij}$, is given as the most recent rotation multiplied by the prior rotation

$$a_{\phi 2} \cdot a_{\Phi} \cdot a_{\phi 1} =$$

$$\begin{vmatrix} \cos\phi 2 & \sin\phi 2 & 0 \\ -\sin\phi 2 & \cos\phi 2 & 0 \\ 0 & 0 & 1 \end{vmatrix} \cdot \begin{vmatrix} 1 & 0 & 0 \\ 0 & \cos\Phi & \sin\Phi \\ 0 & -\sin\Phi & \cos\Phi \end{vmatrix} \cdot \begin{vmatrix} \cos\phi 1 & \sin\phi 1 & 0 \\ -\sin\phi 1 & \cos\phi 1 & 0 \\ 0 & 0 & 1 \end{vmatrix} =$$
$$\begin{vmatrix} \cos\phi 2\cos\phi 1 - \cos\Phi\sin\phi 1\sin\phi 2 & \cos\phi 2\sin\phi 1 + \cos\Phi\cos\phi 1\sin\phi 2 & \sin\Phi\sin\phi 2 \\ -\sin\phi 2\cos\phi 1 - \cos\Phi\sin\phi 1\cos\phi 2 & -\sin\phi 2\sin\phi 1 + \cos\Phi\cos\phi 1\cos\phi 2 & \sin\Phi\cos\phi 2 \\ \sin\Phi\sin\phi 1 & -\sin\Phi\cos\phi 1 & \cos\Phi \end{vmatrix} \tag{2.9}$$

Direction cosines are often defined to relate single directions to coordinate systems. A direction cosine is the cosine of the angle between a new direction and a coordinate system. We will define the direction cosines as $l = a_{11}$, $m = a_{12}$, and $n = a_{13}$ for the direction cosines for an arbitrary orientation of x'.

BIOGRAPHY

LEONHARD EULER (1707–1783)

Euler obtained his master's degree by age 16. Euler is the first person credited with applying concepts of calculus to mechanics. Euler wrote more than 400 papers in the last 6 years of his life. (see Timoshenko, 1953)

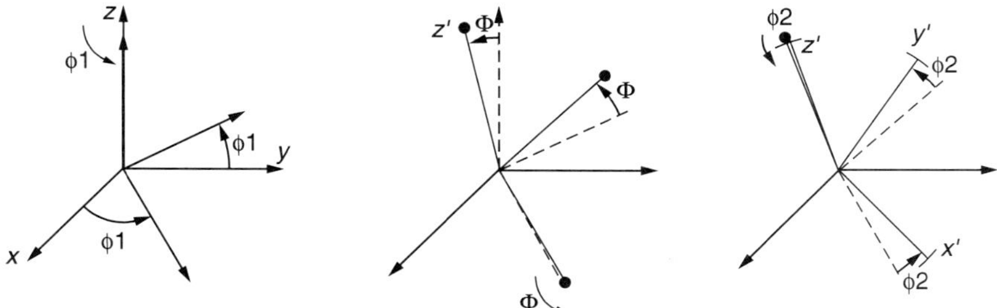

FIG. 2.4 Diagram of the sequence of counterclockwise Euler angle rotations, $\{\phi 1, \Phi, \phi 2\}$.

EXAMPLE 2.1 *Rotations on a Stereographic Projection*

The relationship between orientation and applied stress is critical in describing the mechanical performance of many crystalline materials and composites. The relationship between applied stress and crystal directions is essential in interpreting the microscopic deformation mechanisms operating in deforming crystals. Given an arbitrary dimensionless tensor **T** on the original $\{x, y, z\}$ that has values

$$T_{\mathrm{ij}} = \begin{vmatrix} 2 & 0 & 0 \\ 0 & 3 & 0 \\ 0 & 0 & -1 \end{vmatrix}$$

find **T**′ for the rotation shown in Fig. 2.5 from the initial {[100], [010], [001]} axes of a cubic crystal. The appendix to this chapter (Section 2.8) includes a discussion of stereographic projections.

The rotations applied to a unit cell of a cubic crystal initially oriented as shown in Fig. 2.2 can be written in the following ways:

Direction Indices	Rotation Matrix	Euler Rotations
$\begin{matrix} x' \\ y' \\ z' \end{matrix} \quad \begin{vmatrix} [1\ 1\ 0] \\ [\bar{1}\ 1\ 2] \\ [1\ \bar{1}\ 1] \end{vmatrix} \Rightarrow$	$a_{\mathrm{ij}} = \begin{vmatrix} \frac{1}{\sqrt{2}} & \frac{1}{\sqrt{2}} & 0 \\ \frac{-1}{\sqrt{6}} & \frac{1}{\sqrt{6}} & \frac{2}{\sqrt{6}} \\ \frac{1}{\sqrt{3}} & \frac{-1}{\sqrt{3}} & \frac{1}{\sqrt{3}} \end{vmatrix}$ *or rotations of*	$\phi 1 = 45°$ $\Phi = 54.7°$ $\phi 2 = 0°$

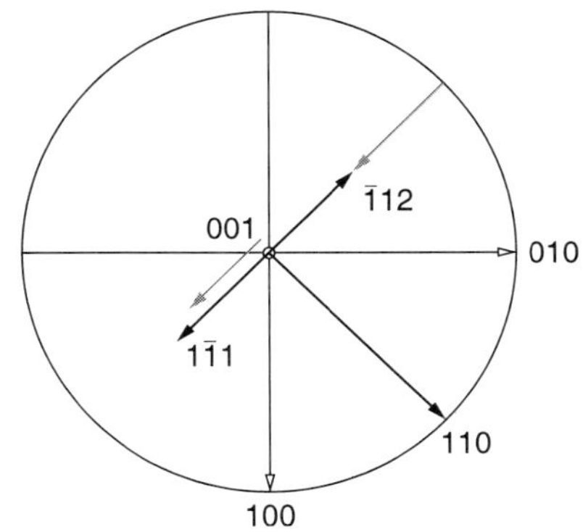

FIG. 2.5 Stereographic projection showing a sequence of two rotations, $\phi 1 = 45°$ and $\Phi = 54.7°$, for an orthogonal coordinate system applied to a cubic lattice.

The matrix multiplication method for tensor rotation given in Eq. 2.8 can be applied to **T** as

$$\mathbf{T}' = \mathbf{a}\cdot\mathbf{T}\cdot\mathbf{a}^{\mathrm{T}} =$$

$$\begin{bmatrix} \frac{1}{\sqrt{2}} & \frac{1}{\sqrt{2}} & 0 \\ \frac{-1}{\sqrt{6}} & \frac{1}{\sqrt{6}} & \frac{2}{\sqrt{6}} \\ \frac{1}{\sqrt{3}} & \frac{-1}{\sqrt{3}} & \frac{1}{\sqrt{3}} \end{bmatrix} \cdot \begin{pmatrix} 2 & 0 & 0 \\ 0 & 3 & 0 \\ 0 & 0 & -1 \end{pmatrix} \cdot \begin{bmatrix} \frac{1}{\sqrt{2}} & \frac{1}{\sqrt{2}} & 0 \\ \frac{-1}{\sqrt{6}} & \frac{1}{\sqrt{6}} & \frac{2}{\sqrt{6}} \\ \frac{1}{\sqrt{3}} & \frac{-1}{\sqrt{3}} & \frac{1}{\sqrt{3}} \end{bmatrix}^{\mathrm{T}} = \begin{pmatrix} 2.5 & 0.289 & -0.408 \\ 0.289 & 0.167 & -1.65 \\ -0.408 & -1.65 & 1.333 \end{pmatrix}$$

Note that the sum of the diagonal in **T′** *is the same as in* **T**. This means that **T** and **T′** are equivalent conditions and that the only difference is the reference coordinate system. ■

2.3 THE SECOND-RANK TENSORS OF STRAIN AND STRESS

2.3.1 Strain Tensors

As shown in Chapter 1, the nonrotational strain tensor for small strains, **ε**, can be expressed as

$$\begin{vmatrix} \varepsilon_{11} & \varepsilon_{12} & \varepsilon_{13} \\ \varepsilon_{21} & \varepsilon_{22} & \varepsilon_{23} \\ \varepsilon_{31} & \varepsilon_{32} & \varepsilon_{33} \end{vmatrix} = \begin{vmatrix} e_{11} & \frac{1}{2}(e_{12}+e_{21}) & \frac{1}{2}(e_{13}+e_{31}) \\ \frac{1}{2}(e_{21}+e_{12}) & e_{22} & \frac{1}{2}(e_{23}+e_{32}) \\ \frac{1}{2}(e_{31}+e_{13}) & \frac{1}{2}(e_{32}+e_{23}) & e_{33} \end{vmatrix}$$

In addition to the tensile, compressive, and shear strains defined in Chapter 1, we can obtain the dilation or volumetric strain from the strains on the diagonal of the strain tensor.

Normally the total volume or dilatant strain is defined as

$$\Delta \approx e_{11} + e_{22} + e_{33} \tag{2.10}$$

Unlike the strains introduced previously, the total volume strain is not altered by the orientation of the coordinate frame. Thus, the sum given in Eq. 2.10 is not altered by a change in the coordinate frame even though the individual values of the nine strain tensor terms do change. For this reason, Eq. 2.10 is called an *invariant* of the second-rank tensor for strain.

The components of either e_{ij} or ε_{ij} can be transformed by means of the standard transformation rules for second-rank tensors:

$$e_{ij}' = a_{ik}a_{jl}e_{kl} \text{ or } \varepsilon_{ij}' = a_{ik}a_{jl}\varepsilon_{kl} \tag{2.11}$$

BIOGRAPHY

AUGUSTIN CAUCHY (1789–1857)

Lagrange encouraged young Cauchy to pursue a career in mathematics. Cauchy is credited with the introduction of the concept of a stress as a pressure on a plane and the concept of strain as a differential quantity. (see Timoshenko, 1953)

The strains can be expressed as principal strains in a similar fashion to principal stresses (see Section 2.3.3).

It is important to remember that only small strains can be used for individual tensors since the tensor form is based on infinitesimal quantities. For problems wherein large strains are employed, displacements retain the real form of the material.

2.3.2 Stress Tensors

The general form of the stress tensor provides a geometric reference to the relationship between applied force and the area across which it is applied. This geometric relationship is credited to Cauchy, hence the stresses are often designated as Cauchy stresses. The stress is the intensity of force acting on an infinitesimal surface. At the limit,

$$\sigma_{\mathrm{ij}} = \lim_{\Delta A_j} \frac{\Delta F_{\mathrm{i}}}{\Delta A_{\mathrm{j}}}$$

which leads to the set of equations

$$F_1 = \sigma_{11}A_1 + \sigma_{12}A_2 + \sigma_{13}A_3$$

$$F_2 = \sigma_{21}A_1 + \sigma_{22}A_2 + \sigma_{23}A_3$$

$$F_3 = \sigma_{31}A_1 + \sigma_{32}A_2 + \sigma_{33}A_3$$

where F_{i} are the force components and A_{j} are the area components describing the stress tensor. For the object to be stationary, the terms $\sigma_{\mathrm{ij}} = \sigma_{\mathrm{ji}}$ for i = j. The general form of the stress tensor is

$$\sigma_{\mathrm{ij}} = \begin{vmatrix} \sigma_{11} & \sigma_{12} & \sigma_{13} \\ \sigma_{21} & \sigma_{22} & \sigma_{23} \\ \sigma_{31} & \sigma_{32} & \sigma_{33} \end{vmatrix}$$

where some of the shear stresses are defined on the cube shown in Fig. 2.6.

Different forms of the stress tensor have well-known definitions. A stress tensor for *simple tension* or compression in the x-direction is given by

$$\sigma = \begin{vmatrix} \sigma_{\mathrm{o}} & 0 & 0 \\ 0 & 0 & 0 \\ 0 & 0 & 0 \end{vmatrix}$$

where $\sigma_{11} = \sigma_{\mathrm{o}} = F_1/A_1$.

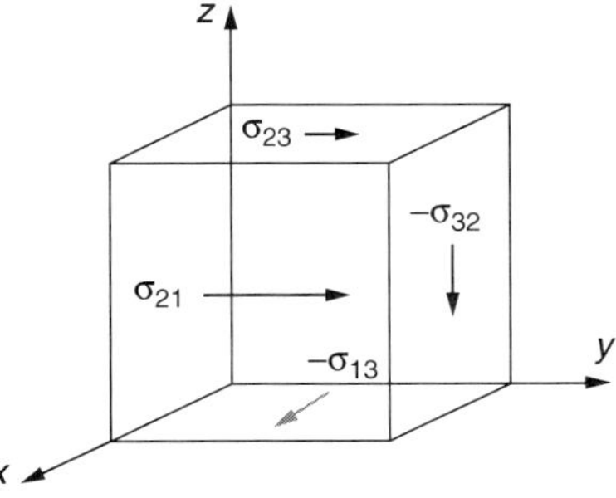

FIG. 2.6 For the unit cube, the stresses, σ_{ij}, are defined by the magnitude of the force in the i-direction and the j-face of the cube. When the direction is positive but the face of the unit cube points in the negative direction, the shear stress is negative. Note the gray arrow for $-\sigma_{13}$, which shows that the force is in the positive x-direction on the negative z-face (the forces on this cube do not sum to zero as they should).

An example of *pure shear* in the z-plane is defined by the stress tensor

$$\begin{vmatrix} 0 & \sigma_o & 0 \\ \sigma_o & 0 & 0 \\ 0 & 0 & 0 \end{vmatrix}$$

where $\sigma_{12} = \sigma_{21} = \sigma_o$, or, if the reference axes are rotated 45° about the z-axis (counter-clockwise),

$$\begin{vmatrix} \sigma_o & 0 & 0 \\ 0 & -\sigma_o & 0 \\ 0 & 0 & 0 \end{vmatrix}$$

A *hydrostatic pressure* can be represented as the stress tensor

$$\begin{vmatrix} P & 0 & 0 \\ 0 & P & 0 \\ 0 & 0 & P \end{vmatrix}$$

where P is compressive. A hydrostatic pressure remains no matter what rotation of axes takes place. By adding a hydrostatic compression (or pressure) or tension and any three pure shear stresses, any stress state can be reproduced. In a way similar to the division of strain into two components, stress can be split into two distinct components as a sum of a hydrostatic component (often called mean stress, σ_m) and a deformation shear-related component called the deviatoric stress, σ_d. The hydrostatic stress

$$\sigma_m = \frac{(\sigma_{11} + \sigma_{22} + \sigma_{33})}{3} \tag{2.12}$$

is independent of orientation or invariant. The deviatoric stress has three components that sum to zero when referred to the principal axes defined in the next section:

$$\begin{aligned} \sigma_{d1} &= \sigma_{11} - \sigma_m = \frac{(2\sigma_{11} - \sigma_{22} - \sigma_{33})}{3} \\ \sigma_{d2} &= \sigma_{22} - \sigma_m = \frac{(2\sigma_{22} - \sigma_{11} - \sigma_{33})}{3} \\ \sigma_{d3} &= \sigma_{33} - \sigma_m = \frac{(2\sigma_{33} - \sigma_{22} - \sigma_{11})}{3} \\ \sigma_{d1} &+ \sigma_{d2} + \sigma_{d3} = 0 \end{aligned} \tag{2.13}$$

2.3.3 Derivation of Principal Stresses

The transformation in plane stress derived in mechanics texts by summation of forces (see Fig. 1.7) can also be determined using tensors. Assuming a state of plane stress—i.e., σ_{13}, σ_{31}, σ_{23}, σ_{32}, and $\sigma_{33} = 0$—the stress tensor becomes

$$\sigma_{ij} = \begin{vmatrix} \sigma_{11} & \sigma_{12} & 0 \\ \sigma_{21} & \sigma_{22} & 0 \\ 0 & 0 & 0 \end{vmatrix}$$

Although we assume a symmetric tensor such that $\sigma_{12} = \sigma_{21}$, to perform the second-rank tensor transformation using Eq. 2.7, we must use the correct subscripts for each term in the transformation

$$\sigma_{ij}' = a_{ik} a_{jl} \sigma_{kl}$$

We can then define the stresses at any orientation in the x-y plane—viz., $\sigma_{ij}' = \sigma_{ij}(\phi 1)$—using the rotation matrix

$$a_{ij} = \begin{vmatrix} \cos\phi 1 & \sin\phi 1 & 0 \\ -\sin\phi 1 & \cos\phi 1 & 0 \\ 0 & 0 & 1 \end{vmatrix}$$

Then solutions for σ_{11}' and σ_{12}' are given by

$$\sigma_{11}' = \sum_{k}^{3} \sum_{l}^{3} a_{1k} a_{1l} \sigma_{kl}$$

$$\begin{aligned} \sigma_{11}' &= a_{11}a_{11}\sigma_{11} + a_{11}a_{12}\sigma_{12} + a_{12}a_{11}\sigma_{21} + a_{12}a_{12}\sigma_{22} \\ &= \cos^2\phi 1\ \sigma_{11} + 2\sin\phi 1\ \cos\phi 1\ \sigma_{12} + \sin^2\phi 1\ \sigma_{22} \end{aligned}$$

$$\begin{aligned} \sigma_{12}' &= a_{11}a_{21}\sigma_{11} + a_{11}a_{22}\sigma_{12} + a_{12}a_{21}\sigma_{21} + a_{12}a_{22}\sigma_{22} \\ &= \sigma_{12}\left(\cos^2\phi 1 - \sin^2\phi 1\right) + \left(\sigma_{22} - \sigma_{11}\right)\sin\phi 1\ \cos\phi 1 \end{aligned}$$

Using trigonometric substitutions and setting $\sigma_{12}' = 0$, the equations for principal stress become

$$s_1, s_2, \text{ or } s_3 = \frac{\sigma_{11} + \sigma_{22}}{3} \pm \sqrt{\left(\frac{\sigma_{11} - \sigma_{22}}{2}\right)^2 + \left(\sigma_{12}\right)^2} \tag{2.14}$$

where s_1 is the algebraically largest principal stress, s_2 the next largest, and s_3 the smallest. These values of stress are the same stresses obtained using Mohr's circle, which is defined in mechanics or strength of materials texts.

EXAMPLE 2.2 *Principal Stresses for Stress Tensors*

Two-Dimensional Stress States

For the Mohr's circle shown in Fig. 2.7, the values are given in tensor form as

$$\begin{vmatrix} 6.66 & 7.66 & 0 \\ 7.66 & -1.66 & 0 \\ 0 & 0 & 0 \end{vmatrix} \text{MPa}$$

where the principal stresses are $s_1 = 10$, $s_2 = 0$, and $s_3 = -5$.

Three-Dimensional Stress States

The principal stresses for the three-dimensional stress state given below are found by obtaining the roots of the determinant shown below. The difference between the diagonal matrix containing SP and the applied stress matrix defines a cubic equation after the determinant is taken.

$$\left| \begin{pmatrix} SP & 0 & 0 \\ 0 & SP & 0 \\ 0 & 0 & SP \end{pmatrix} - \begin{pmatrix} 10 & -3 & 4 \\ -3 & 5 & 2 \\ 4 & 2 & 7 \end{pmatrix} \right| \text{MPa}$$

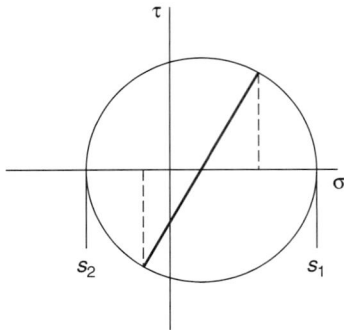

FIG. 2.7 Schematic of a Mohr's circle.

The roots of this determinant are the principal stresses.

$$s = \begin{vmatrix} 13 & 0 & 0 \\ 0 & 7.8 & 0 \\ 0 & 0 & 1.2 \end{vmatrix} \text{MPa}$$

An expression for the determinant is the cubic equation for the principal stress. We can set this equal to zero:

$$s^3 - I_1 s^2 - I_2 s - I_3 = 0$$

where I_1, I_2, and I_3 are taken from the stresses

$$\sigma = \begin{pmatrix} 10 & -3 & 4 \\ -3 & 5 & 2 \\ 4 & 2 & 7 \end{pmatrix} \text{MPa}$$

as
$$I_1 = \sigma_{11} + \sigma_{22} + \sigma_{33}$$
$$I_2 = (\sigma_{12})^2 + (\sigma_{23})^2 + (\sigma_{13})^2 - \sigma_{11} \cdot \sigma_{22} - \sigma_{22} \cdot \sigma_{33} - \sigma_{11} \cdot \sigma_{33}$$
$$I_3 = \sigma_{11} \cdot \sigma_{22} \cdot \sigma_{33} + 2 \cdot \sigma_{12} \cdot \sigma_{23} \cdot \sigma_{13} - \sigma_{11} \cdot (\sigma_{23})^2 - \sigma_{22} \cdot (\sigma_{13})^2 - \sigma_{33} \cdot (\sigma_{12})^2$$

The resulting function can be plotted to show the three roots, as shown in Fig. 2.8. ■

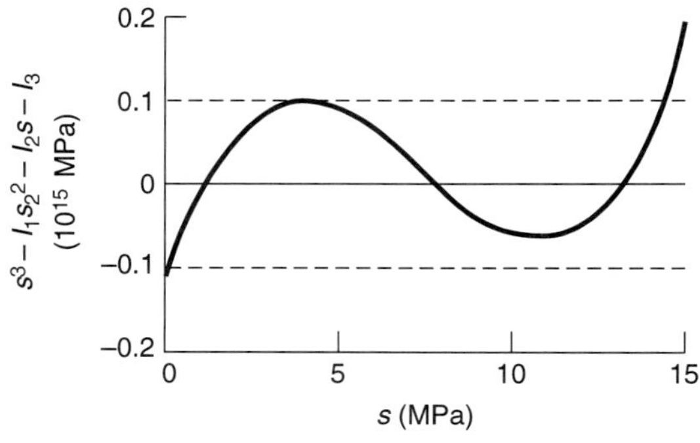

FIG. 2.8 Cubic equation of principal stresses, showing the roots $s_1 = 13$, $s_2 = 7.8$, and $s_3 = 1.2$ MPa.

2.4 DIRECTIONAL PROPERTIES

2.4.1 What Is Symmetry?

symmetry: **1.** correspondence in size, shape, and position of parts that are on opposite sides of a dividing line or center.

2. an arrangement marked by regularity and balanced proportions.

Syn. proportion, balance, harmony

—(Merriam-Webster Dictionary)

A crystallographer might like this one better:

> *"any feature of a body which by a rotation or displacement across or through a defined point or line is repeated."*

Thus, the physical properties of a crystal will be affected by the symmetry inherent to the crystal structure (Neumann's Principle of Crystal Physics). The role that symmetry plays in expressing the properties of a material with tensors is a reduction in the number of independent constants. Of course, the symmetry that is assumed for individual crystals, and in particular for materials produced by deformation processing (e.g., rolling or extrusion) or structural composite materials, includes the assumption that the macroscopic responses of materials will be homogeneous even if the materials are not homogeneous on a microscopic scale. The minimum symmetry operations for each crystal system are given in Table 2.5, along with the relationship between the interatomic distances and the interaxial angles for a lattice of each crystal structure. The types of second-rank tensors required for each crystal structure are given in Table 2.6. The values are defined with respect to the principal (symmetry) axes of each crystal. If the values were not defined on the principal axes, the number of terms in the tensor can be greater than shown. Nevertheless, all of the values would depend on the number of *independent* terms. For centrosymmetric properties, $T_{ij} = T_{ji}$, thus the second-rank tensors for monoclinic and triclinic symmetries can be expressed in three principal components—i.e., the same second-rank tensor as given for orthorhombic (orthotropic) materials.

Whenever it is possible to treat materials that have been worked by means of deformation processing techniques as homogeneous in the macroscopic sense, specific types of symmetry in property anisotropy can be anticipated. If the individual crystals are anisotropic and have a preferred orientation resulting from the deformation processing, the properties will be anisotropic. Axisymmetric forging, extrusion, and drawing processes can result in symmetry that is termed transversely isotropic and can be represented by the same

TABLE 2.5 Crystal Systems and Corresponding Crystal Symmetry

System	Minimum symmetry element	Axes, interaxial angles	Examples
Triclinic	None	$a \neq b \neq c, \alpha \neq \beta \neq \gamma$	$NaMgPO_4$
Monoclinic	1 × 2-fold	$a \neq b \neq c, \alpha = \beta = 90^\circ \neq \gamma$	Pure ZrO_2
Orthorhombic	2 × 2-fold	$a \neq b \neq c, \alpha = \beta = \gamma = 90^\circ$	Ga, $Bi_4Ti_3O_{12}$
Tetragonal	1 × 4-fold	$a = b \neq c, \alpha = \beta = \gamma = 90^\circ$	β-Sn, doped ZrO_2
Rhombohedral	1 × 3-fold	$a = b = c, \alpha = \beta = \gamma \neq 90^\circ$	As, Al_2O_3
Hexagonal	1 × 6-fold	$a1 = a2 = a3 \neq c, \alpha = \beta = 90^\circ, \gamma = 120^\circ$	Zn, Cd, Be, Mg
Cubic	4 × 3-fold	$a = b = c, \alpha = \beta = \gamma = 90^\circ$	Cu, Ag, Fe, Al

TABLE 2.6 Second-Rank Tensor Relations to Crystal Symmetry

System	Number of independent coefficients	Second-rank tensor property
Cubic	1	$\begin{vmatrix} T & 0 & 0 \\ 0 & T & 0 \\ 0 & 0 & T \end{vmatrix}$
Tetragonal Hexagonal Rhombohedral	2	$\begin{vmatrix} T_1 & 0 & 0 \\ 0 & T_2 & 0 \\ 0 & 0 & T_3 \end{vmatrix}$
Orthorhombic	3	$\begin{vmatrix} T_1 & 0 & 0 \\ 0 & T_2 & 0 \\ 0 & 0 & T_3 \end{vmatrix}$
Monoclinic	4	$\begin{vmatrix} T_{11} & 0 & T_{13} \\ 0 & T_{22} & 0 \\ T_{31} & 0 & T_{33} \end{vmatrix}$
Triclinic	6	$\begin{vmatrix} T_{11} & T_{12} & T_{13} \\ T_{21} & T_{22} & T_{23} \\ T_{31} & T_{32} & T_{33} \end{vmatrix}$

tensor expressions used for hexagonal crystals. The same axial symmetry is often seen in thin films grown by evaporation, plating, and sputtering. Rolled materials can be represented by orthotropic symmetry, which provides the same tensor representation as for orthorhombic crystals. Fiber composite materials can also be represented by symmetries that result from the processing approach. Uniaxial fiber composites produce transverse isotropy, and laminated composite materials are orthotropic. Some of these symmetries are shown schematically in Fig. 2.9. For polycrystalline materials with completely random crystal orientations, the properties should be an average of the directional properties of the constituent crystals that relates in isotropy. As mentioned at the beginning of this section, a transformation of axes provides the information necessary to recover the effects of a symmetry element. In many cases, direct inspection can also provide the answer.

2.4.2 Symmetry Effects on Tensor Rotations

Nearly all properties described by second-rank tensors are centrosymmetric. The relationship

$$M_i = T_{ij}N_j$$

implies that a reversal of direction for M_i and N_j to $-M_i$ and $-N_j$ is still satisfied by the same value of T_{ij}. For example, electrical conductivity must be the same if you simply reverse the direction of the current. This is the same action as relating a new set of axes to old axes by the transformation matrix

$$\mathbf{a} = \begin{vmatrix} -1 & 0 & 0 \\ 0 & -1 & 0 \\ 0 & 0 & -1 \end{vmatrix}$$

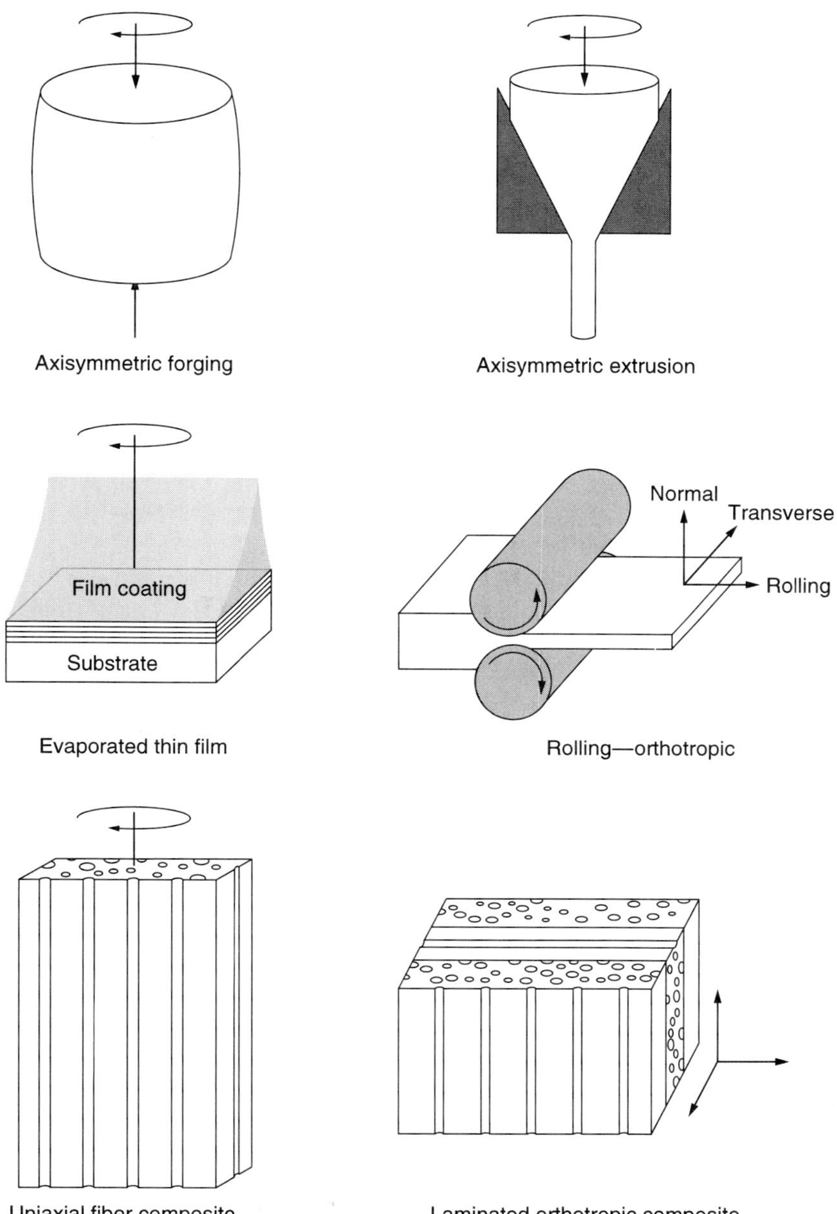

FIG. 2.9 Symmetries for materials with different processing histories.

applying the transformation formula,

$$T'_{ij} = a_{ik}a_{jl}T_{kl}$$

All a_{ij} are equal to zero unless i = j. Then

$$T'_{ij} = a_{ii}a_{jj}T_{ij} = T_{ij}$$

because any multiplication (-1×-1) is of course equal to 1. The same procedure is used to describe any symmetry element.

If a crystal contains a two-fold symmetry operation, then when we measure a certain property along a direction with respect to this axis, we get back the same value for the property after rotating the crystal 180° about this axis. Assume initially that there is a symmetry axis along the y-axis of a crystal oriented with $\{x, y, z\}$. The rotation matrix takes on the form

$$\mathbf{a} = \begin{vmatrix} -1 & 0 & 0 \\ 0 & 1 & 0 \\ 0 & 0 & -1 \end{vmatrix}$$

and gives the following results

$$T'_{11} = a_{11}a_{11}T_{11} = T_{11} \qquad T'_{22} = a_{22}a_{22}T_{22} = T_{22}$$

$$T'_{33} = a_{33}a_{33}T_{33} = T_{33} \qquad T'_{13} = a_{11}a_{33}T_{13} = T_{13}$$

but

$$T'_{23} = a_{22}a_{33}T_{23} = -T_{23} \quad \text{and} \quad T'_{12} = a_{11}a_{22}T_{12} = -T_{12}$$

Because our two-fold symmetry argument requires that $T'_{ij} = T_{ij}$, then the only possible solution is that both T_{23} and T_{12} are equal to zero. The monoclinic crystal structure displays this same symmetry relationship. The same technique is used to transform any second-rank tensor. As shown in Section 2.5, the fourth-rank tensor of elasticity has a similar relationship to crystal symmetry. Examples pertaining to the third-rank tensors for piezoelectricity are given in the problems at the end of this chapter.

2.4.3 Linear Thermal Expansion

One property that demonstrates how second-rank tensor properties are influenced by symmetry is linear thermal expansion. If a material is unconstrained, a change in temperature often results in dimensional changes that can be related to a strain tensor. If constrained, these dimensional changes can also lead to internal elastic stress within materials. An increase in temperature most often results in positive strains, hence the naming of the parameter as the coefficients of thermal expansion (CTE), α_{ij}, related by

$$\varepsilon_{ij} = \alpha_{ij}\Delta T$$

to the strain tensor ε_{ij} and $\Delta T = T_f - T_i$, where T_f is the final temperature and T_i is the initial temperature. Thermal expansion coefficients increase, usually in a fairly linear manner, with temperature. Thus the values most often seen are averages over given ranges of temperature. Table 2.7 provides thermal expansion tensor values for materials with specific symmetries. For convenience, the subscripts α_{11}, α_{22}, and α_{33} are often given as α_1, α_2, and α_3, respectively. For thermal expansion, the second-rank tensor is symmetric, which means that even the lowest symmetries can be expressed in terms of three principal values.

EXAMPLE 2.3 *Thermal Expansion Anisotropy Versus Orientation for Aluminum, Sapphire, and Alumina (Al_2O_3)*

The thermal expansion tensors for aluminum metal, sapphire (single-crystal aluminum oxide), and polycrystalline aluminum oxide with randomly oriented grains are as follows.

$$\alpha_{Al} = \begin{bmatrix} 22 & 0 & 0 \\ 0 & 22 & 0 \\ 0 & 0 & 22 \end{bmatrix} \cdot 10^{-6}, \quad \alpha_{Sapphire} = \begin{bmatrix} 5.7 & 0 & 0 \\ 0 & 5.7 & 0 \\ 0 & 0 & 7 \end{bmatrix} \cdot 10^{-6}, \quad \alpha_{Alumina} = \begin{bmatrix} 6 & 0 & 0 \\ 0 & 6 & 0 \\ 0 & 0 & 6 \end{bmatrix} \cdot 10^{-6}\left(C^{-1}\right)$$

The rotation matrix values are each given with a range of rotation angles about a direction in the plane normal to the z-axis. For the cubic aluminum and the random alumina there should be no dependence of thermal expansion on direction. The first plot (Fig. 2.10) shows this, but the second plot (Fig. 2.11) shows how the values vary with the actual direction—that is, as if the plot were made with respect to a crystal. For the sapphire, this shows how the thermal expansion coefficient varies with crystal direction, assuming that the c-direction or basal normal runs up and down the page. This type of plot is called a *property surface*.

$$a(\theta)=\begin{vmatrix}1 & 0 & 0\\ 0 & \cos\theta & \sin\theta\\ 0 & -\sin\theta & \cos\theta\end{vmatrix}$$

$$\alpha_{\text{prime}}(\theta)=a_{31}\cdot a_{31}\cdot\alpha_{11}+a_{32}\cdot a_{32}\cdot\alpha_{22}+a_{33}\cdot a_{33}\cdot\alpha_{33}$$

The horizontal axis, defined as $\alpha(\theta)\cdot\sin(\theta)$, gives the value of thermal expansion for any direction in the basal plane of a sapphire crystal, and the vertical axis, defined as $\alpha(\theta)\cdot\cos(\theta)$ gives the value of thermal expansion in the basal normal or [001] direction. For directions in between the principal axes, the property value corresponds to the length of line from the origin to the line defining the property surface.

Thermal expansion anisotropy is a factor in the failure of polycrystalline ceramic materials from thermal shock. In some cases, thermal expansion anisotropy is so great that the material fractures with even slow cooling. To demonstrate the development of stresses, we can consider a piece of sapphire bonded at high temperature to a piece of random alumina. The plate of sapphire is oriented with the c-axis normal to the piece of alumina. When the material cools from a bonding temperature of, say, 1400°C, the high-temperature creep that relieves the mismatch strains between the two materials becomes ineffective below about 1000°C. The mismatch in strain can be found as follows using the differences in α and T:

$$\Delta\alpha=\alpha_{\text{alumina}_{11}}-\alpha_{\text{sapphire}_{11}}$$

$$\Delta T=(1000-25)(C)$$

$$\Delta\alpha=0.3\times10^{-6}\left(C^{-1}\right)$$

$$\varepsilon_{\text{mismatch}}=\Delta\alpha\cdot\Delta T=2.9\times10^{-4}$$

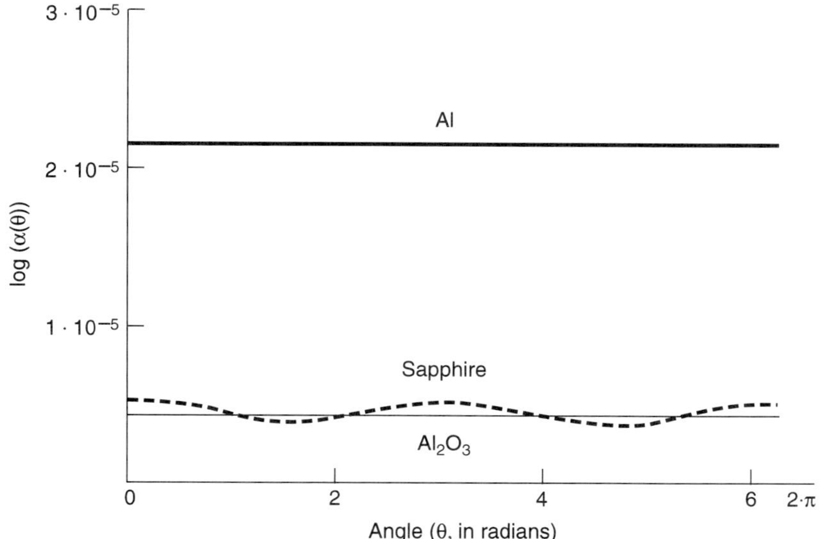

FIG. 2.10 Dependence of thermal expansion on orientation for three materials.

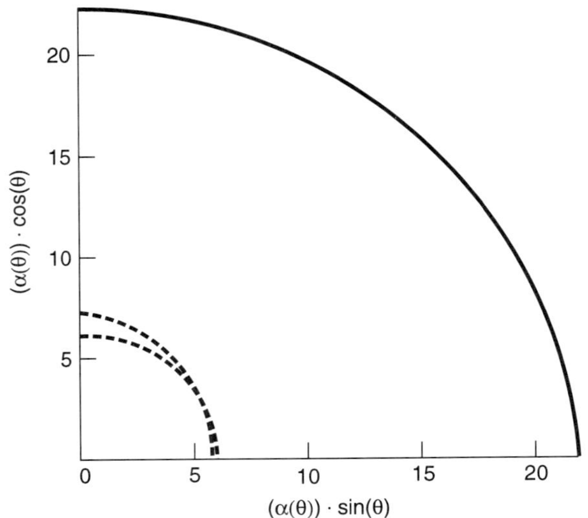

FIG. 2.11 Thermal expansion "surface" for the data plotted in Fig. 2.10 from 0 to π/2.

Then, if we use the properties of alumina (ignoring the elastic anisotropy of the sapphire), we can use Hooke's law for a biaxial stress state in a thin film ($s_1 = s_2$, $s_3 = 0$) to calculate the stresses.

$$E_{\text{alumina}} = 390 \text{ GPa} \qquad \nu_{\text{alumina}} = 0.25$$

$$s_1 = \frac{E_{\text{alumina}} \cdot \varepsilon_{\text{mismatch}}}{(1 - \nu_{\text{alumina}})} = 152 \text{ MPa}$$

TABLE 2.7 Thermal Expansion Coefficients for Materials with Different Symmetries

Material	Symmetry	Temperature °C	α_1 (10^{-6} C^{-1})	α_2 (10^{-6} C^{-1})	α_3 (10^{-6} C^{-1})
Aluminum	Random	25–100	22	22	22
Aluminum*	Cubic	25–100	22	22	22
Iron	Random	25–100	12	12	12
Iron*	Cubic	25–100	12	12	12
Silicon nitride	Random	25–100	1.8	1.8	1.8
Zinc	Random	25–100	20	20	20
Aluminum oxide	Random	25–100	6.0	6.0	6.0
Calcite	Random	25–100	4.5	4.5	4.5
Gypsum	Random	25–100	24	24	24
HDPE	Random	25–100	90	90	90
Epoxy	Random	25–100	60	60	60
Silicon nitride*	Hexagonal	100–200	1.3	1.3	2.8
Zinc*	Hexagonal	25–100	0	0	60
Sapphire*	Rhombohedral	25–100	5.7	5.7	7
Calcite*	Rhombohedral	25–100	–5.6	–5.6	25
Gypsum*	Monoclinic	25–100	1.6	42	29
60 v/o glass fiber in epoxy	Axisymmetric‡	25–100	34	34	6
Kevlar	Axisymmetric‡	25–100	59	59	–2

* Single crystals.

‡ For axisymmetric materials, α_3 is the thermal expansion coefficient along the axis of symmetry.

BIOGRAPHY

ROBERT HOOKE (1635–1703)

Robert Hooke worked with Robert Boyle, of Boyle's gas law, for several years and specialized in precise measurements. Through measurement of the relationship between force and deformation of long wires, he was able to establish that the relationship is linear. As a contemporary of Newton, Hooke was concerned that Newton might get credit for his discovery, so he originally published his law in the form of an anagram, "ceiiinsssttuv" which in latin is "ut tensio sic vis," and translates roughly to "as goes the tension, so goes the stretch." By publishing this relationship in a coded form, he hoped to prevent the law from being understood until the printings were widely disseminated.

The same types of stresses are likely between the individual grains in any noncubic material. Problems with this type of thermal expansion mismatch have been demonstrated for alumina reinforced with sapphire fibers wherein the long axes of the fibers are parallel to the c-direction of the sapphire crystals. ■

2.5 ELASTICITY

As defined in Chapter 1, the linear elastic response of materials is valid for small strains. In very strong materials (high yield strength), elastic strains on the order of 1 percent may begin to show the nonlinearity defined by the interatomic potential. Although we do not have a detailed understanding of atomic or molecular displacements, the fourth-rank tensor of elasticity is a handy tool for describing the macroscopic elastic response of anisotropic materials.

2.5.1 Specification of the Elasticity Tensor

The implications of Hooke's law,

$$\sigma = \mathbf{C}\,\varepsilon \qquad \text{or} \qquad \varepsilon = \mathbf{S}\sigma \tag{2.15}$$

where σ is stress, **C** is stiffness, **S** is compliance, and ε is strain (note that we are excluding rotation here and focusing only on shape change), become more complex in a three-dimensional state of stress and strain:

$$\sigma_{ij} = C_{ijkl}\varepsilon_{kl} \tag{2.16}$$

or

$$\varepsilon_{ij} = S_{ijkl}\sigma_{kl} \tag{2.17}$$

which gives 81 components for the fourth-rank tensors of stiffness and compliance. As long as the body in question is not moving, stress and strain are symmetric tensors so that

$$C_{ijkl} = C_{jikl}$$

and

$$C_{ijkl} = C_{ijlk}$$

taking the tensor of **C** or **S** terms (with only the subscripts given)

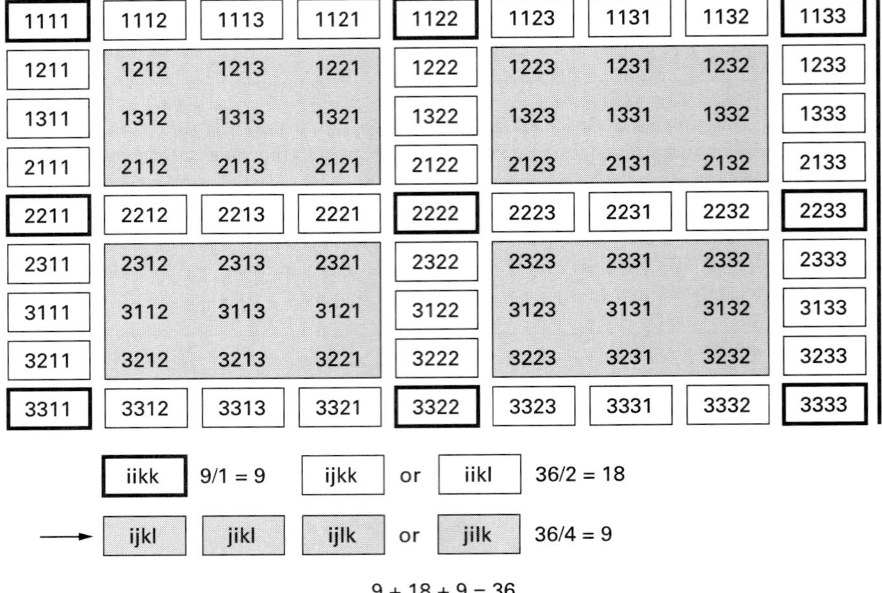

to the matrix form that is often used to describe it,

$$\begin{vmatrix} 11 & 12 & 13 & 14 & 15 & 16 \\ 21 & 22 & 23 & 24 & 25 & 26 \\ 31 & 32 & 33 & 34 & 35 & 36 \\ 41 & 42 & 43 & 44 & 45 & 46 \\ 51 & 52 & 53 & 54 & 55 & 56 \\ 61 & 62 & 63 & 64 & 65 & 66 \end{vmatrix}$$

using the following rules for stress and strain.

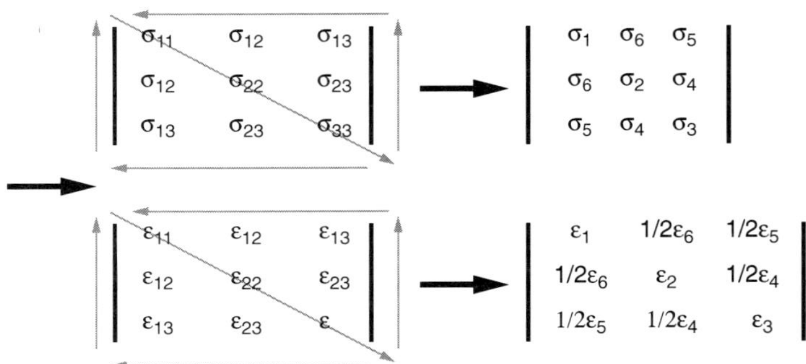

The 36 components in the matrix above can be reduced to 21 by assuming that $C_{1122} = C_{2211}$ or, in the contracted notation above, $C_{12} = C_{21}$. The abbreviation rules for subscripts are

Tensor notation:	11	22	33	23, 32	31, 13	12, 21
Matrix notation:	1	2	3	4	5	6

(2.18)

The factors of 2 that relate

$$\varepsilon_4 = 2\varepsilon_{23} \qquad \varepsilon_5 = 2\varepsilon_{13} \qquad \varepsilon_6 = 2\varepsilon_{12}$$

result in

$$S_{mn} = 2S_{ijkl} \text{ for one of either m or n equal to 4, 5, or 6}$$

$$S_{mn} = 4S_{ijkl} \text{ for both m and n equal to 4, 5, or 6}$$

that is, $S_{11} = S_{1111}$, $S_{15} = 2\ S_{1113}$, and $S_{66} = 4\ S_{1212}$. Using such a definition, the result is

$$\varepsilon_i = S_{ij}\sigma_j$$

No factors of 2 or 4 are encountered in the relationship for stiffness and

$$\sigma_i = C_{ij}\varepsilon_j$$

This reduction of subscripts is a convenience, because some tensor operations must be performed in the full 9×9 tensor. Fortunately, symmetry reduces the complexity of these operations.

2.5.2 Transformations of the Elasticity Tensors

If a specific set of axes is not given for a crystal, then there will 21 different elastic constants. However, by referencing the axis to the crystal structure, symmetry requirements may cause some of the constants to be zero. The effects of symmetry elements can be seen by performing the corresponding rotations of the coordinate system. The transformation rule for a fourth-rank tensor is

$$T_{ijkl}' = a_{im}a_{jn}a_{ko}a_{lp}T_{mnop} \tag{2.19}$$

This can be confirmed by evaluating the equalities

$$\varepsilon_{ij}' = a_{ik}a_{jl}\varepsilon_{kl}$$

$$\varepsilon_{kl} = s_{klmn}\sigma_{mn}$$

$$\sigma_{mn} = a_{om}a_{pn}\sigma_{op}'$$

and substituting into the definition of a compliance to yield

$$S_{ijkl}' = a_{im}a_{jn}a_{ko}a_{lp}S_{mnop} \tag{2.20}$$

The same expression can be applied to the stiffnesses, C_{ijkl}, by replacing S with C in Eq. 2.20.

Tensor Rotations by Matrix Multiplication

For fourth-rank tensor matrices, multiplication is substantially more difficult because a special nine-by-nine rotation matrix must be generated. Hirth and Lothe (1982) describe in detail the procedures for this matrix operation, and similar relations are given in composites texts. Computer and calculator programs that perform matrix multiplication can be very effective for numerical solutions of these problems (see Section 2.7 and Examples 2.5 and 2.6). Symbolic mathematics programs can be used to find generalized expressions.

There are easier techniques for finding the effects of two- and four-fold symmetry. For example, consider the compliance S_{16} in an orthorhombic crystal. S_{16} measures the extension in the x-direction (ε_1) when the crystal is sheared about the z-axis (ε_{12}), as shown in Fig. 2.12*a*.

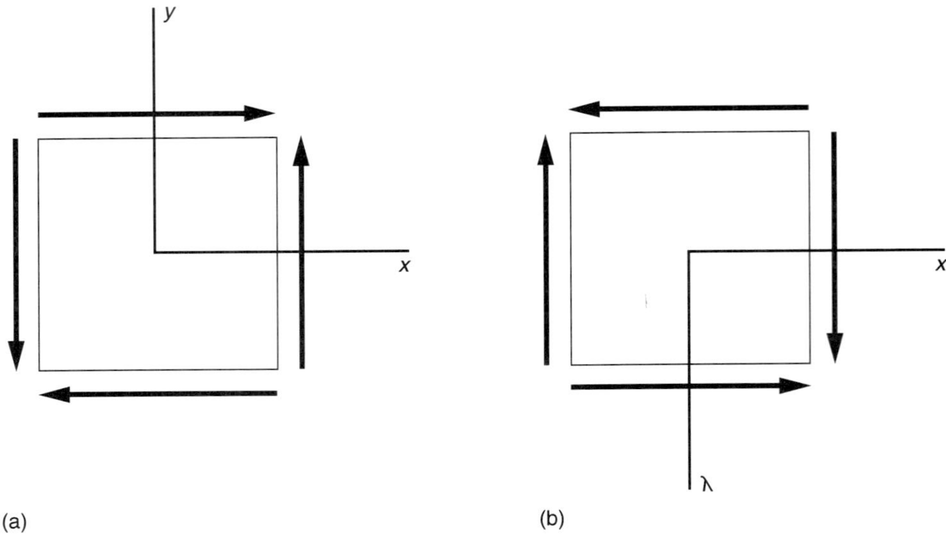

FIG. 2.12 A set of axes and a deformation element (a) before and (b) after a counterclockwise rotation about *x*.

Now apply a two-fold operation about the x-axis, shown in Fig. 2.12*b*. The shear stress in both cases must yield

$$\varepsilon_1' = \varepsilon_1$$

But the rotation of 180° about z changes the sign of the shear stress, producing the same strain. Thus

$$\varepsilon_1 = S_{16}\,\sigma_6$$

$$\varepsilon_1' = S_{16} - \sigma_6$$

and

$$-\varepsilon_1' = \varepsilon_1$$

which means that $\varepsilon_1 = 0$ and $S_{16} = 0$. This, in turn, means that in the presence of a two-fold operation the application of a stress $\sigma_6 = \sigma_{12}$ produces no normal strain in the *x*-direction. Completing a similar process for each of the terms in the compliance or stiffness matrix would lead to the following nonzero terms

$$\mathbf{2^x} = \left|\begin{array}{cccccc} 11 & 12 & 13 & 14 & & \\ 12 & 22 & 23 & 24 & & \\ 13 & 23 & 33 & 34 & & \\ 14 & 24 & 34 & 44 & & \\ & & & & 55 & 56 \\ & & & & 56 & 66 \end{array}\right|$$

Another method is termed *direct inspection*. To demonstrate this technique, we will add a second two-fold axis operating about *y* to the matrix of elastic constants given above for two-fold symmetry about *x*. For a two-fold axis about *y*, we know that the axes make the following changes

$$1 \rightarrow -1 \qquad 2 \rightarrow 2 \qquad 3 \rightarrow -3$$

which means that for nonreduced subscripts (e.g., stress and strain given by two subscripts), the signs of the subscripts follows the same rules of sign as for multiplication

$$11 \rightarrow 11 \qquad 22 \rightarrow 22 \qquad 33 \rightarrow 33$$

$$23 \rightarrow -23 \qquad 13 \rightarrow 13 \qquad 12 \rightarrow -12$$

Then, for reduced subscripts,

$$1 \rightarrow 1 \qquad 2 \rightarrow 2 \qquad 3 \rightarrow 3 \qquad 4 \rightarrow -4 \qquad 5 \rightarrow 5 \qquad 6 \rightarrow -6$$

yielding

$$\begin{vmatrix} 11 & 12 & 13 & -14 & & \\ 12 & 22 & 23 & -24 & & \\ 13 & 23 & 33 & -34 & & \\ -14 & -24 & -34 & 44 & & \\ & & & & 55 & -56 \\ & & & & -56 & 66 \end{vmatrix}$$

Thus, the terms S_{14}, S_{24}, S_{34}, and S_{56} are equal to zero, yielding

$$\mathbf{2^x 2^y} = \begin{vmatrix} 11 & 12 & 13 & & & \\ 12 & 22 & 23 & & & \\ 13 & 23 & 33 & & & \\ & & & 44 & & \\ & & & & 55 & \\ & & & & & 66 \end{vmatrix} \tag{2.21}$$

which is equivalent to a material with three orthogonal symmetry operations $\mathbf{2^x 2^y 2^z}$. This particular matrix, called the Orthotropic Elastic Tensor, is useful for determining the elastic properties of crystals and fiber composite materials.

2.5.3 Elastic Constants for Various Symmetries

Orthotropy or Orthorhombic Crystals If we wish to express the elastic strains for a given set of applied stresses, then the strains are given by equations of the type

$$\varepsilon_1 = S_{11}\sigma_1 + S_{12}\sigma_2 + S_{13}\sigma_3$$

$$\varepsilon_4 = 2\varepsilon_{23} = S_{44}\sigma_{23} = S_{44}\sigma_4$$

and if we wish to express the stress at a given set of elastic strains,

$$\sigma_1 = C_{11}\varepsilon_1 + C_{12}\varepsilon_2 + C_{13}\varepsilon_3$$

$$\sigma_4 = \sigma_{23} = C_{44}\varepsilon_4 = 2C_{44}\varepsilon_{23}$$

Consider a simple tensile test in the x-direction. The resulting stress and strain tensors are

$$\boldsymbol{\sigma} = \begin{vmatrix} \sigma_1 & 0 & 0 \\ 0 & 0 & 0 \\ 0 & 0 & 0 \end{vmatrix}, \quad \boldsymbol{\varepsilon} = \begin{vmatrix} \varepsilon_1 & 0 & 0 \\ 0 & -\nu_{12}\varepsilon_1 & 0 \\ 0 & 0 & -\nu_{13}\varepsilon_1 \end{vmatrix}$$

BIOGRAPHY

S. D. POISSON (1781–1840)

Poisson taught undergraduate calculus and was a student of Laplace and Lagrange. Using the assumptions of Cauchy and Ostrogradsky, ν should be equal to one-fourth in all isotropic solids. Thus, isotropic solids should have only one elastic constant. It was Green that demonstrated that isotropic solids should have two independent elastic constants.

where Poisson's ratio, ν_{ij}, is defined as the negative of the ratio of the strain in the i-direction to the strain in the j-direction owing to an applied stress in the i-direction, yielding

$$\varepsilon_2 = -\nu_{12}\varepsilon_1 \quad \text{or} \quad \nu_{12} = \frac{-\varepsilon_2}{\varepsilon_1} \tag{2.22}$$

Another familiar quantity is Young's modulus, which can be related to principal directions by E_i, the modulus of elasticity in the i-direction. Then

$$\varepsilon_1 = S_{11}\sigma_1 = \frac{\sigma_1}{E_1}$$

$$\varepsilon_2 = S_{21}\sigma_1 = -\nu_{12}\varepsilon_1 = \frac{-\nu_{12}\sigma_1}{E_1}$$

$$\varepsilon_3 = S_{31}\sigma_1 = -\nu_{13}\varepsilon_1 = \frac{-\nu_{13}\sigma_1}{E_1}$$

and

$$S_{11} = \frac{1}{E_1} \qquad S_{21} = \frac{-\nu_{12}}{E_1} \qquad S_{31} = \frac{-\nu_{13}}{E_1}$$

For Poisson's ratio, $\nu_{ij} \neq \nu_{ji}$, because

$$\nu_{12} = \frac{-S_{21}}{S_{11}}$$

and we already know that $S_{12} = S_{21}$.

If we put all of this together, we can write the matrix of values for the compliances as

$$S = \begin{vmatrix} \frac{1}{E_1} & \frac{-\nu_{21}}{E_2} & \frac{-\nu_{31}}{E_3} & 0 & 0 & 0 \\ \frac{-\nu_{12}}{E_1} & \frac{1}{E_2} & \frac{-\nu_{32}}{E_3} & 0 & 0 & 0 \\ \frac{-\nu_{13}}{E_1} & \frac{-\nu_{23}}{E_2} & \frac{1}{E_3} & 0 & 0 & 0 \\ 0 & 0 & 0 & \frac{1}{M_4} & 0 & 0 \\ 0 & 0 & 0 & 0 & \frac{1}{M_5} & 0 \\ 0 & 0 & 0 & 0 & 0 & \frac{1}{M_6} \end{vmatrix} \tag{2.23}$$

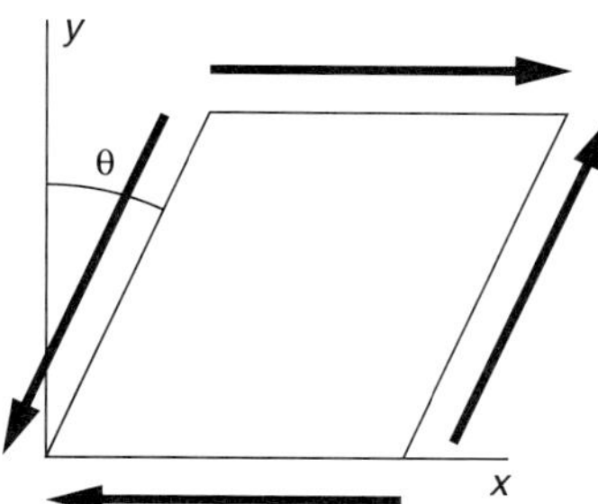

FIG. 2.13 Simple shearing of an element by an angle θ. The strains shown are exaggerated well beyond the linear elastic limit.

Now, consider the simple shearing of the element in Fig. 2.13.

$$\boldsymbol{\sigma} = \begin{vmatrix} 0 & \sigma_{12} & 0 \\ \sigma_{21} & 0 & 0 \\ 0 & 0 & 0 \end{vmatrix}, \qquad \boldsymbol{\varepsilon} = \begin{vmatrix} 0 & \varepsilon_{12} & 0 \\ \varepsilon_{21} & 0 & 0 \\ 0 & 0 & 0 \end{vmatrix}$$

where

$$\mathbf{e} = \begin{vmatrix} 0 & 0 & 0 \\ \dfrac{\sigma_{12}}{\mu_6} & 0 & 0 \\ 0 & 0 & 0 \end{vmatrix}$$

From mechanics of materials, the proportionality constant between the angle θ and the shear stress σ_{12} is μ_6 (possible alternatives include $\mu_6 = 1/S_{66} = C_{66} = C_{1212}$), with the shear modulus in the *x-y* plane referenced to the principal axes.

Remember that

$$\varepsilon_{12} = \frac{1}{2}(e_{12} + e_{21}) = \frac{\sigma_{12}}{2\mu_6} \approx \frac{\tan\theta}{2}$$

so, repeating the same pattern of the definitions (e.g., $\varepsilon_4 = 2\varepsilon_{23} = S_{44}\sigma_{23}$)

$$\varepsilon_{12} = \frac{S_{66}\sigma_{12}}{2} = \frac{\sigma_{12}}{2\mu_6}$$

Cubic Symmetry For cubic crystals, there are four three-fold symmetry axes (along the <111> body diagonals) such that

$$S_{11} = S_{22} = S_{33} \qquad S_{12} = S_{23} = S_{31} \qquad S_{44} = S_{55} = S_{66}$$

This reduces the nine constants for orthotropic symmetry to three: S_{11}, S_{12}, and S_{44}. An anisotropy factor, A, can be defined for cubic materials where

$$A = \frac{C_{44}}{\frac{1}{2}(C_{11} - C_{12})} = \frac{2(S_{11} - S_{12})}{S_{44}} \tag{2.24}$$

In a cubic material, the **S**-terms are related to the **C**-terms by

$$C_{11} = \frac{S_{11} + S_{12}}{(S_{11} - S_{12})(S_{11} + 2S_{12})}$$

$$C_{12} = \frac{-S_{12}}{(S_{11} - S_{12})(S_{11} + 2S_{12})}$$

$$C_{44} = \frac{1}{S_{44}}$$

Using the direction cosines, l, m, and n, for a particular reference direction, one can determine the elastic properties of a cubic single crystal in a particular direction by the relationship

$$\frac{1}{E_{hkl}} = S_{11}' = S_{11} - 2\left[S_{11} - S_{12} - \frac{1}{2}S_{44}\right]l^2m^2 + m^2n^2 + l^2n^2 \tag{2.25}$$

Table 2.8 lists C_{ij} and S_{ij} values for several cubic and hexagonal materials. Figure 2.14 shows Young's modulus surfaces for Eq. 2.25 plotted with reference to a three-dimensional coordinate system.

TABLE 2.8 Stiffness and Compliance Values for Various Crystals (at Room Temperature)

Material	C_{11}	C_{33}	C_{12}	C_{23}	C_{44} (GPa)	S_{11}	S_{33}	S_{12}	S_{23}	S_{44} (10^{-3} GPa^{-1})
Cubic										
Aluminum	108	—	61	—	28	15.7	—	–5.70	—	35.1
Copper	168	—	121	—	75.4	15.0	—	–6.30	—	13.3
Gold	186	—	157	—	42	23.3	—	–10.7	—	24
Lead	47	—	40	—	14.4	93	—	–42.4	—	69
Nickel	247	—	147	—	125	7.3	—	–2.7	—	8.0
Silver	124	—	93.4	—	46	22.9	—	–9.8	—	22
Iron	237	—	141	—	116	8.00	—	–2.8	—	8.60
Molybdenum	460	—	176	—	110	2.8	—	–0.78	—	9.1
Niobium	246	—	134	—	28.7	6.6	—	–2.3	—	34.8
Tantalum	267	—	161	—	82.5	6.9	—	–2.6	—	12.2
Tungsten	500	—	200	—	140	2.6	—	–0.7	—	6.6
C (diamond)	1076	—	125	—	566	1.1	—	–0.22	—	2.1
Germanium	129	—	48.3	—	67.1	9.8	—	–2.7	—	15
Silicon	166	—	63.9	—	79.6	7.7	—	–2.1	—	12.6
Potassium	4.6	—	3.7	—	2.6	82	—	–370	—	380
Zinc sulfide	108	—	72	—	41	20	—	–8.00	—	24
Magnesium oxide	296	—	95	—	156	4.00	—	–1.0	—	6.5
Sodium chloride	49	—	12	—	13	23	—	–4.7	—	79
Lithium fluoride	111	—	42	—	62.8	11.1	—	–3.1	—	15.9
Titanium carbide	513	—	106	—	178	2.1	—	–0.36	—	5.61
Hexagonal										
Cadmium	121	51.3	40	41	20	12.3	35.5	–1.5	–9.3	54
Titanium	160	181	92	69	46.5	9.70	6.9	–4.7	–1.8	2.15
Zinc	161	61	34.1	50	38.3	8.4	28.4	0.5	–7.3	26.1
Magnesium	59.7	61.7	26.2	21.7	16.4	22	20	–8.0	–5.0	61

(Kelly, Groves, and Kidd, 2000, Simmons and Wang, 1971, Hirth and Lothe, 1982, and Nye, 1957)

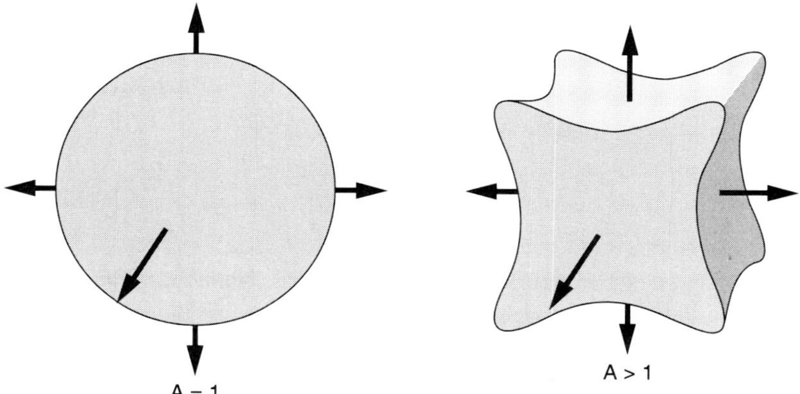

FIG. 2.14 Schematic figures showing the different symmetries for Young's modulus as a function of orientation for an isotropic cubic crystal and a cubic crystal with A > 1. (The true modulus surfaces are smooth in nature and do not have any *edges*.)

Isotropy When the anisotropy factor is equal to 1, there are just two independent components—e.g., C_{11} and C_{12}. In this instance, the rigidity or shear modulus μ is given by

$$\mu = C_{44} = \frac{1}{2}(C_{11} - C_{12}) = \frac{1}{S_{44}} \tag{2.26}$$

and λ is given by

$$\lambda = C_{12} \tag{2.27}$$

These two constants are known as Lamé constants and can be used to describe all the elastic properties of isotropic materials.

Poisson's ratio, defined in Eq. 1.23, can be expressed in terms of two compliances, two stiffnesses, or the two Lamé constants μ and λ:

$$\begin{aligned} \nu &= -\frac{S_{12}}{S_{11}} = -\frac{C_{12}}{C_{11} + C_{12}} \\ &= \frac{1}{2(1 + \mu/\lambda)} \end{aligned} \tag{2.28}$$

An added type of elastic constant has no directionality and by its very definition is a material's elastic response to changes in volumetric strain or hydrostatic pressure. The *compressibility (K) or bulk modulus (B)* relate hydrostatic or mean stress to the volume strain:

$$K = \frac{1}{B} = \frac{\Delta}{\sigma_m} \tag{2.29}$$

which is equivalent to

$$K = \lambda + \frac{2\mu}{3}$$

If we define *Young's modulus* in similar terms, we can write

$$E = \frac{1}{S_{11}} \tag{2.30}$$

which is equivalent to

$$E = \frac{\mu(3 + 2\mu/\lambda)}{(1 + \mu/\lambda)}$$

EXAMPLE 2.4 *Application of the Elasticity Tensor to Determine Elastic Constants*

Suppose we wish to determine the elastic constants of an object with rotational symmetry. Examples of a material with such a macroscopic response could include hexagonal crystals, polycrystalline materials with fiber textures, or fiber composite materials with fibers running only in one direction, as shown in Fig. 2.9. The elasticity tensor of compliances for such materials has the following nonzero terms (with just the subscripts given):

$$\begin{vmatrix} 11 & 12 & 13 & & & \\ 12 & 11 & 13 & & & \\ 13 & 13 & 33 & & & \\ & & & 44 & & \\ & & & & 44 & \\ & & & & & 2(11-12) \end{vmatrix}$$

Imagine that the only available means of making this measurement is a set of weights and a device for analyzing the voltage in surface-mounted resistance strain gages. We can easily find the values of Young's modulus E_i and Poisson's ratio ν_{ij} for any orientation specimen cut from the material by measuring the strains that result from an applied load to find the compliances. Unfortunately, it is essentially impossible to measure the shear modulus directly with this type of equipment. How then do we determine the value of the shear modulus, $\mu_4 = 1/S_{44}$? If we measure Poisson's ratio and Young's moduli at orientations off the symmetry axes defined here, then we have the possibility of getting measurements that we can translate back into the shear modulus in the original coordinate system. So we will perform such a rotation to find μ by cutting a specimen at 45° from the rotational symmetry axis, as shown in Fig. 2.15.

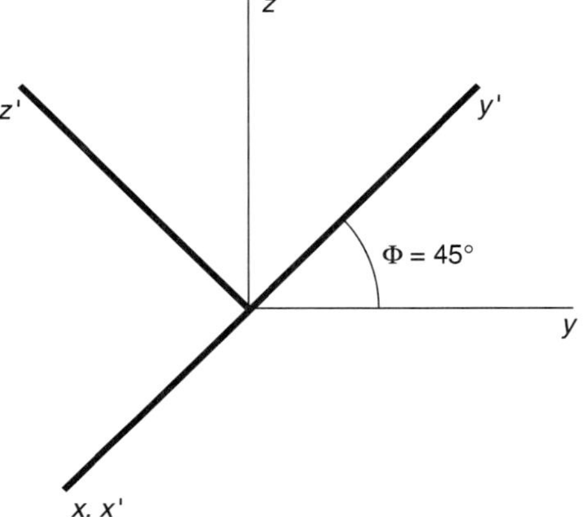

FIG. 2.15 Rotation about x using the Euler angle Φ.

The rotation matrix is then

$$\begin{vmatrix} 1 & 0 & 0 \\ 0 & \cos\Phi & \sin\Phi \\ 0 & -\sin\Phi & \cos\Phi \end{vmatrix}$$

leading to the following sequence of calculations for S'_{44}, where S'_{44} is the inverse shear modulus (or shear compliance) in the original coordinate system, which is related to the specimen cut at 45° by the transformation shown above.

$$S'_{44} = 4S'_{2323} = 4a_{2m}a_{3n}a_{2o}a_{3p}S_{mnop}$$

Note first that any **S′** values with subscripts of 1 will be zero, yielding

$$\begin{aligned} S'_{2323} = {} & a_{22}a_{32}a_{22}a_{32}S_{2222} + a_{23}a_{33}a_{23}a_{33}S_{3333} + \\ & a_{22}a_{32}a_{23}a_{33}S_{2233} + a_{23}a_{33}a_{22}a_{32}S_{3322} + \\ & a_{22}a_{33}a_{22}a_{33}S_{2323} + a_{23}a_{32}a_{23}a_{32}S_{3232} + \\ & a_{22}a_{33}a_{23}a_{32}S_{2332} + a_{23}a_{32}a_{22}a_{33}S_{3223} \end{aligned}$$

$$= \cos^2\Phi \sin^2\Phi\, S_{2222} + \cos^2\Phi \sin^2\Phi\, S_{3333} - 2\cos^2\Phi \sin^2\Phi\, S_{2233} + \cos^4\Phi\, S_{2323} + \sin^4\Phi\, S_{3232} - \cos^2\Phi \sin^2\Phi\, S_{3223} - \cos^2\Phi \sin^2\Phi\, S_{2332}$$

Substituting $\frac{1}{4}S_{44}$ for each of the shear components and $1 - \cos^2\Phi$ for each $\sin^2\theta$ gives the final solution

$$S'_{44} = 4S'_{2323} = 4\cos^2\Phi(1-\cos^2\Phi)\ [S_{22} - 2S_{23} + S_{33}] + (2\cos^2\Phi - 1)^2 S_{44}$$

Using a value of 45° for Φ, the solution is then

$$S'_{44} = [S_{22} - 2S_{23} + S_{33}]$$

which means that by measuring Poisson's ratio and Young's moduli at 45° from the z symmetry axis we can recover the shear modulus of the material. ■

EXAMPLE 2.5 *Stiffnesses for Mo at 25°C in GPa*

Most math software and programming languages are not readily suited to tensors of greater than second rank. This example demonstrates one way to map subscripts from 4 to 2 using the definitions provided by a special matrix **M**. First, we can assign the proper 9 × 9 matrix of 81 terms by setting the magnitudes of the three independent elastic constants for cubic materials on their principal axes.

$$c_{11} = 460 \qquad c_{12} = 176 \qquad c_{44} = 110\ \text{(GPa)}$$

Then the 9 × 9 matrix contains these values at each appropriate position, with

$$\mathbf{C} = \begin{bmatrix} c_{11} & c_{12} & c_{12} & 0 & 0 & 0 & 0 & 0 & 0 \\ c_{12} & c_{11} & c_{12} & 0 & 0 & 0 & 0 & 0 & 0 \\ c_{12} & c_{12} & c_{11} & 0 & 0 & 0 & 0 & 0 & 0 \\ 0 & 0 & 0 & c_{44} & 0 & 0 & c_{44} & 0 & 0 \\ 0 & 0 & 0 & 0 & c_{44} & 0 & 0 & c_{44} & 0 \\ 0 & 0 & 0 & 0 & 0 & c_{44} & 0 & 0 & c_{44} \\ 0 & 0 & 0 & c_{44} & 0 & 0 & c_{44} & 0 & 0 \\ 0 & 0 & 0 & 0 & c_{44} & 0 & 0 & c_{44} & 0 \\ 0 & 0 & 0 & 0 & 0 & c_{44} & 0 & 0 & c_{44} \end{bmatrix}$$

The matrix **M** is defined to translate the positions in stress or strain tensors to rows and columns of the 6×6 matrix version for C that we will use to perform the calculation:

$$\mathbf{M} = \begin{bmatrix} 1 & 6 & 5 \\ 9 & 2 & 4 \\ 8 & 7 & 3 \end{bmatrix}$$

A rotation matrix where $x' = [1\bar{2}1]$, $y' = [111]$, and $z' = [\bar{1}01]$ is given by

$$\mathbf{a} = \begin{bmatrix} \frac{1}{\sqrt{6}} & \frac{-2}{\sqrt{6}} & \frac{1}{\sqrt{6}} \\ \frac{1}{\sqrt{3}} & \frac{1}{\sqrt{3}} & \frac{1}{\sqrt{3}} \\ \frac{-1}{\sqrt{2}} & 0 & \frac{1}{\sqrt{2}} \end{bmatrix}$$

Range variables can be defined for each subscript type as integers 1, 2, or 3 by

$$i = 1 \ldots 3 \qquad j = 1 \ldots 3 \qquad k = 1 \ldots 3 \qquad l = 1 \ldots 3$$

$$m = 1 \ldots 3 \qquad n = 1 \ldots 3 \qquad o = 1 \ldots 3 \qquad p = 1 \ldots 3$$

The double subscripts used in the equation

$$C'_{M_{ij}M_{kl}} = \sum_m \sum_n \sum_o \sum_p a_{ij} \cdot a_{jn} \cdot a_{ko} \cdot a_{lp} \cdot C_{M_{mn}M_{op}}$$

define the same operations for summation over m, n, o, and p on the righthand side to define each of the terms in the new matrix C′ as

$$\mathbf{C}' = \begin{bmatrix} 428 & 197 & 187 & 0 & 0 & -15 & 0 & 0 & -15 \\ 197 & 417 & 197 & 0 & 0 & 0 & 0 & 0 & 0 \\ 187 & 197 & 428 & 0 & 0 & 15 & 0 & 0 & 15 \\ 0 & 0 & 0 & 131 & 15 & 0 & 131 & 15 & 0 \\ 0 & 0 & 0 & 15 & 121 & 0 & 15 & 121 & 0 \\ -15 & 0 & 15 & 0 & 0 & 131 & 0 & 0 & 131 \\ 0 & 0 & 0 & 131 & 15 & 0 & 131 & 15 & 0 \\ 0 & 0 & 0 & 15 & 121 & 0 & 15 & 121 & 0 \\ -15 & 0 & 15 & 0 & 0 & 131 & 0 & 0 & 131 \end{bmatrix} \text{GPa}$$

■

2.6 EFFECTIVE PROPERTIES OF MATERIALS: ORIENTED POLYCRYSTALS AND COMPOSITES

The properties of polycrystals and composites are often calculated as weighted averages of the components expressed in their volume fractions. Consider the uniaxial fiber composite in Fig. 2.9. If we apply a tensile stress along the direction of the fibers and assume that the fibers and the matrix deform together without slipping (i.e., both phases undergo the same elastic strain), then the stress carried by the composite is

$$\sigma = \sigma_{fiber} V_{fiber} + \sigma_{matrix} V_{matrix}$$

where V_{fiber} is the volume fraction of the fiber and V_{matrix} is $(1 - V_{fiber})$. If each of the phases is isotropic, we can write

$$\sigma = E_{fiber}\varepsilon_{fiber}V_{fiber} + E_{matrix}\varepsilon_{matrix}V_{matrix}$$

However, because $\varepsilon_{fiber} = \varepsilon_{matrix} = \varepsilon_{composite}$,

$$E_{composite} = E_{fiber}V_{fiber} + E_{matrix}V_{matrix} \tag{2.31}$$

Eq. 2.31 can be generalized as

$$E_{composite} = \Sigma\, E_i V_i$$

to give the *upper bound* or *Voigt average* of Young's modulus. To include more complex loadings, the upper bound is the integral over all of the stiffnesses of the individual phases or crystal orientations

$$< C_{ijkl} >^{Voigt} = \int C_{ijkl}(i)dV_i \tag{2.32}$$

where the individual anisotropic elastic constants for each phase or crystal orientation are matched with the respective volume fractions of the phase or crystals with a particular orientation.[1] For a fiber composite, Eq. 2.32 can be simplified as follows:[2]

$$\left\langle C_{ijkl} \right\rangle^{Voigt} = C_{ijkl}^{fiber} V_{fiber} + C_{ijkl}^{matrix} V_{matrix} \tag{2.33}$$

The *lower bound* or *Reuss average* for a uniaxial composite includes an assumption of constant stress and would correspond to loading in a direction perpendicular to the fibers. For the uniaxial composite,

$$E_{composite} = (V_{fiber}/E_{fiber} + V_{matrix}/E_{matrix})^{-1} \tag{2.34}$$

For more complex loadings, the lower bound for elasticity is derived from the average over the compliances

$$< S_{ijkl} >^{Reuss} = \int S_{ijkl}(i)dV_i \tag{2.35}$$

Similar to Eq. 2.33, the Reuss average for a fiber composite is

$$\left\langle S_{ijkl} \right\rangle^{Reuss} = S_{ijkl}^{fiber} V_{fiber} + S_{ijkl}^{matrix} V_{matrix} \tag{2.36}$$

In general

$$<C_{ijkl}>^{Voigt} \neq [<S_{ijkl}>^{Reuss}]^{-1}$$

(and vice versa), we can calculate Voigt compliances from

$$<S_{ijkl}>^{Voigt} = [<C_{ijkl}>^{Voigt}]^{-1}$$

and Reuss values from

$$<C_{ijkl}>^{Reuss} = [<S_{ijkl}>^{Reuss}]^{-1}$$

The real values should fall between these two bounds. Hill proposed in 1952 that polycrystalline materials with equiaxed grains could be given by the average

$$<C_{ijkl}>^{Hill} = [<C_{ijkl}>^{Voigt} + <C_{ijkl}>^{Reuss}]/2 \tag{2.37}$$

[1]The <x> symbols give the average (or mean value) of *x*.

[2]Both of these equations would return isotropic elastic constants if the starting values for the fiber and matrix were isotropic. This is not expected, as shown in Example 2.5.

which is approximately equal to

$$<S_{ijkl}>^{Hill} = [<S_{ijkl}>^{Voigt} + <S_{ijkl}>^{Reuss}]/2 \tag{2.38}$$

Because the difference between Eq. 2.37 and Eq. 2.38 is small, either one can be called the *Hill arithmetic mean value* of the elastic constants. A geometric mean can also be calculated. The complexity of finding the inverses of the four subscript elasticity tensors can be avoided by applying the matrix method discussed in the next section.

2.7 MATRIX METHODS FOR ELASTICITY TENSORS

For application to orthotropic or higher symmetry, matrix methods for expressing tensors are very powerful. These matrix methods are often applied to elasticity. Special definitions of the stress and strain tensors as 6 × 1 matrices and the elasticity as a 6 × 6 matrix enable these matrix approaches.

The definition of deformation strain expressed with a single subscript can be used to write strain in the form

$$\varepsilon_i = \begin{vmatrix} \varepsilon_1 \\ \varepsilon_2 \\ \varepsilon_3 \\ \varepsilon_4 \\ \varepsilon_5 \\ \varepsilon_6 \end{vmatrix}$$

where $\varepsilon_1 = e_{11} = \varepsilon_{11}$ and $\varepsilon_4 = 2\varepsilon_{23}$. We will designate ε_i as the strain vector. It does not transform by the same rules as given for tensors. The definition of the stress vector given with a single subscript includes the six components

$$\sigma_i = \begin{vmatrix} \sigma_1 \\ \sigma_2 \\ \sigma_3 \\ \sigma_4 \\ \sigma_5 \\ \sigma_6 \end{vmatrix}$$

where the shear stresses include the assumption of a symmetric tensor such that $\sigma_4 = \sigma_{23} = \sigma_{32}$. With the vector designations for stress and strain, the matrix form of the stiffness constants corresponding to the principal components of an orthotropic material is given by matrix multiplication

$$\begin{bmatrix} \sigma_1 \\ \sigma_2 \\ \sigma_3 \\ \sigma_4 \\ \sigma_5 \\ \sigma_6 \end{bmatrix} = \begin{bmatrix} C_{11} & C_{12} & C_{13} & 0 & 0 & 0 \\ C_{12} & C_{22} & C_{23} & 0 & 0 & 0 \\ C_{13} & C_{23} & C_{33} & 0 & 0 & 0 \\ 0 & 0 & 0 & C_{44} & 0 & 0 \\ 0 & 0 & 0 & 0 & C_{55} & 0 \\ 0 & 0 & 0 & 0 & 0 & C_{66} \end{bmatrix} \begin{bmatrix} \varepsilon_1 \\ \varepsilon_2 \\ \varepsilon_3 \\ \varepsilon_4 \\ \varepsilon_5 \\ \varepsilon_6 \end{bmatrix} \tag{2.39}$$

and the matrix form for the compliance constants is given by

$$\begin{bmatrix} \varepsilon_1 \\ \varepsilon_2 \\ \varepsilon_3 \\ \varepsilon_4 \\ \varepsilon_5 \\ \varepsilon_6 \end{bmatrix} = \begin{bmatrix} S_{11} & S_{12} & S_{13} & 0 & 0 & 0 \\ S_{12} & S_{22} & S_{23} & 0 & 0 & 0 \\ S_{13} & S_{23} & S_{33} & 0 & 0 & 0 \\ 0 & 0 & 0 & S_{44} & 0 & 0 \\ 0 & 0 & 0 & 0 & S_{55} & 0 \\ 0 & 0 & 0 & 0 & 0 & S_{66} \end{bmatrix} \begin{bmatrix} \sigma_1 \\ \sigma_2 \\ \sigma_3 \\ \sigma_4 \\ \sigma_5 \\ \sigma_6 \end{bmatrix} \tag{2.40}$$

Expressed as individual elements, these two matrices yield, as described previously,

$$\sigma_1 = C_{11}\varepsilon_1 + C_{12}\varepsilon_2 + C_{13}\varepsilon_3$$

and

$$\sigma_4 = C_{44}\varepsilon_4$$

or

$$\varepsilon_1 = S_{11}\sigma_1 + S_{12}\sigma_2 + S_{13}\sigma_3$$

and

$$\varepsilon_4 = S_{44}\,\sigma_4$$

In the matrix form, the orthotropic elastic constants are related by the matrix inverse

$$C_{ij} = [S_{ij}]^{-1} \tag{2.41}$$

Rotation operations applied to the matrix form require definition of special rotation matrices that have 6×6 dimensions. This special rotation matrix is

$$b_{ij} = \begin{vmatrix} a_{11}^2 & a_{12}^2 & a_{13}^2 & a_{13}a_{12} & a_{13}a_{11} & a_{11}a_{12} \\ a_{21}^2 & a_{22}^2 & a_{23}^2 & a_{22}a_{23} & a_{21}a_{23} & a_{21}a_{22} \\ a_{31}^2 & a_{32}^2 & a_{33}^2 & a_{32}a_{23} & a_{31}a_{33} & a_{31}a_{32} \\ 2a_{31}a_{21} & 2a_{22}a_{32} & 2a_{23}a_{33} & \left(a_{22}a_{33} + a_{23}a_{32}\right) & \left(a_{21}a_{33} + a_{23}a_{31}\right) & \left(a_{21}a_{32} + a_{31}a_{22}\right) \\ 2a_{11}a_{31} & 2a_{12}a_{32} & 2a_{13}a_{33} & \left(a_{13}a_{32} + a_{12}a_{33}\right) & \left(a_{11}a_{33} + a_{13}a_{31}\right) & \left(a_{12}a_{31} + a_{32}a_{11}\right) \\ 2a_{11}a_{21} & 2a_{12}a_{22} & 2a_{13}a_{23} & \left(a_{12}a_{23} + a_{13}a_{22}\right) & \left(a_{11}a_{23} + a_{13}a_{21}\right) & \left(a_{11}a_{22} + a_{12}a_{21}\right) \end{vmatrix} \tag{2.42}$$

where the terms a_{ij} are from the standard rotation matrix. This matrix can be used to transform ε_i by matrix multiplication

$$[\varepsilon_i'] = [\mathbf{b}][\boldsymbol{\varepsilon}] \tag{2.43}$$

and transform σ_i in a similar fashion by

$$[\sigma_i'] = [\mathbf{b}][\boldsymbol{\sigma}]$$

The compliance matrix can be transformed by

$$[S_{ij}'] = [\mathbf{b}][\mathbf{S}][\mathbf{b}]^T \tag{2.44}$$

and the stiffness matrix is transformed by

$$[C_{ij}'] = [[\mathbf{b}]^T]^{-1}[\mathbf{C}][\mathbf{b}]^{-1}$$

These relations are often used to evaluate the properties of composites.

EXAMPLE 2.6 *Stiffness Coefficients for Bovine Femoral Bone*

The orthotropic elastic constants for bovine (cow) femoral (leg) bone have been reported by Van Buskirk, Cowin, and Ward (1981) from measurements using ultrasound. The values vary on the basis of the position around the bone and along its length. Bone consists of a number of minerals and tissues, and the amounts of these minerals and tissues vary with species and bone type. Different types of bone have very different porosity levels, with increased porosity resulting in lower elastic constants.

The elastic constants of materials can be determined using piezoelectric crystals to propagate and measure the speed of sound in materials. Two types of elastic constants can be determined. Propagation of dilational waves can be used to measure longitudinal stiffnesses (e.g., C_{11}) and propagation of shear waves can be used to measure the shear moduli (e.g., C_{44}) The equation

$$C_{11} = \rho V_{dil}{}^2$$

gives the longitudinal stiffness if the direction of wave propagation is along the x-axis. Moreover, the equation

$$C_{44} = \rho V_{trans}{}^2$$

gives the shear stiffness for wave propagation in the y-direction on the z-plane. The term ρ is the density, and V_{dil} and V_{trans} are the wave speeds of dilational waves and transverse waves, respectively.

Van Buskirk and coworkers used dilational and transverse waves to determine the following elastic constants on sections of bone cut with the length direction of the bone parallel to the z-axis using the following relations

$$\begin{aligned} C_{11} &= \rho v^2_{dil,1/1} & C_{44} &= \rho v^2_{trans,2/3} = \rho v^2_{trans,3/2} \\ C_{22} &= \rho v^2_{dil,2/2} & C_{55} &= \rho v^2_{trans,1/3} = \rho v^2_{trans,3/1} \\ C_{33} &= \rho v^2_{dil,3/3} & C_{66} &= \rho v^2_{trans,1/2} = \rho v^2_{trans,2/1} \end{aligned}$$

where the i/j subscripts are defined with i the direction of travel and j the direction of ultrasound displacements. To find the stiffnesses C_{12}, C_{13}, and C_{23}, they used a tensor rotation that included velocity measurements from 45° sections of C'_{12}, C'_{13} and C'_{23}

$$\begin{aligned} C_{12} &= \sqrt{(C_{11} + C_{66} - 2C'_{12})(C_{22} + C_{66} - 2C'_{12})} - C_{66} \\ C_{13} &= \sqrt{(C_{11} + C_{55} - 2C'_{13})(C_{33} + C_{55} - 2C'_{13})} - C_{55} \\ C_{23} &= \sqrt{(C_{22} + C_{44} - 2C'_{23})(C_{33} + C_{44} - 2C'_{23})} - C_{44} \end{aligned}$$

The approximate reported stiffness values were

$$C_{ij} = \begin{bmatrix} 14 & 6.3 & 4.8 & 0 & 0 & 0 \\ 6.3 & 18.4 & 7 & 0 & 0 & 0 \\ 4.8 & 7 & 25 & 0 & 0 & 0 \\ 0 & 0 & 0 & 7 & 0 & 0 \\ 0 & 0 & 0 & 0 & 6.3 & 0 \\ 0 & 0 & 0 & 0 & & 5.3 \end{bmatrix} \text{MPa}$$

To convert this into the compliance values, we can take the inverse of the matrix to get

$$S_{ij} = \begin{bmatrix} 0.086 & -0.026 & -0.009 & 0 & 0 & 0 \\ -0.026 & 0.07 & -0.014 & 0 & 0 & 0 \\ -0.009 & -0.014 & 0.046 & 0 & 0 & 0 \\ 0 & 0 & 0 & 0.14 & 0 & 0 \\ 0 & 0 & 0 & 0 & 0.16 & 0 \\ 0 & 0 & 0 & 0 & & 0.19 \end{bmatrix} \text{MPa}^{-1}$$

From this compliance matrix, we can see that Young's modulus along the bone length is $1/S_{33} = E_3 = 21.7$ GPa and in the radial direction is $1/S_{11} = E1 = 11.6$ GPa. We can also recover the Poisson's ratios using Eq. 2.23. For example $\nu_{12} = (-11.6)(-0.026) = 0.3$ and $\nu_{13} = (-11.6)(-0.009) = 0.1$. ■

2.8 APPENDIX: THE STEREOGRAPHIC PROJECTION

The use of the stereographic projection is a useful common ground for anyone interested in orientations of crystals. Electronic materials, structural materials, and even coatings depend on crystal orientations. Besides serving as a convenient two-dimensional representation of three-dimensional crystals, the stereographic projection also serves as a convenient approach to consideration of any spatial relationship that involves orientations of objects with definable symmetries.

Projections of the Earth Every map used to describe locations of or distances between points on Earth incorporates the principle of a projection. Figure 2.A.1 is a circular map showing an approximation of the Western Hemisphere of Earth. With the exception of the equator, the lines of latitude are small circles running around the Earth. The lines of longitude are all circles that pass through the North Pole and the South Pole. Each of these circles and the equator is called a great circle. The shortest path between any two points at different latitudes lies along a great circle, as shown in Fig. 2.A.1.

The positions of the small and great circles on the projection can be determined using several strategies. In cartography and analysis of textured polycrystals, the type of projection

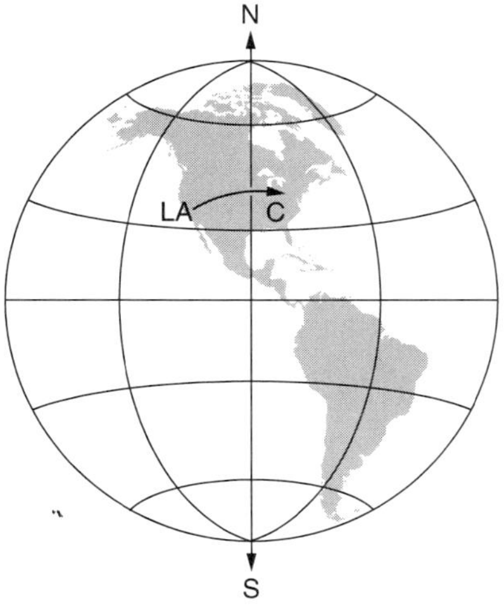

FIG. 2.A.1 Schematic of the Western Hemisphere, showing that the shortest airplane distance between Chicago and Los Angeles follows a curved line.

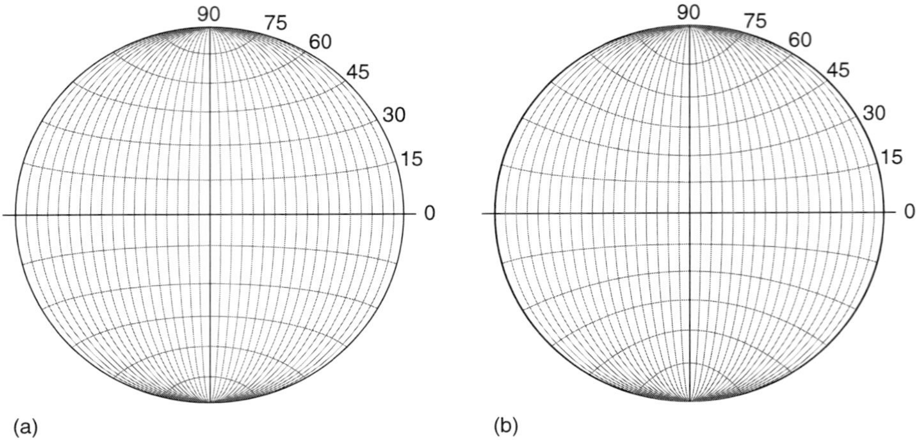

FIG. 2.A.2 (a) The Schmid net for mapping the equal area projection. (b) The Wulff net for mapping the equal angle projection.

preserves, as nearly as possible, the surface area of the sphere onto the projection. An example of this, called a Schmid projection or equal area projection, is shown in Fig. 2.A.2*a*. This projection is obtained by mapping the radius R and the angular arc length σ by the expression

$$AP = \sqrt{2}R\sin\left(\frac{\alpha}{2}\right) \tag{2.A.1}$$

Stereographic Projections of Crystals Stereographic projections describing orientations preserve angular relationships. The angle-preserving stereographic projection is given by a Wulff projection, as shown in Fig. 2.A.2*b*. The Wulff projection is obtained by mapping the radius R and the angular arc length σ:

$$AP = R\tan\left(\frac{\alpha}{2}\right) \tag{2.A.2}$$

Using a Stereographic Projection for Crystallography When a stereographic projection is used for crystallography, directions, plane normals, and planes can be represented. Directions and plane normals are located as though the crystal lies within the sphere, and the crystal direction or plane normal forms a point on the sphere called a pole. To relate a pole to a plane, a simple analogy is given by a standard basketball, which has an air hole at a point that lies on a line normal to the great circle that cuts the ball in half. Each point on the ball represents the intersection of all planes containing the corresponding direction. Using these relationships, and the equation for the Wulff projection, the complete projection for cubic materials in Fig. 2.A.3 can be generated.

Locating a Direction or Pole on the Wulff Projection The easiest strategy for generating the positions of directions or poles on the standard stereographic projection is the use of a Wulff net (or a computer plotting program using Eq. 2.A.2). The standard 001 stereographic projection[3] given in Fig. 2.A.3 shows a right-handed coordinate system with X or the 100 pole pointing toward the bottom of the page and Z or the 001

[3]Note that hkl designated without brackets, parentheses, etc., is intended to represent a pole to a plane. For a cubic material, all poles coincide with directions having the same indices, but this is not the case for directions and planes in noncubic materials.

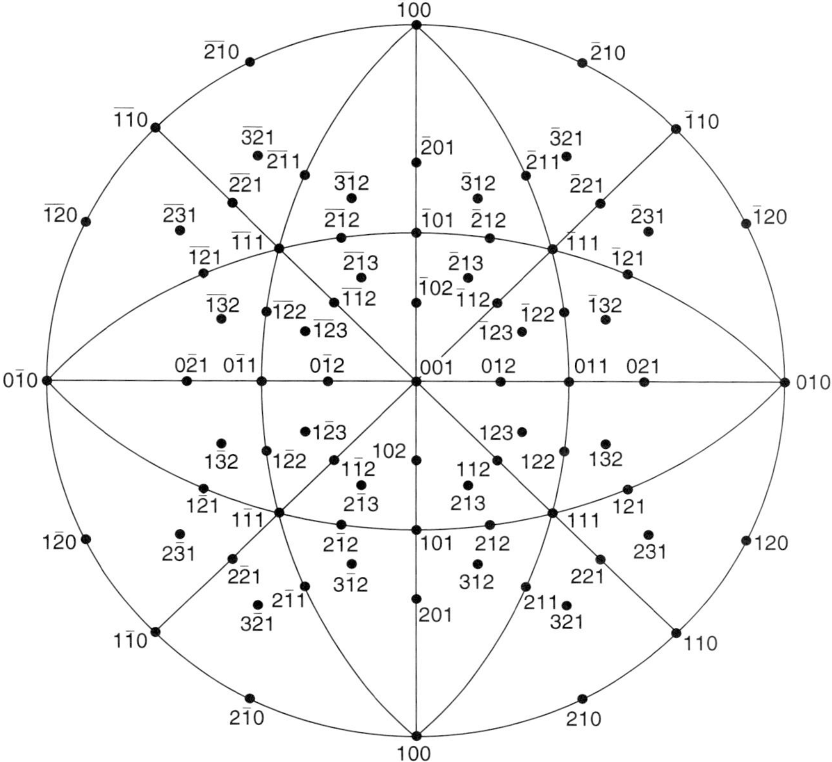

FIG. 2.A.3 The standard cubic stereographic projection.

pole coming out of the page. The cross-product of Z into X gives Y or the 010 pole pointed to the right of the page. The 001 stereographic projection can be reproduced on a transparency by attaching a Wulff net to the transparency with a tack. Then the gradations on the Wulff net can be employed to mark angular distances from the X, Y, and Z poles.

For example, location of the $2\bar{3}1$ pole could proceed in the following manner.

1. Trace a circle onto a transparency, carefully following the outline of a Wulff net.
2. Mark the center, top, bottom, left, and right poles relative to the transparency. Label these 001, $\bar{1}00$, 100, etc.
3. Determine the octant in which the pole lies. Remember that this projection gives only poles that have positive Z-values. The octant for $2\bar{3}1$ will have a positive X-value, thus lying on the lower half of the projection, and a negative Y-value, placing it in the lower left octant.
4. Find the angle between $2\bar{3}1$ and the three poles bounding this octant. The angles are given by the inverse cosines of the dot products between $2\bar{3}1$ and the three bounding poles:

$$\cos^{-1}\left[\frac{100\cdot 2\bar{3}1}{\sqrt{1}\sqrt{14}}\right] = 57.6°$$

$$\cos^{-1}\left[\frac{0\bar{1}0\cdot 2\bar{3}1}{\sqrt{1}\sqrt{14}}\right] = 36.7°$$

$$\cos^{-1}\left[\frac{001\cdot 2\bar{3}1}{\sqrt{1}\sqrt{14}}\right] = 74.5°$$

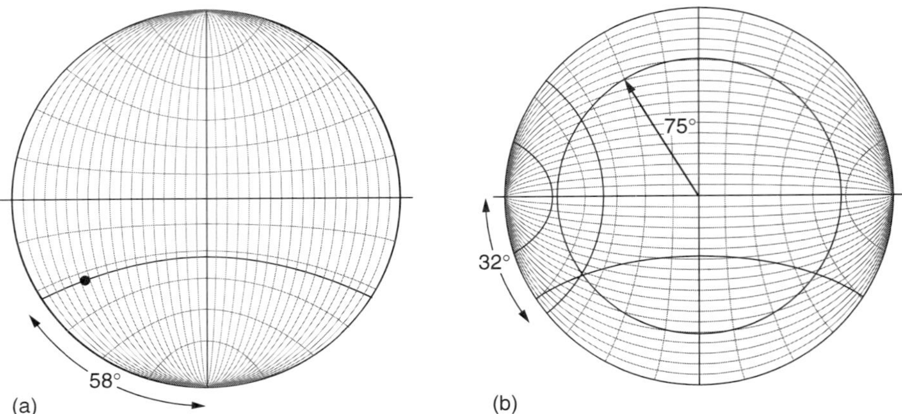

FIG. 2.A.4 (a) Construction of a small circle 58° from the south pole and (b) construction of small circles on the two adjoining axes to define an intersection.

5. Draw small circles the appropriate distances from each of these poles by rotating the Wulff net and counting off the angles. The three small circles intersect at $2\bar{3}1$. This is shown step-by-step in Fig. 2.A.4.

Constructing the Plane Corresponding to a Specific Pole To construct the plane corresponding to a pole, first locate a pole. Rotate the Wulff net with respect to the projection such that the great circle across the Wulff net (the "equator") intersects the pole. Count 90 degrees along this great circle (through the center) and locate the nearest great circle. This great circle is the plane orthogonal to the pole.

2.9 REFERENCES

F. P. BEER AND E. R. JOHNSTON, JR., *Mechanics of Materials,* McGraw-Hill, 1981.

W. C. BUSKIRK, S. C. COWIN, AND R. N. WARD, *J. Biomech. Eng.*, **103**, 67–72, 1981.

G. E. DIETER, *Mechanical Metallurgy*, McGraw-Hill, New York, 1986.

R. W. HERTZBERG, *Deformation and Fracture Mechanics of Engineering Materials*, Wiley, New York, 1988.

J. P. HIRTH AND J. LOTHE, *Theory of Dislocations*, 2nd Ed., Wiley, 1982.

D. HULL AND T. W. CLYNE, *An Introduction to Composite Materials,* Cambridge, 1996.

A. KELLY, G. W. GROVES, AND P. KIDD, *Crystallography and Crystal Defects,* Rev. Ed., Wiley, New York, 2000.

F. A. MCCLINTOCK AND A. S. ARGON (EDITORS), *Mechanical Behavior of Materials,* Addison-Wesley, 1966.

J. F. NYE, *Physical Properties of Crystals,* Oxford, 1957.

G. SIMMONS AND H. WANG, *Single Crystal Elastic Constants,* MIT Press, 1971.

S. P. TIMOSHENKO, *History of Strength of Materials,* Dover, 1953.

J. R. VINSON, *The Behavior of Structures Composed of Composite Materials,* Martinus Nijhoff, 1987.

2.10 PROBLEMS

A.2.1 Write out the SI units for each term of the second-rank tensors given in Table 2.3.

A.2.2 If a stress tensor is given with only σ_{23} as the nonzero term, describe the deformation. (Will the object stay stationary?)

A.2.3 Thermal expansion can also be considered a second-rank tensor. Explain.

A.2.4 A polarization **P** is a vector quantifying charge per unit area. A polarization results when a stress is applied along the axes of some crystals (e.g., quartz). Show how you would write out **d**, the piezoelectric modulus.

A.2.5 The converse piezoelectric effect, which relates electric field to strain, has the same modulus as for the direct piezoelectric effect (see A.2.4). Show that the units must be the same.

A.2.6 Find A, the anisotropy factor for each of the materials in Table 2.8.

A.2.7 Write out the equivalent four subscript notation terms for the following 6 × 6, two-subscript notation compliances and stiffnesses.

$T_{14} = ?T_{????}$___ $T_{34} = ?T_{????}$___

$T_{54} = ?T_{????}$___ $T_{63} = ?T_{????}$___

A.2.8 Section 2.5 describes the 6 × 6, two-subscript tensor. Give rules for a 9 × 9, two-subscript tensor. How does this change the relationship between S_{15} and S_{1113}?

A.2.9 For an isotropic material, write an expression for Hooke's law with principal stresses as a function of G, λ, e_1, e_2, and e_3, where e_i are principal strains.

A.2.10 Show that $S_{66} = 2(S_{11} - S_{12})$ for hexagonal materials.

A.2.11 Which elastic constant is the shear modulus for σ_{12} and ε_{12} (using standard tensor notation)?

(a) S_{44}

(b) C_{11}

(c) C_{12}

(d) S_{11}

(e) C_{66}

A.2.12 Which elastic constants correspond to the symmetry of a hexagonal or axially symmetric material?

(a) $C_{11} = C_{22} = C_{33}$

(b) $C_{11} = C_{22} \neq C_{33}$

(c) $S_{11} = S_{22} = S_{33}$

(d) $C_{44} = C_{55} = C_{66}$

(e) $C_{12} = C_{13} = C_{23}$

A.2.13 Which are the correct equivalencies from two- to four-subscript notation (and vice versa)? Assume orthotropic symmetry and values given on principal axes.

(a) $C_{44} = C_{2323}$

(b) $S_{44} = S_{2323} + S_{2332} + S_{3232} + S_{3223}$

(c) $C_{2211} = C_{21}$

(d) $C_{1313} = C_{13}$

(e) $C_{1313} = C_{55}$

(f) $S_{1122} = S_{21}$

(g) $S_{54} = 0$

(h) $S_{11} = S_{2222}$

A.2.14 The orthotropic stiffness tensor is normally written as

$$C = \begin{vmatrix} C_{11} & C_{12} & C_{13} & 0 & 0 & 0 \\ C_{12} & C_{22} & C_{23} & 0 & 0 & 0 \\ C_{13} & C_{23} & C_{33} & 0 & 0 & 0 \\ 0 & 0 & 0 & C_{44} & 0 & 0 \\ 0 & 0 & 0 & 0 & C_{55} & 0 \\ 0 & 0 & 0 & 0 & 0 & C_{66} \end{vmatrix}$$

When the level of symmetry is increased to cubic, the nine independent constants for orthotropic are reduced to just three. Describe how these stiffnesses become related by writing three sets of equivalencies:

________ = ________ = ________

________ = ________ = ________

________ = ________ = ________

A.2.15 For a cubic material, assume that a compressive stress of 10 MPa is applied in the *y*-direction:

(a) Which of the following expressions would give the corresponding strain in the *z*-direction? Explain.

(i) S_{32} (–10 MPa)

(ii) S_{32} (10 MPa)

(iii) S_{23} (–10 MPa)

(iv) S_{13} (–10 MPa)

(b) Which of the expressions in part (a) would apply for orthotropic symmetry? Explain.

A.2.16 To demonstrate symmetry, we often apply rotations that should reproduce equal values for elastic constants before and after rotations.

(a) A rotation of $\phi_1 = 90°$ could be used to prove which of the following?

(i) Four-fold symmetry about *z*

(ii) Four-fold symmetry about *y*

(iii) Cubic symmetry

(b) If, following a rotation demonstrating a symmetry element, we find

$$C_{1123}' = -C_{1123}$$

what is the only possible value for C_{1123}?

B.2.1 Using the tensor **T′** and {*x′*, *y′*, *z′*} for the material in Example 2.1 find **T″** after a new rotation of $\phi 2 = 30°$. Find the new rotation matrix and direction indices. Show the rotation on a schematic stereographic projection.

B.2.2 For

$$\boldsymbol{\sigma} = \begin{vmatrix} 3 & 0 & 1 \\ 0 & 0 & 0 \\ 1 & 0 & -1 \end{vmatrix}$$

find the principal stresses, **σ′**. What is the rotation matrix required to go from **σ** to **σ′**? What values of the Euler angles will produce this rotation?

B.2.3 The strain tensor

$$\begin{vmatrix} 0.05 & 0 & 0 \\ 0.08 & 0 & 0 \\ 0 & 0 & 0 \end{vmatrix}$$

is applied to a unit cube. Describe the changes, in terms of changes in length and angles, that will take place in lines that were in the [111], [110], [$1\bar{1}0$], and [100] before the strain took place.

B.2.4 Two isotropic crystals, with Poisson's ratios of 0.4 and 0.6 are stretched to an elastic strain of 0.01. Describe the volume change in each material and the origins of the response. Compare the results for the exact volume strain.

B.2.5 If both the crystals in Problem B.2.4 had a Young's modulus of 2 GPa, find the tensile stress that will produce a transverse strain of 0.005. Determine the bulk and shear moduli of the two materials.

B.2.6 Plot values of Young's modulus for all possible directions in the (100), (110), and (123) for MgO. (*Hint:* Find the rotation matrix that puts the plane normal as z' at some rotation of $\phi 1$ and ϕ and then plot Young's modulus as a function of $\phi 1$.)

B.2.7 Write the symbolic expression for C'_{13} for an arbitrary rotation of Φ for orthorhombic and cubic symmetry.

B.2.8 Find the shear modulus for shearing in the [110] on the ($1\bar{1}2$) for iron and aluminum.

B.2.9 Find Young's modulus for Germanium along [110].

(a) Complete the rotation matrix

$$\begin{array}{c|ccc|} & x & y & z \\ x' & \frac{1}{\sqrt{2}} & \frac{1}{\sqrt{2}} & 0 \\ y' & \frac{-1}{\sqrt{2}} & \frac{+1}{\sqrt{2}} & 0 \\ z' & --- & --- & --- \end{array}$$

(b) Given that $S_{11} = 0.98 \times 10^{-11}$/Pa, $S_{12} = -0.27 \times 10^{-11}$/Pa, and $S_{44} = 1.5 \times 10^{-11}$/Pa, find S'_{1111} which is $1/E'$ in the [110].

Remember that there are 21 terms possible on the right-hand side of the equation, although many may have coefficients equal to zero.

B.2.10 Given below are the room-temperature compliances of several refractory metals. All of these metals have BCC crystal structures.

(a) Which of these metals has the highest stiffness along the direction of closest spacing of atoms?

(b) Give the elastic strain expected in the [100] for a Cr crystal compressed by a stress in the [001] of 5 MPa.

(c) If the Cr crystal in part (b) has a square cross section with {001} faces before loading, will it still be square while the compression is applied?

B.2.11 The final equation given in Example 2.4 provides the variation of S'_{33} with orientation (Φ) for axisymmetric composites and hexagonal crystals. Using this equation, plot Young's modulus surfaces for cadmium and titanium single crystals (Table 2.8). Plot the surfaces on the same axes to make a comparison.

B.2.12 Which of the cubic metals in Table 2.8 is most isotropic? Least isotropic?

C.2.1 A cylindrical drawn polymer fiber is reported to have elastic constants (in GPa) of $C_{11} = C_{22} = 0.5$, $C_{33} = 18$, $C_{12} = 1$, $C_{13} = C_{23} = 3$, and $C_{44} = C_{55} = 0.8$. Find C' for a rotation of $\Phi = 30°$. Compare this result with C' for $\phi 1 = 30°$ and $\Phi = 30°$. (Note the axisymmetry.)

C.2.2 Find S_{11}', S_{14}', and S_{44}' for each of the rotations in B.2.9.

C.2.3 For a tensile test on a cylindrical specimen that is deformed by a stress in the z (or 3-direction), what will be Poisson's ratio, ν_{13}? (Write your answer in terms of strains e_{ij}, but with specific choices of i and j.)

C.2.4 A silica glass has Young's modulus of 80 GPa and a shear modulus of 31.5 GPa. We plan to reinforce this glass with 20 v/o particulates of tungsten carbide (WC), which has Young's modulus of 530 GPa and a shear modulus of 219 GPa. Compare the expected composite Young's moduli versus volume fraction for both upper and lower bound estimates assuming that each phase is isotropic.

Refractory metal	Group on periodic chart	Anisotropy factor	S_{11}	S_{12}	$S_{44}(10^{-11}/\text{Pa})$
V	VB	0.78	0.68	−0.23	2.35
Nb	VB	0.51	0.66	−0.23	3.48
Ta	VB	1.56	0.69	−0.16	1.21
Cr	VIB	0.69	0.30	−0.04	0.99
Mo	VIB	0.78	0.28	−0.08	0.91
W	VIB	1.0	0.24	−0.07	0.62

C.2.5 (a) For the properties given in Problem C.2.4, write the compliance matrix as S_{ij} for 100% glass and 100% WC.

(b) Calculate the corresponding effective compliances for an equal fraction composite.

(c) Calculate the corresponding effective stiffnesses, $<C_{ijkl}>^{Reuss}$.

(d) To find the $<C_{ijkl}>^{Voigt}$ for an equal fraction composite, find the C_{ij} matrix for 100% glass and 100% WC and calculate the effective stiffnesses.

(e) Discuss the values found for $<C_{ijkl}>^{Voigt}$ and $<C_{ijkl}>^{Reuss}$.

C.2.6 Find the principal stresses for the stress tensor

$$\sigma = \begin{vmatrix} 10 & 12 & -5 \\ 12 & 4 & 3 \\ -5 & 3 & -5 \end{vmatrix} \text{ MPa}$$

C.2.7 Describe the shape of the three-dimensional Young's modulus surface for cubic materials with anisotropy $A << 1$ and make a sketch.

C.2.8 Find the compliance values for the polymer fiber in Problem C.2.1. What is Young's modulus along the fiber axis?

C.2.9 Estimate the stiffness properties of an aluminum alloy single crystal containing 50 percent cuboidal silicon particles by weight, assuming that the cubic unit cells of both materials are aligned with one another. Find the Voigt prediction and the Reuss prediction.

CHAPTER 3

PLASTICITY

THE TENSORS for stress and strain can also be applied to conditions of plastic deformation. The strengths of materials can be expressed in terms of tensor properties, as can the changes in strength during plastic deformation. To accommodate the small increments necessary for a tensor description of strain, the small increments can be summed to produce large displacements. Finite element analyses of plastic deformation and modeling of grain rotation producing textured polycrystals both employ iterative operations to produce large strains. Plastic deformation is normally defined as a permanent change in shape; however, the classic metallurgical definition of plasticity also includes conservation of volume as an assumption. In materials with well-defined structures, particularly crystalline materials, this is a valid assumption for deformation by slip and twinning. In noncrystalline materials—e.g., most composites, polymers, and ceramic glasses, this assumption is not always valid. As shown in Chapter 1, plastic deformation usually occurs in most materials before the total strain reaches 1 percent of Young's modulus.

EXPERIMENT: Make a small cylinder of a substance like Silly Putty™ (silicon rubber–zinc oxide particulate composite sold as a children's toy). About one-third of the way from one end, mark a circumferential circle around the cylinder. Then compress the cylinder with a book on your desk until it resembles a small pancake. The circle now has rotated by shear deformation from the sides to the flat surface. This demonstrates how flow occurs under conditions of sticking friction. The cylinder very quickly bulges as shear deformation proceeds. The heterogeneous flow is also apparent as the putty in contact with the book and desk undergoes little or no deformation as a result of friction effects.

3.1 CONTINUUM MODELS FOR SHEAR DEFORMATION OF ISOTROPIC DUCTILE MATERIALS

In Chapter 2, tensor rotations transforming shear stresses or strains into principal normal stresses were demonstrated. Similarly, the components of a deformation tensor, ε_{ij}, can be transformed into shear components opposing deformation even though the sum of the diagonal terms will not be changed by rotating the coordinate frame.

3.1.1 Yield Surfaces

Defining a condition under which a particular material will or will not plastically deform for a simple loading state such as tension or compression is as easy as stating the yield stress (ignoring statistical variability). Translating the results of a simple tension test into predictions of more complex loading states involves some assumptions on the response to multiaxial loading. The most common approach employs the principal normal stresses to generate a condition at which the material deforms. By putting the complete set of stresses at which yielding first occurs in Cartesian coordinates, a surface is defined. These surfaces are called *yield surfaces*, and the mathematical expressions describing the surfaces are called *yield criteria*. It is common in the field of mechanics to call these expressions *failure criteria*. Such a designation is often inadequate. Although structural designers use the criteria to prevent failure through plastic distortion of a rigid structure, for designers of metal or plastic forming operations a failure to reach the yield surface and thereby change the shape of a component is a disaster. Although they can be represented in the general form of the tensor, it is quite a bit easier to express the general form for yield criteria as

$$f(s_1, s_2, s_3) = \text{Constant} \tag{3.1}$$

where s_1, s_2 and s_3 are principal stresses. Consider the following assumptions.

1. The yield strength of the material is the same in tension and in compression.
2. The material does not change volume during deformation.
3. The magnitude of σ_m does not influence yielding.

Then we can construct a surface that indicates yielding for each unique combination of principal stresses. As described by Eq. 2.13, the deviatoric stresses provide for the nonhydrostatic components of stress, resulting in modification of Eq. 3.1 to

$$f(s_1 - s_2), (s_3 - s_2), (s_3 - s_1) = \text{Constant} \tag{3.2}$$

The yield criteria defined for these three conditions consist of open-ended surfaces with uniform cross sections for all planes wherein $s_1 = s_2 = s_3$. Two examples of yield criteria that fulfill Eq. 3.2 are shown in Fig. 3.1. Additional requirements for this surface are imposed if these conditions are not met or if strain hardening, strain rate effects, temperature effects, or anisotropy are included.

For most simple calculations, the stresses corresponding to plane stress are shown such that a section of the yield surface is constructed. Figure 3.2*a* shows a section of the von

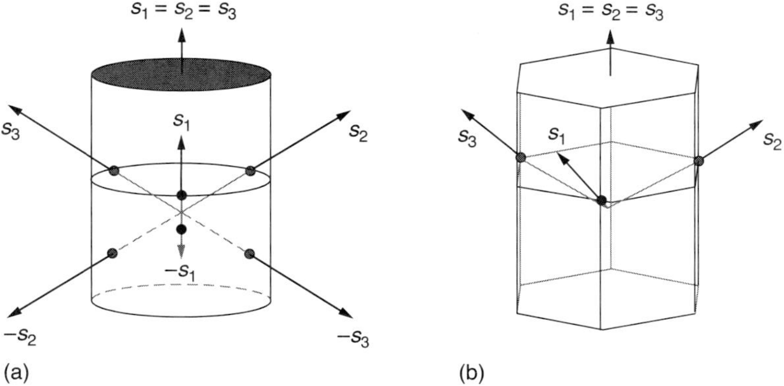

FIG. 3.1 (a) A yield surface as a right cylinder that complies with considerations of volume conservation. This yield surface corresponds to the von Mises criterion. The stress axes, s_i, indicate an orthogonal coordinate system. (b) A yield surface as a hexagonal prism that corresponds to the Tresca criterion. For this example, only the positive portions of the stress are shown. Both s_2 and s_3 have components pointing toward the rear of the von Mises and Tresca surfaces.

BIOGRAPHY

THE VON MISES CRITERION

The traditional attribution of a distortion energy criterion to Richard von Mises (1883–1953) does not give credit to J. C. Maxwell (1831–1879) for suggesting the separation of volume strain from distortion strain. Also, prior to 1904 and independently of von Mises, M. T. Huber developed the approach that results in Eq. 3.4. The paper from von Mises was not published until 1913, but it was published in a more influential journal from Göttingen, Germany. Richard von Mises also authored books on fluid dynamics, aerodynamics, statistics, and human relations (see Timoshenko, 1983).

Mises yield surface corresponding to stresses in the *x-z* plane and zero stress components in the y-direction, or

$$\sigma_{ij} = \begin{vmatrix} s_1 & 0 & 0 \\ 0 & 0 & 0 \\ 0 & 0 & s_3 \end{vmatrix} \tag{3.3}$$

This section is usually shown as a two-dimensional diagram in which s_2 points into the page. The distortion energy or von Mises criterion for isotropic materials is expressed as the condition wherein the square root of the average shear stress reaches a critical value (root-mean-shear stress),

$$\left[\frac{(s_1 - s_2)^2 + (s_3 - s_1)^2 + (s_3 - s_2)^2}{3}\right]^{\frac{1}{2}} = C$$

which, if we eliminate the square root and the factor of 3, can be expressed equivalently as

$$(s_1 - s_2)^2 + (s_3 - s_1)^2 + (s_3 - s_2)^2 = C_{von\ Mises} \tag{3.4}$$

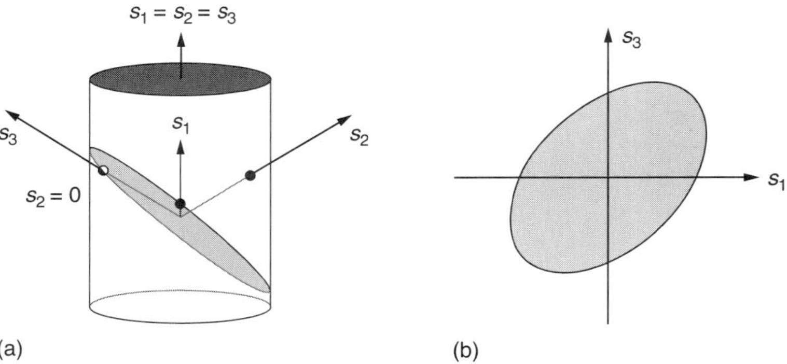

FIG. 3.2 (a) A yield surface shown as a right cylinder with the ellipse equivalent to all stress states wherein $s_2 = 0$ constructed within it. (b) The $s_2 = 0$ section of the yield surface. Because many examples are given in deformation conditions wherein plane stress is considered, this is the common view of the von Mises criterion if the axis representing s_2 is not included. For this construction, right-handed coordinates would require s_2 to point into the page.

BIOGRAPHY

THE TRESCA CRITERION

Henri Tresca (1814–1885) investigated the flow of metals under large hydrostatic pressures and published the basis for the Tresca criterion in 1864. Tresca also was responsible for one of the first designs for the International System of Units standard meter. B. Saint-Venant (1797–1886) used ideas from Tresca's experimental work to devise fundamental equations of plasticity and their applications.

wherein $C_{von\ Mises}$ is a critical value and the s_i values are principal stresses. The von Mises criterion is consistent with the yield surfaces shown in Fig. 3.1*a* and 3.2.

Another yield criterion that is commonly employed, owing partly to a slightly simpler formulation, is the maximum shear stress or Tresca criterion. In contrast with the von Mises criterion, the Tresca criterion is satisfied when the largest of the principal stress differences given in Eq. 3.2 reaches a critical value. It is easiest to state this as

$$s_{max} - s_{min} = C_{Tresca} \tag{3.5}$$

where s_{max} is the most tensile of the s_i values, s_{min} is the most compressive, and C_{Tresca} is the constant that must be satisfied for yielding to occur. The Tresca criterion is consistent with the yield surfaces shown in Fig. 3.1*b* and Fig. 3.3.

We could define any deformation state to decide on the values of the constants in Eq. 3.4 and 3.5, but we will consider two possibilities and their implications related to these two yield criteria. Consider first the condition of uniaxial tension for deformation of a material with $s_1 = Y$, where Y is the magnitude of the tensile yield stress, and s_2 and s_3 are equal to zero. Then the von Mises criterion yields

$$C_{von\ Mises} = 2Y^2 \tag{3.6}$$

and the Tresca criterion yields

$$C_{Tresca} = Y \tag{3.7}$$

The forming operation of rolling (see Fig. 2.9) results in deformation wherein a tensile plastic strain occurs primarily in the rolling direction (RD) and a compressive plastic strain occurs in the normal direction (ND). Almost no plastic deformation occurs in the width or transverse direction (TD). If the material were uniformly loaded in the ND, the

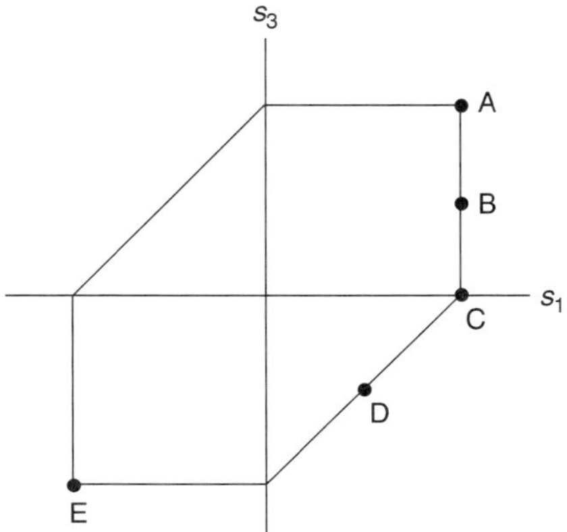

FIG. 3.3 Tresca criterion for a section of the yield surface shown in Fig. 3.1*b* with $s_2 = 0$. The points indicated correspond to the points wherein the stress vectors meet the yield surface with (A) $s_1 = s_3$ and both are tensile; (B) $s_1 = 2s_3$ and both are tensile; (C) $s_1 > 0$, $s_3 = 0$; (D) $s_1 = -s_3$ and s_1 is tensile; and (E) $s_1 = s_3$ and both are compressive.

plastic deformation would be the same in both the RD and the TD. Because the loading is localized along the strip of material in contact with the rolls, the unloaded material before and after the rolls inhibits deformation in the TD. As we learned in Chapter 1, deformation of this type is the plane strain condition called pure shear. Engineers designing mechanical shaping or forming processes often apply plane strain deformation to yield criteria. Under these circumstances, it is convenient to define a magnitude of shear stress, k, at which plastic deformation occurs. Then, rather than using Y as a basis for the size of the yield surface, we can define $s_1 = k$ and $s_3 = -k$. Accordingly, the shear resistances to deformation expressed by these two criteria are given as

$$C_{von\ Mises} = 6k^2 \tag{3.8}$$

and

$$C_{Tresca} = 2k \tag{3.9}$$

The relative yield surfaces for the condition wherein the two yield criteria are equivalent for tension and pure shear are shown in Fig. 3.4*a* and 3.4*b*, respectively. Consider that for Fig. 3.4*a*, if a tensile (or compressive) test is used to determine the yield criterion, the Tresca criterion is more conservative in predicting yielding for complex stress states. Additionally, the greatest difference between the two criteria occurs when the loading is pure shear in nature. The situation is reversed if pure shear is considered as the basis for determining the yield surface (Fig. 3.4*b*).

The most useful way to employ the yield criteria is to express them in terms of an effective stress for deformation, σ_{eff}. We can set the von Mises criterion to

$$\sigma_{\text{eff}} = \left[\frac{(s_1 - s_2)^2 + (s_3 - s_1)^2 + (s_3 - s_2)^2}{2}\right]^{\frac{1}{2}} \tag{3.10}$$

and the Tresca criterion to

$$\sigma_{\text{eff}} = s_{\max} - s_{\min} \tag{3.11}$$

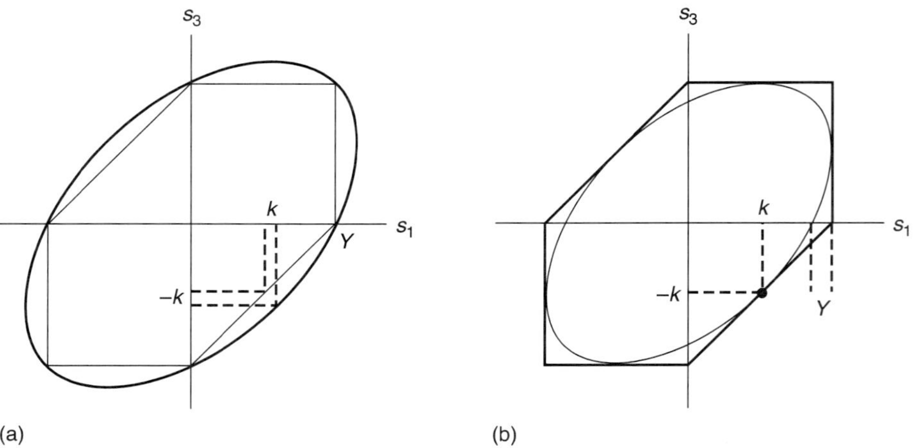

FIG. 3.4 The distinction between setting the yield condition as (a) the uniaxial yield stress Y and as (b) the shear yield stress k for both the von Mises and Tresca yield criteria.

so that either criterion reaches yielding when $\sigma_{eff} = Y$, but then, if pure shear is employed as the condition for deformation, $\sigma_{eff} = \sqrt{3}\,k$ for the von Mises criterion and $\sigma_{eff} = 2k$ for the Tresca criterion.

The von Mises criterion follows well the behavior of essentially isotropic metals, as shown in Fig. 3.5*a*. When materials have a different yield stress in tensile loading than in compressive loading, it is often related to dilatant deformation wherein some volume change from plastic deformation is possible. For some polymers, the rearrangement of organic chain molecules can result in density changes. These density changes result in yield surfaces that deviate from the infinite cylinder shown in Fig. 3.1. Figure 3.5*b* shows yield surfaces for plane stress loading that demonstrate this effect.

3.1.2 Effective Strain

Each criterion has implications for the work that occurs during plastic deformation. If we consider that mean stresses do not cause deformation, the sum of the plastic strains along the strain tensor diagonal should always be zero. Therefore, this should also be the case for the principal plastic strains. Consider the work (per unit volume) done to produce an increment of deformation in a tensile bar

$$dw = (\mathrm{F}dl)/(Al) = \sigma d\varepsilon \tag{3.12}$$

where F is the applied force, dl is the increment of extension for a bar of initial length l, A is the cross-sectional area, σ is the applied stress, and $d\varepsilon$ is the strain increment. The general case can then be expressed in terms of the principal stresses and principal strains as

$$dw = s_1 d\varepsilon_1 + s_2 d\varepsilon_2 + s_3 d\varepsilon_3 \tag{3.13}$$

The *effective strain* $d\varepsilon_{eff}$ provides a way to equate deformation levels for different modes of deformation. The effective strain is defined as the condition wherein $dw = \sigma_{eff}\, d\varepsilon_{eff}$ or

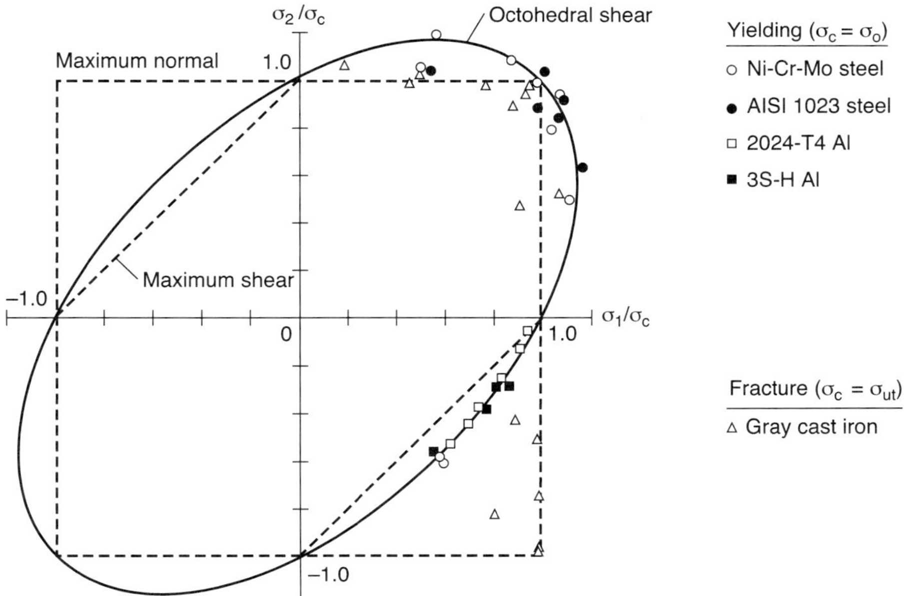

FIG. 3.5 (a) Plane stress yielding for metals normalized to yield stress and corresponding yield criteria (after Dowling, 1999, reprinted by permission).

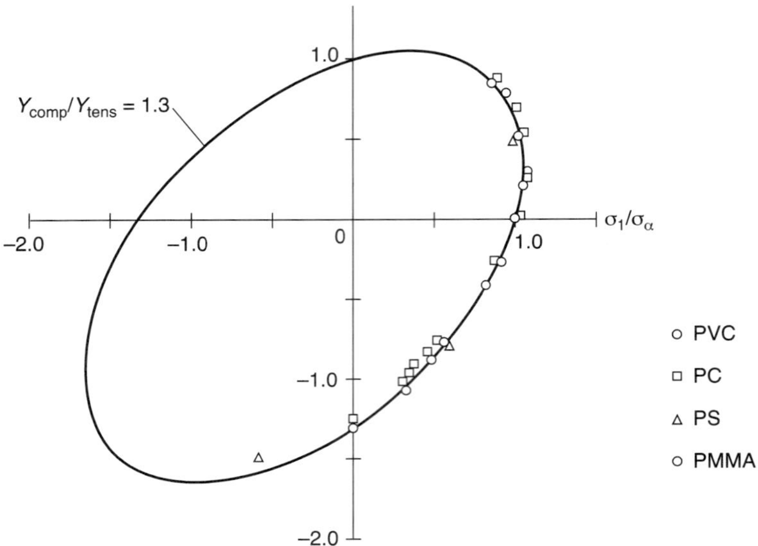

FIG. 3.5 (b) Normalized plane stress yielding for polymers, showing $Y_{comp} > Y_{tens}$ (Raghava, Caddell, and Yeh, 1973, used with permission).

The yield surface for polymers in Fig. 3.5*b* demonstrates that the mean or hydrostatic component of stress affects the yield stress, causing the yield stress to be lower in tension than in compression. This *dilatant* yielding behavior occurs during deformation of noncrystalline materials and partially crystalline materials and also in crystalline materials that undergo phase changes driven by stress. This is discussed in greater detail in Section 3.5.

$$\sigma_{eff} d\varepsilon_{eff} = s_1 d\varepsilon_1 + s_2 d\varepsilon_2 + s_3 d\varepsilon_3 \tag{3.14}$$

This expression, combined with the expression for effective stress using the von Mises criterion and the necessity of volume conservation, leads to the effective strain expressions

$$d\varepsilon_{eff} = \left[\frac{2}{3}\left(d\varepsilon_1^{\,2} + d\varepsilon_2^{\,2} + d\varepsilon_3^{\,2}\right)\right]^{\frac{1}{2}} \tag{3.15}$$

which allows for changes in strain ratios during deformation, and

$$\varepsilon_{eff} = \left[\frac{2}{3}\left(\varepsilon_1^{\,2} + \varepsilon_2^{\,2} + \varepsilon_3^{\,2}\right)\right]^{\frac{1}{2}} \tag{3.16}$$

which applies if the ratio between the strains is kept constant throughout deformation. Equation 3.16 applies to constant loading conditions, whereas Eq. 3.15 can be integrated for varying strains. The von Mises criterion for tensile loading with stress $\sigma_1 = Y$, the plastic strain tensor for isotropic materials, has the form

$$\varepsilon_i = \begin{vmatrix} \varepsilon_1 & 0 & 0 \\ 0 & \varepsilon_2 & 0 \\ 0 & 0 & \varepsilon_3 \end{vmatrix} \tag{3.17}$$

where $\varepsilon_2 = \varepsilon_3$ and both are equal to $-\varepsilon_1/2$. If we substitute these values in Eq. 3.16, we get

$$\varepsilon_{\text{eff}} = \left[\frac{2}{3}\left(\varepsilon_1^2 + \frac{\varepsilon_1^2}{4} + \frac{\varepsilon_1^2}{4}\right)\right]^{\frac{1}{2}} = \left[\frac{2}{3}\cdot\frac{3}{2}\varepsilon_1^2\right]^{\frac{1}{2}} = \varepsilon_1 \tag{3.18}$$

Thus, the effective stress and strain that we have defined correspond to a uniaxial true stress–true strain curve. Therefore, the general expression for describing the plastic portion of the stress-strain curve

$$\sigma_{\text{eff}} = K\varepsilon_{\text{eff}}^{\,n} \tag{3.19}$$

where n is the strain hardening coefficient, can be employed to approximate deformation behavior for isotropic materials at large plastic strains. As pointed out in Chapter 1, such an expression should be applied only to the plastic portions of stress-strain curves.

Equivalent shear expressions can also be defined using the condition wherein $\tau = k$ such that the effective shear stress is given as

$$\tau_{\text{eff}} = \left[\frac{(s_1 - s_2)^2 + (s_3 - s_1)^2 + (s_3 - s_2)^2}{6}\right]^{\frac{1}{2}} \tag{3.20}$$

and the effective shear strain is given as

$$\gamma_{\text{eff}} = \left[2\left(\varepsilon_1^2 + \varepsilon_2^2 + \varepsilon_3^2\right)\right]^{\frac{1}{2}} \tag{3.21}$$

These expressions demonstrate that $\tau_{\text{eff}} = \sigma_{\text{eff}}/\sqrt{3}$ and $\gamma_{\text{eff}} = \sqrt{3}\,\varepsilon_1$ for simple tension or compression. Equation 3.21 written as the time-dependent function of strain rates, $d\varepsilon/dt$, is

$$(d\gamma/dt)_{\text{eff}} = \dot{\gamma}_{\text{eff}} = \left[2\left(\dot{\varepsilon}_1^2 + \dot{\varepsilon}_2^2 + \dot{\varepsilon}_3^2\right)\right]^{\frac{1}{2}} \tag{3.22}$$

Equations 3.20 and 3.22 can be used for the development of the deformation mechanism maps discussed in Chapter 8.

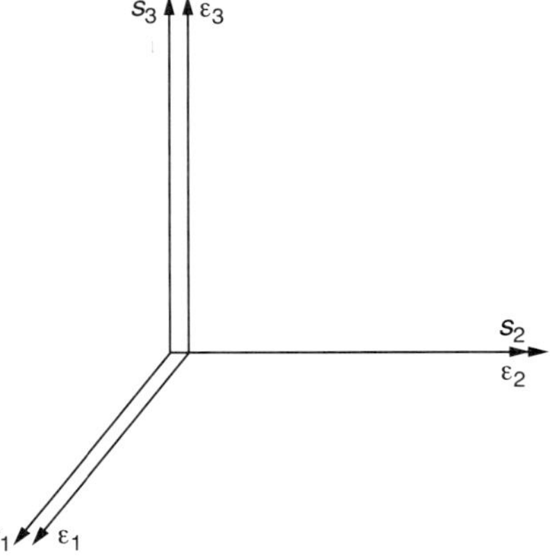

FIG. 3.6 Parallel coordinate systems for the yield surface and the strain vector.

3.1.3 Flow Rules

The change in dimensions anticipated from the assignment of yield criteria can also be described by employing the *Levy–von Mises flow rules* to describe the relationship between applied stresses and strains. These flow rules take the same form as the multiaxial expression for Hooke's law, but we employ a resistance to plastic deformation or plastic potential to balance this relationship. For example, the magnitude of incremental strain in the x-direction should be

$$d\varepsilon_1 = \frac{d\varepsilon_{\text{eff}}}{d\sigma_{\text{eff}}}\left[s_1 - \frac{1}{2}\left(s_2 + s_3\right)\right] \tag{3.23}$$

if the expression $d\varepsilon_{\text{eff}}/d\sigma_{\text{eff}}$ is considered to be the resistance to plastic deformation and we require that the volume change is zero. Keeping the pattern given in Eq. 3.23 for $d\varepsilon_2$ and $d\varepsilon_3$, the ratio of the strains should then correspond to the ratios of the driving force for each deformation as

$$d\varepsilon_1 : d\varepsilon_2 : d\varepsilon_3$$

coinciding with

$$\left[s_1 - \frac{1}{2}\left(s_2 + s_3\right)\right] : \left[s_2 - \frac{1}{2}\left(s_1 + s_3\right)\right] : \left[s_3 - \frac{1}{2}\left(s_2 + s_1\right)\right] \tag{3.24}$$

The implications of the flow rule given in Eq. 3.24 are that if the principal strain axes and the principal stress axes were drawn to coincide, as shown in Fig. 3.6, the vector corresponding to the strain ratios must have no component along the $s_1 = s_2 = s_3$ axis. Or in other words, if such a strain vector were drawn at the surface of the yield criterion, it must lie normal to it, as shown in Fig. 3.7*a*. Of course, this means that if the projection of the vector is drawn in a planar section of the yield stress, as in Fig. 3.7*b*, the vector will not appear to be normal to the elliptical section. This will be true whenever $s_3 \neq 0$. The principle

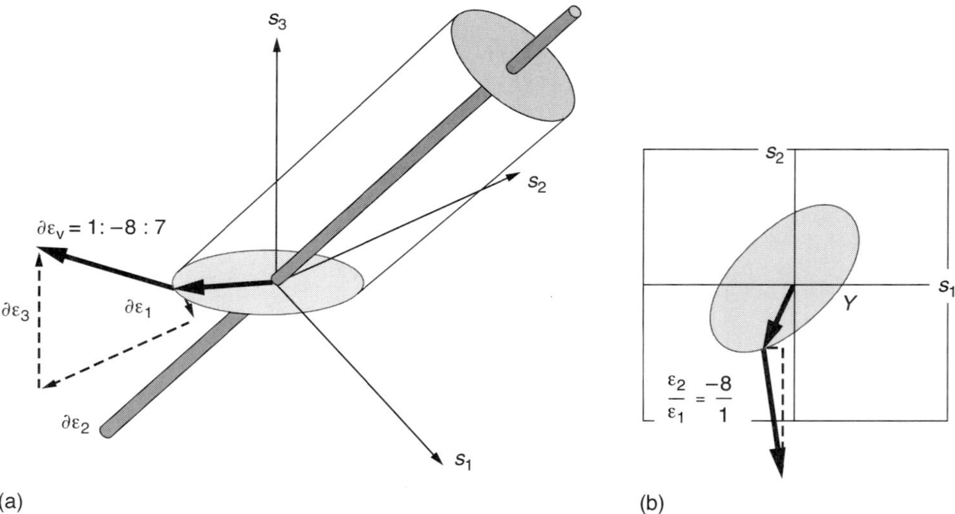

FIG. 3.7 Demonstration of the principle of normality, showing the incremental strain vector $\partial\varepsilon_v$ as the ratio of the three strains $\partial\varepsilon_1{:}\partial\varepsilon_2{:}\partial\varepsilon_3$ (a) in three dimensions and (b) in a projection on the plane wherein $s_3 = 0$. For this example, $s_1 = 0.4s_2$ and s_1 and s_2 are compressive (after Hosford and Caddell, 1993).

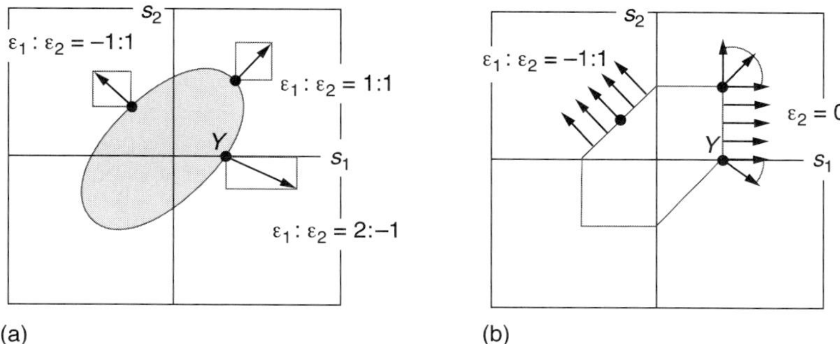

FIG. 3.8 Strain vector projections for the (a) von Mises and (b) Tresca yield criteria. For each of these vectors, any component of strain along ε_3 is not apparent. Also, for the Tresca criterion in (b), the corners represent a range of possible deformations and the straight sides have fixed deformation imposed. This is one reason that the von Mises criterion is considered more meaningful in describing deformations (after Hosford and Caddell, 1993).

described here is often termed in plasticity as the *Principle of Normality*. This principle is difficult to apply to the Tresca criterion because of the flat facets and the corners of the yield surface. For the Tresca criterion, the flat facets suggest that even though the stress ratios are changed, the strain ratios are fixed for a facet. Even worse, the corners do not have a fixed definition of the strain ratios. Projections of the strain vector for the von Mises and Tresca yield criteria are shown in Fig. 3.8.

EXAMPLE 3.1 *Channel Die Compression*

Let us consider what is the *easiest* way to change the shape of an object. The definition of "easiest" in this instance is the one that requires the lowest applied stress. We will consider flattening of a piece of aluminum from an 8-cm^3 cube to a square plate 1 cm in height. We will ignore any friction effects and assume that the deformation is *perfectly plastic*—i.e., proceeds without strain hardening. To date, your new employer has used a two-step process of *channel die compression,* as shown in Fig. 3.9.

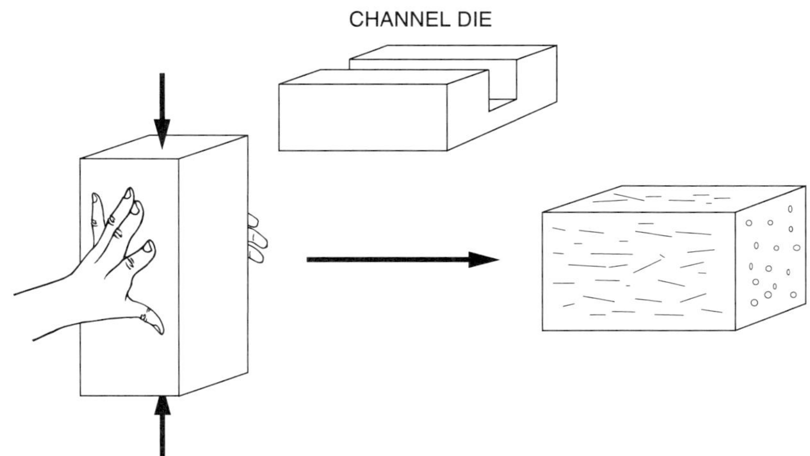

FIG. 3.9 Cartoon showing channel die compression. The sample is compressed in one direction and allowed to flow freely in another. The hands show how a channel constrains the sides of the sample from deforming.

In channel die compression, the material is deformed in plane strain by placing the cube in a rectangular slot that fits snuggly against two sides. The cube is free to deform in the other direction as it is compressed from the normal direction. The deformation is analogous to that in rolling with compressive deformation in the normal direction **(ND)**, extension in the free direction or rolling direction **(RD)**, and near-zero deformation in the confined direction or transverse direction **(TD)**. To reach the final dimensions, the plastic deformation in this first step is given in true strain as

$$\varepsilon^{(1)} = \begin{vmatrix} \ln\frac{\sqrt{8}}{2} & 0 & 0 \\ 0 & 0 & 0 \\ 0 & 0 & \ln\frac{\sqrt{2}}{2} \end{vmatrix} = \begin{vmatrix} 0.347 & 0 & 0 \\ 0 & 0 & 0 \\ 0 & 0 & -0.347 \end{vmatrix}$$

where $\sqrt{8}$ cm is the x-dimension, the y-dimension is unchanged at 2 cm, and the width of the die, $\sqrt{2}$ cm, is the z-dimension after the first deformation step.

In the second deformation step, the x-dimension is held at $\sqrt{8}$ cm and the material is free to deform in the y-direction as it is again in the z-direction. This time the deformation is

$$\varepsilon^{(2)} = \begin{vmatrix} 0 & 0 & 0 \\ 0 & \ln\frac{\sqrt{8}}{2} & 0 \\ 0 & 0 & \ln\frac{\sqrt{2}}{2} \end{vmatrix} = \begin{vmatrix} 0 & 0 & 0 \\ 0 & 0.347 & 0 \\ 0 & 0 & -0.347 \end{vmatrix}$$

Assume that the stress applied to produce this deformation was $s = 100$ MPa. Using the flow rules for the first deformation step, the strain ratios are

$$0.347:0:-0.347$$

Applying the flow rules given in Eq. 3.24, if the magnitude of the stress in the confining direction is one-half that of the deformation stress, the deformation will proceed with equal and opposite strains in the x- and z-directions, respectively. This stress level can be used in both the von Mises and Tresca criteria to predict deformation under other conditions. Using the von Mises criterion from Eq. 3.4,

$$\left(0-\frac{1}{2}s\right)^2+(s-0)^2+\left(s-\frac{1}{2}\right)^2 = C_{von\ Mises}$$

$$\frac{3}{2}s^2 = 15{,}000(\text{MPa})^2$$

This fixes the yield criterion at one point. Because the yield criterion is simply an infinite cylinder, it also fixes all other values. The yield stress Y for simple compression predicted by the von Mises stress is then

$$(0-0)^2+(Y-0)^2+(Y-0)^2 = 15{,}000\ (\text{MPa})^2$$

so we can solve for the yield stress as follows:

$$Y = \sqrt{\frac{15{,}000}{2}} = 86.6\ \text{MPa}$$

Consequently, the von Mises criterion predicts that the applied stresses for the simple compression will be less than for the channel die compression by

$$Y = \left(\sqrt{3}/2\right)s$$

Using the Tresca criterion,

$$s - 0 = C_{Tresca}$$

yielding

$$s = Y = 100 \text{ MPa}$$

The second step requires the same stresses on different axes. The addition of $\varepsilon^{(1)}$ and $\varepsilon^{(2)}$ gives the final shape specified. In other words, the Tresca criterion predicts no difference in stress level for the two methods. Careful evaluation of the loading path employed here shows that channel die compression loading described here is given in Fig. 3.4*b*. At $s_3 = \frac{1}{2}s_1$, the values for the von Mises and Tresca criteria can be equated. When this is done, the yield stress predicted by the von Mises criterion is lower. ■

To this point, we have considered how the yield surface is approached, but not what the implications are when we get there. If the material strain hardens, such that the resistance to deformation increases as we deform the material consistent with Eq. 3.19, then the yield surface must change in size to correspond to this condition. Thus, the yield condition is not exceeded, but follows the strain hardening. The yield surface might not change size uniformly along all directions, as shown in Fig. 3.10*a*, because the imposed deformation is directional, but it can distort as suggested by Fig. 3.10*b*. Nonetheless, if we believe that the deformation should respond to the direction involving the least energy, the strain vector should still lie normal to the distorted yield surface. Each of the effects not considered in this section—anisotropy, strain hardening, strain rate, temperature, and volumetric change—can dramatically alter the material's response. Single crystals provide an example of strong plastic anisotropy because slip is confined to specific planes and directions.

EXAMPLE 3.2 *The Strains Resulting from Isotropic Hardening*

Engineers have estimated that a stress state of

$$\sigma = \begin{vmatrix} 120 & 80 & 0 \\ 80 & 240 & 0 \\ 0 & 0 & -120 \end{vmatrix} \text{ MPa}$$

has apparently caused unexpected modest deformation in a critical region of an aluminum alloy component. You have been asked to determine if the material was fulfilling the expected specifications with particular emphasis on the required yield stress of 250 MPa. You have been assured that the expected flow curve according to Eq. 3.19 is reliable for small strains with $K = 634$ MPa and a strain hardening exponent of $n = 0.15$ (although this is hardly typical). You have also been asked to describe

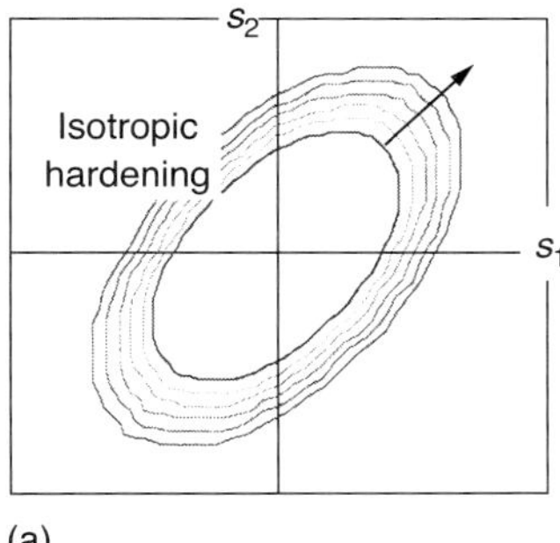

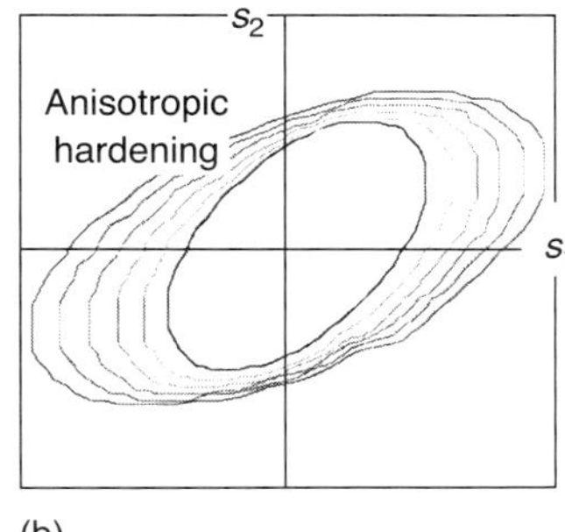

FIG. 3.10 (a) Isotropic strain hardening and (b) anisotropic strain hardening for a material that deforms by the von Mises criterion. For (b), the principal stresses must be referenced to specific principal stress directions within the material.

the nature of the shape change that may have occurred so as to help understand any repairs that might be necessary.

First, we will need to determine the principal stresses as shown in Chapter 2. We know that $\sigma_{33} = -20$ MPa is already a principal stress. We can find the other two principal stresses using Eq. 2.14:

$$\frac{120+240}{2} \pm \sqrt{\left(\frac{120-240}{2}\right)^2 + (80)^2} = 280 \text{ and } 80 \text{ MPa}$$

Then, if we use the von Mises criterion, we can write

$$\sqrt{\frac{(80-280)^2 + (80-(-120))^2 + (280-(-120))^2}{2}} = 346 \text{ MPa}$$

which is clearly greater than the specified yield stress. To estimate the amount of deformation that occurred locally, we first assume that hardening is isotropic. If we do so, we can use Eq. 3.19 to roughly estimate the amount of effective shear strain. Using the given values, we can calculate

Effective strain	Effective stress (MPa)
0.002	250
0.005	286
0.01	318
0.015	338
0.02	353
0.025	365

The values above suggest that the total amount of deformation expected is about 2 percent effective strain. If we want to predict the shape change, we can use the Levy–von Mises relations of Eq. 3.23 and 3.24.

Using these relations yields the following principal strains.

$$\varepsilon_1 = \frac{0.02}{346}\left[280 - \frac{1}{2}(80 + (-120))\right] = 0.017$$

$$\varepsilon_2 = \frac{0.02}{346}\left[80 - \frac{1}{2}(280 + (-120))\right] = 0.00$$

$$\varepsilon_3 = \frac{0.02}{346}\left[-120 - \frac{1}{2}(280 + 80)\right] = -0.017$$

We can check our answer to see that the strains sum to zero, as they should for plastic deformation.

Whenever deformation is driven by a given stress state that meets a criterion for shear deformation, we can express this in a single term called the effective shear stress. Related to this effective shear stress is a single term that we can also use to represent the amount of shear deformation taking place. Both of these terms are used to describe deformation in Chapter 8. Using the values from this example, the effective shear stress is

$$\tau_{\text{eff}} = \sqrt{\frac{(80-280)^2 + (80-(-120))^2 + (280-(-120))^2}{6}} = 200 \text{ MPa}$$

and the effective shear strain is

$$\gamma_{\text{eff}} = \sqrt{2\left(0.017^2 + (-0)^2 + (-0.017)^2\right)} = 0.024$$

■

3.2 SHEAR DEFORMATION OF CRYSTALLINE MATERIALS

At one time, recognition that plastic deformation of crystals had taken place on specific planes and in specific directions required the knowledge that the materials were crystalline. Most materials readily identified as crystalline prior to the start of the twentieth century were transparent materials with strong ionic bonding that inhibited observation of large-scale plasticity. Glide had first been reported by Reusch in 1867 from observations of surface steps on rock salt crystals. The first investigators to call these surface steps slip lines were Ewing and Rosenhain, who published research on metal crystals in 1899. Once it was recognized that most metals were crystalline and large metal crystals were made, scientists were able to observe slip bands or markings at specific orientations in deformed crystals. It was in the intervening years that slip between planes of atoms, similar to the card example given in Chapter 1, became the interpretation. Using this information along with recently developed techniques for determining crystal orientation and structure by x-ray diffraction, scientists were able to determine the planes and directions that became displaced during tensile or compressive deformation. Through experiments on aluminum, iron, and brass single crystals, Taylor and Elam demonstrated slip behavior related to crystal planes and directions in several papers in the 1920s. Some examples from Elam on deformation of crystals are shown in Fig. 3.11. Deformation twinning was also ascribed to a crystallographic origin over this same period.

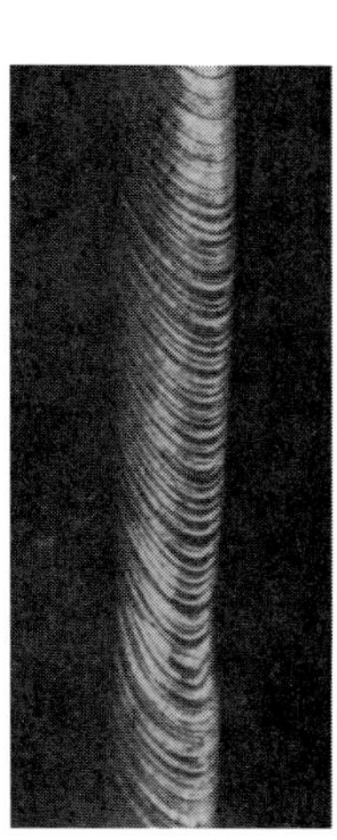

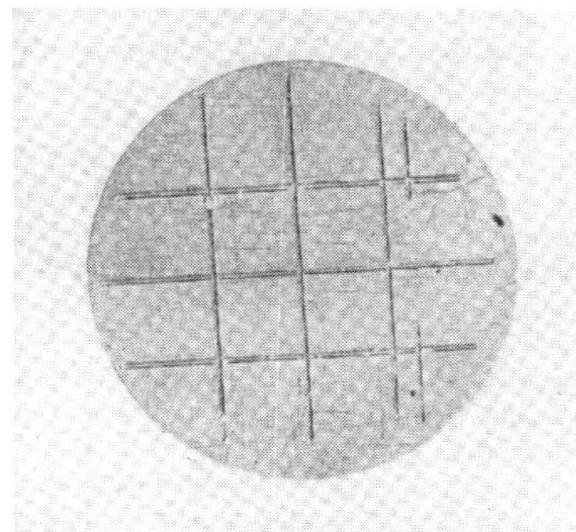

(a) (b)

FIG. 3.11 (a) Slip bands in a copper-aluminum crystal (Fig. 3 [Plate 1] from *Distortion of Metal Crystals* by C. F. Elam, 1935, used with permission of Oxford University Press) and (b) elliptical aluminum crystal after compression (Fig. 2, p. 532, G. I. Taylor and W. S. Farren, *Proc. Roy. Soc. A.*, 111, 759, 1926, used with permission).

BIOGRAPHY

GEOFFREY I. TAYLOR (1886–1975) AND CONSTANCE (F. ELAM) TIPPER (1894–1995)

Sir G. I. Taylor made major contributions to experimental approaches to testing of metal crystals and to models of deformation of polycrystals and is one of the three scientists credited with the discovery of dislocations in 1934. His collaboration with C. F. Tipper dramatically advanced our understanding of crystal plasticity. Taylor also made significant contributions in meteorology, oceanography, aeronautics, metal physics, and chemical engineering. He was knighted in 1944.

In 1949 C. F. Tipper became Reader and the first woman engineering faculty member at Cambridge University following pivotal activities during the Second World War, wherein she led efforts on understanding how the steel in Liberty Ships caused their catastrophic failures. She provided a full perspective on this in her book, *The Brittle Fracture Story* (Cambridge University Press), which was published in 1962. (See also Batchelor, 1996, Cambridge, 2001, and Taylor, 1965.)

3.2.1 Deformation of Single Crystals

For a specific glide plane and direction, the geometric relationship to applied stress provides information on the shear stress producing the shear translation and the ratios of the strain along the deformation axis and the shear strain. Stresses producing plastic deformation and plastic shear strains can be represented by tensors with the special proviso that the strains should be given in small increments. Slip systems, as shown in Fig. 3.12, can be defined with the slip plane as **SP**, the normal to the glide or slip plane as **SPN**, and the glide or slip direction as **SD**. Glide on the negative or opposite-sign slip plane, **–SP**, in the negative or opposite sign direction, **–SD**, produces the same deformation because shear always consists of a pair of force couples (see Chapters 1 and 2). For a given crystal, the number of crystallographically equivalent slip systems is a result of the specific system and the crystal symmetry. We will designate slip systems as we have tensor subscripts for stress and strain: a slip system is then a combination of directions and planes, <uvw> {hkl}, where the **SD** = [uvw] and the **SP** = (hkl) for a single slip system. Of course, in cubic materials, all **SPN** (hkl or <hkl>) are orthogonal to the **SP** ((hkl) or {hkl}) with the same indices. Table 3.1 gives slip systems observed for a large number of crystalline materials. For the noncubic materials, specific indices are used for directions and planes that are unique as a result of differences in axis length or interaxis angles. For most metals, slip occurs before deformation mechanisms at temperatures less than half the absolute melting temperature. For some ceramic and intermetallic materials, fracture occurs at room temperature before plasticity. For these materials, slip may occur only at high temperatures and under compressive loading. Polymer crystals can also undergo plastic deformation; however, the deformation is complicated by the chain or network structures of large molecules, and producing fully crystalline polymers is quite difficult.

The observed slip directions, as discussed in more detail in the next chapter, are normally the shortest interatomic spacings found within the crystal. The combination of this direction **(SD)** with the plane of slip (**SP**) usually requires the smallest amount of shear stress to produce deformation. As shown in Fig. 3.11, the **SD** must lie within the **SP**. This restricts the number of slip systems possible.

TABLE 3.1 Favored Slip Systems for Crystals

Crystalline material	Family of directions	Family of planes	Independent slip systems
FCC metals and solid solutions (Al, Cu, Ni, Ag, Au)	$\langle 1\bar{1}0\rangle$	$\{111\}$	5
Diamond cubic (Si, Ge, C)	$\langle 1\bar{1}0\rangle$	$\{111\}$	5
BCC metals and solid	$\langle 1\bar{1}1\rangle$	$\{110\}$	5
solutions	$\langle 1\bar{1}1\rangle$	$\{21\bar{1}\}$*	5
(Fe, Nb, Mo, Ta, W)	$\langle 1\bar{1}1\rangle$	$\{\bar{1}23\}$*	5
HCP metals and alloys	$\langle 11\bar{2}0\rangle$	(0001)	2
(Zn, Cd, Mg)	$\langle 11\bar{2}0\rangle$	$\{10\bar{1}1\}$	4
(Ti, Zr)	$\langle 11\bar{2}0\rangle$	$\{10\bar{1}0\}$	2
	$\langle 11\bar{2}0\rangle$	$\{10\bar{1}1\}$	4
	$\langle 11\bar{2}0\rangle$	(0001)	2
β-Sn, tetragonal	$[001]$	$\{110\}$	
	$[001]$	$\{100\}$	
	$\langle 1\bar{1}1\rangle$	$\{110\}$	
Ga, orthorhombic	$[010]$	(001)	1
	$[010]$	(102)	1
	$\langle 0\bar{1}1\rangle$	$\{011\}$*	
U, orthorhombic	$[100]$	(010)	1
	$[100]$	(001)	1
Rock salt structures	$\langle 1\bar{1}0\rangle$	$\{110\}$	2
(NaCl, LiF, MgO)	$\langle 1\bar{1}0\rangle$	$\{001\}$	3
Cesium chloride structures (CsCl, NiAl, FeAl)	$\langle 001\rangle$	$\{110\}$	3
Sapphire-α-Al_2O_3	$\langle 11\bar{2}0\rangle$	(0001)	2
β-Si_3N_4, hexagonal	$\langle 0001\rangle$	$\{10\bar{1}0\}$	2
Fluorite structures	$\langle 1\bar{1}0\rangle$	$\{001\}$	3
(CaF_2, UO_2, ThO_2)	$\langle 1\bar{1}0\rangle$	$\{110\}$	2

*Slip systems for which the sign of the slip direction is important. Reversing the slip direction produces a different arrangement of the crystal.

EXAMPLE 3.3 *Specific Slip Systems in Cubic Materials*

Consider the list of paired directions and planes for FCC metals with regard to slip on $\langle 1\bar{1}0\rangle\ \{111\}$.

(a) $[110]\ (1\bar{1}1)$
(b) $[110]\ (\bar{1}1\bar{1})$
(c) $[\bar{1}\bar{1}0]\ (1\bar{1}1)$
(d) $[011]\ (1\bar{1}1)$
(e) $[110]\ (\bar{1}11)$
(f) $[110]\ (\bar{1}\bar{1}1)$

We want to establish the nature of the specific slip systems listed above. Systems (a) and (b) represent slip in the same direction on the same type of plane. The only difference is the definition of the normal given by the sign. In most slip systems, this has no importance, although in specifically defining dislocations and dislocation interactions (Chapter 4), this can be important. Systems (a) and (c) represent slip on the same plane in two different directions. They are really the same slip system; only the direction of slip is different. Thus (a), (b), and (c) are all representations of the same individual slip system. System (d) is a slip system on the same plane as the prior three, but with slip in a separate direction. This slip system is physically distinct from the first three. System (e) represents slip in the

TABLE 3.2 Physically Distinct Slip Systems for FCC Metals

Direction	Plane	Direction	Plane
1. $[1\bar{1}0]$	(111)	7. $[110]$	$(1\bar{1}1)$
2. $[0\bar{1}1]$	(111)	8. $[011]$	$(1\bar{1}1)$
3. $[10\bar{1}]$	(111)	9. $[\bar{1}01]$	$(1\bar{1}1)$
4. $[110]$	$(\bar{1}11)$	10. $[011]$	$(11\bar{1})$
5. $[101]$	$(\bar{1}11)$	11. $[101]$	$(11\bar{1})$
6. $[0\bar{1}1]$	$(\bar{1}11)$	12. $[1\bar{1}0]$	$(11\bar{1})$

The operative slip systems are determined by the loading geometry. For BCC metals that slip on $<1\bar{1}1>\{110\}$ systems, the reversal of the designations in Table 3.2 gives the possible slip systems. In noncubic systems, the physically distinct slip systems may include a mixture of designations, as discussed in Example 3.4.

same direction as for (a) to (c), but on a different plane. This is the second physically distinct plane of the $\{111\}$ family that can produce slip in the $[110]$ direction. The direction and plane pair in (f) *is not* a slip system since $[110]$ does not lie in $(\overline{1}\overline{1}1)$. It is always important to check that the direction of slip lies in the plane before identifying a direction–plane combination as a slip system. The physically distinct slip systems for FCC metals can be expressed as three specific $<1\bar{1}0>$ directions on four specific $\{111\}$ planes or two planes of the family $\{111\}$ that can produce slip in six distinct $<1\bar{1}0>$directions. Table 3.2 gives a set of 12 physically distinct slip systems for FCC metals. ■

EXAMPLE 3.4 *Slip Systems in Tetragonal Tin*

Consider the reported slip systems for tetragonal tin, which has a body-centered tetragonal unit cell with dimensions $a = b = 0.583$ nm and $c = 0.318$ nm. This phase is the equilibrium phase for tin at temperatures above 14°C. Below this temperature, the equilibrium phase is cubic, although the transformation is difficult to achieve and requires significant undercooling. The designation of [001] as a specific slip direction is a result of the uniqueness of this direction. The spacing of atoms in the [001] direction is much shorter than in the [100] and [010] directions, which might be grouped as the $<100>$ family with the recognition that [001] would not be included in this family in a tetragonal material. The slip planes matched with the [001] slip direction are given the family designations $\{110\}$ and $\{100\}$. The physically distinct slip systems that can be gleaned include the (110) and $(1\bar{1}0)$ planes for $\{110\}$ because the spacing of the four distinct $\{101\}$ planes—(101), $(\bar{1}01)$, (011), and $(0\bar{1}1)$—have different interatomic spacings and do not contain the [001] direction. The corresponding case for $\{100\}$ includes only the (100) and (010) planes. Although this format is often used for reports on slip in noncubic materials, designations are not always given, nor are they always consistent in published works. ■

The relationship between applied uniaxial stress and strain and glide on individual slip systems is related to a quantity usually defined as the Schmid factor

$$\tau = \sigma_{\text{applied}} \cos\theta \cos\lambda \tag{3.25}$$

where τ is the shear stress on the slip plane in the slip direction, σ_{applied} is the applied tensile or compressive stress along a defined tensile axis **TA**, λ is the angle between the slip direction and the direction of the applied stress, and θ is the angle between the slip plane normal and the direction of the applied stress. The Schmid factor can be derived quite concisely by tensor rotations, as shown below.

The Schmid Factor The relationship between the applied stress σ_{applied} and the shear stress τ is simply an example of a tensor rotation wherein

BIOGRAPHY

ERICH SCHMID (1896–1983)

Along with coauthor W. Boas, Erich Schmid authored the first important book on crystal plasticity (*Kristallplastizität*) in 1935. This critically important work was translated into english in 1950. This book is complementary to the first book authored by C. F. (Elam) Tipper, *Distortion of Metal Crystals*, which was published in 1936.

$$\sigma = \begin{vmatrix} 0 & 0 & 0 \\ 0 & 0 & 0 \\ 0 & 0 & \sigma_{\text{applied}} \end{vmatrix}$$

and

$$\sigma' = \begin{vmatrix} \sigma_{11} & 0 & \tau \\ 0 & \sigma_{22} & 0 \\ \tau & 0 & \sigma_{33} \end{vmatrix}$$

We can then write

$$\tau = \sigma_{13}{}' = a_{13} a_{33} \sigma_{33}$$

and employ the $a\{\phi 1, \Phi, \phi 2\}$ rotation matrix to yield

$$\tau = \sin\Phi \sin\phi 2 \cos\Phi \; \sigma_{\text{applied}}$$

Because a_{13} is the cosine of the angle between x' and z, we can then set $\cos\lambda = \sin\Phi \sin\phi 2$ and $\theta = \Phi$ to arrive at the expression for Eq. 3.25.

The relationship of the Schmid factor to axial strain and shear strain is given by

$$\varepsilon = \gamma \cos\theta \cos\lambda \tag{3.26}$$

where ε is the strain along the direction of applied stress, γ is the glide strain expressed as a simple shear, and θ and λ are as defined previously.

The strains for a specific slip system can be expressed in a simple way by using the unit vectors for the slip direction, $|\mathbf{SD}| = [uvw]$, and slip plane normal, $|\mathbf{SPN}| = hkl$. The individual shape-change strains referenced to the unit cell of a cubic material can then be expressed as

$$\varepsilon_{ij} = \gamma \begin{bmatrix} u \cdot h & \frac{1}{2}(v \cdot h + u \cdot k) & \frac{1}{2}(u \cdot l + w \cdot h) \\ \frac{1}{2}(v \cdot h + u \cdot k) & v \cdot k & \frac{1}{2}(w \cdot k + v \cdot l) \\ \frac{1}{2}(u \cdot l + w \cdot h) & \frac{1}{2}(w \cdot k + v \cdot l) & w \cdot l \end{bmatrix} \tag{3.27}$$

where γ is the magnitude of the simple shear on the slip system.

EXAMPLE 3.5 *Shear Strain Transformations to Normal Strains*

Consider slip on the [110] $(1\bar{1}0)$ system with a shear strain magnitude of 0.02. How much shape change will this cause with respect to the unit cell orientation for a cubic material?

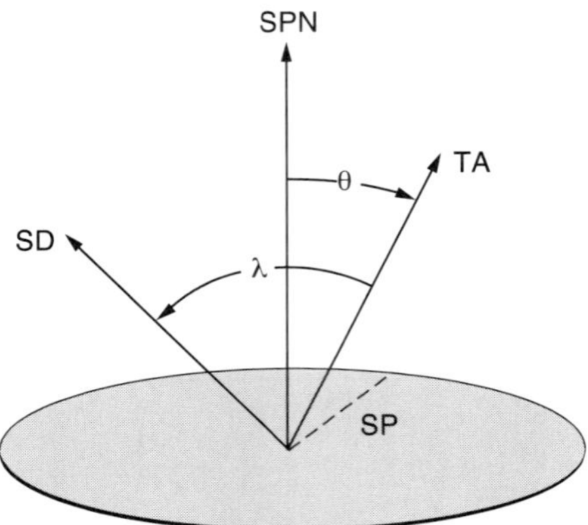

FIG. 3.12 Diagram showing the relationship among **SD**, **SP**, **SPN**, **TA**, λ, and θ for slip of a single crystal. The shadow in SP is the projection of TA on this plane. **SD** lies in **SP**.

First, we need to express the slip system in terms of unit vectors,

$$|\mathbf{SD}| = \left[\frac{1}{\sqrt{2}} \cdot \frac{1}{\sqrt{2}} \; 0\right] \quad \text{and} \quad |\mathbf{SPN}| = \frac{1}{\sqrt{2}} \cdot \frac{-1}{\sqrt{2}} \; 0$$

The resulting strains given by Eq. 3.27 are

$$\varepsilon_{ij} = \gamma \begin{bmatrix} \frac{1}{\sqrt{2}} \cdot \frac{1}{\sqrt{2}} & \frac{1}{2}\left(\frac{1}{\sqrt{2}} \cdot \frac{1}{\sqrt{2}} + \frac{1}{\sqrt{2}} \cdot \frac{-1}{\sqrt{2}}\right) & \frac{1}{2}\left(\frac{1}{\sqrt{2}} \cdot 0 + 0 \cdot \frac{1}{\sqrt{2}}\right) \\ \frac{1}{2}\left(\frac{1}{\sqrt{2}} \cdot \frac{1}{\sqrt{2}} + \frac{1}{\sqrt{2}} \cdot \frac{-1}{\sqrt{2}}\right) & \frac{1}{\sqrt{2}} \cdot \frac{-1}{\sqrt{2}} & \frac{1}{2}\left(0 \cdot \frac{-1}{\sqrt{2}} + \frac{1}{\sqrt{2}} \cdot 0\right) \\ \frac{1}{2}\left(\frac{1}{\sqrt{2}} \cdot 0 + 0 \cdot \frac{1}{\sqrt{2}}\right) & \frac{1}{2}\left(0 \cdot \frac{-1}{\sqrt{2}} + \frac{1}{\sqrt{2}} \cdot 0\right) & 0 \cdot 0 \end{bmatrix}$$

$$= \begin{bmatrix} \frac{\gamma}{2} & 0 & 0 \\ 0 & \frac{-\gamma}{2} & 0 \\ 0 & 0 & 0 \end{bmatrix} = \begin{bmatrix} 0.01 & 0 & 0 \\ 0 & -0.01 & 0 \\ 0 & 0 & 0 \end{bmatrix}$$

For noncubic systems, the unit vector designations must be scaled to the axis lengths and the interaxial angles. ■

The critical resolved shear stress τ_{crss} corresponds to the condition (Eq. 3.25) wherein slip occurs for this particular crystal orientation, or the yield stress Y, as

$$\tau_{crss} = Y \cos\theta \cos\lambda \tag{3.28}$$

As the single crystal undergoes tensile or compressive plastic deformation with slip on a single slip system, the crystal must rotate as shown in Fig. 3.13. This rotation also leads to a change in the crystal width along the projection of the slip direction. Figure 3.11*b* shows such a change in cross section for an aluminum crystal. The rotation takes place partly because of the rotation of the crystal relative to the confining grips or platens and partly because of the simple shear component of the deformation. As discussed by Hosford (1993), a mathematical expression for the rotation would have the rotation take place as a rotation about the axis that is mutually orthogonal to the **SD** and the **SPN**, or **SD** × **SPN**, where × is the cross product. The rotation tensor increments can be expressed as the rotation tensor

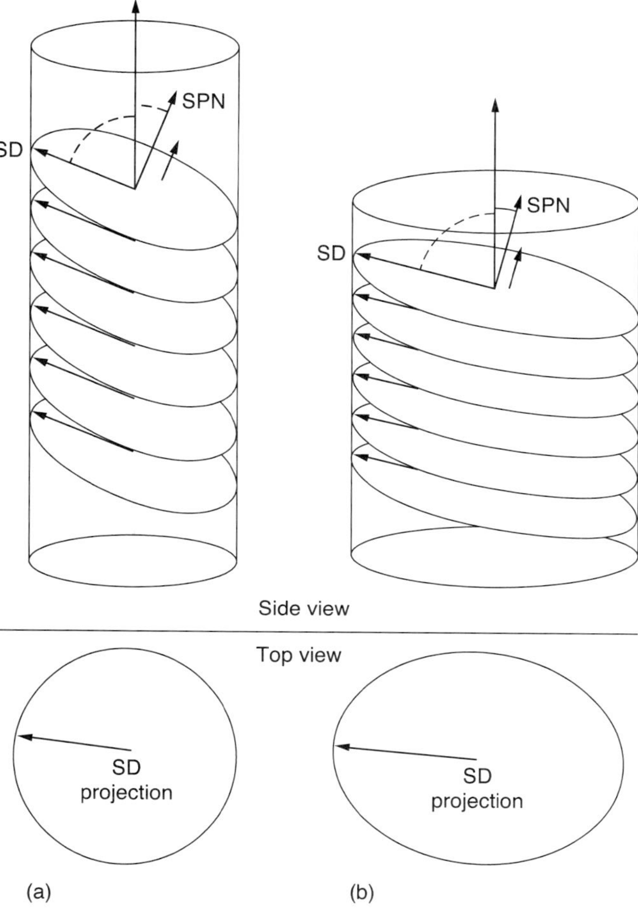

FIG. 3.13 Rotation of the **SPN** toward the compression axis, as occurred in Fig. 3.11*b*, resulting in the change from a circular to an elliptical cross section. (a) Before compression; (b) after compression.

$$\omega_{ij} = \gamma \begin{bmatrix} 0 & \frac{1}{2}(u \cdot k - v \cdot h) & \frac{1}{2}(u \cdot l - w \cdot h) \\ \frac{1}{2}(v \cdot h - u \cdot k) & 0 & \frac{1}{2}(v \cdot l - w \cdot k) \\ \frac{1}{2}(w \cdot h - u \cdot l) & \frac{1}{2}(w \cdot k - v \cdot l) & 0 \end{bmatrix} \tag{3.29}$$

EXAMPLE 3.6 *Shear Strain Transformations to Rotations*

For the slip system, [110] $(1\bar{1}0)$, and strain level used in Example 3.5, the rotations would be about the mutual normal direction [001] and would be given as

$$\omega_{ij} = \begin{vmatrix} 0 & \frac{-\gamma}{2} & 0 \\ \frac{\gamma}{2} & 0 & 0 \\ 0 & 0 & 0 \end{vmatrix} = \begin{vmatrix} 0 & -0.01 & 0 \\ 0.01 & 0 & 0 \\ 0 & 0 & 0 \end{vmatrix}$$

If we use the relation given in Chapter 1, the angle of rotation about [001] is $\tan^{-1}(0.01) = 0.57°$.

Crystallographic rotations during plastic deformation and the initiation of secondary slip systems are discussed in greater detail in the next section. ■

3.2.2 Stereographic Projections and Slip Systems

Interactions between slip systems and demonstrations of the effects of deformation on crystal orientations are frequently shown using stereographic projections. The Appendix to Chapter 2 contains a brief tutorial on their uses. As an example, the square-cross-section cubic crystal shown in Fig. 3.14*a* designates a (111) slip plane and a slip direction [$\bar{1}$01] for tensile deformation along the [$\bar{1}$23]. If slip takes place on the parallel planes and directions corresponding to this specific slip system, then the **SD** should rotate toward the tensile axis. In normal tension or compression testing, this axis is fixed by the testing machine. The rotation from slip then changes the direction within the crystal that is parallel to the tensile or compressive axis defined by the testing machine. However, because the stereographic projection is drawn as fixed relative to the frame of the crystal, the tensile axis should rotate as shown in Fig. 3.15. This rotation is drawn as though it results from a counterclockwise rotation about **SPN** × **SD** = [1$\bar{2}$1].

The triangle highlighted in Fig. 3.14*b* is often called the standard stereographic triangle, although single slip can occur for orientations defined within any triangle because they are all equivalent. The implication of rotating to the boundaries shown on the triangle is that the slip process can occur on additional slip systems that by symmetry have the same value of the Schmid factor given above. Plastic slip was observed to occur on particular slip systems even before the discovery of dislocations (described in greater detail in Chapter 4). The easiest slip system implies the system with the lowest value of τ_{crss}. For FCC metals, this combination of closest atomic spacing, <110>, and closest packed planes, {111}, provides for easy slip.

Fig. 3.15*a* shows designations of the slip system for a given tensile axis in the standard stereographic triangle. Figure 3.15*b* gives the multiple slip systems for one of the three boundaries of the triangle for slip on the <1$\bar{1}$0> {111} slip system, which is the easiest slip

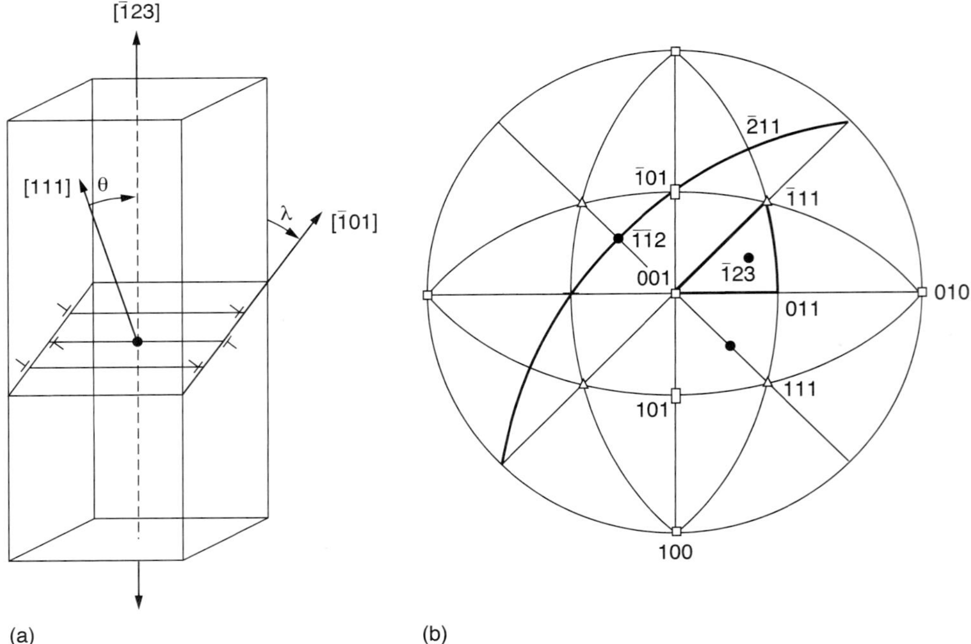

FIG. 3.14 (a) A square-cross-section crystal showing the slip direction and slip plane normal of a cubic crystal. (b) The same orientation shown on a standard stereographic projection.

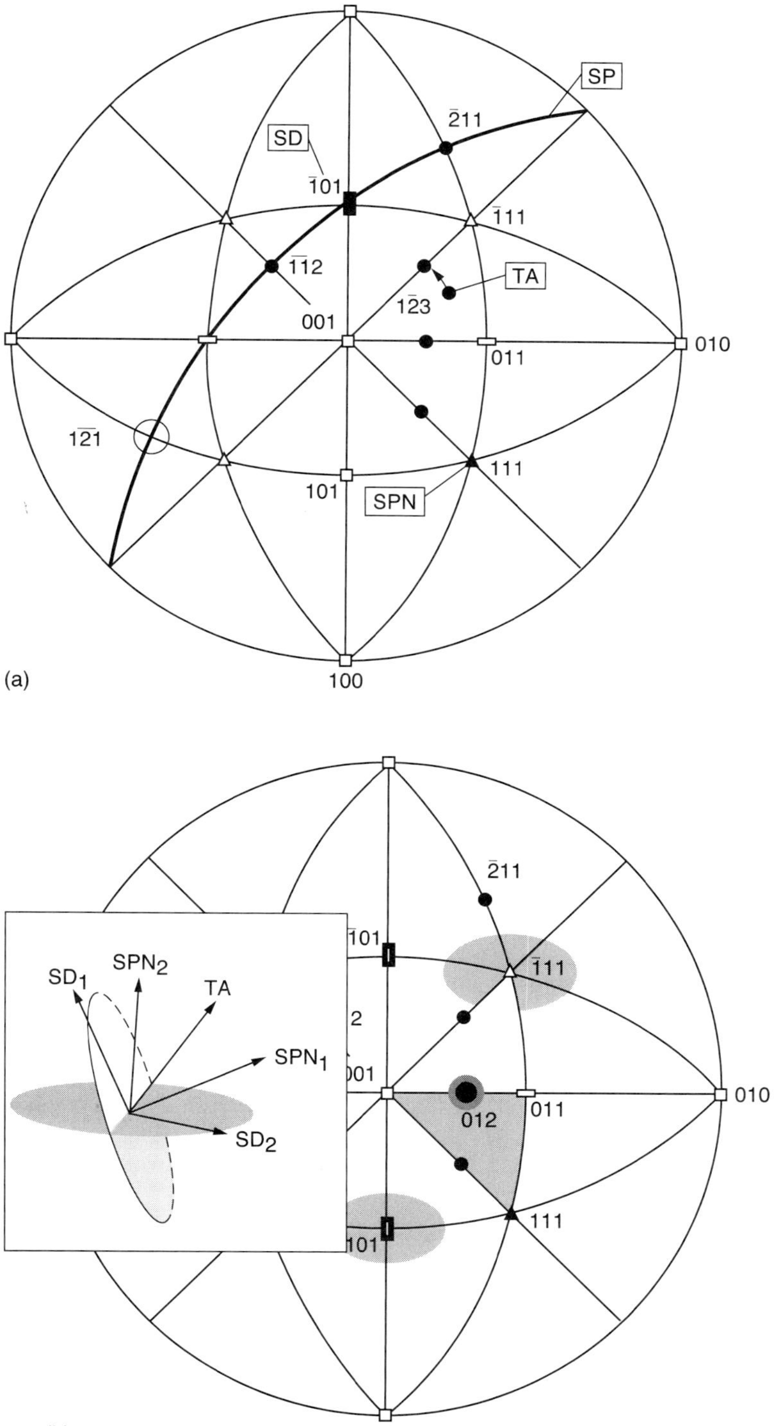

FIG. 3.15 (a) Rotation of the tensile axis during testing of a cubic crystal oriented for single slip relative to the crystal axis. In compression, the rotation would have the opposite sense. The rotation related to simple shear occurs about the [$1\bar{2}1$]. (b) Slip on two equivalent slip systems at the boundary of the standard stereographic triangle.

system in FCC metals. As discussed in Chapter 5, slip on multiple systems, which occurs at these boundaries, and in most polycrystals, results in higher yield stresses and higher rates of work hardening.

3.2.3 Deformation by Twinning

Twins (or domains in ordered compounds) can produce an abrupt transition within a crystal such that one portion of the crystal is a mirror image of another portion, as shown in Fig. 3.16 for compressive deformation of an aluminum crystal. The process occurring in twinning is shown in Fig. 3.17*a*. Other types of twins also occur wherein two adjoining portions of crystal are oriented with other types of symmetry. Some twins arise during annealing, but in most materials that twin, the process of twin formation and the motion of the boundaries between these regions result in deformation. Although only some of the parallel planes within a material undergo slip, in twinning every layer of twin undergoes a sudden displacement relative to the planes above and below it. Twinning can produce abrupt deformation that is often audible. In many materials, twinning in one sense is not the same as in the opposite sense (see the discussion of slip in BCC metals in Chapter 4). That is, the sign of the shear stress may determine whether or not twinning can take place. In most metals, twinning and the motion of twin boundaries are difficult when compared with slip by dislocation motion. In materials with few available slip systems—in particular, HCP metals—and at low temperatures when slip is difficult, the shear stress required to produce mechanical twinning can be reached before deformation occurs by other mechanisms.

The crystallographic definitions of twins can be very complex. Fortunately, an elementary type of twinning is found in FCC metals. As shown in Fig. 3.17*a*, this twin is formed as a mirror image across {111} planes with a shear relationship in <112>. The displacement of one-sixth of a <112> applied to the crystal shears the crystal, producing a reversal in stacking from ABCABC to ABCBAC. The twin results by applying this $\frac{a}{6}[11\bar{2}]$ on each successive (111) plane. This type of twin can be found in FCC metals that have been grown by vapor or electrolytic deposition. Twinned grains are also found in FCC metals following recrystallization or deformation.

The geometry of twins is classified as shown in Fig. 3.17*b*. The crystal above the twin plane K^1 is displaced by

$$u_1 = \gamma d_2$$

where u_1 is the displacement proportional to the distance d_2 above the twin plane multiplied by γ, the strength of the simple shear. The plane of shear, S, contains the direction of shear, η^1, and the normal to the K^1. The vector η^2 in S will be unchanged in length if it makes an angle α that is the same before and after the shear. The magnitude of simple shear is then given as

$$\gamma = 2 \tan \alpha$$

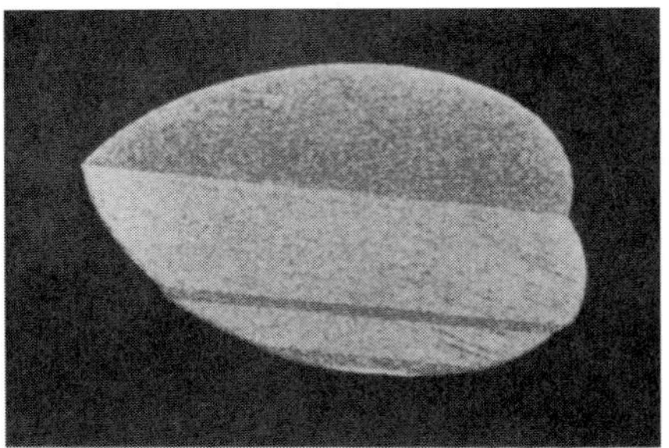

FIG. 3.16 End view showing twinning of an aluminum crystal tested in compression. The crystal was originally cylindrical and was compressed in the same manner as shown in Fig. 3.11b. (Fig. 5, Plate 3, C. F. Elam, *Proc. Roy. Soc. A.*, 121, 787, 1928, used with permission)

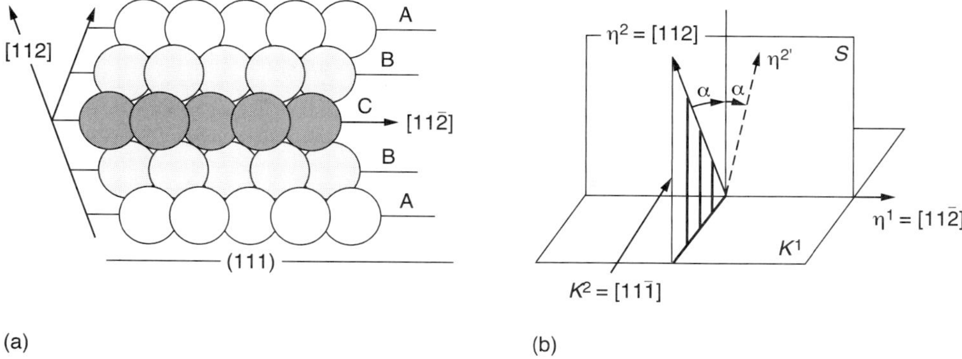

FIG. 3.17 (a) Twin boundary showing the mirror images of planes within an FCC grain and (b) definition of twinning elements for an FCC metal (after Kelly, Groves, and Kidd, 2000).

The normal to the S plane and η^2 define a plane K^2 that is undistorted by the twinning. For the twin in Fig. 3.17*b*,

$$K^1 = (111) \qquad \eta^1 = [11\bar{2}]$$
$$K^2 = (11\bar{1}) \qquad \eta^2 = [112]$$

For this twin, the magnitude of simple shear is $g = 0.707$. Table 3.3 lists twinning elements found in several materials.

Propagation of a twin is generally much easier than twin initiation. The propagation is carried out by dislocations (discussed in Chapter 4) at the interface between the twinned and untwinned regions. The initiation requires very high stresses, because large displacements are associated with twinning.

Examples of domains in materials with tetragonal and lower symmetries are prevalent, and many can be put to use for useful applications. Piezoelectric materials take advantage of coupled electronic and elastic domain character to serve as sensors and actuators, and some ceramics and minerals undergo deformation by domain boundary motion. An example of domain motion in a hypothetical tetragonal material is shown in Fig. 3.18. The 90° misorientation between the regions provides that if the domains are driven to move by an imposed stress, strain will result from the transition of unit cells at the boundary from one orientation to another. This property is called ferroelasticity. The strains are typically

TABLE 3.3 Twinning Elements for Various Crystals*

Material	K^1	η^1	K^2	η^2	γ
FCC metals	(111)	$[11\bar{2}]$	$(11\bar{1})$	[112]	0.707
BCC metals	(112)	$[\bar{1}\bar{1}1]$	$(\bar{1}\bar{1}2)$	[111]	0.707
Cd, HCP, c/a = 1.89	$(10\bar{1}2)$	$[\bar{1}011]$	$(10\bar{1}\bar{2})$	$[10\bar{1}1]$	0.171
Zn, HCP, c/a = 1.86	$(10\bar{1}2)$	$[\bar{1}011]$	$(10\bar{1}\bar{2})$	$[10\bar{1}1]$	0.140
Mg, HCP, c/a = 1.632	$(10\bar{1}2)$	$[\bar{1}011]$	$(10\bar{1}\bar{2})$	$[10\bar{1}1]$	–0.130
	$(10\bar{1}1)$	$[\bar{1}012]$	$(\bar{1}013)$	$[30\bar{3}2]$	0.137
C, hexagonal graphite	$(11\bar{2}1)$	$[\bar{1}\bar{1}26]$	(0001)	$[11\bar{2}0]$	0.367
Al_2O_3 hexagonal (sapphire)	$(10\bar{1}1)$	$[\bar{1}012]$	$(\bar{1}012)$	$[10\bar{1}1]$	0.202

* The examples are given in specific indices; however, corresponding symmetric planes and directions for the given material could also be used. The negative values for g correspond to the sense of the shear relative to the given coordinates.

(See Kelly, Groves, and Kidd, 2000, and Reed-Hill, 1964)

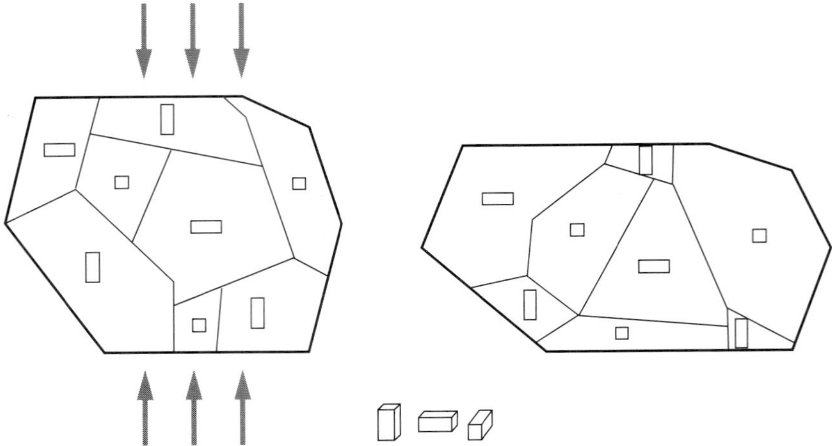

FIG. 3.18 Compression of a tetragonal crystal containing domains. All three possibilities for the domains are shown in projection within the grain.

low, because large strains would involve high resistances to boundary motion. In materials that do not otherwise deform by slip, such deformation mechanisms can enable these materials to be more resistant to crack propagation.

3.2.4 Deformation of Polycrystalline Materials

Polycrystalline materials consist of crystallite aggregates differing in orientation. If a polycrystalline material is considered to consist of randomly oriented grains, then all properties, including plasticity, should be nondirectional or isotropic (see Hosford, 1993).[1] To evaluate the problem of a polycrystal, Sachs considered that if the problem is treated for the average value of the Schmid factor, for randomly oriented crystals the value of slip in an FCC metal would be 2.238, so that $\sigma = 2.238\tau$. Taylor noted that by allowing single slip within each grain, Sachs' model would not allow the grains to deform while maintaining intact grain boundaries. Taylor developed a new model that required that slip take place with the grains deforming uniformly and thereby maintaining all boundaries. Because each grain in the material would then undergo the same strains as those imposed externally on a macroscopic basis, five independent terms of the nine in a plastic strain tensor must be identical ($\varepsilon_{12} = \varepsilon_{21}$, $\varepsilon_{13} = \varepsilon_{31}$, $\varepsilon_{23} = \varepsilon_{32}$, and $\varepsilon_{11} + \varepsilon_{22} + \varepsilon_{33} = 0$).

Taylor's analysis of slip was based on the increment of work per unit volume, dw, that is produced by slip on all participating slip systems inside one grain as

$$dw = \tau \sum_i \tau_i |d\gamma_i| \tag{3.30}$$

where τ_i is the magnitude of shear stress required for slip on slip system i. The term $|d\gamma_i|$ is the absolute value of the slip increment on the same slip system. For a single type of slip system—e.g., <110> {111} in FCC metals—the magnitude of shear stress required to produce deformation is initially equal for all specific slip systems of this type at the start of deformation. As deformation proceeds, the work hardening on one slip system may proceed at a different rate than on other slip systems. Also, in materials with multiple <uvw>{hkl} possibilities for slip systems, the differences in the shear stress required for deformation and

[1]A good rule of thumb for deformation studies is that the stress is applied over a cross section containing at least 10 grains.

the crystal orientation control deformation. The favored slip system is, of course, the one that requires the lowest shear stress for deformation.

For the 12 slip systems of FCC metals, Taylor calculated that the magnitude of what is now called the Taylor factor, <*M*>. The value of <*M*> can be seen as a ratio of applied uniaxial deformation to shear deformation. For FCC metals, Eq. 3.30 can be simplified to

$$dw = \tau \sum_i |d\gamma_i| \tag{3.31}$$

For a simple tensile test in the z-direction, the work per unit volume can be written from Eq. 3.12 as $\sigma_z d\varepsilon_z$. We can express the work from slip on multiple slip systems as

$$\sigma_z d\varepsilon_z = \tau \sum_i |d\gamma_i|$$

which then allows for definition of M as

$$\mathrm{M} = \frac{\sum_i |d\gamma_\mathrm{i}|}{d\varepsilon_z} = \frac{\sigma_z}{\tau} \tag{3.32}$$

The average value of M is given as <*M*>, which can be attained by averaging the value for Eq. 3.32 over the distribution of orientations in the material.

Selecting 5 cases from 12 on an arbitrary basis results in 792 combinations. Consequently, it was not a small task to do such calculations before computers were available. In addition, the independent nature of the five slip systems means that no two or more of the slip systems can be summed to produce the same strain as another (see Appendix to this chapter). Thus, the relationship between the yield stress and the critical resolved shear stress for FCC metals is given as <*M*> = 3.06. For BCC metals assumed to be slipping in <111>, but on any of the possible {hkl}, the value of <*M*> is 2.73. This type of slip is called *pencil glide* and is discussed in more detail in Chapter 4.

The rotation process defined for the slip of single crystals also occurs within polycrystalline materials, but because each grain undergoes slip on several specific slip systems, there are rotational contributions from each participating mechanism according to the magnitude of slip on that system. For crystalline materials, this rotation occurs on specific paths corresponding to the appropriate easiest slip systems. An initially random material subjected to plastic deformation by extrusion, drawing, rolling, or similarly directional processes eventually acquires a preferred orientation. Conversely, the final intensification of the preferred orientation of the material provides a historical account of the processing history and the deformation mechanisms that participated. If the nature or degree of slip is changed, the preferred orientation will also change in type and magnitude. As the crystals in the material become oriented, each of the properties that are affected by this orientation can become anisotropic, including the Taylor factor. Some processes designed to produce particular textures—e.g., transformer steels oriented for ideal magnetic properties—include low-temperature deformation orientation during forming operations followed by recrystallization that produces a different texture. Even without a deformation history, solidification or other processing steps often include directionality that leads to preferred orientation.

Pole Figures A pole figure is a representation of specific directions for a set of polycrystals plotted relative to the external deformation coordinates. Thus, the stereographic projection discussed previously, which shows an external coordinate relative to crystal

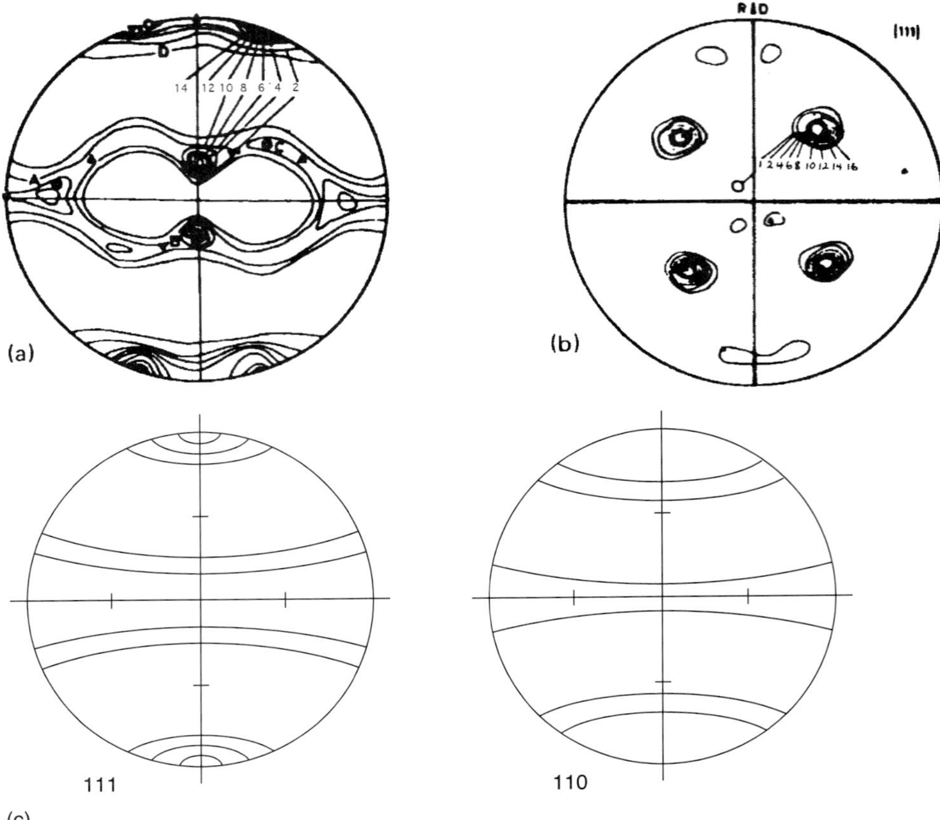

FIG. 3.19 (a) 111 pole figure for rolling of copper with contours of 1, 2, 4, 6, 8, 10, 12, and 14 for the texture intensity. The contour of 14 indicates that 14 times the number of grains has a 111 pole parallel to that direction than there would be in a random polycrystal. (b) 111 pole figure showing subsequent recrystallization of copper, indicating a stronger texture with tightly spaced contours up to a texture intensity of 16. Contours indicate prevalence of the 111 pole relative to the **RD** (top of each pole figure). The **ND** is normal to the page (Mecking, 1985, used with permission of Elsevier Limited). (c) Schematic of 111 and 110 pole figures for a cubic material with an axisymmetric <111> extrusion texture. The extrusion axis runs north to south on the pole figures. The 111 pole figure shows an alignment of 111 directions nearly parallel to the extrusion axis.

axes, is the inverse of a pole figure with one specimen axis plotted on it. Figure 3.19*a* shows an example of pole figure data for rolling and recrystallization textures in copper. Both methods of representing orientation information for a material provide part of the entire set of orientation relationships between the specimen or external coordinate system and the coordinated systems of the crystals within the polycrystal. Because plotting of the discrete orientation of each crystal in most polycrystals would be impossible to measure and show, functional representations of what is called the orientation distribution function (ODF) are often given to provide a statistical accounting of orientation as a function of the Euler angle relationship between each set of crystal axes and the specimen axes. If we can write the orientation of an individual crystal in terms of Euler angles as $g = (\phi_1, \Phi, \phi_2)$, then we can write the ODF as

$$\frac{\Delta V}{V} = \int f(g)dg \tag{3.33}$$

where $\Delta V/V$ is the volume of grains under consideration and $f(g)$ is a function describing grain orientations. The ODF is commonly normalized such that a random sample would have an ODF of 1 at all points. If the magnitude of a point within a contour in Fig. 3.19*a* is greater than 1, that orientation relationship occurs more frequently than it would in a random sample. If the magnitude is less than 1, that orientation relationship is depleted when compared with the frequency of its occurrence in a random sample. A single crystal represents an infinite degree of preferred orientation. Preferred orientations that result from processing history in materials can be advantageous or deleterious depending on the final application of the material. Preferred orientations in rolled steels are often designed to fulfill specifications for formability in later shaping operations. In aluminum alloys, the plastic anisotropy of the rolled sheet hinders the efficiency of the deep drawing process for manufacture of aluminum beverage cans, as demonstrated in Fig. 3.20.

The texture of metals from rolling is expressed in terms of the rolling direction <uvw> and the plane of the rolled sheet {hkl} (see Fig. 2.9). For extruded materials, the expected textures give pole figures that are axisymmetric about the extrusion axis, as shown in Fig. 3.19*c*. The schematic 111 and 110 pole figures both correspond to a cubic material with a <111> extrusion direction. The 111 pole figure shows a large number of orientations in the extrusion direction (top and bottom). The two symmetric bands also shown on the 111 pole figure represent the positions of the three <111> that lie approximately 70° from the <111> that lie concentrated in the extrusion direction. Typical textures in engineering alloys expressed in this way are given in Table 3.4.

TABLE 3.4 Typical Textures for Various Ductile Metals

	Rolling
Material	**Rolling texture**
	<RD> {RP}
FCC metals, Ag, brass, stainless steel	$\langle\bar{1}12\rangle\{110\}$, <001> {110}
FCC metals, Cu, Ni, Al	$\langle 11\bar{1}\rangle\{112\}$, $\langle 112\rangle\{\bar{4}21\}$
BCC metals	<110> {001}, $\langle 1\bar{1}0\rangle\{111\}$, <112> {111}
HCP metals, Ti, Zr	$[10\bar{1}0]$ (0001) tilted ≈ 30° from plane
HCP metals, Mg	$[11\bar{2}0]$ (0001)
	Extrusion
Material	**Extrusion direction**
FCC metals, Ag, brass	Strong <100>, weak <111>
FCC metals, Cu, Au, Ni, Al	Strong <111>, weak <100>
BCC metals	Strong <110>
HCP metals	Strong <hk0>
	Axisymmetric Compression
Material	**Compression direction**
FCC metals	<110> strong
BCC metals	Strong <111>, weak <100>
HCP metals	Strong [0001]

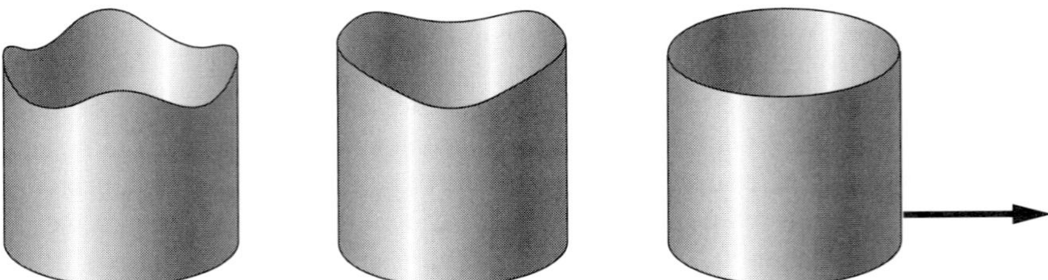

FIG. 3.20 Types of earing behavior during deep drawing of cups from FCC metal sheets with different starting preferred orientations. The values of the anisotropy ratio R are dependent on the orientation within the sheet. Substantial earing leads to unnecessary waste material and thickness variations.

Plastic Anisotropy Plastic anisotropy can be expressed as a nonuniform yield surface wherein the yield stress in a direction within a material may be different from another. The most extreme cases would be the single crystal examples discussed in Section 3.2. Deformation of a thin sheet of rolled stock has been considered in the greatest detail. The plastic anisotropy factor R is defined as

$$R = \varepsilon_w/\varepsilon_t \tag{3.34}$$

where ε_w is the width of a tensile specimen cut from a sheet of a ductile material and ε_t is the thickness of the sheet, as shown in Fig. 3.21. If a material is isotropic, the value of R will be 1 no matter in what direction the specimen is cut out of the sheet. In a material with plastic anisotropy, R can be greater or less than 1. Because the accuracy of measurements on the thinnest dimension of a tensile specimen might be difficult to obtain, R is often obtained by measuring the axial and width strains and assuming that plastic deformation is volume conservative.

Consider the problem of deforming a thin sheet material that shows plastic anisotropy. If $R < 1$, then the material strains faster through the thickness than across the width. If thinning of the material can lead to failure of the forming operation, this is not desirable. On the other hand, if $R > 1$, then most of the strain takes place across the width. For simple tensile deformation, this may be fine, but in biaxial tension the stresses required for deformation could be quite high because plastic deformation of the thickness dimension is required for stretching. In addition, unless the symmetry of the plastic anisotropy is uni-

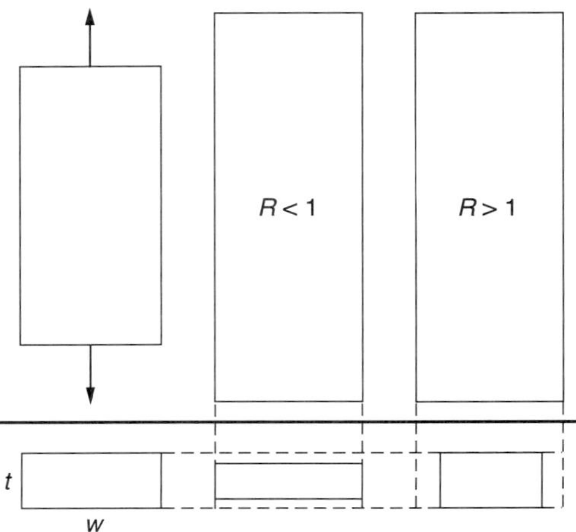

FIG. 3.21 Values of R for tensile deformation of specimens, with original dimensions shown at the left.

form in all directions parallel to the sheet surface (axisymmetric), the value of R will be different for each orientation within the sheet. In this instance, the plastic anisotropy may be expressed as an angle within the sheet from the rolling direction so that the plastic anisotropy in the rolling direction is R_0 and the plastic anisotropy in the transverse direction across the width of the sheet is R_{90} (see Fig. 2.9). To relate the plastic anisotropy as a single value, the mean plastic anisotropy is used from tensile tests at 0°, 45°, and 90° from the rolling direction as

$$\overline{R} = \frac{R_0 + 2R_{45} + R_{90}}{4} \tag{3.35}$$

The differential in the R-values with direction is often used to characterize the degree of in-plane anisotropy that results in the earing behavior shown in Fig. 3.20.

EXAMPLE 3.7 ***Plastic Anisotropy in Sheet Materials***

Your employer has requested that you find a sheet material that when extended by 20 percent in one direction will be reduced in thickness by less than 5 percent. What R-value will this sheet material be required to have?

First, we should convert to true strains

$$\ln(1.2) = 0.182$$

$$\ln(0.95) = -0.05$$

and then, assuming that volume is conserved, we can calculate the width strain: $-(0.182 - 0.05) = -0.132$. Then the R-value that will be required is $R = -0.132/-0.05 = 2.64$. This is a fairly large value and probably only likely in noncubic materials that are strongly textured. Some materials with hexagonal crystal structures can be processed to produce R-values of 3 to 6. ■

3.3 NECKING AND INSTABILITY

The final strain to failure for most ductile materials is determined by the point at which the material begins to deform nonuniformly or to neck, as shown in Fig. 3.22. Necking can occur whenever the continued resistance to deformation is decreasing. If the increase in force necessary for continued deformation of the material reaches zero, then necking will occur. Necking can take place as a result of insufficient work or strain hardening or as a result of strain rate effects. We will consider the first case, with strain rate effects discussed in a later chapter.

In tensile deformation of a ductile material, the maximum force that can be applied to the material coincides with the engineering definition of the ultimate tensile stress (UTS), defined as the maximum force on the stress-strain curve over the original cross-sectional area. Thus, as shown in Fig. 3.22, necking occurs at the maximum value of the engineering stress-strain curve. In terms of the true stress–true strain curve, the rate of change of force expressed in terms of effective stress σ_{eff} and instantaneous cross-sectional area A can be given as

$$dF = \sigma_{eff} dA + A d\sigma_{eff} \tag{3.36}$$

and

$$d\sigma_{eff}/\sigma_{eff} = -dA/A = d\varepsilon_{eff} \tag{3.37}$$

Then,

$$d\sigma_{eff}/d\varepsilon_{eff} = \sigma_{eff} \tag{3.38}$$

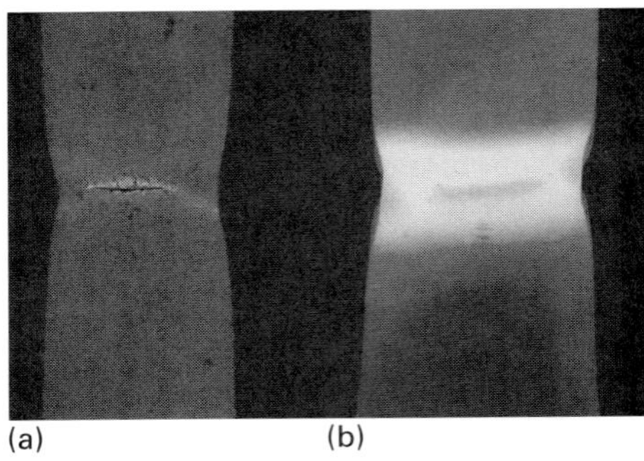

FIG. 3.22 Necking of (a) a 1020 steel specimen and (b) an initially transparent, amorphous polyethylene terephthalate (PET) specimen in tension. At the later stages of necking, the center of the necked region begins to show the effects of a local condition approximating hydrostatic tension. The steel specimen cracks first at the center, with shear on the edges of the necked zone. The PET specimen shows similar behavior. This hydrostatic tensile condition suppresses plastic deformation by shear and may enable the opening of small voids within the center portions of the eventual fracture surface. More discussion is given on these phenomena in Chapter 7.

If this result is substituted in Eq. 3.19, then $\varepsilon_{eff} = n$. For strains below the magnitude of the strain hardening exponent, the deformation is uniform. If a region of the specimen were to have a slightly smaller diameter, the deformation that resulted in that smaller diameter would cause a sufficient amount of hardening to maintain the applied loads until the larger and softer regions could "catch up." In other modes of deformation, the magnitude of strain before necking is different than for uniform tension. Obviously, the potential for necking is not present in compression, although, as discussed above for shear band formation in polymers, some types of plastic instability are possible if local heating takes place. This is a contributing factor to the strain rate sensitivity of PET shown in Fig. 1.22.

3.4 SHEAR DEFORMATION OF NONCRYSTALLINE MATERIALS

As is clear from the preceding discussion, if a volume change does not occur, the deformation can always be resolved into shear components. For the deformation mechanisms of grain-boundary sliding and diffusion creep discussed in Chapter 6, the deformation can be modeled as fluid flow that deviates from ideal behavior of liquids. For noncrystalline materials, which are normally noncrystalline owing to cooling rates from a liquid state more rapid than would allow the formation of crystals, complex molecular structures are often the case. In both inorganic and organic noncrystalline materials or glasses, a difference in the bond strength between bonds within the molecular unit and between molecular units can allow flow of the individual molecules past one another whenever the strength of the weaker bonds is exceeded. The resulting flow and alignment of linear macromolecules result in one type of orientation strengthening, as will be discussed in Chapter 5.

3.4.1 Fluid Flow Analogy

A solid can be defined in terms of physical or mechanical properties easier than almost any other characteristic. The property that clearly distinguishes a liquid from a solid is that you cannot stir a solid. The property that determines the difficulty of stirring a liquid is called viscosity, designated by η, the proportionality constant between a shear stress τ and the rate of the liquid velocity change with distance from a surface dv_x/dy, where fluid flows in the x-direction over the y-plane with

$$\tau = -\eta \frac{dv_x}{dy} \tag{3.39}$$

For a solid, in the equivalent expression we consider shear strain rate (or rate of deformation) and

$$\tau = \eta \frac{d\gamma}{dt} = \eta\dot{\gamma} \tag{3.40}$$

where $\dot{\gamma}$ is defined as the shear strain rate in the x-direction, dx/dy. The sign difference in these two expressions is from conventions employed in the respective fields of fluid mechanics and solid mechanics. Both cases are depicted in Fig. 3.23.

If Equation 3.40 is expressed in terms of the effective stress and effective strain rate,

$$\sigma_{\text{eff}} = \kappa\dot{\varepsilon}_{\text{eff}} \tag{3.41}$$

For deformation described by the von Mises criterion, $\kappa = 3\eta$, and for the Tresca criterion, $\kappa = 4\eta$. The difference between a solid and a liquid can be defined as a distinct change in viscosity. So, as most materials solidify, the viscosity increases by several orders of magnitude. This abrupt transition is a result of crystallization. If, as for many glasses, the viscosity decreases but crystallization does not occur, a threshold viscosity or the maximum change in viscosity with temperature ($d\eta/dT$) can be used to determine the transition temperature from a liquid to a solid. This approximation is not a simple one, because many polymers and ceramics that readily form glasses do not demonstrate the simple linear relationship shown in Eq. 3.39. Figure 3.24*a* shows the effective viscosity changes versus temperature for several materials. In Fig. 3.24*a*, the viscosities in the solid state are estimated from lattice diffusion creep data for low-stress deformation of large crystals (grain diameter d greater than 0.1 mm).

The size of the crystals in a polycrystalline material does indeed influence the effective viscosity of the solid (see Chapter 6). An example of the *estimated* grain-size effect for the viscosity of solid aluminum is shown in Fig. 3.24*b*. As grain size decreases, the viscosity just below the solidification temperature decreases dramatically. The change in viscosity predicted for solid aluminum is based on the consideration that the strain rate for creep of polycrystalline materials scales to between d^{-2} and d^{-3}. The overall viscosity–temperature relationship for aluminum approaches that shown for the glasses in Fig. 3.24*a* as the grain size becomes very small. The data for grain-size dependence in aluminum do not include consideration of the grain growth that takes place in aluminum over time at elevated temperatures, making experimental measurements of these properties nearly impossible.

Although the diffusive deformation mechanisms that occur near the melting temperature in many crystalline solids have a linear relationship, the relationship between stress and strain rate is often nonlinear. In noncrystalline materials consisting of large molecules,

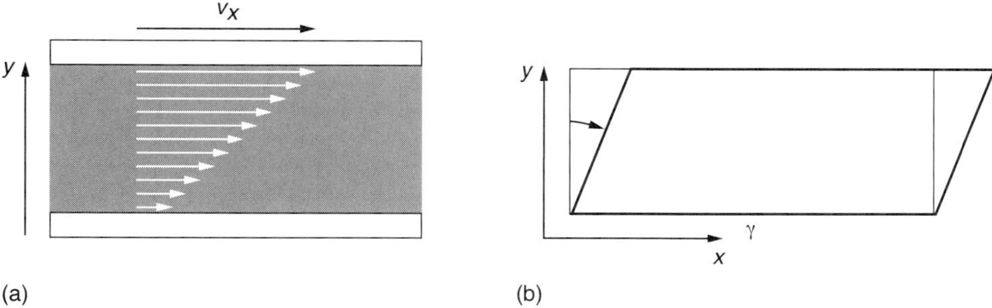

FIG. 3.23 (a) Fluid flow across a stationary y-plane for a flat plate using stress $\tau = F_x/A_y$, where F_x is the force in the x-direction on the top plate. The gradient v_x is shown. (b) Shear strain γ, demonstrated by the distortion of a rectangle into a parallelogram.

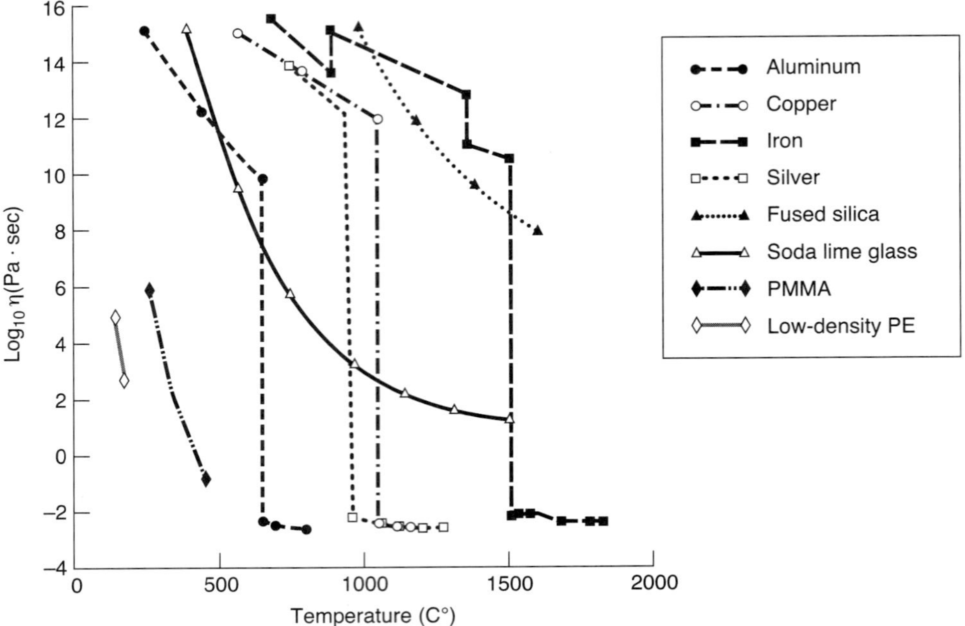

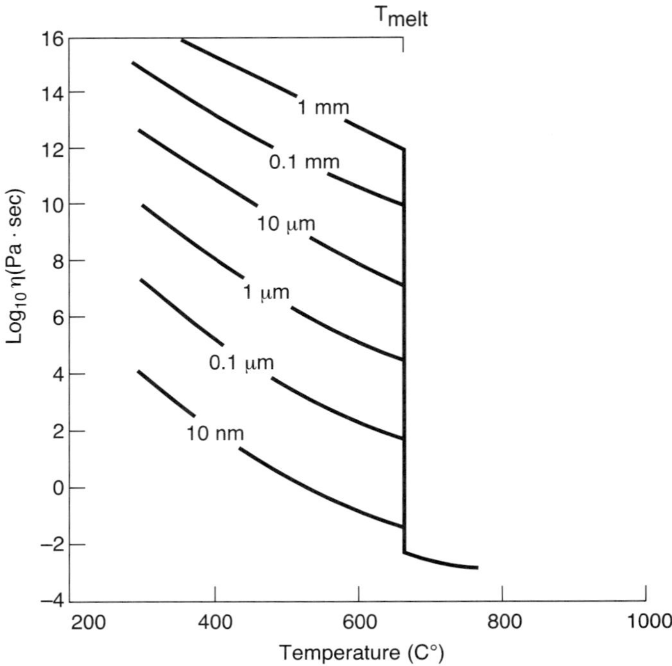

FIG. 3.24 (a) Viscosity versus temperature for several metals, glasses, and polymers. (b) Viscosity versus temperature and grain size for aluminum using models for diffusion creep discussed in Chapter 6. These plots are hypothetical, because rapid rates of grain growth at high temperatures would preclude measurement of this property for the given grain sizes.

a nonlinear relationship for stress and strain rate is also quite common. Nonlinear expressions for Eq. 3.41 can be defined as

$$\sigma_{\text{eff}} = C\dot{\varepsilon}_{\text{eff}}{}^{m} \tag{3.42}$$

where C is a constant, with units that depend on the magnitude of m, which is called the strain rate sensitivity.

A combination of strain rate effects and strain hardening can be written as

$$\sigma_{\text{eff}} = C'\dot{\varepsilon}_{\text{eff}}^{m}\varepsilon_{\text{eff}}^{n} \tag{3.43}$$

The form of this expression shows the potential for plastic instabilities from either an insufficient strain sensitivity or a low strain hardening exponent.

3.5 DILATANT DEFORMATION OF MATERIALS

Throughout the earlier portions of this chapter, the assumption was made that deformation takes place without any permanent change in volume. For crystal plasticity and fluid flow, this is often a reasonable consideration. However, in some materials there are pronounced effects of imposed hydrostatic stresses because the deformation process of the material involves a change in volume. If there is a change in the volume of the material during deformation, the cross section of the yield surface (i.e., the cylinder diameter) does not have to stay constant along the equal stress axis. In fact, the yield surface is not even necessarily open on either end. As shown in Fig. 3.5*b*, many polymers show yield behavior that includes dilatant effects that result in higher yield stresses in compression than in tension. This results in part from the large values of volumetric strain that occur because of the low elastic moduli of most polymers and the effects of rearranging polymer molecules on specific volume or density. In polymers, this hydrostatic effect can be extreme enough to result in different deformation mechanisms for tension and compression. Another instance wherein hydrostatic stress may affect the size of the yield surface is in porous materials. Hydrostatic compression of a porous metal enables densification by plastic deformation of a macroscopic part. So, in this instance the deformation on a microscopic basis consists of slip processes and particle rearrangement that do not involve a change in the volume of metal, but rather a removal of porosity. Permanent changes in volume also accompany many phase changes. Some special types of phase transformations involve changes in crystal structure that are triggered mechanically. Martensitic and similar types of phase changes provide a good opportunity for our first exploration of dilatant effects on deformation. In later chapters, other mechanisms involving volume change will be discussed in more detail.

3.5.1 Martensitic Transformations

Diffusionless phase transformations accompanied by volume changes and shear typify martensitic transformations. In addition to the effective distribution of carbon and other alloying agents, microstructural refinement, and increased dislocation densities that can accompany martensitic transformation strengthening of steels, these transformations are also deformation processes. In transformation-induced plasticity (TRIP) steels, shear deformation and volume changes associated with transformation from a face-centered cubic to a body-centered tetragonal structure are initiated by permanent deformation.

The metastable phase called martensite was originally discovered to harden steel. Similar phase transitions enable shape memory effects and toughening of ceramics (see next section). As shown in Fig. 3.25, it forms within a phase that has been cooled quickly

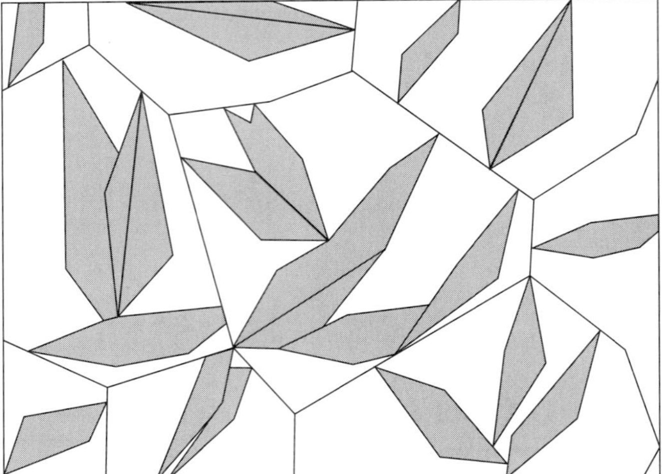

FIG. 3.25 Schematic of lenticular martensite formed during the phase transformation from retained austenite to the body-centered tetragonal metastable phase. The lenticular-shape regions in gray indicate the martensitic phase.

below its equilibrium transformation temperature. The formation of martensite within quenched austenite (FCC iron) can be initiated by deformation, suggesting that strain energy can drive the process. The formation of martensite from austenite crystals involves volume strains and simple shear strains of approximately

$$e_{\text{mart}} = \begin{vmatrix} 0 & 0 & 0 \\ 0 & 0 & \pm 0.20 \\ 0 & 0 & 0.05 \end{vmatrix} \tag{3.44}$$

which are much greater than the allowable elastic strains without plastic deformation. The actual magnitudes of these strains depend on the composition of the steel. The phase transformation causes enough plastic distortion in adjoining untransformed material to cause strain hardening through dislocation motion. The amount of transformation and the orientations that are transformed determine the strain contribution of martensite to deformation of a polycrystal.

The temperature of the solid-state allotropic transformation from the FCC austenite (γ) structure to the BCC ferrite (α) structure depends on carbon content. In pure iron, this $\gamma \rightarrow \alpha$ transformation occurs on cooling through 912°C. The transformation temperature can be as low as 727°C at the eutectoid shown in Fig. 3.26. The dramatic difference in carbon solubility between γ and α can also be seen in Fig. 3.26. If γ containing more than 0.02 percent carbon by weight is cooled slowly, diffusive processes enable the formation of BCC iron and the cementite or Fe_3C phase. If γ is cooled too rapidly for precipitation of Fe_3C, carbon atoms can be trapped in interstitial positions that prevent the complete change to the BCC structure. Instead, the BCT structure of martensite can result. The tetragonality, or c over a ratio, of the martensite depends on the carbon content, as shown in Fig. 3.27. This variation in lattice parameters between the phases changes the value of the transformation strains given in Eq. 3.44.

EXAMPLE 3.8 *Stress Tensor Effects on Martensitic Transformation*

Consider the stress-driven transformation of a single crystal of austenite to martensite with the strains given in Eq. 3.44. We will give the crystal an orientation, $g = (0, 45°, 0)$ or $\Phi = 45°$, and we will apply a tensile stress in the z-direction, σ_z. Using Eq. 3.13, the deformation work per volume transformed can be written as

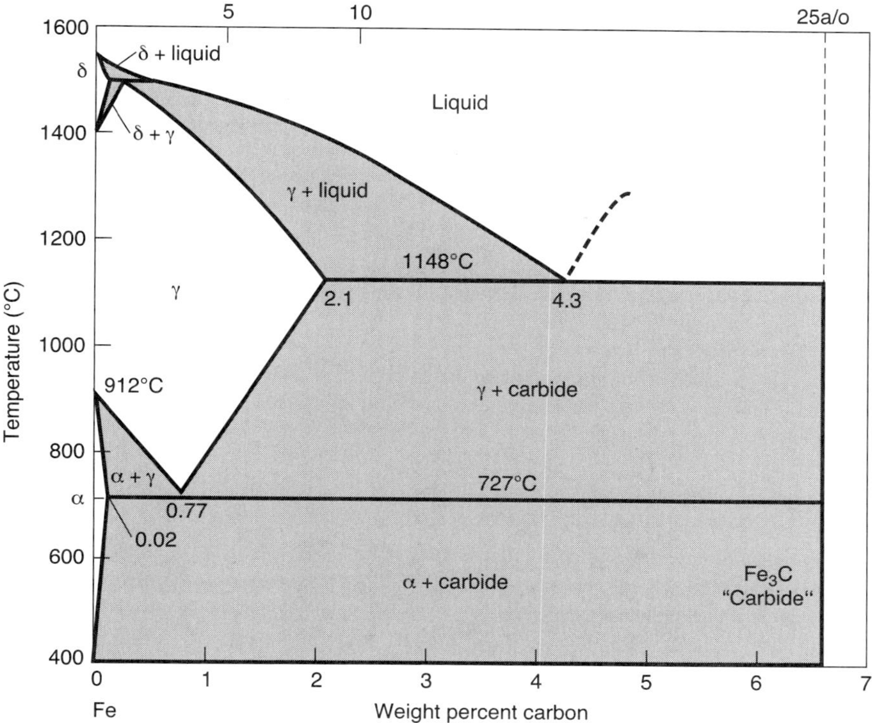

FIG. 3.26 Section of the Fe-Fe_3C phase diagram, showing the range of compositions for steels.

$$W_{\text{trans}} = \sigma_z \varepsilon_z$$

By analogy to Eq. 3.30, it should be possible to write this in terms of the transformation strains derived in Eq. 3.44 through a tensor rotation. Then, if we ignore rotation and use the symmetric form of e_{mart} with positive shears

$$\varepsilon_{\text{mart}(+)} = \begin{vmatrix} 0 & 0 & 0 \\ 0 & 0 & 0.1 \\ 0 & 0.1 & 0.05 \end{vmatrix}$$

we can expect that

$$\begin{aligned} \sigma_z \varepsilon_z &= \sigma_z \varepsilon'_{\text{mart}} \\ \varepsilon'_{\text{mart}(+)} &= a_{32} a_{33} \varepsilon_{23} + a_{33} a_{32} \varepsilon_{32} + a_{33} a_{33} \varepsilon_{33} \\ &= \frac{-1}{\sqrt{2}} \cdot \frac{1}{\sqrt{2}} \cdot (0.1) + \frac{1}{\sqrt{2}} \cdot \frac{-1}{\sqrt{2}} \cdot (0.1) + \frac{1}{\sqrt{2}} \cdot \frac{1}{\sqrt{2}} \cdot (0.05) \\ &= -0.1 + 0.025 = -0.075 \end{aligned}$$

Because this result would give us a negative or compressive strain and thereby would produce negative W_{trans}, it should be clear that

$$\varepsilon'_{\text{mart}(-)} = \begin{vmatrix} 0 & 0 & 0 \\ 0 & 0 & -0.1 \\ 0 & -0.1 & 0.05 \end{vmatrix}$$

would represent the sense of shear that would allow for tensile strain along the z stress axis.

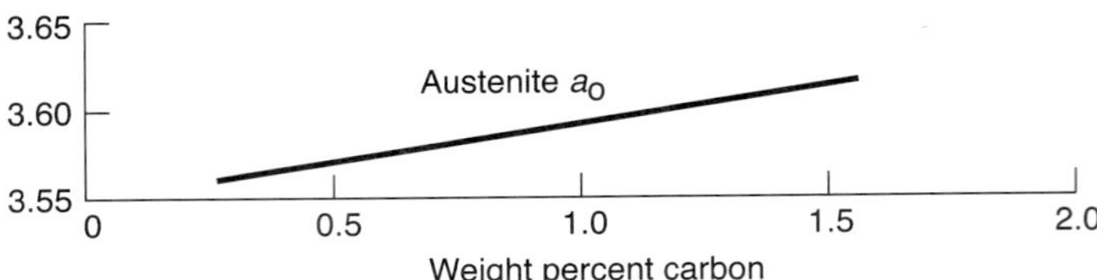

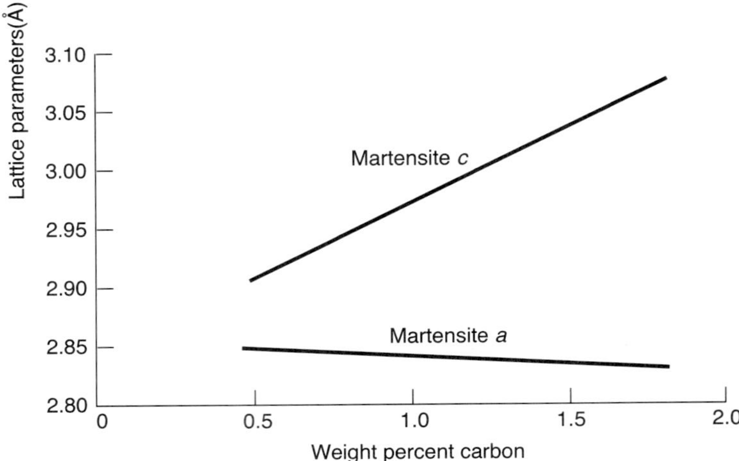

FIG. 3.27 The lattice parameters of austenite and martensite versus carbon content. (Based on data from Roberts, 1953, used with permission.)

$$\varepsilon'_{\text{mart}(-)} = \frac{-1}{\sqrt{2}} \cdot \frac{1}{\sqrt{2}} \cdot (-0.1) + \frac{-1}{\sqrt{2}} \cdot \frac{1}{\sqrt{2}} \cdot (-0.1) + \frac{1}{\sqrt{2}} \cdot \frac{1}{\sqrt{2}} \cdot (0.05)$$
$$= 0.125$$

which is a large strain well beyond the elastic limit of almost any material.

The difference in strain values for the two $\varepsilon'_{\text{mart}}$ calculations shows that the shear deformation can overcome the dilatant strains to produce compressive deformation, but this depends on the crystal orientation. If a critical value of work is required to produce transformation, it will be reached at a lower stress in tension than in compression. If the critical transformation stress is 10 MPa in tension, then

$$W_{\text{trans, crit}} = 10 \cdot 0.125 = 1.25 \text{ MPa} \left(\text{or } \frac{\text{MJ}}{\text{m}^3} \right)$$

To reach this amount of work in compression would require an applied compressive stress of

$$\frac{1.25 \text{ MPa}}{0.075} = 16.7 \text{ MPa}$$

For this example, we constrained the transformation strain to the value given in Eq. 3.44. In fact, the transformation can also occur at any symmetry-related orientation permitted by the cubic crystal structure, which means that, by analogy to slip systems, twelve possible transformation tensors are possible. ■

3.5.2 Transformation Toughening in Ceramics

Transformation toughening is a process wherein a change in the phase of a material triggered near the tip of a crack inhibits the propagation of the crack. The process of permanent deformation called transformation toughening is also often labeled as "transformation plasticity." It consists of features that are common to crystallographic slip processes and phase transformations (see Chen and Reyes-Morel, 1986). In several alloys based on the ceramic oxide zirconia (ZrO_2), alloying of the material makes it possible to maintain the tetragonal phase below the equilibrium transformation temperature to a monoclinic structure. If a stress state is applied to the zirconia initiating the phase transformation, the tetragonal zirconia transforms to the more stable monoclinic phase (see Fig. 3.28). One mode of transformation involves changes in the a, b, and c crystallographic axes corresponding to tensors of the type

$$e_{\mathrm{ij}} = \begin{vmatrix} e_{11} & 0 & e_{13} \\ 0 & e_{22} & 0 \\ 0 & 0 & e_{33} \end{vmatrix} \tag{3.45}$$

where the magnitude of the shear strain can be as high as ± 0.16 (if we include the rotation component) and a volumetric strain (sum of the diagonals) of 0.04. The rotational shear component is manifested as the change in the orthogonal axes of the tetragonal crystal to the single nonorthogonal (clinic) angular relationship between the b- and c-axes. In addition to a simple shear of a fixed amount per unit of transformed crystal, there is a volume change corresponding to a fixed amount per unit of transformed crystal. If the condition for yielding is considered to correspond to some amount of work done to initiate the transformation in a given grain, both shear and volumetric components will have an influence. Also, if we consider simple loading conditions such as tension and compression, for the same orientation of an individual crystal the stress required to initiate the transformation should be different. Because the phase transformation involves an increase in size, the tensile hydrostatic component of a tensile stress should make it easier for the transformation to happen.

Inversely, a compressive stress should suppress transformation. Therefore, the yield stress in compression should be higher than in tension because the resolved shear stress for the same crystal orientation should be equivalent. In fact, if a certain energy or work per unit volume is all that is required to initiate transformation, it should be possible to initiate transformation in just hydrostatic tensile loading (a condition that is approached near the tip of a crack). For compressive loading, the magnitude of the shear strain resolved in the

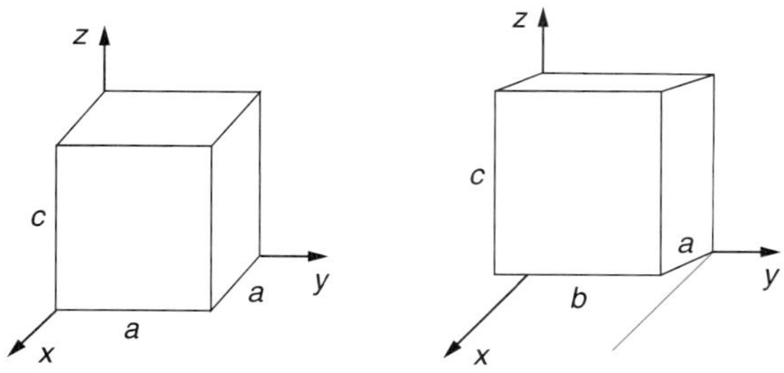

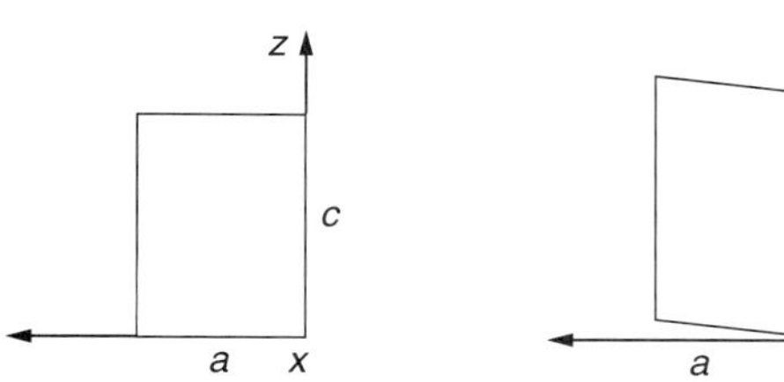

FIG. 3.28 Schematic diagram of the unit cell changes on going from the tetragonal to the monoclinic crystal structure in some zirconia ceramics. Note that the sense of the angular change can be positive or negative. The changes include a volume change and a simple shear.

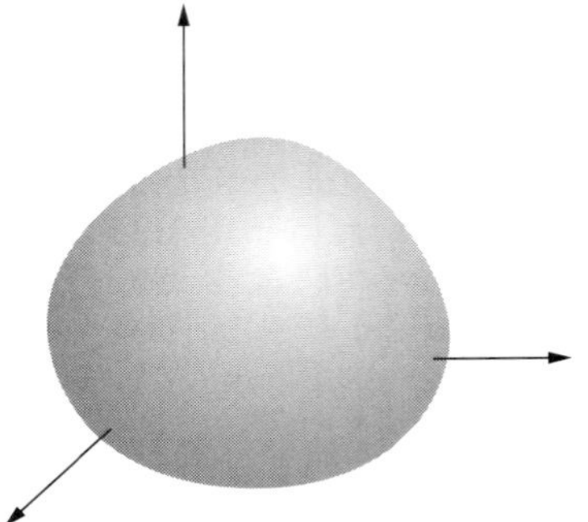

FIG. 3.29 Hypothetical domelike yield surface for a material that undergoes a positive volume change during deformation. Unlike the yield surfaces shown in Fig. 3.1, a tensile hydrostatic stress reaches the yield surface to enable permanent deformation.

direction of compression must exceed the magnitude of the volumetric strain component in that direction, or the specimen will not reduce in height. If it were possible to apply a compressive (negative) stress and get a tensile (positive) strain, we would be producing negative work, which is impossible. Thus, in this case the yield surface can be closed in the positive or tensile stresses, but it must increase in size, as suggested by Fig. 3.29. Furthermore, the principal of normality, given that the strain ratios should result in a deformation vector normal to the yield surface, is anticipated to hold for deformation of polycrystals.

Both the tetragonal and monoclinic phases of zirconia can undergo formation of twins or domains and thereby domain boundary motion that results in deformation. If only hydrostatic strains result from the transformation, then the volume fractions of the domains formed would sum such that the net shear stress would be zero.

3.6 APPENDIX: INDEPENDENT SLIP SYSTEMS

Independent slip is plastic deformation that cannot be produced by motion of any other slip system or combination of slip systems. Cotton, Kaufman, and Noebe (1991) have shown a simple approach for resolving the number of independent slip systems using the matrix operation called "row echelon reduction"—a technique that is easily accomplished using most math software. Consider slip on the slip systems

$$\text{I. } [0\bar{1}1]\ (011)$$

and

$$\text{II. } [110]\ (\bar{1}10)$$

By Eq. 3.27, slip system I produces strains of

$$\varepsilon^{\mathrm{I}} = \begin{vmatrix} 0 & 0 & 0 \\ 0 & -\gamma/2 & 0 \\ 0 & 0 & \gamma/2 \end{vmatrix}$$

Slip system II produces independent strains of

$$\varepsilon^{\mathrm{II}} = \begin{vmatrix} -\gamma/2 & 0 & 0 \\ 0 & \gamma/2 & 0 \\ 0 & 0 & 0 \end{vmatrix}$$

At first glance, the only other physically distinct slip system,

$$\text{III. } [\bar{1}01]\ (101)$$

which gives the strain

$$\varepsilon^{\mathrm{III}} = \begin{vmatrix} -\gamma/2 & 0 & 0 \\ 0 & 0 & 0 \\ 0 & 0 & \gamma/2 \end{vmatrix}$$

may seem independent. It is not. If equal shears take place on slip system I and II, then

$$\varepsilon^1 + \varepsilon^2 = \varepsilon^3$$

For this reason, Table 3.A.1 gives the number of independent specific slip systems for $\langle 1\bar{1}0\rangle\ \{110\}$ as only two. When multiple slip system families are possible, these combinations can be quite complex.

The row echelon technique allows quick conversion from physically distinct slip systems to independent systems. If we remember that volume is conserved in plastic deformation by shear ($e_{11} + e_{22} + e_{33} = 0$), we can set up five columns for strains as $e_{11} - e_{33}$, $e_{22} - e_{33}$, e_{12}, e_{13}, and e_{23}. For slip in the intermetallic compound NiAl, which has the ordered cesium chloride structure, one atom at the body center and one at each corner of the cubic cell, slip is easiest on <100> {011}. If we find the shear strain components for each physically distinct slip system using Eq. 3.27 for $\gamma = 1$, the values in Table 3.A.1 results.

This leads to the matrix for <100> {011} slip

$$\begin{vmatrix} 0 & 0 & 0.5 & 0.5 & 0 \\ 0 & 0 & -0.5 & 0.5 & 0 \\ 0 & 0 & 0.5 & 0 & 0.5 \\ 0 & 0 & -0.5 & 0 & 0.5 \\ 0 & 0 & 0 & 0.5 & 0.5 \\ 0 & 0 & 0 & 0.5 & -0.5 \end{vmatrix}$$

This matrix contains values that are dependent on one another; there are more equations than unknowns. Row rank form is reached by multiplying by constants and adding or subtracting rows, as is common in linear algebra. This form defines the minimum number of rows required to describe the calculated strains. Following these procedures, the matrix becomes

$$\begin{vmatrix} 0 & 0 & 1 & 0 & -1 \\ 0 & 0 & 0 & -1 & 1 \\ 0 & 0 & 0 & 1 & 0 \\ 0 & 0 & 0 & 0 & 0 \\ 0 & 0 & 0 & 0 & 0 \\ 0 & 0 & 0 & 0 & 0 \end{vmatrix}$$

TABLE 3.A.1 Matrix Solution Example

System	$\varepsilon_{11} - \varepsilon_{33}$	$\varepsilon_{22} - \varepsilon_{33}$	ε_{12}	ε_{13}	ε_{23}
[100](011)	0	0	0.5	0.5	0
[100]($0\bar{1}1$)	0	0	−0.5	0.5	0
[010](101)	0	0	0.5	0	0.5
[010]($\bar{1}01$)	0	0	−0.5	0	0.5
[001](110)	0	0	0	0.5	0.5
[001]($1\bar{1}0$)	0	0	0	0.5	−0.5

TABLE 3.A.2 Physically Distinct Slip Systems and Resultant Independent Slip Systems

Slip	Physically distinct	Independent
<100> {011}	6	3
<1$\bar{1}$0> {110}	6	2
<1$\bar{1}$0> {110} + <100> {011}	12	5
<110> {001}	6	3
<100> {011} + <110> {001}	12	3

which has a rank of 3. Thus, only three of the six physically distinct slip systems are independent. Cotton, Kaufman, and Noebe (1991) used this technique to demonstrate that <100> {011} and <110> {001} slip systems each give three independent systems when they are considered alone. When considered to act simultaneously, these slip systems give a 12×5 matrix that, following row echelon reduction, has a rank of 3. The number of independent slip systems is not necessarily additive, but depends on interactions between slip systems. Examples of independent slip determinations are given in Table 3.A.2.

3.7 REFERENCES

G. BATCHELOR, *The Life and Legacy of G. I. Taylor,* Cambridge University Press, 1996.

F. P. BEER AND E. R. JOHNSTON, JR., *Mechanics of Materials,* McGraw-Hill, 1981.

CAMBRIDGE, *the Magazine of the Cambridge Society,* **47**, 11–18, 2001.

I-W. CHEN AND P. E. REYES-MOREL, *J. Am. Ceram. Soc.*, **69**, 181–189, 1986.

J. D. COTTON, M. J. KAUFMAN, AND R. D. NOEBE, *Scripta Met. & Mater.*, **25**, 2395–2398, 1991.

T. COURTNEY AND G. E. DIETER, *Mechanical Metallurgy,* McGraw-Hill, New York, 1986.

N. E. DOWLING, *Mechanical Behavior of Materials,* 2nd Ed., Prentice-Hall, 1999.

C. F. ELAM, *Proc. Roy. Soc. A,* **121**, 787, 237–243, 1928.

C. F. ELAM (MRS. G. H. TIPPER), *Distortion of Metal Crystals,* Oxford University Press, 1935.

R. W. HERTZBERG, *Deformation and Fracture Mechanics of Engineering Materials,* Wiley, New York, 1996.

W. F. HOSFORD, *The Mechanics of Crystals and Textured Polycrystals,* Oxford University Press, 1993.

W. F. HOSFORD AND R. M. CADDELL, *Metal Forming: Mechanics and Metallurgy,* 2nd Ed., Prentice-Hall, 1993.

A. KELLY, G. W. GROVES, AND P. KIDD, *Crystallography and Crystal Defects,* Rev. Ed., Wiley, 2000.

U. F. KOCKS, C. N. TOMÉ, AND H.-R. WENK, *Texture and Anisotropy,* Cambridge University Press, 1998.

H. MECKING, in *Preferred Orientation in Deformed Metals and Rocks: An Introduction to Modern Texture Analysis,* edited by H. R. Wenk, Elsevier, 1985.

J. F. NYE, *Physical Properties of Crystals,* Oxford, 1957.

R. RAGHAVA, R. M. CADDELL, AND G. S. Y. YEH, *J. Mater. Sci.*, Chapman & Hall, London, Vol. 8, pp. 225–232, 1973.

R. E. REED-HILL, *Deformation Twinning,* Gordon and Breach, New York, 1964.

C. S. ROBERTS, *Trans. AIME–Journal of Metals,* 203–204, 1953.

E. SCHMID AND W. BOAS, *Plasticity of Crystals* (Kristallplastizitaet), F. A. Hughes, 1950, English Translation.

G. I. TAYLOR, in *The Sorby Centennial Symposium on the History of Metallurgy,* edited by C. S. Smith, Gordon and Breach, 355–358, 1965.

G. I. TAYLOR AND W. S. FARREN, *Proc. Roy. Soc. A.*, **111,** 759, 529–551, 1926.

S. P. TIMOSHENKO, *History of Strength of Materials,* Dover, 1983 (originally McGraw-Hill, 1953).

3.8 PROBLEMS

A.3.1 Which of the following strain tensors shows deformation without volume change?

(a) $\varepsilon = \begin{vmatrix} 0.01 & 0.01 & 0 \\ 0.01 & 0.02 & 0 \\ 0 & 0 & -0.03 \end{vmatrix}$ **(b)** $\varepsilon = \begin{vmatrix} 0.03 & 0 & 0 \\ 0 & 0.02 & 0 \\ 0 & 0 & -0.04 \end{vmatrix}$ **(c)** $\varepsilon = \begin{vmatrix} 0.01 & 0 & 0 \\ 0 & 0.01 & 0 \\ 0 & 0 & -0.02 \end{vmatrix}$ **(d)** $\varepsilon = \begin{vmatrix} 0 & 0 & 0.01 \\ 0 & 0 & 0 \\ 0.01 & 0 & 0 \end{vmatrix}$

A.3.2 Calculate the effective strain for part (c) of Problem A.3.1. For a 1-cm^3 cube, what would be the equivalent degree of deformation in simple compression?

A.3.3 Plot the Tresca criterion for a tensile yield stress of 5 MPa, defining each of the lines or facets of the yield surface by a separate equation.

A.3.4 Plot the von Mises criterion for a yield stress of 5 MPa.

A.3.5 An individual crystal is deformed such that slip planes 1 μm apart are displaced by 0.1 μm. What is the glide strain? If both θ and λ are 45°, what is the axial strain?

A.3.6 Write the general tensor expression of the type shown in Eq. 3.44 for transformation between the following crystal structures.

(a) Cubic to tetragonal

(b) Tetragonal to orthorhombic

(c) Orthorhombic to monoclinic

(d) Monoclinic to triclinic

B.3.1 Plot a projection of the Tresca criterion onto an s_1–s_3 section of the yield surface at a hydrostatic pressure of 10 MPa for a tensile yield stress of 5 MPa.

B.3.2 Repeat Problem B.3.1 for the von Mises criterion.

B.3.3 For the following stress states (in MPa), determine if the Tresca or von Mises criterion has been exceeded, assuming a tensile yield stress of 10 MPa.

(a) $\begin{vmatrix} 0 & 0 & 4 \\ 0 & 8 & 0 \\ 4 & 0 & 0 \end{vmatrix}$ **(c)** $\begin{vmatrix} 10 & -2 & 0 \\ -2 & 5 & 0 \\ 0 & 0 & 3 \end{vmatrix}$

(b) $\begin{vmatrix} 0 & 0 & 0 \\ 0 & 10 & 12 \\ 0 & 12 & -10 \end{vmatrix}$ **(d)** $\begin{vmatrix} 12 & 0 & 1 \\ 0 & 0 & 0 \\ 1 & 0 & 3 \end{vmatrix}$

B.3.4 Calculate the resolved shear stress for slip on SP = (100) and the SD = $[0\bar{1}1]$ for a stress of 10 MPa applied along the stress axes.

(a) [100] **(d)** [123]

(b) [110] **(e)** $[2\bar{1}1]$

(c) $[\bar{1}23]$ **(f)** $[\bar{1}21]$

B.3.5 Calculate the axial strain ε along $[\bar{1}23]$ for a glide shear strain of γ = 0.05 on the following proposed slip systems for a cubic material (make sure the combinations give valid slip systems).

(a) [110] $(\bar{1}01)$ **(e)** [111] $(0\bar{1}1)$

(b) [111] $(\bar{1}01)$ **(f)** [111] $(\bar{1}\bar{1}2)$

(c) $[\bar{1}01]$ (111) **(g)** [110] $(\bar{1}12)$

(d) [111] $(\bar{1}10)$ **(h)** [101] $(\bar{1}\bar{1}\bar{1})$

B.3.6 Which of the following are likely slip systems in a tetragonal material with $c/a = 2$?

(a) [110] (001) **(d)** $[10\bar{1}]$ (101)

(b) [100] (011) **(e)** [111] $(\bar{1}01)$

(c) $[1\bar{1}0]$ (110) **(f)** $[\bar{1}01]$ (111)

B.3.7 Find and sketch in an $s_1 - s_3$ section of the von Mises yield surface the value of the plastic strain ratio for the following stress states (in MPa). Assume that only yielding takes place in each case, and also calculate the value of Y.

(a) $\begin{vmatrix} 0 & 0 & 4 \\ 0 & 8 & 0 \\ 4 & 0 & 0 \end{vmatrix}$ **(c)** $\begin{vmatrix} 12 & 0 & 1 \\ 0 & 0 & 0 \\ 1 & 0 & 3 \end{vmatrix}$

(b) $\begin{vmatrix} 0 & 0 & 0 \\ 0 & 10 & 12 \\ 0 & 12 & -10 \end{vmatrix}$ **(d)** $\begin{vmatrix} 10 & -2 & 0 \\ -2 & 5 & 0 \\ 0 & 0 & 3 \end{vmatrix}$

B.3.8 Plot the Schmid factor for a set of FCC crystals lying at each of the boundaries of the standard stereographic triangle (pick at least eight for each boundary, including the symmetric points at the corners).

B.3.9 Determine how the factor of 3η comes into being if we consider that Eq. 3.41 relies on an assumption that the von Mises criterion holds. How would this equation change if the Tresca criterion were assumed?

B.3.10 Chen and Reyes-Morel (1986) were able to deform a tetragonal zirconia polycrystal (TZP) in compression. Explain which features of the transformation strain tensor (Eq. 3.44) would be required to make deformation in compression possible.

B.3.11 (a) A BCC crystal with a $[\bar{1}23]$ orientation undergoes single slip on a $(\bar{1}\bar{1}2)$ plane. Find the slip direction that is most likely.

(b) Sketch a standard stereographic triangle, showing clearly the direction of rotation of the stress axis for compression.

B.3.12 Write the physically distinct slip systems for slip in BCC metals on the slip system $\langle 1\bar{1}1 \rangle$ $\{21\bar{1}\}$.

B.3.13 Write the physically distinct slip systems for slip on both types of slip systems given for rock salt materials in Table 3.1.

B.3.14 Find the number of slip systems in Table 3.2 that have the same Schmid factor for the following positions on a standard stereographic triangle: (a) the [001]–[011] boundary; (b) the [011]–$[\bar{1}11]$ boundary; (c) the [001]–$[\bar{1}11]$ boundary.

B.3.15 For a material with Young's modulus of 100 GPa, a yield strength of 50 MPa, and parabolic work hardening behavior, write the stress for a strain of 2%.

For this material, work hardening occurs at $s_{plastic} = 50 \sqrt{\varepsilon}$ MPa.

C.3.1 Find the simplified expression relating the effective shear stress and the following stress states in terms of σ.

(a) $\begin{vmatrix} \sigma & 0 & 0 \\ 0 & 0 & 0 \\ 0 & 0 & 0 \end{vmatrix}$ (d) $\begin{vmatrix} 0 & \sigma & 0 \\ \sigma & 0 & 0 \\ 0 & 0 & 0 \end{vmatrix}$

(b) $\begin{vmatrix} \sigma & 0 & 0 \\ 0 & \sigma & 0 \\ 0 & 0 & 0 \end{vmatrix}$ (e) $\begin{vmatrix} \sigma & 0 & 0 \\ 0 & 0 & 0 \\ 0 & 0 & -\sigma \end{vmatrix}$

(c) $\begin{vmatrix} 2\sigma & 0 & 0 \\ 0 & 0 & 0 \\ 0 & 0 & 0 \end{vmatrix}$ (f) $\begin{vmatrix} \sigma & \sigma & 0 \\ \sigma & \sigma & 0 \\ 0 & 0 & 0 \end{vmatrix}$

C.3.2 Calculate the Schmid factor of the dominant slip system for a cubic crystal pulled in a direction with cosines of [0.81 –0.46 0.36], assuming <111> {110} slip.

(a) Write the Euler rotations (ϕ1, Φ, ϕ2) required to reach this orientation.

(b) Write the integer direction indices for a direction that lies within 5°.

(c) Show this tensile direction and highlight the cube triangle on a stereographic projection.

(d) Give the specific easiest slip system and the Schmid factor value.

(e) Assuming that slip occurs only as a simple shear rotation on the slip system, what angle of rotation would take place before a second slip system would have an equivalent resolved shear stress? Which slip system(s) would that be?

C.3.3 Repeat Problem C.3.2 using <110> {111} slip.

C.3.4 Repeat Problem C.3.3 using <110> {110} slip.

C.3.5 Consider the Rankine criterion, wherein yielding is predicted if any principal stress exceeds *Y* (see Fig. 3.5). Does this criterion fulfill the conditions for shear deformation given at the beginning of this chapter? Sketch the yield surface in three dimensions and with an $s_2 = 0$ section.

C.3.6 From a macroscopic viewpoint, hot pressing, which combines deformation with sintering to produce densification, is a deformation process. Describe the type of yield surface you would expect for hot pressing, including the case for isostatic pressure. Sketch both a three-dimensional section and an $s_2 = 0$ section of the yield surface.

C.3.7 Consider the yield condition for transformable zirconia materials in tension and compression. Perform a calculation to find the crystal orientation wherein the transformation should be easiest for each case. (*Hint:* using the resolved tensor components, relate the stresses and strains such that the most work is produced for a given applied stress.)

C.3.8 Derive Eq. 3.15.

C.3.9 Find that the necking strain is twice the strain hardening exponent, *n,* for balanced biaxial tension.

C.3.10 Discuss the implications of Fig. 3.8 on the value of the anisotropy factor *R*.

C.3.11 Find ε and ω relative to a unit cube using Eq. 3.27 and 3.29 for slip on $[\bar{1}10]$ (111).

C.3.12 The actual transformation from γ to α iron involves a change in the defined crystal axes. Describe this change in the coordinate system by a Euler angle rotation. Is Eq. 3.44 defined in terms of the original γ axes or in terms of the axes for FCC iron or martensite?

CHAPTER 4

DISLOCATIONS IN CRYSTALS

THE IMPORTANCE of defects in enabling of plasticity was not recognized until the understanding of crystals had been fully developed. The technological requirements for this understanding were x-ray diffraction and techniques for growing large crystals. Dislocations and their motions are responsible for the impressive ductilities demonstrated by well-annealed cubic metals. The ductility of polycrystalline metals is even more impressive when one considers that individual neighboring crystals of different orientations must maintain intact mutual grain boundaries throughout high-strain deformation processes such as rolling, extrusion, and deep drawing. The processes for strain hardening, recovery, and recrystallization rely on changes in the dislocation content within a material. The attributes of dislocation motion, formation, and multiplication control the performance and the prior processing of many engineering materials.

4.1 DISLOCATION THEORY

The understanding of dislocations requires recognition that dislocations provide stepwise motion scaled to the spacing of atoms in crystals. This stepwise motion makes it possible to deform materials to large strains by accumulation of these small displacements.

In 1934, three papers were published on the discovery of a new defect called a dislocation ("Versetzung" in German). These three papers describe the first justifications for the tremendous difference between the theoretical shear strength of a material and that actually measured in experiments. Each of the authors, Orowan, Taylor, and Polanyi, provide insight into the types of crystalline defects that could explain why the measured yield stresses for deformation of single crystals fall several orders of magnitude below those calculated by the best available models. Without dislocations, scientists of the time could not provide reasonable models for bonding of materials that simultaneously explained the elastic constants and the plastic deformation properties. (For more on the discovery of dislocations, see Hirth and Lothe, 1982, and the personal account of Orowan, 1965.)

4.1.1 The Case for Dislocations

An important fundamental problem facing researchers at the beginning of the twentieth century was the calculation of the theoretical stress for slip on particular crystallographic planes. These models were required to explain the threshold character of a yield stress that begins at a particular stress, but does not result in an instability wherein the material instantaneously fractures. Initial models of slip were based entirely on the recently documented periodicity of atoms within the crystalline material. As shown in Fig. 4.1, a simple two-dimensional

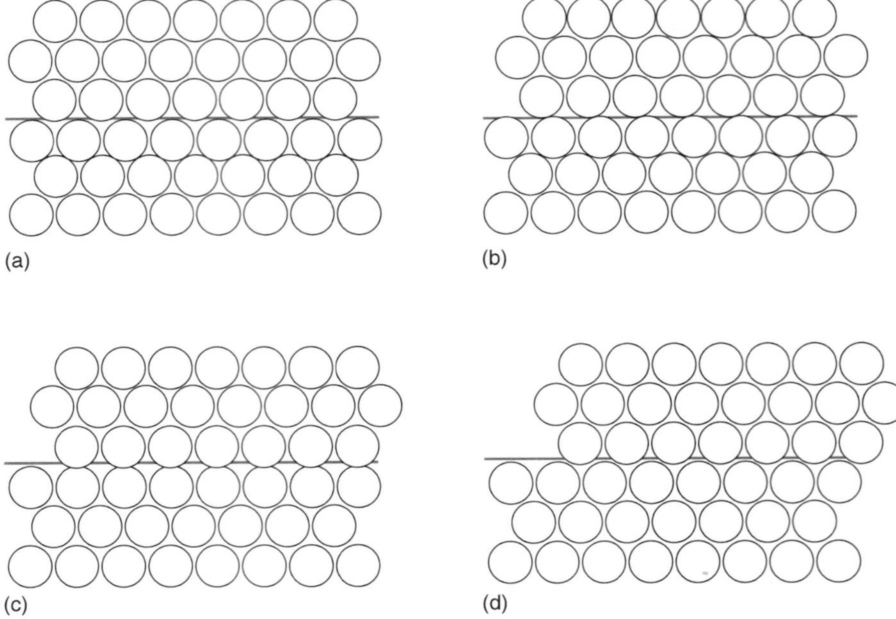

FIG. 4.1 An idealized model showing slip in the absence of dislocations. (a) Schematic planes of atoms with a line of slip shown in gray. To produce shear deformation, the bonds must be (b) stretched, (c) broken, and (d) finally reformed.

model for slip behavior from Frenkel can be proposed wherein the variation of shear stress with the relative sliding of two crystals has the periodic form of the sine function

$$\tau = \tau_{max} \sin \frac{2\pi x}{a} \tag{4.1}$$

where τ_{max} is the maximum stress before deformation begins, a is spacing of the atoms, and x is the specific slip of one plane relative to another. Using this function results in a stress for simultaneous breaking of all bonds of $\tau_{max} = \mu/(2\pi)$, where μ is the shear modulus (G is also often used for this quantity). If more reasonable expressions are employed for the periodic function, this value becomes as low as $\mu/30$. As shown in Table 4.1, the values of deforma-

TABLE 4.1 Comparison of Theoretical and Experimental Shear Strengths of Crystalline Materials (Average or Polycrystalline Value)

	Theoretical strength, $\mu/2\pi$ (MPa)	Experimental strength (MPa)
Aluminum	4200	0.7–0.8
Silver	4800	0.4–0.5
Copper	7700	0.5–3
Nickel	12,000	3–5
Iron	13,000	25–30
SiC*	23,000	11,000
Al_2O_3*	20,000	19,000
Diamond (C)*	46,000	21,000

*Estimated from deformation under hydrostatic compression (e.g., indentation hardness testing).

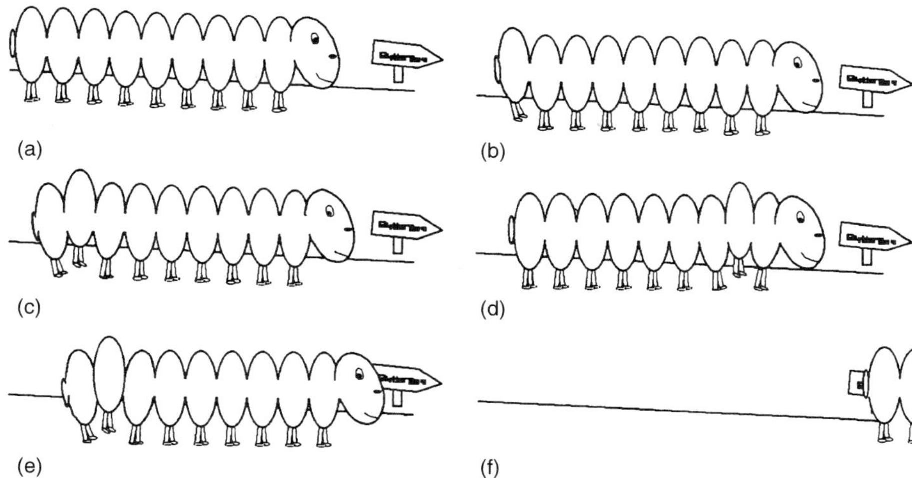

FIG. 4.2 To provide locomotion, you must somehow translate your feet, legs, and the rest of your body relative to the ground. If, as a caterpillar, you would like to move forward and attempt to do so by breaking all bonds with earth (jumping with all 18 legs), it will take a tremendous amount of energy and coordination. (a) If, instead, you lift just the rear segment and plant it closer to the second-last segment, you have moved at least your last segment forward. (b) If now you lift the next-to-last segment, relieving the pinching between the last and next-to-last segments, you have moved two segments forward and have translated the pinched region to a point between segments farther forward. (c, d) By continued motion of this segment, you can produce a single forward step without lifting all of the segments from the ground simultaneously (e, f).

tion of ductile single crystal materials fall well below these theoretical ones. Moreover, this mechanism of slip allows for neither stable plastic deformation nor strain hardening. Only in the three brittle materials at the bottom of the list does the resistance to deformation approach the theoretical values. All of these materials fracture before undergoing plastic deformation.

Consider a problem similar to dislocation slip. You are a small caterpillar on your way to meet some butterflies. As illustrated in Fig. 4.2, your body is comprised of multiple segments, each with two legs that are attached to the segment (you have no knees). The ripple that passes along your body provides motion. A similar analogy is the moving of a large carpet by forming a wrinkle at an edge and then pushing the wrinkle from one end to the other.

4.1.2 Observation of Dislocations

The linear defect called a dislocation enables stable crystal deformation to occur in deformation of single and polycrystalline materials. Furthermore, the nature of dislocations provides for all aspects of the creation, motion, multiplication, annihilation, and strengthening effects observed in low-temperature ($\leq \frac{1}{2}T_m$) deformation of crystalline materials. Surprisingly, direct evidence for dislocations was not available until about 20 years after their discovery, when the new instrument called a transmission electron microscope (TEM) was employed to image the elastic distortions around the dislocation lines in thin crystals. A thin sheet of tungsten is seen in silhouette by passing high-energy electrons through it, as shown in Fig. 4.3. At about the same time as the first TEM images, dislocation motion was inferred from the inspired investigation of etch pits in lithium fluoride (LiF) crystals (see Fig. 4.4) by Gilman and Johnston (1957). Etch pits form as a result of the high elastic energy surrounding the dislocation core.

FIG. 4.3 Transmission electron micrograph of straight <111> screw dislocations in tungsten following plastic deformation (by K. J. Bowman and R. Gibala, unpublished). Three sets of long <111> screw dislocations are shown. The fourth set lies nearly perpendicular to the thin foil of material. The other markings and contrast are mostly surface relief and strain fields in the TEM foil (marker is in microns).

As evaluation of deformation processes has progressed, the influences of the presence or absence of dislocations in the deformation and fracture of materials has become a topic of great importance. Single crystals made from materials that are normally readily deformed become extremely resistant to plastic deformation if they are made with very few dislocations. To define the quantity of dislocations within a material, the dislocation density $\rho_\perp$ is given as

$$\rho_\perp = \frac{L_\perp}{V} \approx \frac{\#_\perp}{A} \tag{4.2}$$

where $L_\perp$ is the total length (m) of the dislocations within a given volume V (m^3), $\#_\perp$is the number of dislocations, and A is the area. An approximation of dislocation density that also fits these units is dislocation number per unit area. When dislocation densities are determined from micrographs such as those shown in Fig. 4.3 and 4.4, edge effects often cause

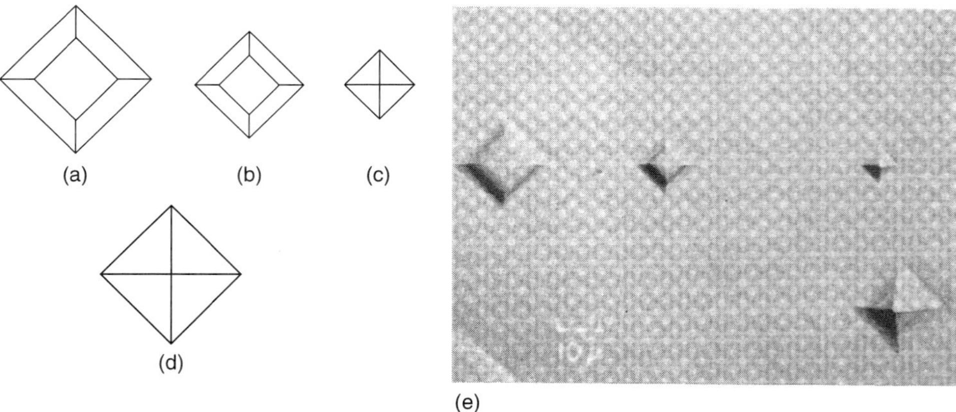

FIG. 4.4 Schematic showing motion of a dislocation as tracked by repeated etching of lithium fluoride (LiF) crystals. (a) The largest smooth bottom etch pit was formed during the first etching treatment. (b) The medium smooth bottom etch pit was formed after dislocation motion and the second etch treatment. (c) The third sharp bottom etch (indicating the final dislocation position) was formed after additional dislocation motion and the third etch treatment. (d) The large sharp bottom pit shows dislocations that did not move from the applied stresses. Part (e) shows the actual images of the dislocation etch pits. (Gilman and Johnston, 1957, used with permission)

some errors. In the TEM sample shown in Fig. 4.3, the preparation of the thin sample can result in dislocations lost to the surfaces in order to relax strain energy. As shown in Fig. 4.5, nearly perfect crystals have very high resistances to plastic deformation but deform at relatively low stresses after dislocation multiplication takes place. If these crystals are stressed to levels at which dislocations increase in number, the resistance to deformation of these crystals decreases. Thus, sufficient dislocations are present for the dislocation multiplication mechanisms discussed in Section 4.4 to occur. Since most crystals within polycrystalline materials are not so perfect, they contain significant numbers of dislocations and fall somewhere within the middle regions in Fig. 4.6. However, as plastic deformation continues, dislocation multiplication leads to dislocations becoming obstacles to the motion of each other, resulting in strain or *work hardening*.

EXAMPLE 4.1 *Dislocation Density in Transmission Electron Micrographs*

The dislocations shown in Fig. 4.3 can be measured and used to calculate the dislocation density. The thickness of the foil was approximately 200 nm, so this will enable us to compare estimates of dislocation density in terms of length per unit volume and number per unit area. The lengths of the dislocations in Fig. 4.3 are about 15 μm for the 20 or so visible dislocations. The area of the micrograph is 1.5 μm by 1.25 μm. The calculated dislocation densities are then

$$\rho_\perp = \frac{L_\perp}{V} = \frac{15\times10^{-6}\ \text{m}}{\left(1.5\times10^{-6}\ \text{m}\right)\left(1.25\times10^{-6}\ \text{m}\right)\left(200\times10^{-9}\ \text{m}\right)} = 4\times10^{13}\ \text{m}^{-2}$$

$$\rho_\perp \approx \frac{\#_\perp}{A} = \frac{20}{\left(1.5\times10^{-6}\ \text{m}\right)\left(1.25\times10^{-6}\ \text{m}\right)} = 1.1\times10^{13}\ \text{m}^{-2}$$

This result shows that there is often a small difference in the two approaches for documenting dislocation density. One problem with using TEM for estimating dislocation densities is that edge dislocations can slip out of the thin foils (and reduce the strain energy in the foil) because they are so close to the surfaces. ■

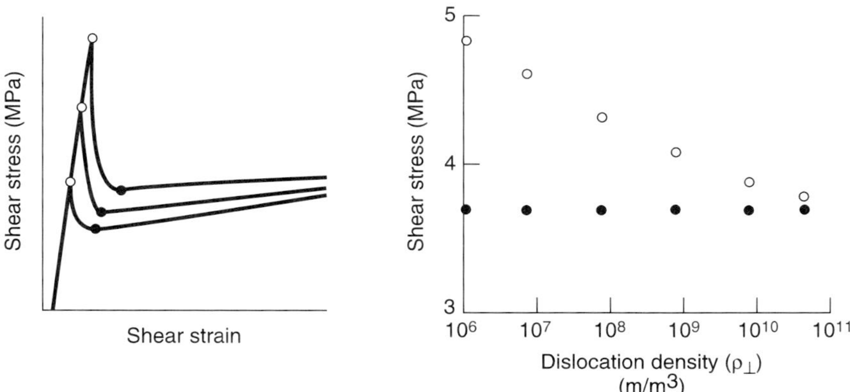

FIG. 4.5 (a) Schematic shear stress–shear strain curve with upper and lower shear yield stresses indicated. (b) Effect of initial dislocation density on initial upper shear yield point (open circle) and shear flow stress (closed circle) at a plastic strain of approximately 1 percent in compressive deformation of LiF crystals (using data from Johnston, 1962).

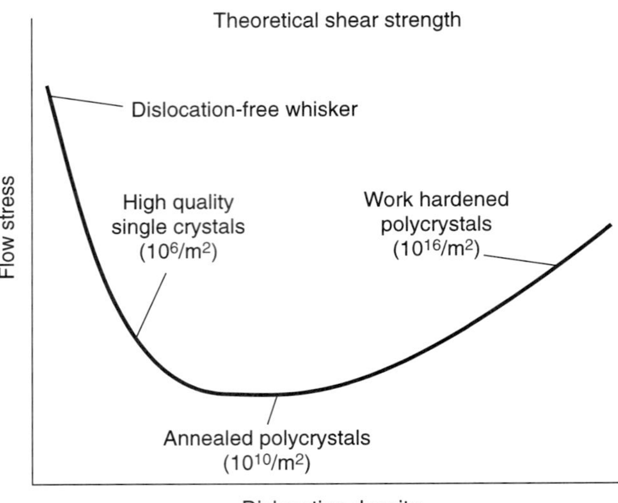

FIG. 4.6 Schematic representation of yield strength versus dislocation density.

4.2 SPECIFICATION OF DISLOCATION CHARACTER

To understand the linear defects called dislocations, one must consider that any local deviation in the perfection of a crystal changes the elastic strain energy of the crystal. The strain energy is a positive quantity that is related to the increase in energy produced by forcing the atoms to be away from their equilibrium average positions. To minimize the nature and extent of elastic defects in materials, the elastic strain energy must be balanced between being so localized that it causes singularities and being so extensive that it increases the energy level of the entire crystal. For this reason, simple geometric examples of cutting the crystal, and either adding a half-plane in the case of an edge dislocation or displacing the crystal into a spiral in the case of a screw dislocation, provide comforting geometric similarities without the details of a crystalline lattice. The nature of any dislocation can be described using a Burgers circuit, which, as is the case for many tools revealing the physical nature of crystals, involves a sign convention. Sign conventions can frustrate understanding if the focus is on the convention and not on the phenomena, so we will focus more on understanding.

4.2.1 Edge Dislocations

The edge dislocation is the simplest form of dislocation and follows quite readily from the description of a linear defect given in 1934. The insertion or removal of a so-called half-plane of crystal, as shown in Fig. 4.7, provides a straight defect through the material that goes from one end to the other. The line describing the core or center of the defect lies orthogonal to the slip or Burgers vector, **b**, thereby defining a slip or glide plane. Burgers described a tool, now called a Burgers circuit, that compares the dislocated crystal with an otherwise identical perfect (dislocation-free) crystal. To construct a Burgers circuit, we must answer the following questions.

1. Will the circuit be constructed in the clockwise or counterclockwise direction?
2. Will the Burgers vector be constructed from the start to the finish or from the finish to the start of the circuit?
3. Does the line direction, **t**, of the dislocation core point out of the face on which the Burgers circuit is constructed or into it?

The Burgers circuit relies on a closure failure to describe the direction and magnitude of the Burgers vector (as shown in Figure 4.8); however, the complete description of the dislocation requires a combination of the Burgers vector and the line direction.

Let us evaluate these questions by means of an example to find a reason for having a sign convention in the first place. As shown in Fig. 4.7*a*, the ledge produced if we move the dislocation from right to left creates an overhang on the right-hand side of the crystal from

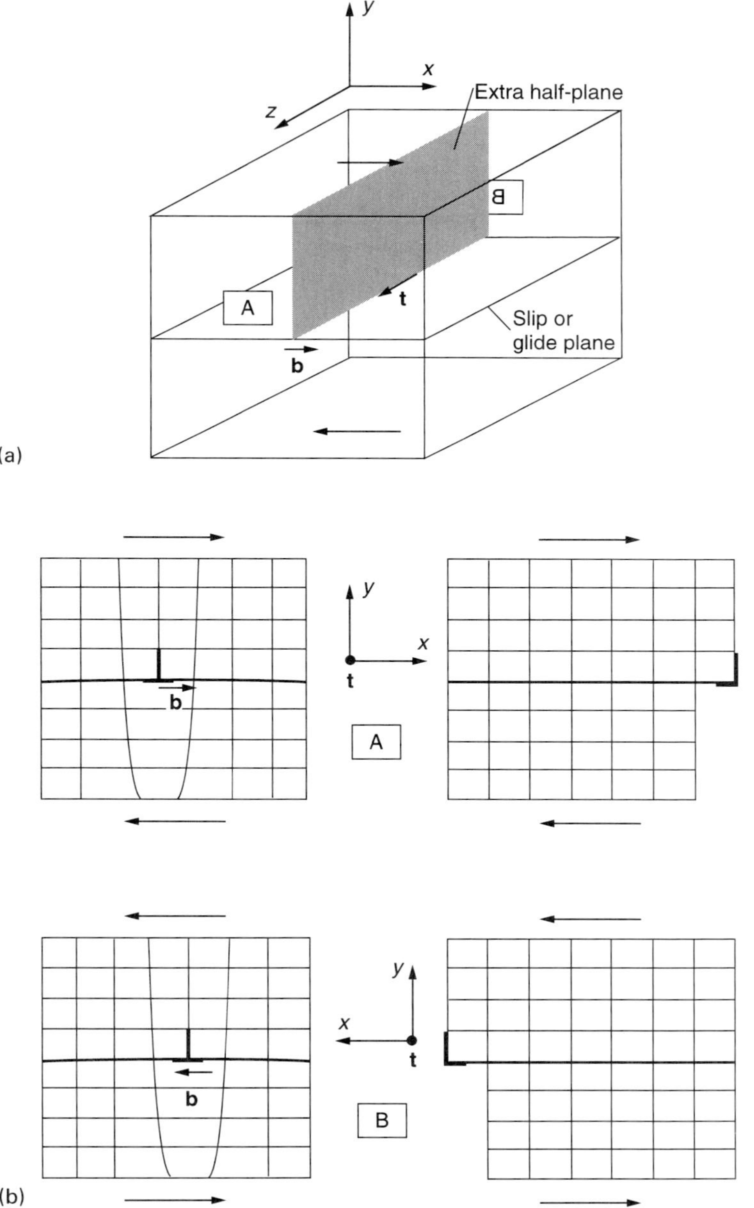

FIG. 4.7 (a) The three-dimensional diagram shows an extra half-plane in the top half of a schematic crystal. (b) The two-dimensional diagrams show crystal planes viewed from positions A and B before (left) and after (right) shear deformation.

the reader's vantage point at A. If, however, the reader had a partner viewing the crystal from behind the page, at point B, the partner would see the dislocation moving to the left, resulting eventually in a ledge on the left-hand side (from the partner's perspective) of the crystal (Fig. 4.7*b*). If our goal is for both viewers of this dislocation to provide the same definition of the edge dislocation, they should be able to adapt the rules above so that they are describing the same dislocation. Obviously, a clockwise circuit from one side of the crystal moves in the opposite direction from the other side of the crystal. Therefore, if we employ a clockwise circuit on both sides, and draw the Burgers vector from start to finish, we get the opposite direction for the apparent Burgers vector if we keep the relative positions of you and your partner on opposite sides of the crystal. The only difference between your position and your partner's is that if you define the line direction of the dislocation as coming out of the crystal, your partner must define it in the opposite way. Since your partner inverted one of the three rules of the sign convention, the discrepancy is solved by reversing the direction of the Burgers vector on your side or your partner's. Then you can agree that you are observing the same type of dislocation.

To ascertain the nature of the region surrounding an edge dislocation, we must first establish a format for presenting expressions relating the nature of the dislocation. This is necessary to provide mathematical expressions for the elastic stresses and strains around dislocations. If we employ the coordinate system expressed in Fig. 4.7, the line direction for the edge dislocation is given in the positive z-direction and the Burgers vector is given in the positive x-direction for the extra half-plane. At any given point surrounding the dislocation, we can define the strain and stress states using expressions based on elasticity theory.

The elastic displacements around edge dislocations in isotropic materials include all three normal strains, ε_{11}, ε_{22}, and ε_{33}, and the shear strains in the x-y plane, ε_{12} and ε_{21}. Because accommodating an extra half-plane causes relative states of compression above the glide plane and tension below the glide plane, we expect that at least at some positions around the dislocation core there will be stresses on the diagonal of the stress tensor. In addition, symmetric simple shears are present on either side of the half-plane and above and below the glide plane. Thus, only the shear stresses with z-components—i.e., σ_{13}, σ_{31}, σ_{32}, and σ_{23}—will be nonzero everywhere surrounding the core of the edge dislocation in an isotropic material. The solution for the elastic stresses surrounding an edge dislocation in an isotropic material is very complex, but follows a process similar to the simpler derivation available for a screw dislocation in the next section. Expressions defining the elastic stresses relative to the core of an edge dislocation in an isotropic material (see Fig. 4.7) can be given in cartesian coordinates as

$$\sigma_{11} = \frac{-\mu b}{2\pi(1-\nu)} \frac{y\left(3x^2+y^2\right)}{\left(x^2+y^2\right)^2} \tag{4.3a}$$

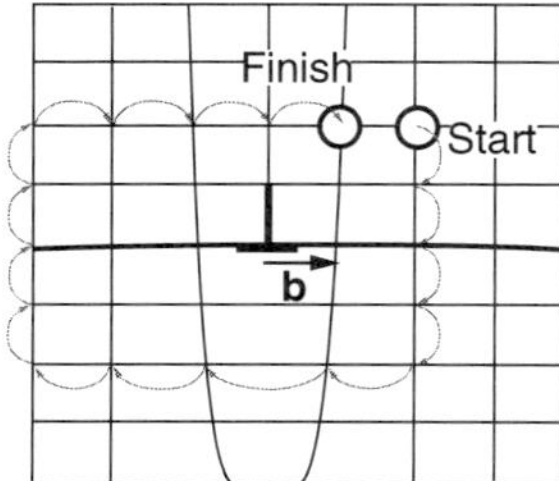

FIG. 4.8 Edge dislocation from Fig. 4.7 viewed from position A with a clockwise Burgers circuit that steps four lattice sites in both vertical and horizontal translations. If the Burgers vector is defined to go from finish to start and the line direction of the dislocation is consistent with our defined rule, then the closure vector has the direction and magnitude given for **b**.

$$\sigma_{22} = \frac{\mu b}{2\pi(1-\nu)} \frac{y(x^2 - y^2)}{(x^2 + y^2)^2} \tag{4.3b}$$

$$\sigma_{12} = \sigma_{21} = \frac{\mu b}{2\pi(1-\nu)} \frac{x(x^2 - y^2)}{(x^2 + y^2)^2} \tag{4.3c}$$

where b is the magnitude of the Burgers vector. Because for particular positions the expressions above include normal stresses, there must also be a Poisson effect given by Hooke's law such that $\sigma_{33} = -\nu(\sigma_{11} + \sigma_{22})$. Figure 4.9 shows the relative magnitude of σ_{11} and the mean or hydrostatic stress as the dislocation core is approached.

At small distances from the core of the dislocation, the magnitudes of the stresses become quite large owing to large strains. Since the strains are large near the dislocation core, deviations from linear elasticity prevent the possibility of "infinite" stresses at the singularity.

EXAMPLE 4.2 *Stresses Near Edge Dislocation Cores*

Using the values for an aluminum alloy, we will calculate the magnitudes of the stresses σ_{11} given by Eq. 4.3*a* for positions along the *y*-axis, which means that the *x*-value is zero.

$\mu = 25$ GPa $\qquad \nu = 0.35 \qquad b = 2.9 \times 10^{-10}$ m

$$\sigma_{11} = \frac{-\mu b}{2\pi(1-\nu)} \frac{y^3}{y^4} = \frac{-25\text{ GPa}(2.9 \times 10^{-10}\text{ m})}{2(3.14)(1-0.35)} \frac{1}{y} = \frac{1.8 \times 10^{-9}\text{ GPa}\cdot\text{m}}{y}$$

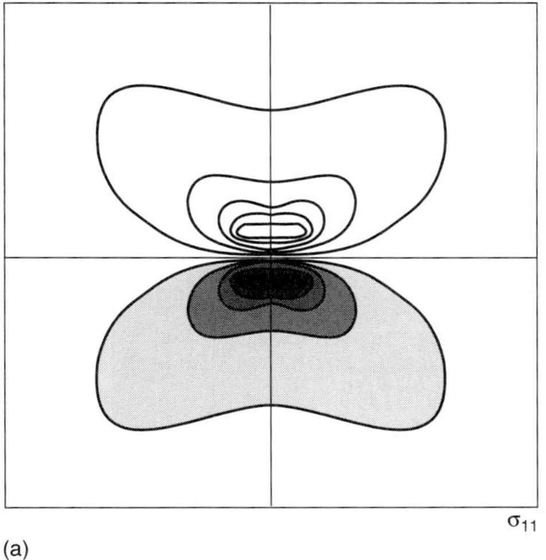

(a)

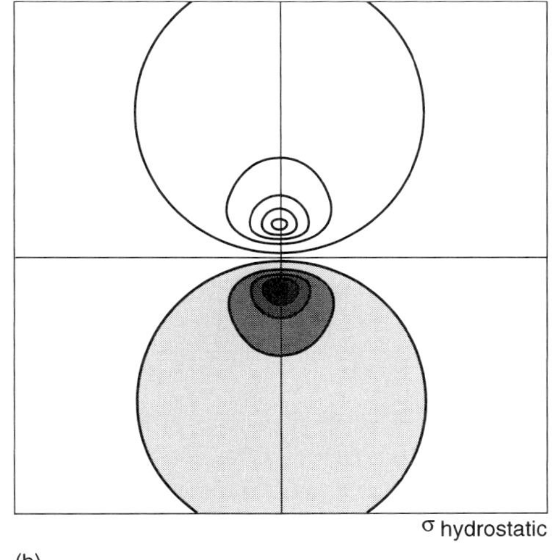

(b)

FIG. 4.9 Gray scale plots of the stresses (a) σ_{11} and (b) hydrostatic stress = $(\sigma_{11} + \sigma_{22} + \sigma_{33})/3$ from Eq. 4.3 as a function of position around the centered edge dislocation core from Fig. 4.7. The extra half-plane lies in the top half of each figure. White represents compressive stress, and gray represents tensile stress.

Distance (10^{-10} m)	Stress (GPa)
40	–0.4
30	–0.6
20	–0.9
10	–1.8
0	∞
–10	1.8
–20	0.9
–30	0.6
–40	0.4

As shown in the calculation results, the magnitudes of the stresses become quite large, with this linear elastic relation predicting infinite stresses at the dislocation core, large tensile stresses just below the glide plane, and large compressive stresses just above the glide plane. At the large strains near the dislocation core, an assumption of linear elasticity is not a good one. ■

The elastic distortion of the material outside the dislocation core (wherein the elastic calculation would give an infinite result) can be calculated as a function of distance from the dislocation using

$$\begin{aligned} E_{\text{elastic}} &= \frac{\mu b^2 \ell}{4\pi(1-\nu)} \int_{r_o}^{R} \frac{dx}{x} \\ &= \frac{\mu b^2 \ell}{4\pi(1-\nu)} \ln \frac{R}{r_o} \end{aligned} \tag{4.4}$$

The core radius r_o and the outer radius R can then be used to calculate the energy of the dislocation, and ℓ is the dislocation length. The magnitude of r_o should be chosen such that the deformation is still small enough to be considered elastic. The magnitude of the outer radius can extend to the crystal surfaces if only one dislocation is present in the crystal. For crystals containing many dislocations, half the spacing between dislocations is often employed to provide an estimate of the elastic energy surrounding each dislocation. Equation 4.4 also shows that the energy of the dislocation is linearly dependent on the shear modulus and the dislocation length. The dislocation energy has a squared dependence on the magnitude of the Burgers vector.

The plane of slip for an edge dislocation is completely determined by the plane defined by the Burgers vector and line direction. With reference to Chapter 3, it is then clear that the cross product of the slip plane normal, **SPN**, and the Burgers vector gives the line direction for the edge dislocation.

EXAMPLE 4.3 *The Elastic Energy of Edge Dislocation*

We can calculate the energy of an edge dislocation using Eq. 4.4. Consider 1 cm^3 of a cold-worked material with $\mu = 50$ GPa, $b = 2.5 \times 10^{-10}$, and $\nu = 0.3$. We will use a cutoff of $4b = 1 \times 10^{-9}$ m for the magnitude of r_o. With a dislocation density of $\rho_\perp = 1 \times 10^{14}\ \text{m}^{-2}$, we can estimate an average spacing of $L = (1 \times 10^{14})^{-0.5} = 1 \times 10^{-7}$ m. The outer radius can then be set to half this value, $R = 5 \times 10^{-8}$ m. The total length of dislocation in our 1 cm^3 of material is $l = (1 \times 10^{-6}\ \text{m}^3)(10^{14}\ \text{m/m}^3) = 1 \times 10^8$ m. Then the elastic energy is calculated as

$$E_{\text{elastic}} = \frac{(50\times10^9\ \text{N/m}^2)(2.5\times10^{-10}\ \text{m})^2(1\times10^8\ \text{m})}{4\pi(1-0.3)} \ln\left(\frac{5\times10^{-8}\ \text{m}}{1\times10^{-8}\ \text{m}}\right) = 0.14\ \text{Nm}$$

■

BIOGRAPHY

JOHANNES MARTINUS BURGERS (1895–1981)

This specialist in fluid mechanics was appointed as professor at the Technical University of Delft (Netherlands) at the age of 23. He is credited with the first real description of a screw dislocation in 1939. He also helped establish early forms of the expressions for dislocation displacements and dislocations in subgrain boundaries.

4.2.2 Screw Dislocations

A screw dislocation is a spiral defect within a material. The nature of the screw dislocation is similar to a simple parking garage wherein one circuit translates you and your auto up or down one level. Etching of surfaces intersected by screw dislocations can show the spiral character of the screw dislocation (see Fig. 4.10). For consideration of a screw dislocation in a crystal, the relationship between the screw dislocation and an edge dislocation provides some insight, as shown in the three views of the same dislocation in Fig. 4.11. Of course, the most distinctive difference between edge and screw dislocations is the relationship between the line direction of the dislocation and the Burgers vector. For the edge dislocation, these vectors are orthogonal; in the case of screw dislocations, the Burgers vector and the line direction are either parallel or antiparallel. Screw dislocations with the same Burgers vector, but opposite line directions, have the opposite sign of the twist. Unlike the case of the edge dislocation, the screw dislocation does not define a single glide plane in which it can move, but it is possible for it to move on any plane that contains its Burgers vector if the shear stress required for motion on that plane is exceeded.

The distortion of a cylindrical element of material by a screw dislocation in the material is shown in Fig. 4.12*a*. The screw dislocation of the unrolled element is equivalent to a simple shear of the unrolled cylindrical element, as shown in Fig. 4.12*b*. When translated back into the coordinate system of the dislocated cylinder, the only shear strains possible are those consisting of shear strains with z-components. These strains are given in cartesian and cylindrical coordinates as

$$\varepsilon_{13} = \varepsilon_{31} = \frac{-b}{4\pi} \frac{y}{\left(x^2 + y^2\right)} = \frac{-b}{4\pi} \frac{\sin\theta}{r} \qquad (4.5a)$$

$$\varepsilon_{23} = \varepsilon_{32} = \frac{b}{4\pi} \frac{x}{\left(x^2 + y^2\right)} = \frac{b}{4\pi} \frac{\cos\theta}{r} \qquad (4.5b)$$

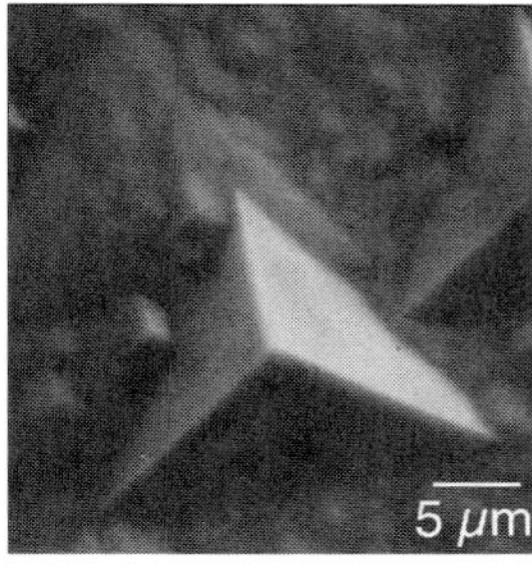

(a)

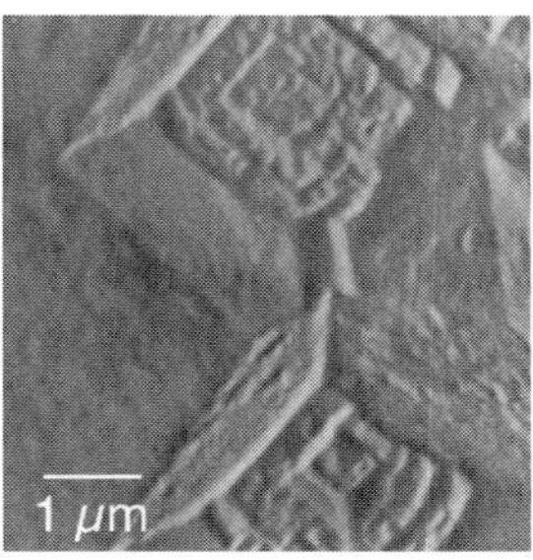

(b)

FIG. 4.10 Dislocation etch pits of <111> screw dislocations in tungsten (a BCC metal). (a) Pits for screw dislocations that lie nearly orthogonal to the surface. (b) Etch pits for dislocations at an angle to the surface, showing the spiral ledges associated with the screw dislocations (K. J. Bowman and R. Gibala, unpublished).

All other strains should be zero in an isotropic material. The associated stresses are given by

$$\sigma_{11} = \sigma_{22} = \sigma_{33} = \sigma_{12} = \sigma_{21} = 0 \tag{4.6a}$$

$$\sigma_{13} = \sigma_{31} = \frac{-\mu b}{2\pi} \frac{y}{\left(x^2 + y^2\right)} = \frac{-\mu b}{2\pi} \frac{\sin\theta}{r} \tag{4.6b}$$

$$\sigma_{23} = \sigma_{32} = \frac{\mu b}{2\pi} \frac{x}{\left(x^2 + y^2\right)} = \frac{\mu b}{2\pi} \frac{\cos\theta}{r} \tag{4.6c}$$

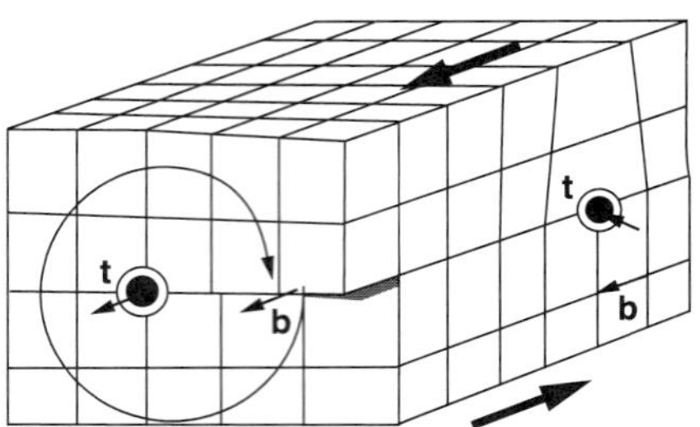

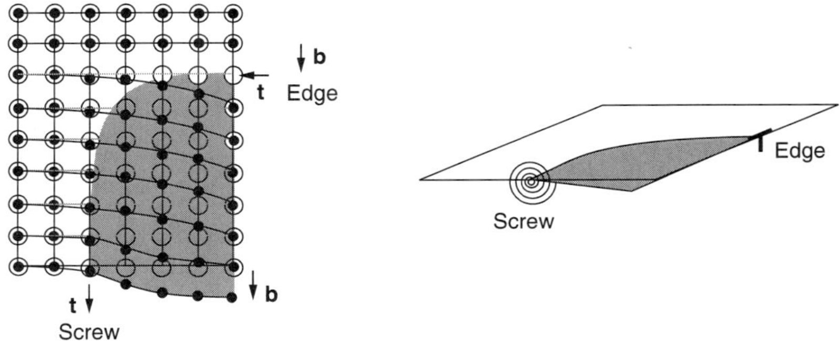

FIG. 4.11 Screw dislocation viewed on a cube of material giving a screw dislocation connected to an edge dislocation to form a quarter loop. Top view is a three-dimensional view with an overhang of magnitude b on the front right portion of the crystal. Open arrows denote a shear stress that if high enough would cause expansion of the loop through motion of the edge dislocation to the back of the crystal and the screw dislocation to the left. Figure at lower left shows how the atoms in the top half of the crystal (solid black circles) are shifted forward from their positions above atoms in the lower half of the crystal (open gray circles). Schematic at lower right shows a perspective view of the quarter loop.

BIOGRAPHY

F. R. N. NABARRO (1916–)

F. R. N. Nabarro is well known for developing physical formulations to describe the resistance to dislocation motion and mechanisms of diffusional creep. Nabarro has made many important contributions in elastic theory applied to dislocation.

As mentioned for the edge dislocation, the magnitude of the stresses at the dislocation core of a screw dislocation cannot be predicted by these expressions; however, the predictions for stresses are reasonable at distances of 3b to 4b from the dislocation center.

Similar to an edge dislocation, the elastic distortion surrounding the core of a screw dislocation is given by

$$\begin{aligned} E_{\text{elastic}} &= \frac{\mu b^2 \ell}{4\pi} \int_{r_o}^{R} \frac{dx}{x} \\ &= \frac{\mu b^2 \ell}{4\pi} \ln \frac{R}{r_o} \end{aligned} \quad (4.7)$$

The terms are defined the same as in Eq. 4.4. For common values of Poisson's ratio, the energy of the screw dislocation in an isotropic material is approximately two-thirds that of an edge dislocation.

Mixed dislocations are dislocation segments wherein the angle between the Burgers vector and the line direction is neither 90° (edge) nor 0° (screw). Each mixed dislocation can be resolved into edge and screw components to relate the energy of such dislocations as

$$E_{\text{mixed}} = \frac{\mu b^2 \ell \left(1 - \nu \cos^2 \theta\right)}{4\pi(1-\nu)} \ln \frac{R}{r_o} \quad (4.8)$$

where θ is the angle between the Burgers vector and the line direction. The elastic energy of a dislocation can be generalized as

$$E_{\text{dislocation}} = \alpha \mu b^2 \ell \quad (4.9)$$

where α is a dimensionless factor based on dislocation arrangement that is between 0.5 and 1, and ℓ is the dislocation length. To minimize the energy of a dislocation, it is then clear that the smallest possible Burgers vector is essential to keeping the dislocation energy small. Minimization of dislocation length, though not as strong a term, is then important in determining the shapes of dislocations or dislocation loops with a given Burgers vector. The slip directions described in Chapter 3 normally correspond to the shortest Burgers vector for a particular crystal structure.

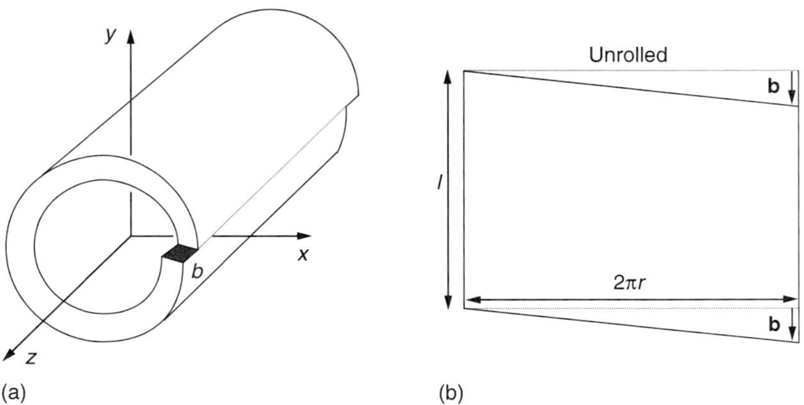

FIG. 4.12 (a) Distortion of a cylinder of material in a model for a screw dislocation. (b) Shear distortion shown after unrolling a cylinder of material containing a screw dislocation.

4.2.3 Dislocation Loops and Nodes

Dislocation loops provide the essential features to prevent a dislocation from ending within a crystal. A dislocation loop provides closure of the dislocation onto itself. The quarter loop shown in Figure 4.11 provides all that is required to construct a dislocation loop that can glide through a crystal and produce shear deformation. By combination with a mirror image of the side containing the edge dislocation in Fig. 4.11, one-half of a glide loop can be constructed as shown in Fig. 4.13. A complete glide loop is then constructed by putting together matching halves to form the full glide loop shown in Fig. 4.14.

Evaluation of the glide loops shows that the Burgers vector of the loop lies in the plane of the loop. In this way, the opposite sides of the loop are defined as opposites by the inversion of the line direction for opposite sides of the loop. For this glide loop, any applied shear stress that would cause motion of edge, screw, or mixed segments to expand the loop would similarly tend to move the other portions of the dislocation loop to produce expansion of the loop. Consider expansion of the planar dislocation loop shown in Fig. 4.14. The edge dislocation at the front of this plane has an extra half-plane on the bottom half of the crystal. To move this segment forward requires a resolved shear stress pointing in the direction of the defined Burgers vector below the plane of the loop and opposite to the defined

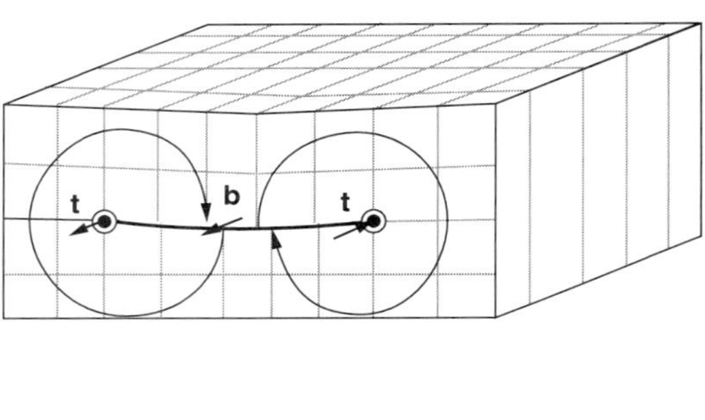

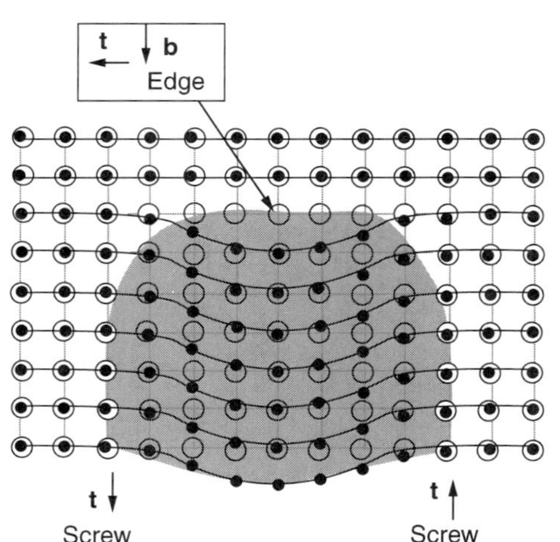

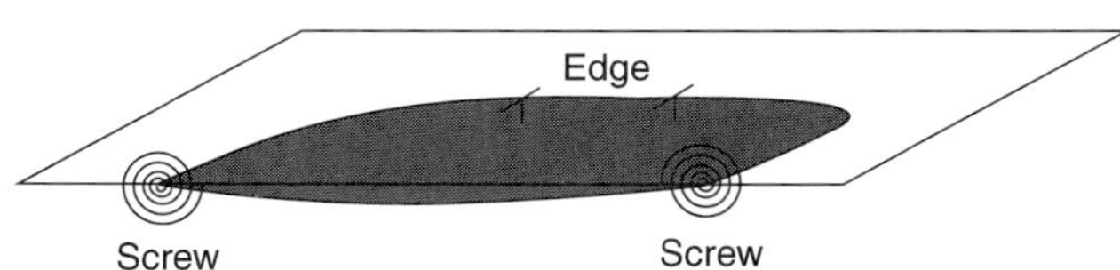

FIG. 4.13 By adding two quarter-loops together, a half-loop can be created. Top view is a three-dimensional view with two opposite line direction screw dislocations and the same Burgers vector. Middle view shows the displacement of atoms in the top half of the crystal (solid black circles) versus the atoms in the lower half (open circles). Only the top atoms are shown as displaced, whereas the atoms above and below the slip plane would actually be displaced. The gray area is the slipped region due to formation of the half dislocation loop defined in the lower schematic view in perspective.

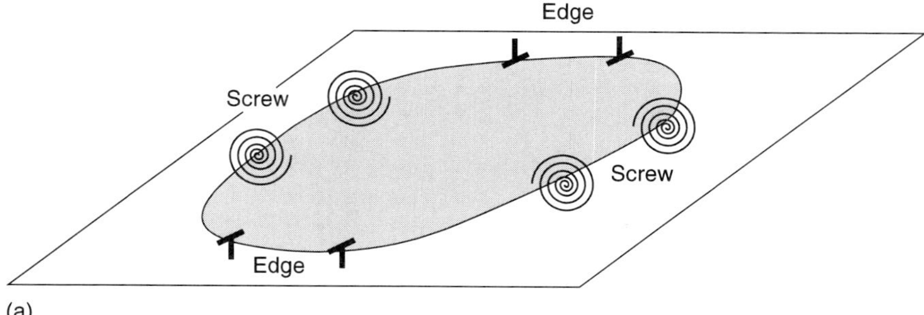

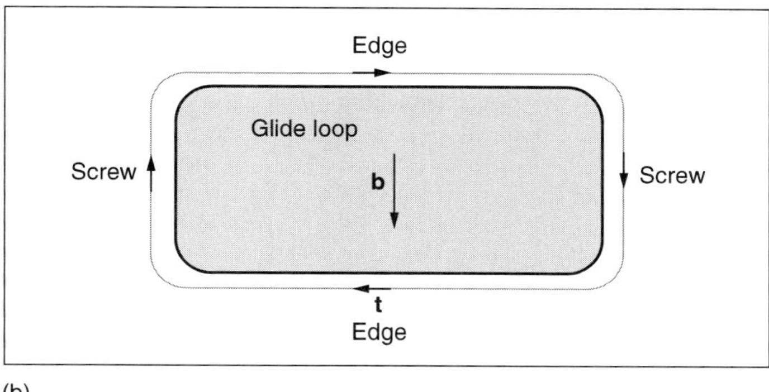

FIG. 4.14 Two views of a complete glide loop shown (a) in perspective and (b) with the Burgers vector and the line direction defined.

Burgers vector above the plane of the loop. The same stress will cause the other segments to expand. If the opposite-sign resolved shear stress is applied, the loop will contract and the opposite-sign segments may mutually annihilate one another to eliminate this loop. Differences in the energy of edge, screw, and mixed segments will result in larger displacements of the lower-energy portions of the glide loop. Dislocation motion is further described in Section 4.3.

Nonglide or sessile dislocation loops consist of dislocation loops with a Burgers vector that do not lie in the loop plane. When the Burgers vector is orthogonal to the dislocation loop, there are two types of dislocation loops, *vacancy* and *interstitial* (by analogy with the point defects). As shown in Fig. 4.15, a vacancy loop consists of a missing plane of atoms that ends within a crystal. An interstitial loop consists of an extra plane of atoms that

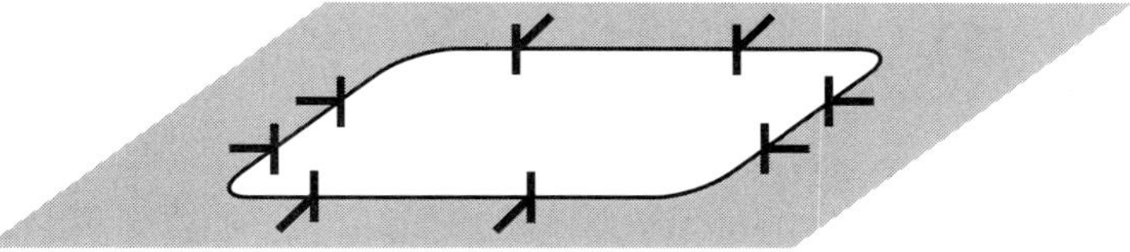

FIG. 4.15 A prismatic loop of vacancy character showing that the Burgers vector is orthogonal to the line direction around the entire loop. The vacancy loop shown consists of a missing square of atoms within a crystal.

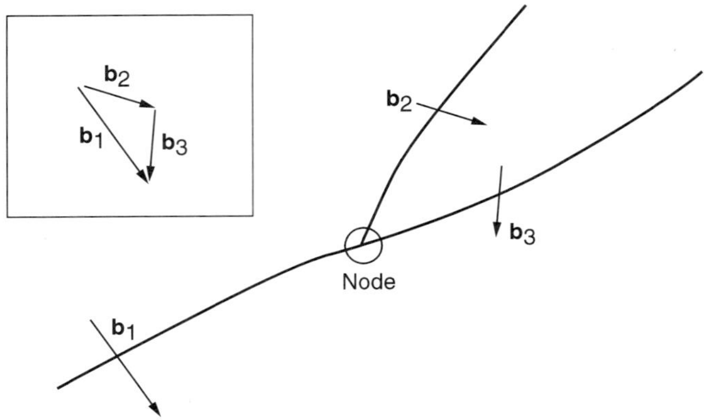

FIG. 4.16 Two dislocations, $\mathbf{b}_2$ and $\mathbf{b}_3$, combine to yield a single dislocation $\mathbf{b}_1$ at the node.

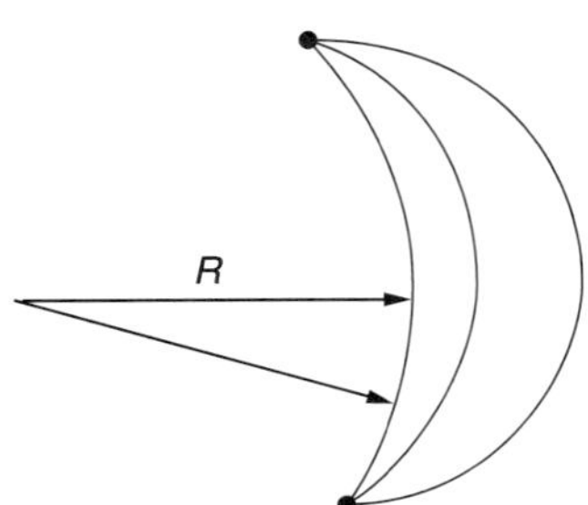

FIG. 4.17 A curved segment that shows bowing between two obstacles wherein R is used to define the radius of curvature.

ends within a crystal. Dislocations with different Burgers vectors can meet at *a dislocation node*, as shown in Fig. 4.16. At this *node* $\mathbf{b}_1 = \mathbf{b}_2 + \mathbf{b}_3$.

The initial formation of a half-loop is often observed when a dislocation is pinned (or held) at two points by obstacles lying across the glide plane. These obstacles can be changes in the plane on which the dislocation glides due to a pair of jogs or cross-slip of a screw dislocation segment, intersections with other dislocations, or second-phase particles. If the dislocation has no other mechanism for overcoming these obstacles and the material is isotropic, the dislocation can glide until it forms a semicircle that has the radius of curvature defined by the half-distance between the obstacles. Figure 4.17 shows a curved segment bowing between two obstacles. For this dislocation segment, the Burgers vector is already fixed, so extension of the dislocation results in an energy increase determined by the change in length defined by Eq. 4.9. If differences in the energies of edge, mixed, and screw dislocations are ignored, the shear stress required to bow a dislocation to a radius of R is

$$\tau_{\text{bowing}} = \frac{\mu b}{R} \tag{4.10}$$

This expression is used in Chapter 5 to describe how some strengthening mechanisms rely on pinning dislocations using closely spaced obstacles to dislocation motion.

4.3 DISLOCATION MOTION

The result of dislocation motion in response to shear stress is shear strain. Orowan defined this strain as

$$\gamma = \rho_{\perp} b x_{\text{avg}} \tag{4.11}$$

where γ is the simple shear strain resulting from deformation of a material with dislocation density $\rho_{\perp}$, the Burgers vector magnitude b, and the average dislocation displacement x_{avg}.

4.3.1 Glide

The process of dislocation glide is reliant on the sequential stretching, breaking, and reforming of bonds as each portion of the dislocation moves to respond to an applied shear stress. Figure 4.18*a* shows the formation of *kinks* as an edge dislocation moves. Part of the dislocation's resistance to glide is related to the size of the dislocation core, often termed the dislocation width.

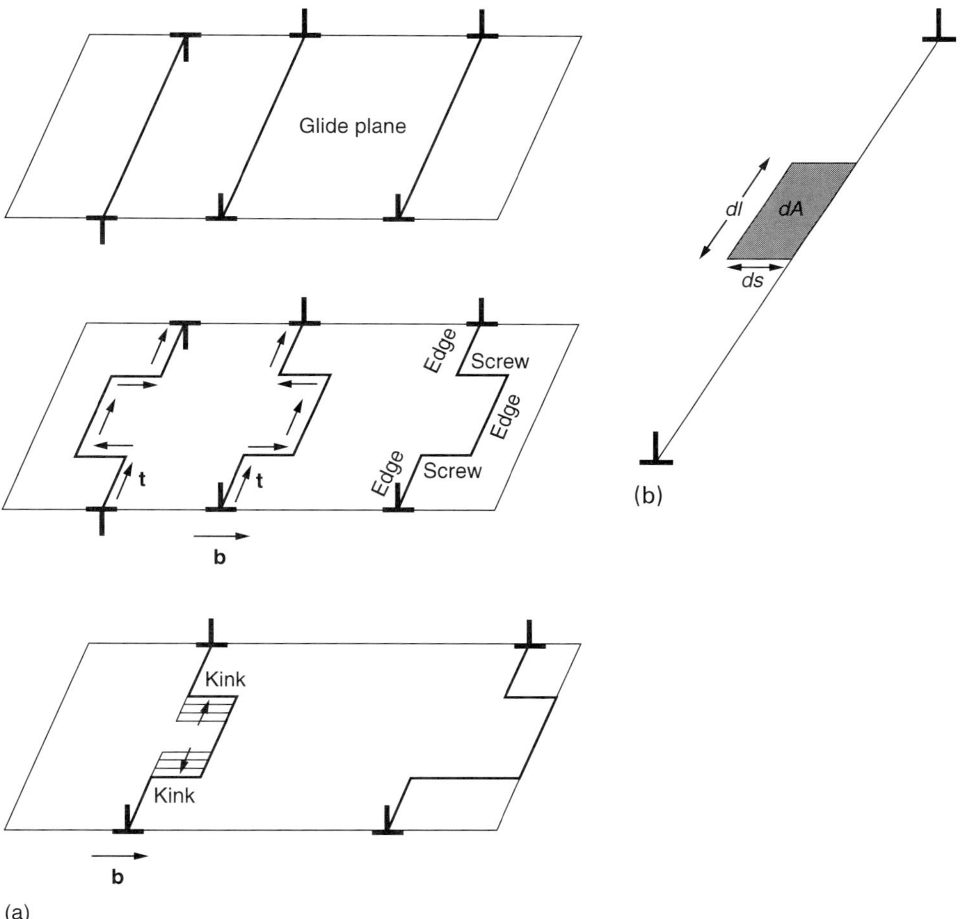

FIG. 4.18 (a) Glide plane showing schematic edge dislocations that form kinks under applied shear. Top view shows straight edge dislocations. Middle view shows the same dislocations after applied shear stresses have caused kinks to form on these dislocations. Lower view shows the kinks propagating under the same shear stress that causes this dislocation to move to the right. (b) The slipped area, displaced by the magnitude of the Burgers vector *b*, consists of components *ds* and *dl*. For this exaggerated example wherein dislocations of mixed character are not shown, $dA = ds\, dl$. A dislocation would have such sharp corners only if the material had very large elastic anisotropy.

Dislocations with widely distributed dislocation move more readily than dislocations with narrow cores. Dislocation width can be expressed in terms of the Burgers vector magnitude for a dislocation. The width of the edge dislocation can be defined as a fixed difference in displacement, Δu, of atoms above and below the slip plane. The width of the dislocation is then defined in terms of the distance over which the displacements are large enough so that linear elasticity theory does not apply. If, for example, we use bounds of $b/10$ for the strain level defining the extent of the dislocation's width, then the width of the dislocation core is much wider in Fig. 4.19*a* than in Fig. 4.19*b*. The plots of $|\Delta u|$ versus x show the disregistry (or deviation from perfect stacking) between the atoms above and below the plane.

The examples in Fig. 4.19 are expressed in terms of a stationary dislocation resting in its equilibrium, lowest-energy position. When a shear stress interacts with one of these dislocations, it imparts changes in the disregistry curves that change the energy of the dislocation. Peierls and Nabarro related the dislocation energy to the dislocation position by what is called the Peierls–Nabarro energy, E_{PN}

$$E_{PN} = \frac{\mu b^2 \ell}{\pi(1-\upsilon)} \exp\left(\frac{-2\pi w}{b}\right) \tag{4.12}$$

From this we can obtain the force or, even better, the stress required to move a dislocation by finding the maximum slope of E_{PN}. This is called the Peierls–Nabarro stress, τ_{PN}

$$\tau_{PN} = \frac{2\pi}{b^2} E_p = \frac{2\mu\ell}{(1-\upsilon)} \exp\left(\frac{-2\pi w}{b}\right) \tag{4.13}$$

The magnitude of τ_{PN} is several orders of magnitude lower than predictions based on the theoretical shear strength and is often close to the observed critical shear stress values for deformation.

As expected, the Peierls–Nabarro model predicts slip between the planes with the widest separations. Extended, broad planar dislocation cores are found in materials with low τ_{CRSS}. The advantages of broad planar disregistry are one reason that edge dislocations are often more mobile than screw dislocations. For FCC and HCP metals, large dislocation

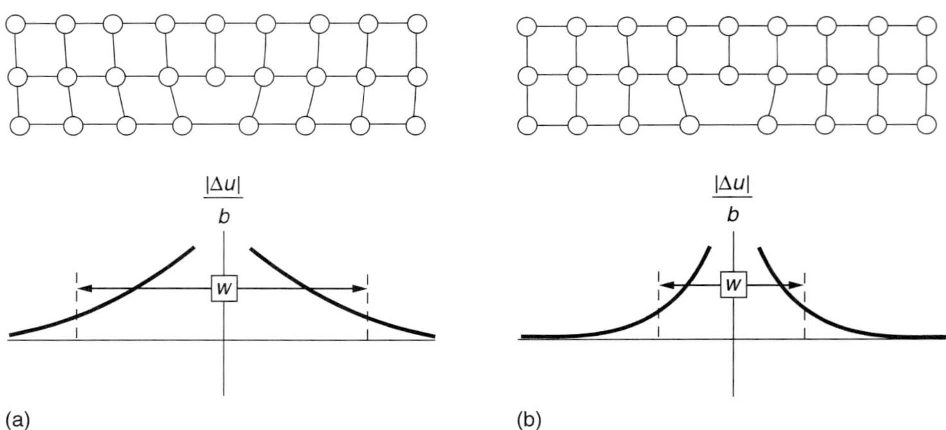

FIG. 4.19 (a) Edge dislocations with (a) wide and (b) narrow dislocation cores (after Hull and Bacon, 1984), with diagrams below showing the absolute value of the displacement $|\Delta u|$ from equilibrium in fractions of the Burgers vector magnitude *b*.

BIOGRAPHY

EGON OROWAN (1902–1989)

Egon Orowan was one of the founders of dislocation theory. As an undergraduate, he was inspired to think of dislocations from the deformation of zinc single crystals. Orowan also described expressions for strain from dislocation motion and mechanisms for dislocations to climb over obstacles (see Orowan, 1965).

---

widths result in low resistances to dislocation motion and even the possibility of dissociation of dislocations to produce stacking faults (see Section 4.5).

Figure 4.8 provides a somewhat misleading view of an edge dislocation since the atoms shown are really representative of lines of atoms lying parallel to the dislocation. The individual stretching and bond breaking must take place in coordination with the stretching and distortion of bonds along the dislocation line. For the edge dislocation line to move from one position to another, screw segment kinks form and propagate to enable translation of the dislocation line. As shown in Fig. 4.18, kinks formed on edge or screw dislocations to facilitate glide must have a line direction sense consistent with the glide process.

To this point we have suggested that the applied shear stresses of the type shown in Fig. 4.7 produce motion of a dislocation. To be consistent with the development of stress fields and the elastic energy associated with dislocations, it is essential to have an understanding of how a shear stress produces the virtual force that results in dislocation motion.

Although there are differences in the energies of dislocations that are of edge, screw, and mixed character, the differences are small compared to the affects of changing the length of the dislocation. Formation of a kink or glide step on a glide plane results in a displacement of the material above and below the glide plane at the kink. The magnitude of the force on the element in Fig. 4.18*b* is simply $\tau\mathbf{b}$, where τ is the shear stress resolved on the glide plane in the direction of the Burgers vector $\mathbf{b}$. This force arises from doing work W on a unit length of dislocation to move it a unit distance, where

$$\boldsymbol{F} = \frac{dW}{(ds dl)} = \frac{dW}{dA} = \tau\mathbf{b} \tag{4.14}$$

with the terms in this equation[1] defined in Figure 4.3.1(b).

Thus, we have an energy (or force) corresponding to the length of the dislocation and a force (or energy) required to move a portion of the dislocation that is resisted by the magnitudes of E_{PN} and τ_{PN}.

The examples showing kinked dislocations in Fig. 4.18 exaggerate the nature of a kink by showing sharp corners. Since the energy of a dislocation depends on its length, the energy is often reduced by eliminating those sharp transitions with the inclusion of mixed dislocation segments. The Peierls–Nabarro energy also affects the shapes of kinked dislocations. The minimum energy positions defined by Eq. 4.12 are energy hills that must be overcome to move the dislocation from one valley to the next. Because the dislocation must be continuous and glides by the formation of kinks, the shape of the dislocation will be determined by the magnitude of the energy barrier (see Fig. 4.20). We also know from Eq. 4.9 that increasing the length of a dislocation adds to its energy. Consequently, the shape of the dislocation is controlled by a balance between the position of the dislocation line and minimization of its length. By forming a pair of kinks, a process that is assisted by thermal

[1]Remember that by defining this expression "per unit length," this force is scaled to the length or "force per unit length."

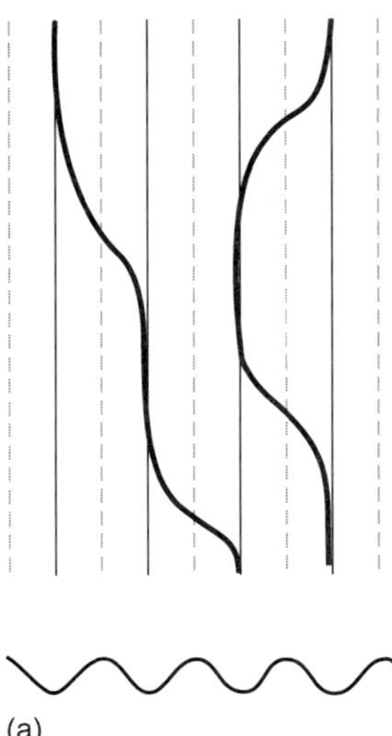

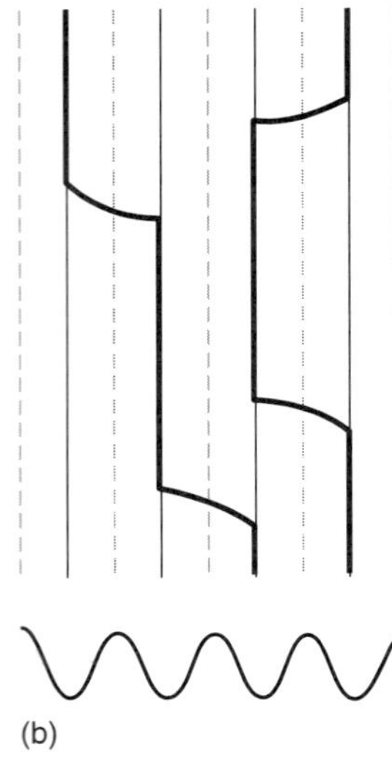

FIG. 4.20 (a) The shape of a dislocation and the magnitude of E_{PN} with position for a low-energy barrier. (b) The shape of a dislocation and the magnitude of E_{PN} with position for a high-energy barrier.

vibration, the dislocation can be carried across a barrier by the propagation of the kinks, which requires much less energy than moving the entire dislocation over the barrier at one time. These kink pairs consist of opposite-sign dislocations that will annihilate if the kink pair is too small. For this reason, the nucleation of kink pairs to stable spacings of 20b or so requires thermal activation. The activation energy for stable kink pair nucleation scales with the magnitude of E_{PN}.

The shear stresses driving dislocation motion result in dislocation velocities that are observed experimentally to have a power law relationship

$$\mathrm{v} = A\left(\frac{\tau}{\tau_o}\right)^m \tag{4.15}$$

where v is the velocity of dislocation motion, τ is an applied shear stress greater than τ_o, the stress below which no dislocation motion takes place, and A is a constant of approximately 1 m/s for pure metal crystals at room temperature. The stress exponent m is near 1 for pure metals and typically 2 to 5 for alloys and compounds. The exponents tend to increase with lower temperatures as damping from thermal vibration weakens at lower temperatures. Figure 4.21 shows the shear stress dependences of velocity measured for various materials.

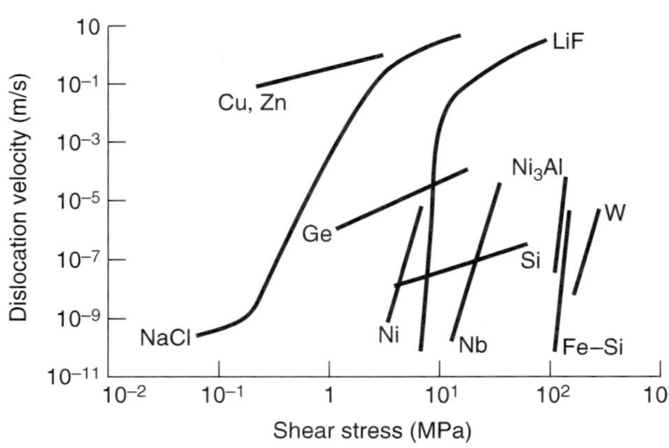

FIG. 4.21 The dependence of dislocation velocity on applied shear stress for several materials at room temperature (after Haasen, 1987). Data for Ge and Si are at 700°C (Siethoff, 2002). Data for Ni_3Al are from Jiang et al. (2001).

4.3.2 Cross Slip

Edge dislocations are restricted to glide within a plane defined by the cross product of the Burgers vector and the line direction of the dislocation. The same restriction does not apply to a screw dislocation, because **b**||**t**. Screw dislocations tend to glide in crystallographic planes from certain families with low critical resolved shear stresses; however, they can switch from one specific plane to an intersecting plane from the same family. Figure 4.22 shows that the screw dislocation can alternate from one plane to another through *cross slip*. In FCC metals, the $\frac{a}{2}$<110> screw dislocations can alternate between the two {111} planes that can contain those screw dislocations. In BCC metals, slip can occur on several of the planes that can contain $\frac{a}{2}$<111> screw dislocations, including three each of {110}, {112}, and even {123} (see Fig. 4.23). This easy cross slip in <111> directions on almost any plane containing this direction is called *pencil glide*.

4.3.3 Climb

Another process for dislocation motion involves diffusion to extend or shorten the extra half-planes of edge dislocations. At high temperatures, diffusion of vacancies and interstitials combined with applied stress enable the deformation shown in Fig. 4.24 called *climb*.

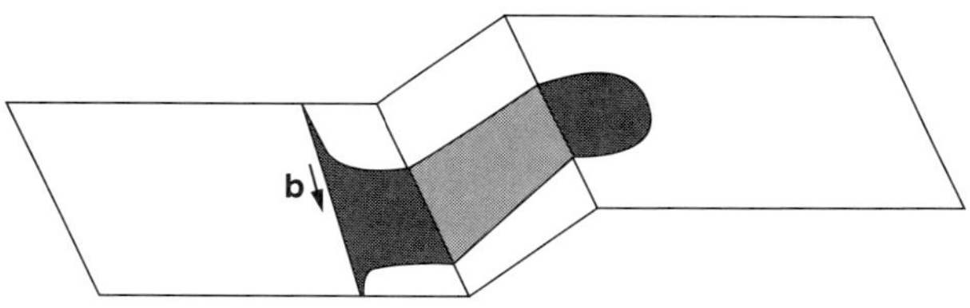

FIG. 4.22 Intersecting planes for a cross-slipping screw dislocation in an FCC metal with a Burgers vector of *a*/2 [110] slipping from a $(\bar{1}11)$ plane onto an intersecting $(1\bar{1}1)$ plane and back to another layer of $(\bar{1}11)$ plane.

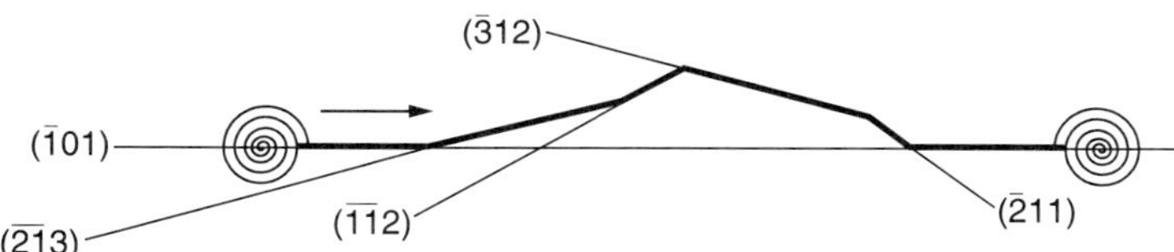

FIG. 4.23 Cross slip on multiple planes for an $\frac{a}{2}$[111] screw dislocation in BCC metal. The thick gray line shows the path of the cross-slipping screw dislocation.

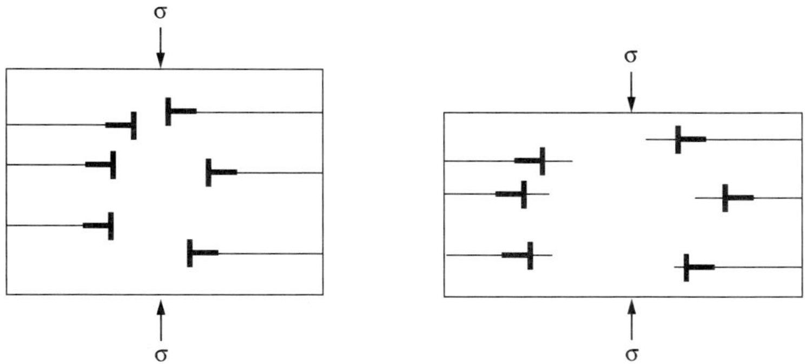

FIG. 4.24 Climb of dislocations from applied stress. The deformation of a material through climb results from normal stresses and is not volume conservative.

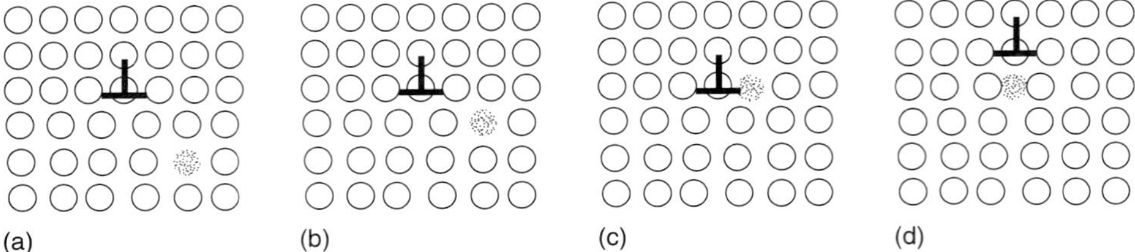

FIG. 4.25 The initial distance between the dislocation core and the vacancy decreases from (a) through (c). If the atom at the core in this layer changes places with the vacancy, the extra half-plane raises up one atom distance as shown in (d).

Motions of vacancies to produce climb in an individual plane are shown in Fig. 4.25. *Jogs* can form on both edge and screw dislocations. The jogs on screw dislocations are steps out of a favored glide plane that move by diffusive processes. Unlike slip through dislocation glide, the stresses that drive climb are not driven by shear but by tensile and compressive components of stress. For the edge dislocation shown in Fig. 4.7 stresses in the y-direction would produce climb. Similar to dislocation glide, climb takes place in segments called jogs shown in Fig. 4.26. Climb is a critical process for high-temperature deformation of crystalline materials. The phenomenon of power law or dislocation creep discussed in Chapter 6 consists of dislocation motion enabled by diffusion.

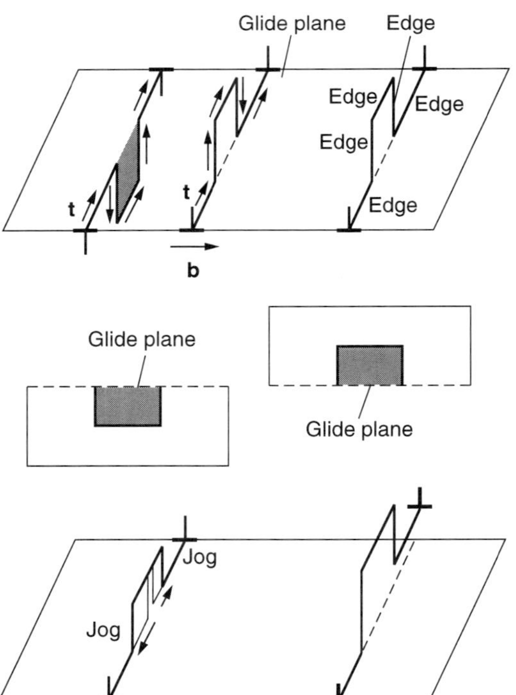

FIG. 4.26 Jog formation and motion in schematic dislocations. The top view shows the same dislocations originally identified in Fig. 4.18 undergoing glide by the formation of jogs. The middle view shows climb down (left) and climb up (right) from the glide plane. The shaded region indicates the material that must diffuse away to produce climb. The lower view shows an example of jog propagation and a dislocation that has already climbed partway out of the original glide plane.

4.4 DISLOCATION CONTENT IN CRYSTALS AND POLYCRYSTALS

The characteristics of dislocations and their mobility are directly related to the symmetry and stacking within a specific crystal structure. We are fortunate that the simplest crystal structures are also the ones that have the most technological applications. On the other hand, increasing interest in more complex dislocation motion is an important topic in development of new engineering materials.

4.4.1 Dislocation Formation

Dislocations can come about in a number of ways. Because dislocations are not equilibrium defects, their formation occurs through (a) errors in stacking during solidification or crystallization, (b) stresses from solidification shrinkage or thermal gradients, (c) mismatch at interfaces, (d) concentration gradients, and (e) coalescence of vacancies or interstitials to form prismatic dislocations. All of these processes can contribute to the dislocation density found in a polycrystalline material prior to any working processes.

4.4.2 Dislocation Multiplication

The multiplication of dislocations may seem unlikely considering that the increase in length of a dislocation is itself an increase in free energy taking the crystal away from equilibrium. During the application of shear stress, the lengthening of an existing dislocation is a response that lowers the free energy of a dislocation. When the stress is removed, the path for reversing the dislocation back to its original length is complex and not readily reversible.

The multiplication of dislocations is accomplished by a self-perpetuating process defined by Frank and Read. Figure 4.27 shows three different depictions of Frank–Read source operation. The left sequence shows an edge dislocation with two jogs and arrows indicating applied shear stress that would act on the horizontal segment to move it toward the front of the crystal. The two jog segments would not be moved by the shear stress shown. As the edge segment moves forward to form a loop, the area swept out by the dislocation loop is shown in gray. The middle sequence shows a more realistic portrayal with curved segments whereas the right sequence shows the dislocation without the mixed components. The edge dislocation, defined by the Burgers vector **b** and the line direction **t**, moves forward in the crystal, generating two screw dislocation segments of opposite sign. As the shear stress is increased, these two screw segments should move outward away from the defined pinning points. As these two screw segments move outward, they leave trailing edge segments on each end. These edge segments are opposite in sign and should move apart under the applied stress. The new segments, which are attached to the pinning points, should glide toward the rear of the crystal, leaving opposite-sign screw segments trailing back to the pinning points. These new screw segments will move together, each trailing behind it a new pinned edge dislocation. They then mutually annihilate to close the loop.

Other types of dislocation multiplication are possible, including the same process shown in Fig. 4.27 with a pinned screw dislocation. The multiplication of screw dislocations also enables the multiplication of sources by cross slip. When dislocations produced by a source on one plane cross slip onto another plane, edge segments are left on a different plane. The screw segment can reach a parallel plane to that of the original source to create multiple new sources, as shown in Fig. 4.28.

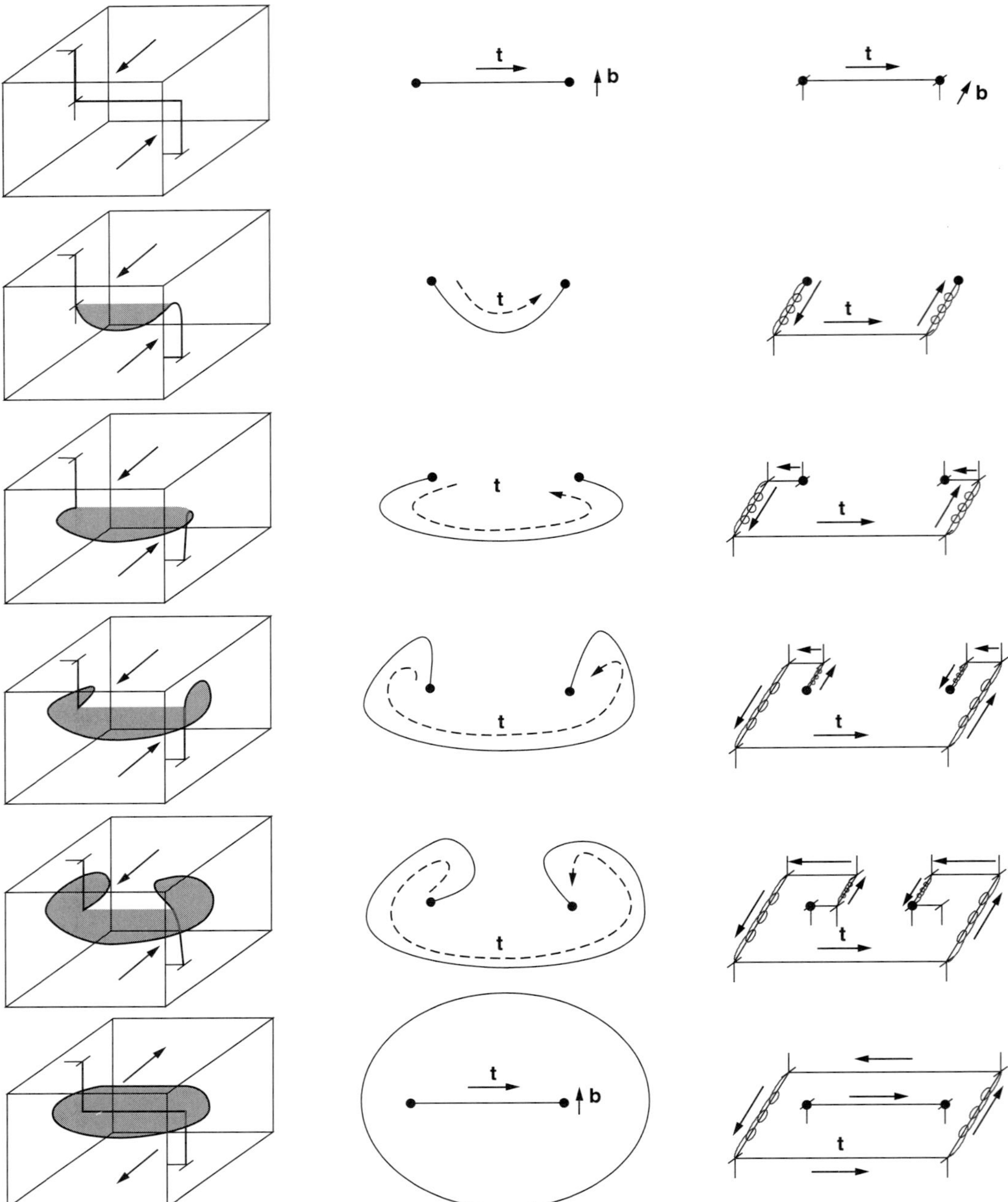

FIG. 4.27 Frank–Read source. The left set of images shows the entire dislocation in a three-dimensional crystal with the sense of simple shear defined by the arrows (the complementary shear stresses are not shown). The middle set of images shows the defined line direction and Burgers vector forming a realistic curved loop. The right set of images shows a perspective view with straight edge and screw dislocations defining the extra half-planes and sense of spiral for each type.

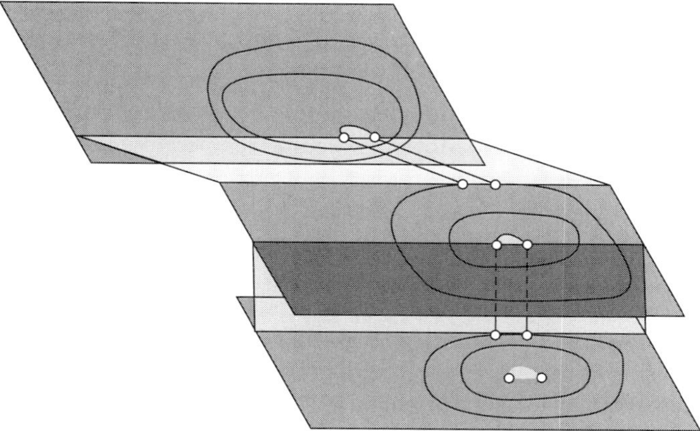

FIG. 4.28 Double cross slip multiplication of dislocation sources (after Low and Guard, 1959).

4.5 DISLOCATIONS AND DISLOCATION MOTION IN SPECIFIC CRYSTAL STRUCTURES

4.5.1 Face-Centered-Cubic (FCC) Metals

FCC metals have low yield stresses (relative to the magnitudes of the elastic constants) and high ductilities when compared to deformation of most any other polycrystalline material at low homologous temperatures (T < 0.5 T_m). The combination of close packing, which provides short Burgers vectors, and cubic symmetry, which provides abundant slip systems, enables this easy plasticity. Figure 4.29 shows schematic shear stress–shear strain behaviors for tensile testing of FCC metals. For these plots, the single crystal shear values are determined from the Schmid factors (see Eq. 3.25) to translate tensile stresses and strains into their shear components. The stage I behaviors, often called easy glide, show the slip behavior on a single slip system. As the rotation of the slip direction toward the tensile axis proceeds (discussed in Chapter 3) until the boundary of a stereographic triangle is reached, additional slip systems become active and the higher rate of strain hardening shown in stage II begins. The start of the stage II region is dependent on the initial crystal orientation. Stage I behavior depends on the angular distance from a stereographic triangle boundary. If the

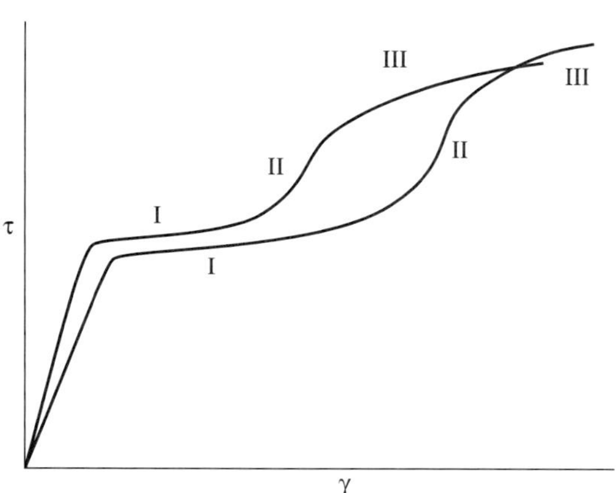

FIG. 4.29 Single crystal shear stress–shear strain curves for FCC metals with different initial crystal orientations.

initial crystal orientation is on a stereographic triangle boundary, no stage I behavior is observed. The multiple slip of polycrystals also shows no stage I behavior. Stage III behavior begins when the dislocations can cross slip or climb to overcome obstacles created by other dislocations slowing the strain hardening rate. For deformation of polycrystalline materials, the rate of strain hardening at low homologous temperatures is governed by the ability to cross slip.

The ABCABC stacking of {111} in FCC metals is shown in Fig. 4.30. Close-packed planes stacked in this arrangement differ from hexagonal-close-packed (HCP) materials, which have ABAB stacking. If we take a single layer from Fig. 4.30, we can show the close-packed <110> directions that lie in this plane in Fig. 4.31*a*. The $[1\bar{1}0]$ direction is shown, and the $[\bar{1}01]$ and $[0\bar{1}1]$ directions (and their opposites) also lie in this A plane. To first define the Burgers vector, consider that the length of the $[1\bar{1}0]$ vector drawn between two neighboring atoms is one-half the vector length for a <110> direction that forms a face diagonal on a unit cube. The convention employed in the field of dislocation theory is to designate this vector as $\frac{a}{2}[1\bar{1}0]$, where a represents the unit cell parameter. The $\frac{a}{2}[1\bar{1}0]$ vector drawn in Fig. 4.31*a* shows the motion when an atom is moved from one position in the B layer to the next. An atom moving along this path passes near a C position as it crosses over the A layer. In fact, for a hard ball model of atoms, a shuffle motion with a brief stay in the C position involves less motion of the atom normal to the plane. The two vectors connecting these B and C positions are both in <112> directions. Then, two vectors of the type $\frac{a}{6}$<112> can be summed to produce the net motion in the <110> direction. For Fig. 4.31, the vectors are $\frac{a}{6}[1\bar{2}1]$ and $\frac{a}{6}[2\bar{1}\bar{1}]$. For many FCC metals, the dislocation width is sufficiently broad that the $\frac{a}{2}$<110> dislocation can split into two partial dislocations separated by some distance, as shown in Fig. 4.31*b*. These two partial dislocations are called Shockley partial dislocations. They are not unit dislocations. They do not displace an atom from one perfect

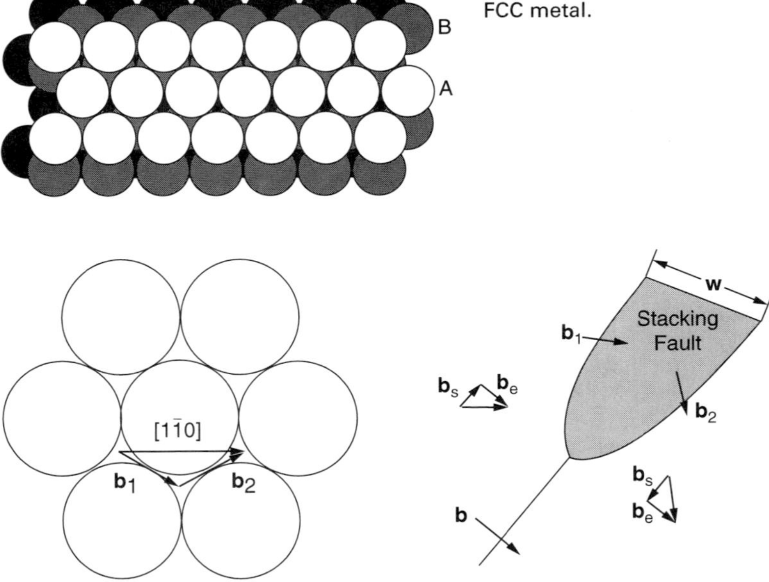

FIG. 4.30 Stacking of three {111} layers in an FCC metal.

FIG. 4.31 (a) Designation of two partial dislocations with Burgers vectors in <112> directions for **b** = $\frac{a}{2}[1\bar{1}0]$ lying in a (111) plane. (b) A stacking fault formed by the dissociation of a unit edge dislocation into two partial dislocations.

lattice site to another. Rather, the displacement results in an error in stacking called a stacking fault. The region of crystal between the two partial dislocations deviates from perfect repetition of the ABCABC stacking that occurs outside of the fault. These Shockley partial dislocations have elastic distortions surrounding them that are both repulsive and attractive in nature. The separations for a stacking fault, designated as the stacking fault width, are related to a quantity called the stacking fault energy (SFE). The SFE is the resistance *per unit length* for a material to form a wider stacking fault. When this resistance is balanced with the repulsive and attractive stress fields from the partial dislocations, the equilibrium width w_{eq} is related to the SFE as

$$\text{SFE} = \frac{\mu b^2}{4\pi w_{eq}} \tag{4.16}$$

where b is the magnitude of the Burgers vector for the unit $\frac{a}{2}$<110> dislocation. Equation 4.16 shows that the SFE is inversely proportional to w_{eq}.

The splitting of the unit dislocation into two separate components may seem unlikely if we consider that the energy of a dislocation was given as proportional to $\mu b^2 l$. Examination of this relationship for splitting into $\frac{a}{6}$<112> partials can be made by calculating the dependence on b^2 for each case as

$$\text{Unit dislocation} \rightarrow \frac{a}{2} < 110 >: b^2 = \frac{a^2}{4}\left(1^2 + 1^2 + 0^2\right) = \frac{a^2}{2}$$

$$\text{Two partials} \rightarrow 2\frac{a}{6} < 112 >: 2b_p^2 = 2\frac{a^2}{36}\left(1^2 + 1^2 + 2^2\right) = \frac{a^2}{3}$$

These relations provide an energetic basis for formation of the partial dislocations since the energy of the two partials is less than that of the unit dislocation. When $\frac{a}{2}$<110> edge dislocations in FCC metals separate into partials, the Burgers vectors for the two partials each retain half of the net edge displacements $\mathbf{b}_e$ in <110> directions and acquire small opposite-sign screw components $\mathbf{b}_s$, as shown in Fig. 4.31*b*. The same-sign edge components have mutually repulsive elastic stress fields, and the opposite-sign screw components have mutually attractive elastic stress fields.

Figure 4.32*a* shows a schematic hard ball model for stacking of (111) planes projected onto a $(11\overline{2})$ projection. In addition to the ABCA layers of (111) planes shown in projection, the ABAB stacking of $(1\overline{1}0)$ planes is also shown. The $[1\overline{1}0]$ vector shown is of length $\frac{a}{2}[1\overline{1}0]$ and shows that an atom-to-atom displacement of this type goes from one A layer to another. If the same planes are shown with a perfect dislocation inserted, the extra half-plane must consist of an AB layer, as shown in Fig. 4.32*b*. It is this AB layer of the {110} extra half-plane that separates for formation of partial dislocations and a stacking fault. Figure 4.33 shows the splitting of such an extra half-plane across a glide plane. Although it is not shown in this figure, the stacking of {111} planes between the partials is for one layer the AB stacking of close-packed planes that is expected in an HCP material. Outside of the partials and above and below this plane, the close-packed planes have the expected ABC stacking.

The splitting of edge dislocations in FCC metals does not have a strong influence on their glide mobility, although the elastic distortions at a stacking fault are believed to offer favorable locations for point defects and interstitial and substitutional solutes. Alloying usually results in a reduction in the SFE. As discussed in Chapter 5, these solutes are important in solid solution strengthening. Climb of edge dislocations is a bit more complex since the stacking fault must then conform to jogs. On the other hand, screw dislocation motion by cross slip is suppressed by wide stacking faults.

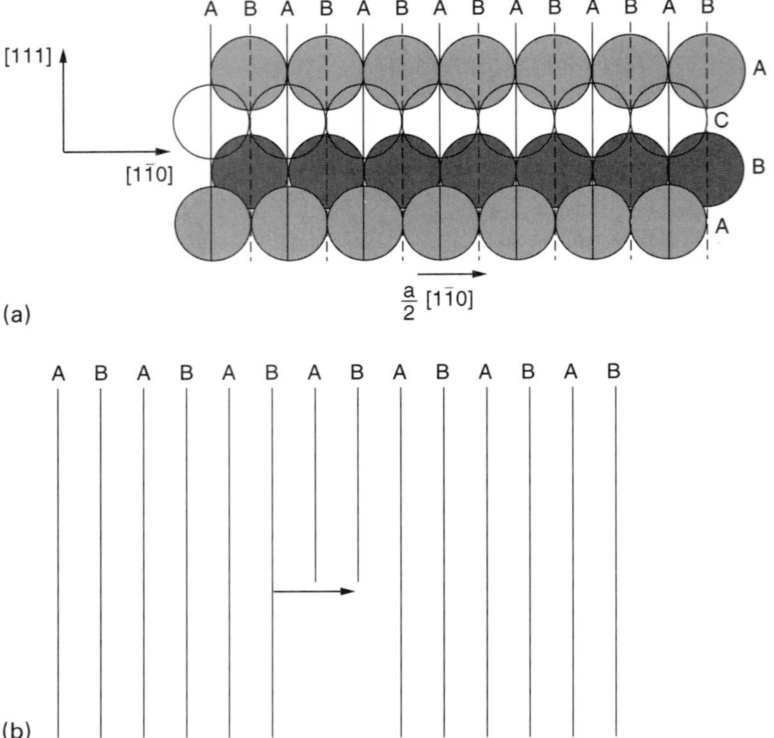

FIG. 4.32 (a) View of a (11$\bar{2}$) plane in projection, showing stacking of the corresponding (1$\bar{1}$0) and (111) planes. The stacking of the (1$\bar{1}$0) planes is ABAB, and the stacking of the (111) planes is ABCABC. (b) Section of the {110} planes from Fig. 4.31, showing that an extra half-plane for the easy glide <110> dislocations consists of two {110} layers.

The formation of partial dislocations from a unit screw dislocation results in repulsive screw components that sum to produce the unit dislocation and attractive, opposite-sign edge components. These edge components confine the dissociated screw dislocation to the plane defined by the edge components. Consequently, dissociated screw dislocations must be restored to their unit configuration before cross slip takes place. This means that shear stresses must bring the two partials together before cross slip can occur. Without cross slip, multiplication of dislocation sources is suppressed and overcoming obstacles to dislocation motion becomes very different. SFE can influence ductility, strain hardening, fatigue, and fracture processes in FCC metals. Table 4.2 gives SFE values at room temperature for a

TABLE 4.2 Approximate Stacking Fault Energies of FCC Metals and Alloys

Metal	SFE (mJ/m^2)
70–30 brass	<10
Stainless steel (austenitic)	<10
Ag	20–25
Au	50–75
Cu	80–90
Ni	130–200
Al	200–250

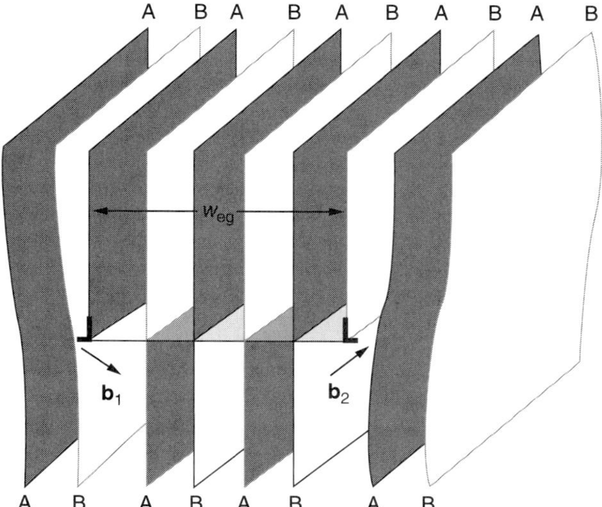

FIG. 4.33 An edge dislocation that has split into two Shockley partials separated by a stacking fault (after Seeger, 1957).

number of FCC metals and alloys. The low-SFE materials have wide stacking faults and inhibited cross slip. Slip in these materials is often coarse, with slip confined to a few planes. The high-SFE alloys toward the bottom of the list cross slip easily and demonstrate finely distributed slip.

EXAMPLE 4.4 *The Widths of Stacking Faults*

We can use Eq. 4.16 to calculate the expected equilibrium widths of stacking faults. We will make a comparison of Ag and Ni using the lowest values given in Table 4.2. We have the following data:

Ag $\quad \text{SFE} = 20 \ \text{mJ/m}^2 \quad \mu = 26 \ \text{GPa} \quad \text{b} = 2.9 \times 10^{-10} \ \text{m}$

$$w_{eq} = \frac{26 \times 10^9 \left(\text{N/m}^2\right)\left(2.9 \times 10^{-10} \ \text{m}\right)^2}{4\pi\left(20 \times 10^{-3} \ \text{Nm/m}^2\right)} = 8.7 \ \text{nm (about 30b)}$$

Ni $\quad \text{SFE} = 130 \ \text{mJ/m}^2 \quad \mu = 79 \ \text{GPa} \quad \text{b} = 2.5 \times 10^{-10} \ \text{m}$

$$w_{eq} = \frac{79 \times 10^9 \left(\text{N/m}^2\right)\left(2.5 \times 10^{-10} \ \text{m}\right)^2}{4\pi\left(130 \times 10^{-3} \ \text{Nm/m}^2\right)} = 3.0 \ \text{nm (about 12b)}$$

■

4.5.2 Body-Centered-Cubic (BCC) Metals

Slip in BCC metals is easiest in <111> slip directions and has been observed on planes including {110}, {112}, and {123}. The easy slip Burgers vector is $\frac{a}{2}$<111>, which is the closest spacing between atoms in the structure. In contrast with FCC metals wherein the extra half-plane represents two AB plane layers of {110}, the extra half-plane in BCC metals can involve three layers of {111}, as shown in Fig. 4.34. This does allow for some planar spreading of edge dislocations. The nature of the planar spreading differs depending on the plane of slip. The combination of the Burgers vector and the slip plane then specifies the line direction **t** for edge dislocations lying on different planes. The wide edge dislocation cores for BCC metals enable easy motion of edge dislocations in these materials.

In contrast, $\frac{a}{2}$<111> screw dislocations have cores spread over the threefold symmetry inherent about <111> directions. The spiral of the screw dislocation shows relaxations

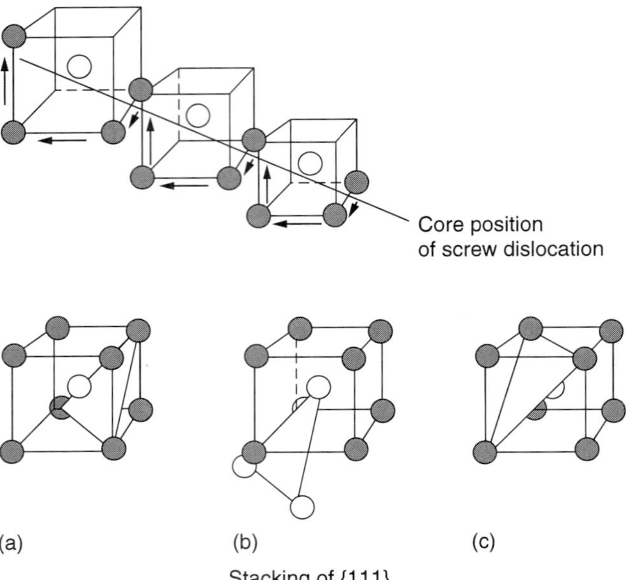

FIG. 4.34 The core position for the easy glide dislocation and the stacking of {111} planes for a BCC metal.

of the dislocation core onto {110} or {112} planes. Rather than a continuous spiral, the displacement of atoms along the spiral is higher at some positions along the Burgers circuit around the screw dislocations. Figure 4.35 shows the arrangement of atoms in a perfect BCC crystal. The layers are labeled 0, 1, and 2 to designate the difference in position along <111> where each layer is separated by $\frac{a}{6}$<111>. Figure 4.35*a* and *b* show that the structure of BCC metals along <111> directions consists of alternating spirals before a screw dislocation is introduced. As shown in Fig. 4.35*b*, superimposing a same-sign spiral for a screw dislocation onto the BCC structure produces a possible arrangement of atoms with the local arrangement changed into an opposite-sense spiral. Superimposing an opposite-sign spiral would produce an impossible arrangement. The shaded regions about the dislocation shown in Fig. 4.36 show the regions of high displacement. This results in a fairly narrow core for the screw dislocation and also affects dislocation motion, multiplication, and cross slip. The difficult motion of screw dislocations relative to edge dislocations in

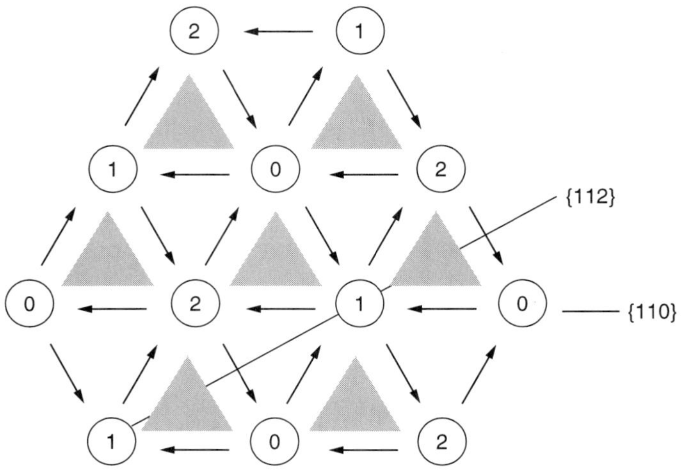

FIG. 4.35 (a) Three layers of {111} planes in a BCC metal. A ternary numbering system is used to designate the positions of spirals that can be found in the perfect crystal structure. The unshaded triangles are counterclockwise spirals coming forward on the page, and the shaded triangles are clockwise spirals on the page (see Duesbery, 1984).

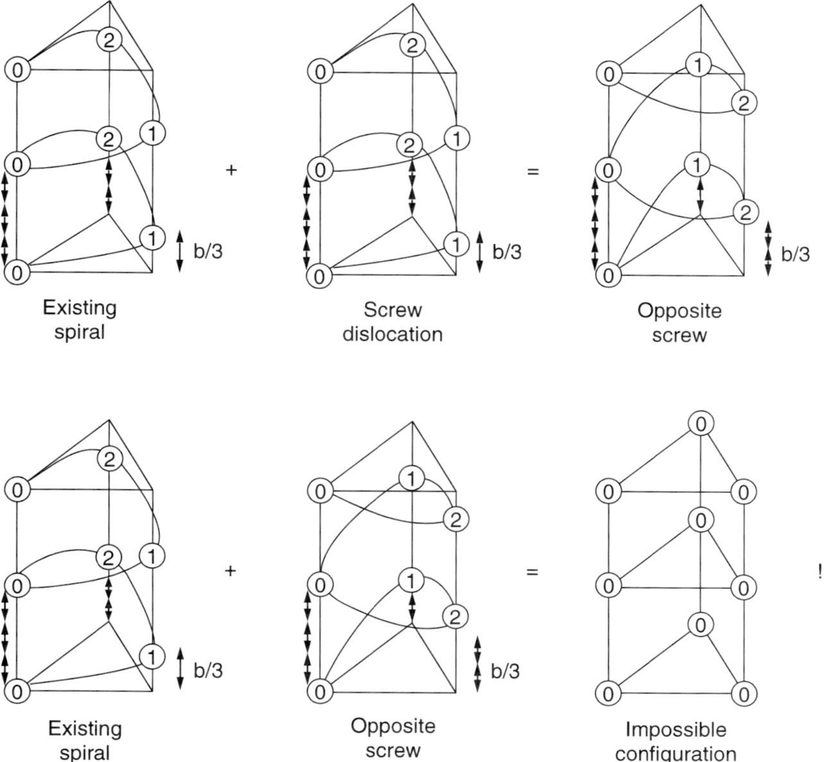

FIG. 4.35 (b) The top set of three-dimensional representations shows that the existing counterclockwise spiral (assuming it is infinite) added to a counterclockwise screw dislocation results in a reversal of the spiral character along the dislocation core. The bottom set shows that when a clockwise spiral or dislocation is added to the counterclockwise site in the perfect crystal, the atoms are placed in the same plane, which is not possible without very large internal displacements (remember that the atoms are much larger than shown and touch along <111>).

BCC metals at low temperatures, $\leq 0.15\ T_m$, has been observed on almost any plane that can contain <111> directions and is often connected to the high increase in yield strength for deformation of BCC metals at low temperatures. For screw dislocation motion on {112} planes, the crystallography of BCC metals also leads to asymmetric slip. Asymmetric slip is defined as slip that requires a different critical shear stress for slip in one direction versus another. Thus, motion of the screw dislocation shown in Fig. 4.36 onto $(\bar{1}\bar{1}2)$ planes by cross slip is easy moving to the right, but difficult moving to the left. This also can be seen in Fig. 4.35. When dislocations move on {112}, the rearrangements of dislocations are different for moving to the left versus moving to the right.

Figure 4.37 shows a stereogram with angular positions within the standard unit triangle defined by the angles χ and ξ. If the Schmid factor is calculated for χ near 30°, the favored slip system is $\frac{a}{2}[111](\bar{2}11)$. At χ values near 0°, the favored slip is on $(\bar{1}01)$ planes. At χ near negative 30°, slip on $(\bar{1}\bar{1}2)$ is expected. The asymmetric slip that is possible for slip on {112} planes is manifested by the unusual phenomenon of yield stresses that differ for tension and compression. For slip on $(\bar{2}11)$ planes, tensile deformation is more difficult than compressive deformation, and for slip on $(\bar{1}\bar{1}2)$ planes, the reverse is true. Figure 4.37*b* shows asymmetric slip in tension and compression.

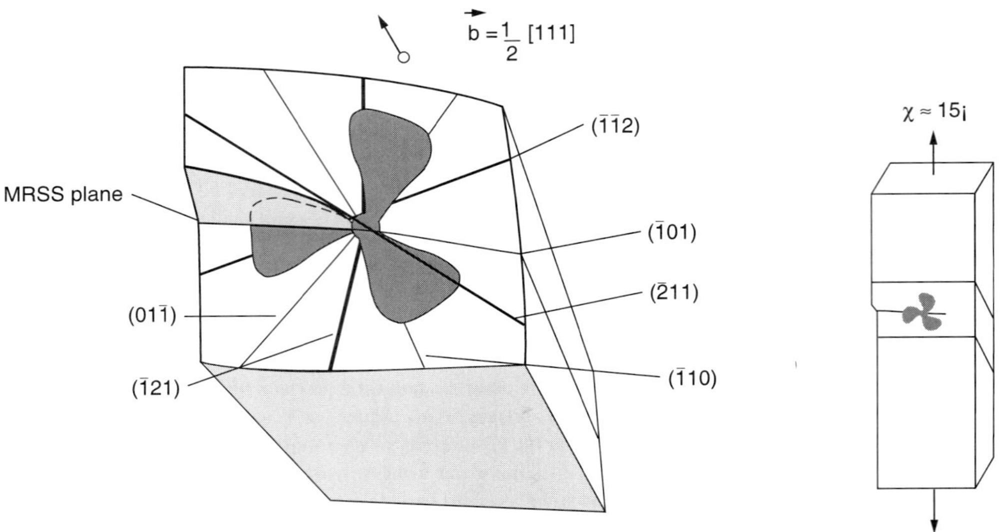

FIG. 4.36 Glide of a screw dislocation in a BCC metal from left to right. The resolved shear stress drives the counterclockwise dislocations, which have the greatest forward displacements in the dark shaded regions, along a path that would probably consist of cross slip on the (1̄01) and (1̄1̄2) planes (see Vitek, 1985).

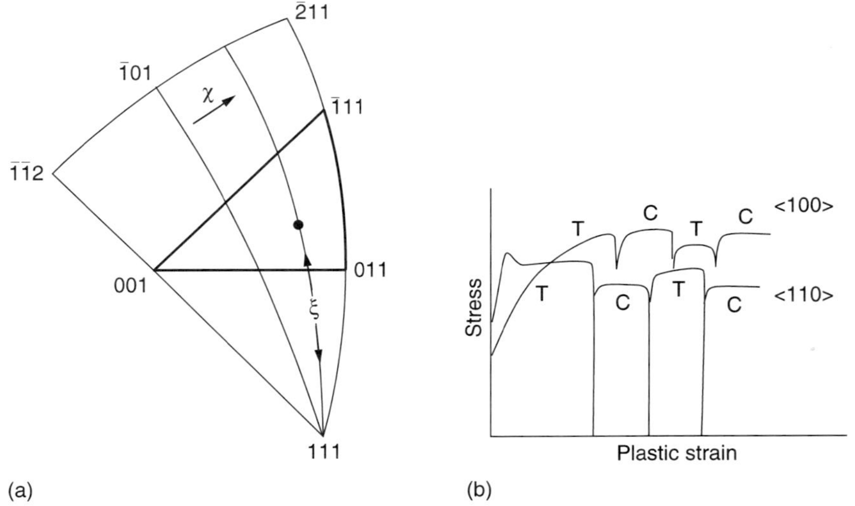

FIG. 4.37 (a) Stereogram showing [111] slip direction and corresponding {110} and {112}. (b) Argon and Maloof showed that slip in tension and compression could be asymmetric such that tension is greater than compression, or vice versa (Argon and Maloof, 1966, reprinted with permission).

4.5.3 Slip in Other Materials

Dislocations in low-symmetry crystals and compounds are even more complex than those we have already discussed. This complexity also results in significant dislocation motion being strongly rate-dependent and occurring mostly at high fractions of the melting temperature. For that reason, more significant discussion will occur in Chapter 6, where the

BIOGRAPHY

JOHN "JACK" CHRISTIAN (1926–2001)

Christian led efforts in the 1960s and 1970s to understand the special nature of the lattice resistance in BCC metals and its relation to dislocation mobilities. He also is well known for *The Theory of Transformations in Metals and Alloys*, the authoritative work on phase transformations since its first publication in 1965.

focus is on high-temperature deformation mechanisms. Lower symmetries result in a large number of possible dislocation types and often an insufficient number of slip systems for ductile deformation of polycrystals. In ordered compounds, including intermetallics, ionic solids, and ceramics, maintaining the ordered structure often results in very large unit Burgers vectors. The Burgers vector becomes large because the unit dislocation must have a Burgers vector that repeats the structure for all atoms or molecules in the compound. Many brittle compounds possess very high Peierls–Nabarro stresses, which inhibit the motion and multiplication of dislocations. If dislocation motion does not maintain the perfect stacking of atoms, defects called antiphase boundaries are formed, which involve pairs of dislocations, as shown in Fig. 4.38. Consequently, the resistance to dislocation motion can be very high.

In polymer crystals, slip by dislocation motion is limited because of low symmetry of crystal structures despite the weak bonding between planes in many materials. In addition, the integrity of long-chain molecules restricts the amount of slip possible. Crystal orientation and orientation effects on strengthening are essential to produce many high-performance plastics. This is discussed in the section on orientation strengthening in Chapter 5.

The dislocation pairs bound a region called an antiphase domain (see Vitek, 1985). The antiphase domain is created when one dislocation interrupts the periodic structure of the compound, changing the local nearest-neighbor relationships. The second dislocation in the pair returns the compound to the arrangement in the bulk crystal. Because these dislocations must travel in pairs, they have very restricted possibilities for climb and cross slip. This compound is often called γ' and serves as the principal reinforcement phase in nickel-base superalloys.

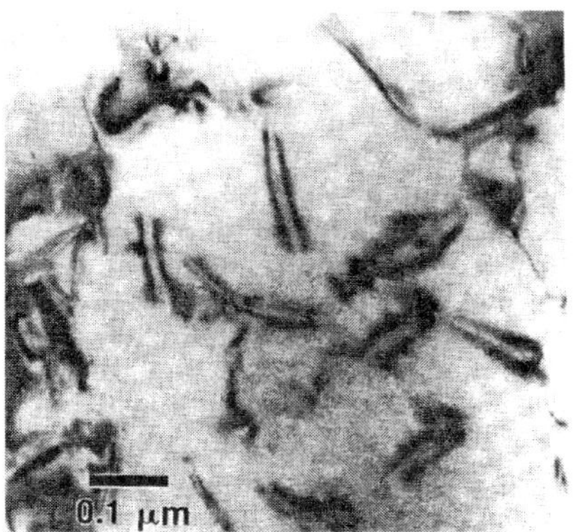

FIG. 4.38 TEM micrograph showing dislocation pairs in Ni_3Al. These pairs of dislocations are called superdislocations and they define the boundaries of a region in the ordered crystal that is stacked out of phase (from Hertzberg, 1996, micrograph by M. Khobaib).

4.6 REFERENCES

Dislocations

A. H. COTTRELL, *Dislocations and Plastic Flow in Crystals*, Oxford, 1953.

Dislocations and Properties of Real Materials, The Institute of Metals, 30–50, 1985.

J. FRIEDEL, *Les Dislocations*, Gauthier-Villars, 1956.

J. J. GILMAN AND W. G. JOHNSTON, in *Dislocations and Mechanical Properties of Crystals*, edited by J. C. Fisher, W. G. Johnston, R. Thomson, and T. Vreeland, Jr., Wiley, New York, 116–161, 1957.

P. HAASEN, *Physical Metallurgy*, 2nd Ed., Cambridge, 1987.

R. W. HERTZBERG, *Deformation and Fracture Mechanics of Engineering Materials*, 4th Ed., Wiley, 1996.

J. P. HIRTH AND J. LOTHE, *Theory of Dislocations*, 2nd Ed., Wiley, 1982.

D. HULL AND D. J. BACON, *Introduction to Dislocations*, 3rd Ed., Pergamon Press, 1984.

C. B. JIANG ET AL., *Intermetallics*, **9**, 355–360 (2001).

W. G. JOHNSTON, *J. Appl. Phys.*, **33**, 2716, 1962.

U. F. KOCKS, A. S. ARGON, AND M. F. ASHBY, *Thermodynamics and Kinetics of Slip, Progress in Materials Science*, Vol. 19, Pergamon Press, 1975.

R. LOW AND R. W. GUARD, *Acta Met.*, **7**, 171 (1959).

F. R. N. NABARRO, *The Theory of Crystal Dislocations*, Oxford University Press, 1967.

E. OROWAN, in *The Sorby Centennial Symposium on the History of Metallurgy*, edited by C. S. Smith, Gordon and Breach, 359–376, 1965.

A. SEEGER, in *Dislocations and Mechanical Properties of Crystals*, edited by J. C. Fisher, W. G. Johnston, R. Thomson, and T. Vreeland, Jr., Wiley, New York, 243–329, 1957.

H. SIETHOFF, *Philosophical Magazine*, **82**, 1299–1316, 2002.

J. WEERTMAN AND J. R. WEERTMAN, *Elementary Dislocation Theory*, Macmillan, 1964.

BCC Metals and other Materials

A. S. ARGON AND S. R. MALOOF, "Plastic deformation of tungsten single crystals at low temperatures," *Acta Metallurgica,* **14**, 1449–1462, 1966.

J. W. CHRISTIAN, *Metall. Trans.,* **A14**, 1233, 1983.

M. DUESBERY ET AL., *Proc. Roy. Soc.,* **A332**, 85–111, 1973.

M. DUESBERY, *Proc. Roy. Soc.,* **A392**, I 145-173, II 175-197, 1984.

V. VITEK, in *Dislocations and Properties of Real Materials,* The Institute of Metals, 30–50, 1985.

4.7 PROBLEMS

A.4.1 Distinguish between a kink and a jog on an edge dislocation (a sketch may help).

A.4.2 Give a first-order estimate of the dislocation density for the crystal shown in Fig. 4.4 in number per square meter.

A.4.3 Describe three examples of dislocations observed in daily life. For each case, provide a sketch and define a Burgers vector. Does each example have a line of dislocation?

A.4.4 Explain the shape of the dislocation etch pits in Fig. 4.10*a* and *b*.

A.4.5 Consider shear stresses applied to the dislocation loop that (a) point to the right above, and to the left below, the loop plane and (b) point up on the left side of the loop and down on the right side of the loop.

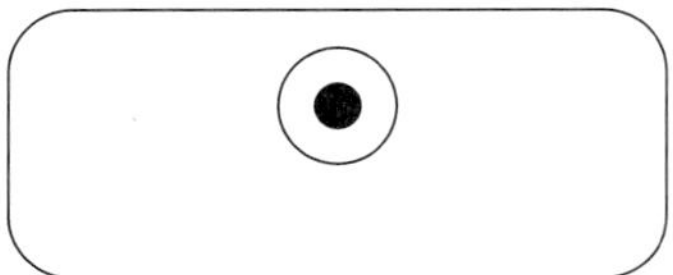

A.4.6 The Burgers vector in the figure below is normal to the plane containing the dislocation loop. Identify with an E or S the character of the dislocation around the loop on all sides. Show the line direction.

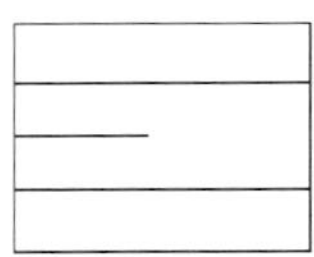

A.4.7 For the dislocation shown in the figure above, show what forces should be applied to the crystal to move the dislocation upward.

B.4.1 Give a numerical estimate with correct units for the energy of a 1-meter-long edge dislocation in MgO.

B.4.2 Single crystals of two different FCC alloys were tested in tension to the same strain. If both specimens were originally the same orientation, explain why one might have slip steps much larger than the other but with a larger spacing between the larger slip steps.

B.4.3 **(a)** A straight screw dislocation in a BCC metal appears to be 2 μm in length in a TEM micrograph. The direction of the electron beam (normal to the TEM micrograph) was a <113>. What was the foil thickness?

(b) Is there only one possible thickness?

B.4.4 Two Frank–Read sources are operating on the same plane. If a shear stress is applied to operate these two sources, describe the circumstances under which the loops meeting between the sources would create a larger loop. Assume that cross slip is very difficult. Use detailed sketches. Does it matter if the pinned dislocation segments acting as Frank–Read sources are of the same sign?

B.4.5 (a) An FCC crystal has been deformed by shock loading, resulting in formation of high-energy unit dislocations. Rank order the likely unit dislocations in FCC metals.

$\frac{a}{2}$<100>	a<100>
$\frac{a}{2}$<111>	$\frac{a}{2}$<112>
a<112>	a<111>

(b) Consider a hypothetical $a[100]$ unit edge dislocation in an FCC metal lying on a (010). What is the line direction?

(c) Write a simple partial dislocation reaction for the dislocation in (b), giving a specific example.

(d) Is the partial dislocation dissociation in (c) energetically favorable?

B.4.6 Estimate the force (N) between two parallel straight edge dislocations of opposite sign each 1 m in length at 20b away from one another. Explain your answer. (Use μ = 100 GPa and b = 2.5 Å.)

B.4.7 If it is assumed that the Burgers vector and elastic properties are about the same for 70–30 brass and copper, which is more likely to undergo cross slip? Justify your answer with a calculation.

C.4.1 (a) Define a $[\bar{1}34]$ crystal with square cross section wherein the front face contains the easy slip direction and the side face is also identified. (b) Find the angle made by the easy slip system on both faces. (c) Find the line direction of edge dislocations for this slip system. (d) Give the line direction of edge dislocations in the second easiest slip system(s). (e) Can screw dislocations of the first slip system cross slip onto planes of the second?

C.4.2 (a) A BCC crystal contains a dislocation lying at 30° from the ideal line direction for $\frac{a}{2}$<111> dislocations that lie on a {112} plane. Calculate the energy of this mixed dislocation assuming that $R/r_o \approx 1000$.

(b) Why might this dislocation lie along this direction?

C.4.3 Describe the intersection of a $\frac{a}{2}[111]$ dislocation gliding on a $(\bar{2}11)$ plane with a stationary $\frac{a}{2}[\bar{1}11]$ gliding on a $(1\bar{1}2)$ plane in a BCC metal.

C.4.4 A prismatic dislocation with Burgers vector $a[100]$ is present in an NiAl crystal (ordered intermetallic with cubic structure having Ni atoms on corners and Al atoms at the body-centered position). The line segments preferentially lie along <110>. Sketch the loop and give the planes on which these segments can glide. Show how an applied stress could lead to operation of two Frank–Read sources.

C.4.5 A single crystal of aluminum with a $[\bar{1}45]$ compression direction is deformed such that the average dislocation velocity is 1×10^{-7} ms^{-1}. The dislocation density is 10^8 m^{-2}. Estimate the compressive strain rate that would produce such a strain rate. Also give the strengths and weaknesses of any assumptions you have made.

C.4.6 Express Eq. 4.3a–c in terms of stiffness C_{ij}.

C.4.7 Express Eq. 4.3a–c in terms of compliances S_{ij}.

C.4.8 Consider the Orowan equation (Eq. 4.11). If $\rho_\perp$ is considered to increase linearly during straining and the resistance to slip is parabolic with strain, does $\bar{x}$ change?

C.4.9 Sketch a process for climb-based dislocation multiplication for compression.

CHAPTER 5

STRENGTHENING MECHANISMS

NEARLY ALL strengthening mechanisms in materials can be placed into three categories. The first category is based on elastic effects suppressing deformation over some distance such that, despite high stress levels, the deviatoric or von Mises effective stress driving shear deformation is not high enough to cause deformation. The apparent "strengthening" is not an increase in the resistance to deformation, but rather a lack of driving force for shear deformation. This apparent strengthening is found in thin films, composites, and materials subjected to strain gradients (e.g., rolling or hardness testing). In the second category, the particular mobile defect that drives a deformation mechanism is slowed by either operating over a more limited or a more circuitous path or is not available because of ineffective self-perpetuation. The third category is related to crystal or molecule orientation wherein the resolved stresses driving deformation are not aligned with the easiest orientation for operation of the available deformation mechanism. Unlike the concentration on crystalline materials in Chapter 4, this chapter provides examples for both crystalline and noncrystalline materials.

5.1 "CONSTRAINT"-BASED STRENGTHENING

Constraint-induced effects are possibly the most misunderstood aspects of mechanical behavior. Constraint effects play a role in plasticity of thin films attached to substrates and metal or polymer matrices in composites, grain size effects in ductile metals, frictional effects in forging, and many other deformation conditions. In composites, constraint fulfills the complementary purpose to that of load transfer in elastic loading. Constraint can be so great in some cases—for example, in the case of epitaxial films grown on mismatched substrates—that elastic strains of several percent can be supported without plastic deformation (another factor can be very low densities of mobile dislocations).

EXAMPLE 5.1 *Constraint in a Rod with a Cylindrical Shell Covering*

Consider the problem shown in Fig. 5.1—a soft rod of material bonded to a shell of a much harder material. When will yielding take place if we assume that both materials have the same elastic properties and we ignore end and friction effects while loading the specimen in compression? It is clear that the softer inner material cannot begin to deform differently from the hard outer shell without some significant problems at the interface, but let us consider the processes of deformation applied to this case.

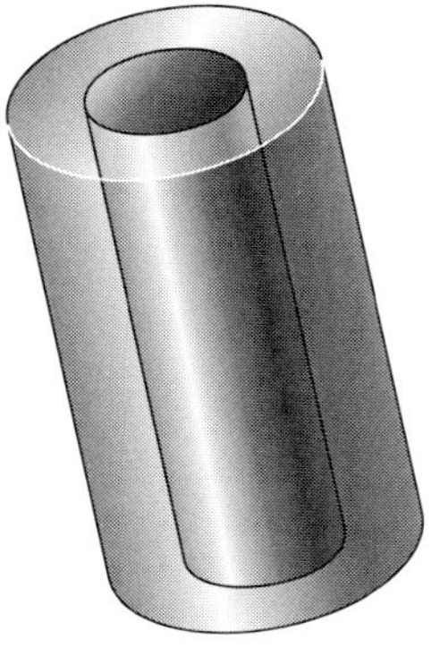

FIG. 5.1 A rod and shell combination wherein the rod is made from a soft material and the shell is made from a harder material.

Since both materials have the same elastic constants, we should be able to evaluate the general case by assigning the following properties:

	Rod	Shell
Young's modulus	100 GPa	100 GPa
Poisson's ratio	0.3	0.3
Yield strength	50 MPa	500 MPa
Radius	r	$2r$

If we take a simple rule of mixtures approach, the deformation stress should be, at most,

$$0.25(50 \text{ MPa}) + 0.75(500 \text{ MPa}) = 387.5 \text{ MPa}$$

which is only about 3 percent greater than if the center core were empty.[1] The key property in the list is Poisson's ratio. If the deformation process within the soft rod occurs only through shear, then the transverse strains are half the magnitude of the axial strain. However, for the soft material of the rod to undergo such a deformation, it would require deformation that would push against the outer shell. A similar difference could occur if the hard and soft materials were made of the same material, but the material had anisotropic plastic properties. Aspects of this idealized situation occur in many composites and two-phase materials whether the deformation is by slip or by some other mechanism. The key element is an abrupt change or a steep gradient in the stresses from one location to another. The effect of specimen thickness on the resistance to fracture discussed in Chapter 7 and the plane strain deformation during rolling discussed in Chapter 10 are related examples of constraint. ■

For most composites of materials, differences in yield strength and elastic constants are expected. As discussed in Chapter 2, different rules of mixtures can be applied depending on the composite and the loading direction. The difference in elastic constants leads to stresses across and near the interface if the strains are totally elastic. In this case, deformation can occur near the interface despite the fact that the macroscopic tensor would not result in deformation of the material if the material were homogeneous. The tensile properties of a composite consist of a transition from initially elastic behavior in both phases to plastic deformation in the phase with a lower yield stress (see Fig. 5.2). For entirely elastic deformation in a composite with fibers aligned in the tensile direction, we have already shown in Chapter 2 that the composite modulus is

$$E_{\text{composite}} = E_{\text{fiber}} V_{\text{fiber}} + E_{\text{matrix}} V_{\text{matrix}} \tag{5.1}$$

[1]Consider a rule of mixtures with $r^2 h Y_{\text{soft}} + [(2r)^2 - r^2]\, h Y_{\text{hard}} = (2\text{r})^2 h Y_{\text{composite}}$

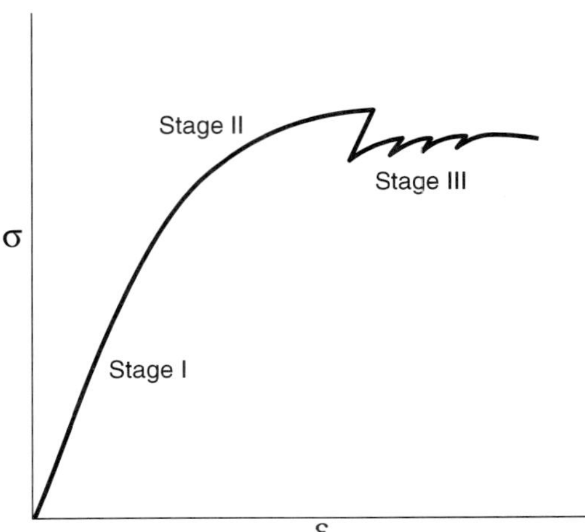

FIG. 5.2 Tensile deformation for an aligned fiber composite showing rules of mixtures behavior in an elastic region, stage I, before the start of plastic deformation of the matrix in stage II. Stage III shows the unloading effects expected once fiber fracture starts.

with V_{fiber} and V_{matrix} representing the volume fractions of fiber and matrix where $V_{\text{fiber}} + V_{\text{matrix}} = 1$. The elastic portion of the tensile behavior shown in Fig. 5.2 is given by $\sigma = E_{\text{composite}}\varepsilon_{\text{composite}}$.

Once plastic deformation begins in the matrix phase, the tensile deformation designated as stage II in Fig. 5.2 might be modeled by

$$\sigma = E_{\text{fiber}}\varepsilon_{\text{comp}}V_{\text{fiber}} + \sigma_{\text{matrix}}V_{\text{matrix}} \tag{5.2}$$

where σ_{matrix} is the resistance to plastic deformation of the matrix. This flow stress of the matrix, $\sigma_{\text{matrix}} = f(\varepsilon_{\text{matrix}})$, is often obtained from a uniaxial tensile test of the matrix material without composite reinforcements. Such a simple model for the matrix contribution, of course, ignores the constraint imposed by the fibers on matrix deformation. The presence of the fiber will suppress deformation of matrix material adjacent to the fibers. It also ignores the possibility for the properties of the matrix material to be altered by processing history differences between a matrix with and without reinforcements or for residual internal stresses to develop as a result of thermal expansion differences. In addition to mismatch between the tensile deformation of the fiber and matrix, the Poisson strains of the elastic fibers will not match the transverse matrix strains, which should be

$$\varepsilon_{\text{transverse}} = -\frac{\varepsilon_{\text{axial}}}{2}$$

if matrix deformation occurs entirely by shear.

Constraint effects are relevant to deformation of polycrystals. The necessity for multiple slip in polycrystals comes from the effects of neighboring grains, as discussed later in Section 5.2.5. In single- and multiple-phase materials, some degree of local deformation may occur at phase boundaries and grain boundaries. For grain boundaries in a single phase material, this can occur in part as a result of elastic anisotropy. Since the local strains are most likely fairly complex, in crystalline materials slip near the interface occurs on multiple slip systems, with the contribution of the slip systems depending on the local stress state and the orientation of the crystal. Two-phase materials wherein each phase can have a similar grain size also show constraint-based strengthening. Dual-phase steels rely on the presence of the harder martensite phase to inhibit deformation of the softer ferrite (or BCC iron) phase.

Frictional effects in mechanical forming can also provide apparent strengthening by constraining deformation. For this reason, direct comparison of tensile tests using reduced

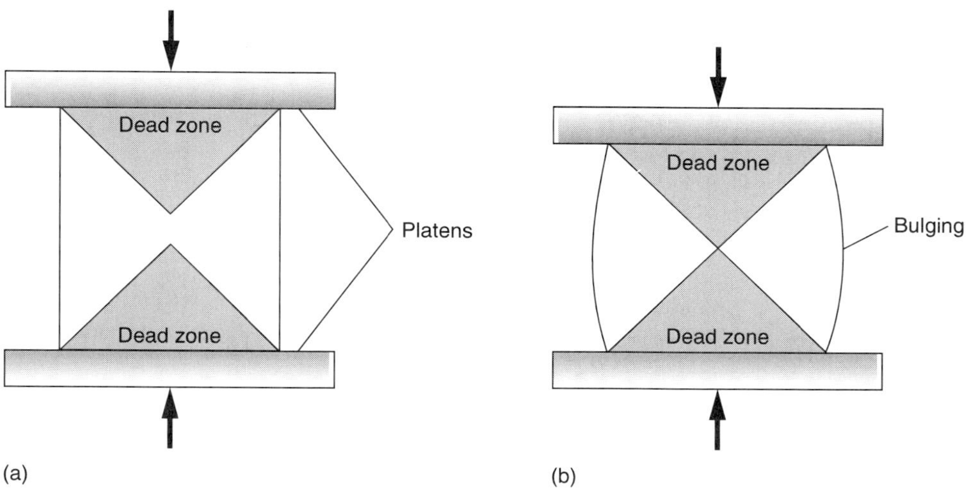

FIG. 5.3 (a) Cross section of compression of a simple right cylinder showing nondeforming dead zones due to friction at the platens. (b) Cross section of the same cylinder after some compression, showing bulging.

gage sections to compression tests can be quite challenging. If frictional coefficients are known, inhibition of deformation by frictional stresses can be modeled using approaches shown in Chapter 10. Friction coefficients usually vary from beginning to end of the deformation. In the ideal case, plastic deformation occurs throughout a part by uniform shear deformation. As shown in Fig. 5.3, friction can result in "dead zones" near the specimen surfaces in contact with the platens. The height-to-diameter ratio of the sample then controls the volume of material deforming to produce a particular magnitude of strain. If the dead zone is a large fraction of the sample volume, the stresses to produce plastic deformation will be exaggerated.

EXAMPLE 5.2 *Constraint During Compression Testing*

The Watts and Ford method (1955) provides an empirical approach for finding a true stress–true strain curve for compression without the influence of end effects. The data obtained by this method also allow for measurement of stress-strain behavior to very large strains without the influence of necking that would occur in a tensile test. Cylinders of equal diameter and different heights are deformed to identical loads. Then, as shown in Fig. 5.4*a*, the reduction in height can be extrapolated to a diameter-to-height ratio, d/h, with infinite height ($d/h = 0$). Figure 5.4*b* shows that a reduction in friction through lubrication dramatically affects the strain for a given load. This is not surprising, since the coefficient of friction can be reduced by orders of magnitude by lubrication (see Chapter 9). Problem B.5.6 demonstrates the "strengthening effect" caused by frictional constraint. ■

The localized stresses of a hardness test are another example wherein constraint can apparently strengthen a material. Figure 5.5 shows a simple model for indentation of a material with a flat indenter. Flow paths of the material in each triangular element are shown by the arrows. If we first consider the deformation as a two-dimensional flow problem, the flow process of the material under the indenter can be separated into the flow regions shown in Fig. 5.5. A dead zone exists below the indenter where the hydrostatic stresses are high. Regions 1a and 1b consist of deformation 30 degrees from the direction of compression. Regions 2a and 2b consist of deformation rotated 60 degrees until parallel

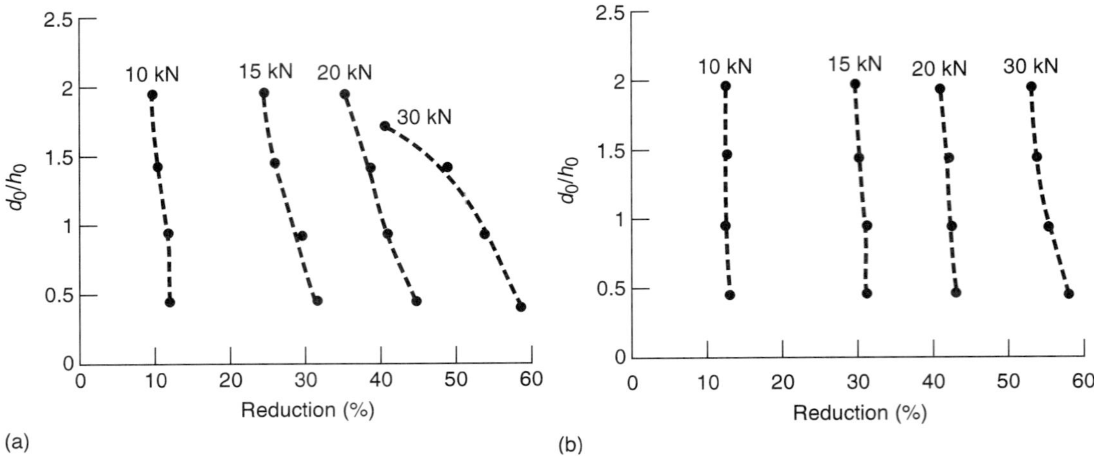

FIG. 5.4 Watts and Ford method (1955) showing constraint effects on testing of aluminum cylinders 7.9 mm in diameter of various heights. The data in (a) show the results for unlubricated testing and (b) shows the results for lubricated tests (Spoo and Bowman, unpublished data). The unlubricated results in (a) show a greater variation in the percent reduction for loading to the force level. The lubricated results in (b) show that with less constraint the diameter-to-height ratio (d_o/h_o) has less effect on the actual reduction.

to the surface of the indented material. Regions 3a and 3b are further rotated 60 degrees so that some uplift of the specimen surface will occur outside the indenter. The abrupt shear deformation between regions 1 and 2, between regions 2 and 3, and between these regions and the nondeforming material outside these regions is much more complex than simple compression and requires a higher stress to get deformation. Using calculations described further in Chapter 10, we can show that the yield stress for the deformation shown in Fig. 5.5 is nearly three times higher than the yield stress for simple frictionless compression. The highly stressed material below the indenter is constrained from deforming by the adjacent material that is under lower stresses.

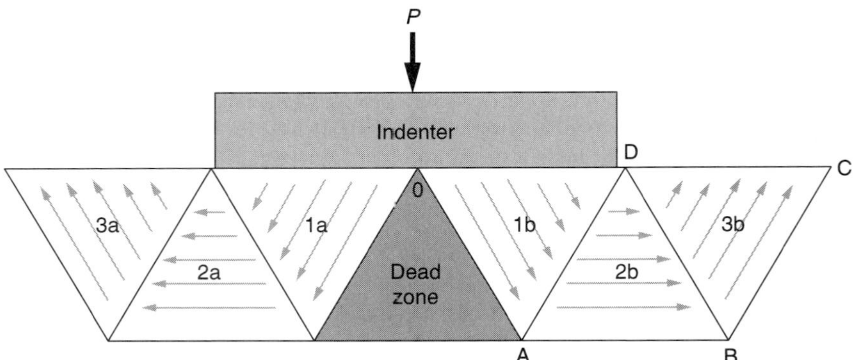

FIG. 5.5 Two-dimensional model for indentation of a plastic material showing flow of material modeled as equilateral triangles. A dead zone in the high hydrostatic region beneath the indenter is also shown. For such a model, assume that the deformation takes place abruptly at any point in which the direction of deformation changes. This model predicts that the stress P required for plastic deformation will be nearly three times the uniaxial yield strength.

5.2 STRENGTHENING MECHANISMS IN CRYSTALLINE MATERIALS

Control of dislocation motion is the primary method for changing the yield strengths of crystalline materials. As described in the next section, how dislocations overcome obstacles to motion controls the effectiveness of different strengthening mechanisms. Each of the strengthening mechanisms described below is expressed in terms of either shear or axial stresses. For strengthening mechanisms that occur within the crystals, the strengthening is given in terms of shear stress. To convert this to an increase in tensile yield strength, the appropriate geometric considerations (defined in Chapter 3) must be made. For strengthening in polycrystals—e.g., grain refinement—the strengthening is given in terms of a uniaxial applied stress.

5.2.1 Obstacle Strengthening

Obstacle strengthening is the effect that impediments to dislocation motion introduce by interacting with the strain field of the dislocation. The critical aspects are that the length of the dislocation directly corresponds to the magnitude of strain energy associated with the dislocation and that dislocations move under the influence of an applied shear stress. Although there is a difference in the energy of a dislocation that is of edge, screw, or mixed character, the difference is small compared to the affects of changing the length of the dislocation. The elastic energy *per unit length* introduced into the materials by the presence of a dislocation was defined in Chapter 4 as

$$E_{\perp} = \frac{\mu b^2}{4\pi} \ln \frac{R}{r_o} \tag{5.3}$$

where μ is the shear modulus, b is the magnitude of the Burgers vector, R is the radius around the dislocation that the strain energy influences, and r_o is the core radius of the dislocation. In an isotropic solid, the edge dislocation has a greater energy than the screw dislocation by a factor of $1/(1 - \nu)$, where ν is Poisson's ratio. If R is half the distance to the next dislocation and r_o is taken to be on the order of 3b to 5b, then

$$E_{\perp} = \alpha \mu b^2 \tag{5.4}$$

where α is a dimensionless factor affected by the arrangement of dislocations (typically, 0.5 to 1) and the values of the two radii employed. Formation of a kink or glide step on a glide plane results in a displacement of the material above and below the glide plane at the kink. The magnitude of the force on the element in Fig. 5.6 is simply $\tau\mathbf{b}$, where τ is the shear stress resolved on the glide plane in the direction of the Burgers vector $\mathbf{b}$. This force *per unit length* arises from doing the work on a unit length of dislocation to move it a unit distance, where

$$\boldsymbol{F} = \frac{dW}{(ds\,dl)} = \frac{dW}{dA} = \tau\mathbf{b} \tag{5.5}$$

with the terms in this equation defined as in Fig. 5.6 (which is reproduced from Chapter 4).

Thus, we have an energy (or force) corresponding to the length of the dislocation and a force (or energy) required to move a portion of the dislocation.

To put our dislocation into motion, we must get part of the dislocation to move forward. We will consider the bowing of the dislocation as shown in Fig. 5.7. We will provide two pinning points that prevent the motion of the dislocation at that location. These pinning

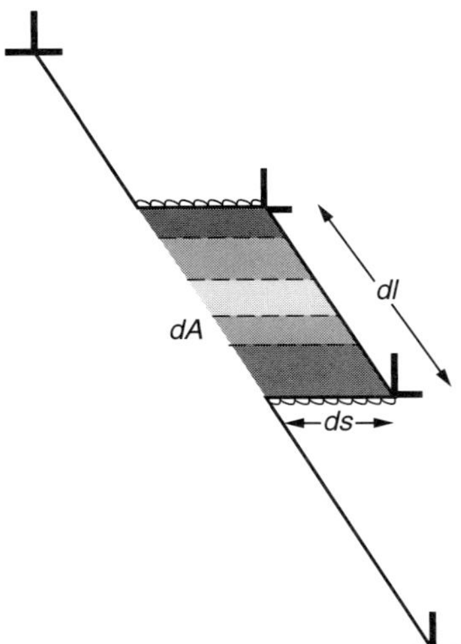

FIG. 5.6 The slipped area, displaced by the magnitude of the Burgers vector **b**, consists of components *ds* and *dl*. For this exaggerated example, wherein dislocations of mixed character are not shown, $dA = ds\,dl$. A dislocation would have such sharp corners only if the material had very large elastic anisotropy.

points are fixed obstacles or points where the dislocation line either leaves the glide plane (e.g., an immobile jog) on which the shear stress is resolved or consists of material into which the dislocation is unable to pass. Possible reasons why the dislocation might not be able to pass by these points are (1) the material at these points consists of a second phase with crystal planes that do not match, (2) the planes near these points are distorted with elastic stresses between a second phase and the matrix material, and (3) the elastic stress fields for dislocations operating on other slip systems inhibit dislocation motion.

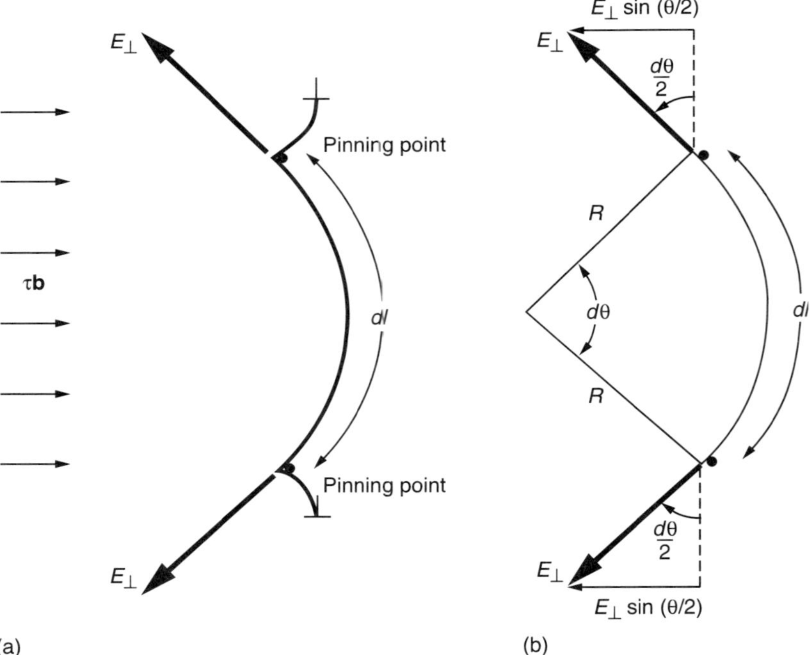

FIG. 5.7 (a) Dislocation bowing past two pinning points wherein a balance between the applied force $\tau\mathbf{b}$ and the components of $E_\perp$ must be balanced for an equilibrium radius R. (b) Relationship between the radius of curvature R and the angular range $d\theta$, equivalent to the length of the dislocation segment dl to the components of $E_\perp$ that must balance $\tau\mathbf{b}$.

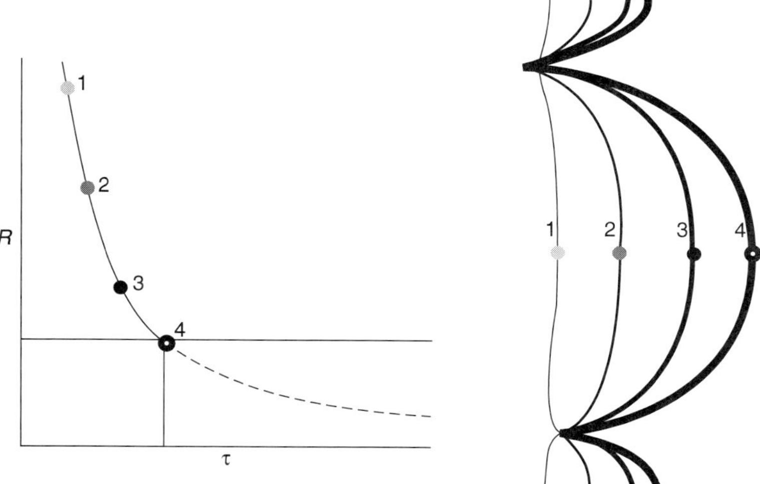

FIG. 5.8 As the applied shear in Eq. 5.7 increases, the radius of curvature for a given obstacle spacing decreases until the critical value of stress is reached wherein the loop segment has an angular range $\theta \geq \pi$. At this point the loop begins to bow around the obstacle. Of course, if other mechanisms of bypassing the obstacles intervene, such as cutting of pinning particles or dislocation segments, climb, or cross slip, this stress level will not be required to produce motion of the dislocation past the obstacle.

Using Eq. 5.4 and 5.5, we can define equilibrium for the dislocation at radius R wherein the forces imposed by the pinning points balance the applied force with

$$E_{\perp} d\theta = \tau b dl \tag{5.6}$$

where $2E_{\perp} \sin(\theta/2) = E_{\perp} d\theta$ for small values of $d\theta$ and $\tau b dl$ is the force on the dislocation element dl. By substitution,

$$\tau = \frac{\alpha \mu b}{R} \tag{5.7}$$

providing an expression for the stress required to bend a dislocation to a value of the radius equal to half the obstacle spacing, which means that the critical shape is a semicircle. This also is the stress required to operate the Frank–Read sources defined in Chapter 4. At lower stresses, the radius of curvature for the segment dl is higher. Therefore, closer spacings of pinning points, or obstacles, that induce smaller radii of curvature on the dislocation result in higher critical resolved shear stresses, as shown in Fig. 5.8.

5.2.2 Dislocation Multiplication Strengthening: Strain Hardening

Strengthening by multiplication of dislocations and an increase in the dislocation density take place in materials containing enough dislocations for operation of dislocation multiplication sources. Commercial materials, such as those used for the stress-strain curves in Fig. 1.20 and 1.21, typically have enough mobile dislocations so that dislocation multiplication occurs stably. In the stress-strain curve for 1020 steel shown in Fig. 1.19, the upper yield point near the vertical axis results from the dislocations present in the material being pinned in place by interstitial carbon (for more discussion, see Section 5.2.3). As discussed in Chapter 4, an insufficient number of dislocations, as in high-perfection, single-crystal electronic materials or whiskers, requires high stresses to nucleate dislocation sources, and lower stresses are sufficient for continued plastic flow. Subsequently, the increase in strength associated with multiplication of dislocations is manifested in several ways. The first is that under applied stress the dislocations bow around obstacles, resulting in a lengthening of the dislocations and thereby smaller radii of curvature. The second is that the spacing between dislocations decreases. This interaction is demonstrated by the example shown in Fig. 5.9, wherein the dislocation energy put forth in Eq. 5.4 is used to show that at the closest possible spacing between two dislocations, when the dislocation consists of a Burgers vector twice that of a unit dislocation, the elastic energy associated with the same length dislocation is four times higher, $\alpha\mu(2b)^2$. This same consideration becomes even more important when considering dislocation pile-ups, as discussed in Section 5.2.5. Even when dislocations do not share the same glide plane, the elastic stress fields surrounding dislocations interact (see Fig. 5.9). Both repulsive and attractive configurations can cause interactions that hinder dislocation motion. Repulsive forces between dislocations can inhibit the approach of dislocations, and attractive forces between dislocations can inhibit motions separating adjacent dislocations.

The third aspect of increasing dislocation density is the effect of mobile and nonmobile dislocations that intersect the path of the gliding dislocations on a given plane. Although the stress fields of some dislocation types do not interact with one another at special relative orientations in isotropic materials, neither one of these two instances occurs with any significant frequency. Thus, it is probable that meeting and intersection of one dislocation with another set of dislocations will require stresses that constitute a significant fraction of those required to bow around pinning obstacles.

Taylor also considered a somewhat simpler dislocation interaction wherein same-sign dislocations interact on parallel slip planes of spacing l. Then this elastic interaction stress depends on the spacing between dislocations in a form similar to the expression for bowing given in Eq. 5.7,

$$\tau = \frac{\beta\mu b}{l} \tag{5.8}$$

but the fitting parameter β is generally small ($\approx$0.1 or less). The smaller magnitude of strengthening given in Eq. 5.8 comes about from the potential for highly stressed dislocations to cut through inclined dislocations, as shown in Fig. 5.10.

We can consider the relationship of this expression to the Orowan equation (Eq. 4.11), which gives us the strain from dislocation motion as

$$\gamma = \rho_{\perp} b x_{avg}$$

where $\rho_{\perp}$ is the dislocation density, b is the Burgers vector magnitude, and x_{avg} is the average dislocation displacement. If the average spacing of dislocations, l, can be approximated

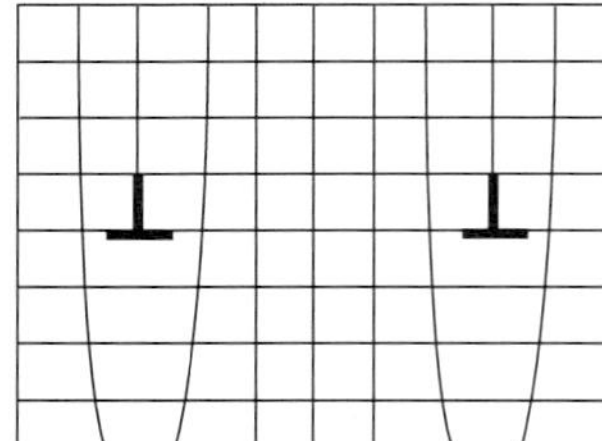

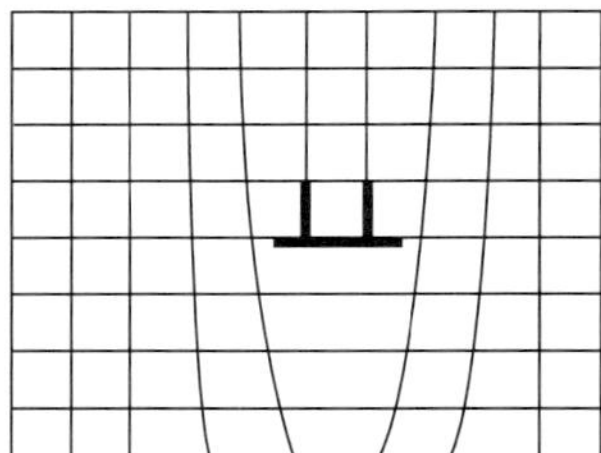

FIG. 5.9 Increase in elastic energy as two identical dislocations on the same glide plane are brought into such close proximity that the dislocation is effectively one consisting of a Burgers vector that is two unit b in magnitude.

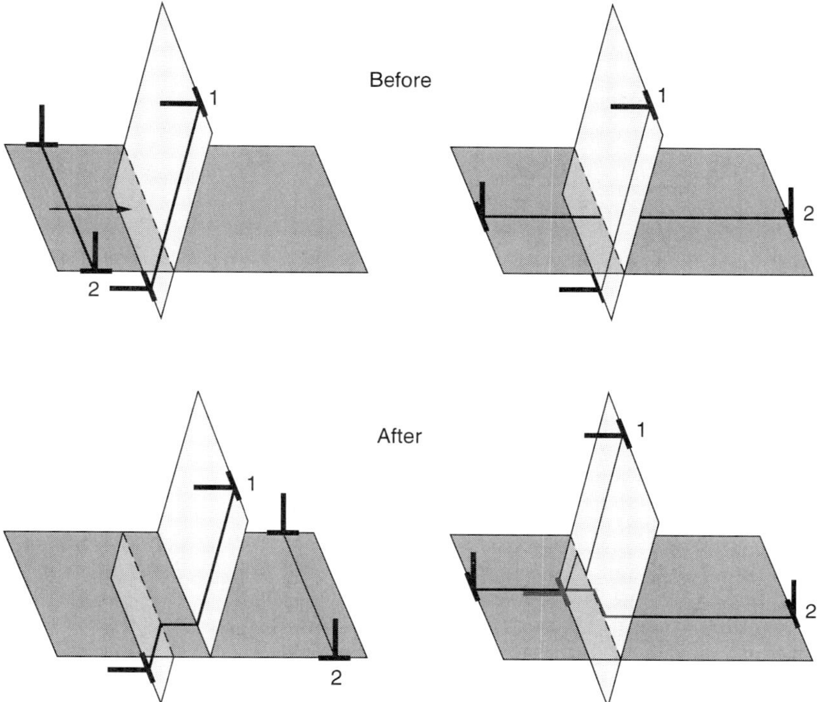

FIG. 5.10 Dislocation intersections. The left-hand set of figures shows the step made in dislocation 1 by the passage of 2. Because the Burgers vector of 1 lies along 2, no step is made in 2. The right-hand set of figures shows that if the Burgers vectors of 1 and 2 are normal to the line direction of the other dislocation, a step will be formed in each.

by $\rho_{\perp}^{-1/2}$, then it is possible to write an expression for the dislocation density dependence of strength as

$$\tau = \tau_{ini} + \beta\mu b\rho_{\perp}^{\frac{1}{2}} \tag{5.9}$$

In Eq. 5.9, τ_{ini} is the shear stress required to move the dislocation without obstacles. By substituting dislocation density for plastic strain via the Orowan equation, the parabolic strain hardening expression defined by Taylor can be written as

$$\tau = \tau_{ini} + \beta\mu\left(\frac{b\gamma}{x}\right)^{n} \tag{5.10}$$

if the value of n is $\frac{1}{2}$. To translate Eq. 5.10 into terms of a yield stress for tensile or compressive deformation, a new expression can be written:

$$\sigma_{yield} = \sigma_{ini} + \kappa\mu\left(\frac{b\varepsilon}{x}\right)^{n} \tag{5.11}$$

where κ compensates for the geometric changes from shear stress to a uniaxial tensile or compressive stress. The increase in yield strength is $\Delta Y(\varepsilon) = \sigma_{yield} - \sigma_{ini}$. Expressions of the type shown in Eq. 5.11 quite readily describe the deformation behavior of polycrystalline materials, but such an expression fails to describe the behavior of single crystals. Consequently, despite the geometric derivation employed, this equation is really applied empirically as an increase in yield strength, $\Delta Y \propto \varepsilon^n$. In real polycrystalline materials, the changes in grain shape, dislocation density, and dislocation arrangement would be required components of a strain hardening model based on fundamental considerations. The same functional dependence should already be familiar as the empirical expression in Eq. 3.19.

$$\sigma_{\text{eff}} = K\varepsilon_{\text{eff}}^n$$

which includes the same exponent n. Remember that this expression applies only to significant levels of plastic deformation.

5.2.3 Solid Solution Strengthening

Alloying of crystalline materials results in several types of strengthening. Solid solution strengthening occurs by either substitutional elements (or ions in the case of ionic materials) or elements that are soluble as interstitials. Since in most cases there is a limit to the solubility of a particular species in another, second phase effects, and in particular precipitation strengthening, become relevant if alloy additions are carried beyond the solubility limit for a given material. If the elements are mutually soluble, as are silver and gold, you can see strengthening effects, as shown in Fig. 5.11*a*. Just as stresses can be assigned to the elastic strains introduced by dislocations, substitutional and interstitial solutes are often considered to function as local elastic distortions within alloys. In some materials, alloying elements can assist in generation or motion dislocations and thereby soften the material, but in most cases the effect of solutes is to make the motion of dislocations more difficult. The elastic strain fields of edge and screw dislocations are profoundly different. As discussed in Chapter 4, the strain field surrounding the core of a screw dislocation consists of a helical shear that can be represented as

$$\boldsymbol{\varepsilon} = \begin{vmatrix} 0 & 0 & \varepsilon_{13} \\ 0 & 0 & \varepsilon_{23} \\ \varepsilon_{31} & \varepsilon_{32} & 0 \end{vmatrix} \tag{5.12}$$

where each value of ε_{ij} depends on the x-y position for dislocations lying along the z (or 3 axis). For an edge dislocation with the dislocation core along the z-axis and the Burgers vector in the positive x-direction, the nonzero terms of the strain field for an isotropic material are given by

$$\boldsymbol{\varepsilon} = \begin{vmatrix} \varepsilon_{11} & \varepsilon_{12} & 0 \\ \varepsilon_{21} & \varepsilon_{22} & 0 \\ 0 & 0 & \varepsilon_{33} \end{vmatrix} \tag{5.13}$$

The mutually exclusive character of the two strain fields provides that ideal screw dislocations should interact only with other defects that cause shear distortions along the dislocation core. On the other hand, edge dislocation strain fields interact with defects that cause either dilatant (volumetric) strain components or shear strain components. On this basis, interstitial defects would be anticipated to be more effective in strengthening than substitutional defects since the distortions associated with introducing a smaller or larger substitutional solute are primarily dilatant in nature. The insertion of solute into most octahedral or tetrahedral interstitial sites is often associated with both dilation and shear, but this depends on the crystal structure. Stress fields that repel or resist the approach of dislocations obviously affect deformation, but stress fields that attract dislocations strengthen the crystal by making motion of dislocations away from those positions more difficult.

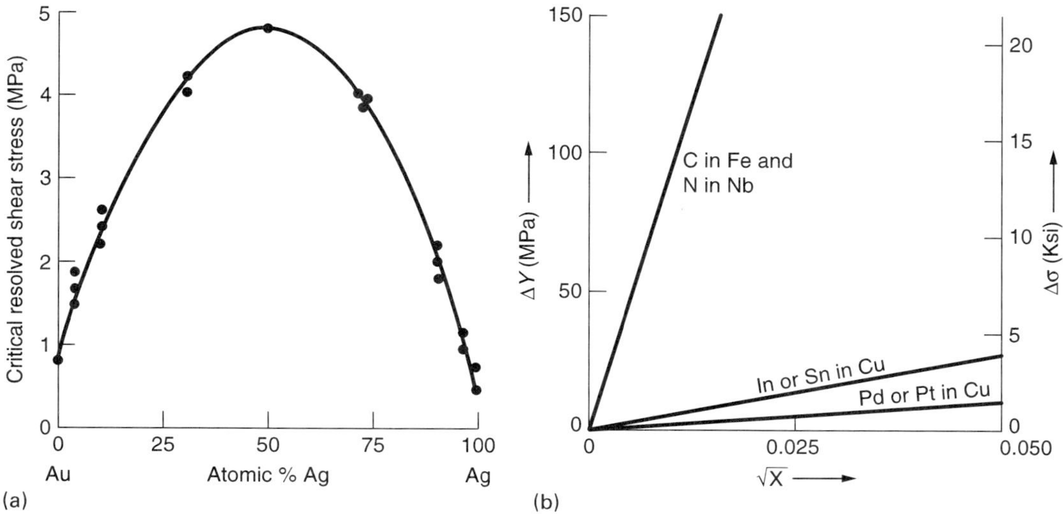

FIG. 5.11 (a) Critical resolved shear stress of gold and silver alloys given as functions of composition. Gold and silver form a continuous solid solution, which leads to the largest strengthening effects at nearly 50 atomic percent mixtures [after Sachs and Weerts (Fig. 7, p. 480), 1930, used with permission of Springer-Verlag]. If the near 50 atomic percent mixtures are heat treated, the alloy is susceptible to some ordering, which results in even greater amounts of strengthening. (b) Alloy strengthening as an increase in flow stress with solute content raised to the one-half power (Ralls, Courtney, and Wulff, 1976, Copyright © This material is used with permission of John Wiley & Sons, Inc.).

Since screw dislocation motion through cross slip is an important mechanism for circumventing other obstacles and screw dislocation motion is integral to dislocation multiplication, the strengthening potency of interstitials is usually much greater than that of substitutional solutes. Although the solubility of interstitials is normally very small, the strengthening potency, expressed as $\Delta\tau/c$, where $\Delta\tau = \tau - \tau_{ini}$ is the strengthening increment and c is the fraction of the interstitial solute, is generally at least an order of magnitude greater than that for substitutional strengthening, as can be seen in Fig. 5.11*b*. In addition, the magnitude of the misfit strain also has a role in solution strengthening, leading to strengthening expressions of the form

$$\tau = \tau_{ini} + \xi \mu \varepsilon_{misfit}^{f} c^{g} \tag{5.14}$$

In Eq. 5.14, τ_{ini} is the shear yield strength of the pure material, ξ is a proportionality constant, μ is the shear modulus, ε_{misfit} is the misfit strain $\Delta a/a$ (with Δa the difference in lattice parameter between solvent and solute or the displacement to accommodate an interstitial solute), and c is the solute fraction. The misfit strain exponent is typically $f = 3/2$ and, as mentioned above, g, the concentration exponent, is 1/2 to 1. For very dilute solutions, the effects are nearly linear. Note that for all solutes the strain associated with their presence is only part of the story, because even interstitial solutes that fit within the interstitial sites in the pure material affect the local bonding and thereby can influence dislocation motion.

EXAMPLE 5.3 *Bake-Hardened Steels for Automobile Body Panels*

Low-temperature annealing can cause interstitials to diffuse to the dislocations, making it difficult for the dislocations to move without reaching some critical stress level. This is perhaps best demonstrated in bake-hardened steels that can be readily shaped for automobile body panels and then given a light heat treatment (170 to 200° C) that increases their strength dramatically and enhances their resistance

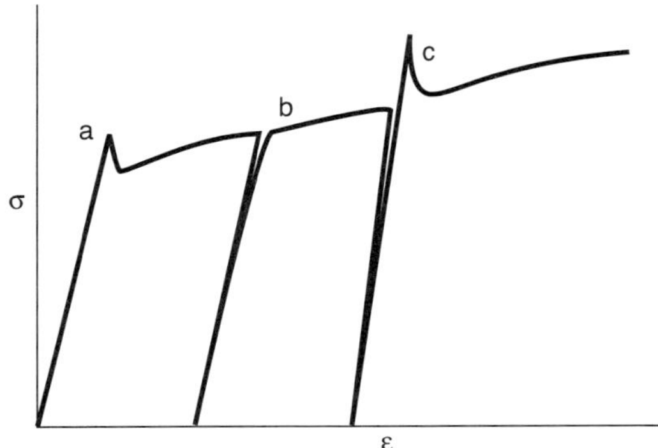

FIG. 5.12 Bake-hardened steels have C and N interstitials that provide strengthening. This strengthening can be enhanced by a heat treatment wherein the C and N diffuse to the dislocations stabilizing their strain fields. This results in a large stress being required for yielding to occur. Three steps of deformation are shown, with (a) representing the initial plastic deformation of a strain aged steel. This material is unloaded and immediately reloaded, as shown at (b). The second loading shows no yield point since the interstitial C and N have had no time to segregate to the dislocations. Strain hardening occurs through multiplication of dislocations. During the next unloading, long times or light heat treatments at 170 to 200°C (bake hardening) can allow for a return of the yield point due to the pinning of the dislocations by strain aging. For each loading, some plastic deformation may precede the upper yield point due to motion of edge dislocations, leading to the impression that the steel has a lower than expected Young's modulus. The lower mobility of screw dislocations and the more effective pinning of them by interstitials leads to formation of the upper yield points. (See the sections on BCC metals in Chapters 4 and 9.)

to denting (Shi, McCormick, and Fekete, 1998). This "strain aging" process can produce increases in yield stress of 20 percent even though the total interstitial alloying content (usually C and/or N) may be less than 0.01 weight percent. The effect of the interstitials is to reduce the mobile dislocation content. To once again get dislocation motion, we must go to a high enough stress level for the dislocations to "break away" from the interstitials. This strengthening is dependent on the time and temperature of the heat treatment. Once they do, the dislocations can move at a much lower flow stress. Tensile stress-strain curves for these materials after strain aging show the type of upper yield point behavior shown for 1020 steel in Fig. 1.19. Figure 5.12 describes the strain aging process as it is applied in bake-hardened steels. ■

5.2.4 Particulate Strengthening

Precipitation Strengthening Precipitation strengthening (or hardening) is normally much more potent than solution strengthening methods. An understanding of precipitation strengthening incorporates appreciation for nonequilibrium phase changes with dislocation mechanics. Precipitation strengthening is particularly effective in circumstances under which the alloy can be worked into the near final shape before the strengthening treatment. Materials can be shaped at relatively low stresses and then strengthened when at or near the desired dimensions. As shown in Fig. 5.13, the potential increase in strength from controlled precipitation of a second phase can be fivefold or greater. A similarly dramatic example was shown for an aluminum alloy in Fig. 1.21.

As might be anticipated by evaluation of Eq. 5.7, the pinning point spacing along the dislocation is critical to strengthening, as is the strength of the pinning obstacle. If these obstacles have a different crystal structure, or just a higher resistance to slip, they can func-

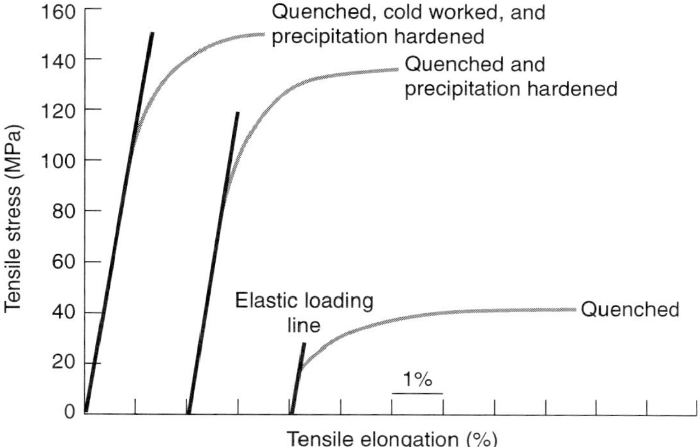

FIG. 5.13 Strengthening of Cu–2.2% Be alloys (after Smith and van Wagner, 1941, used with permission of ASTM). The extreme increase possible through precipitation strengthening is accomplished by placing potent obstacles to dislocation motion at short distances from one another. The added strength from work hardening is small compared to the gain for precipitation strengthening.

tion as pinning points. The strengthening from precipitation can be more than doubled if in addition to physically blocking dislocation motion these obstacles produce elastic strain fields in the alloy. This strengthening will follow the same form as Eq. 5.7, but the effective spacing is not just the spacing between particles but the effective distance between the elastic strain fields around the particles. If we assume that the critical value of R is a semicircle, then the spacing, L, is $2R$. This gives the increase in strength as

$$\Delta\tau \approx \frac{2\alpha\mu b}{L} \tag{5.15}$$

Unfortunately, accomplishing the fine spacing of a second phase throughout the individual grains of a polycrystalline material is not that easy. The heat treatment process for strengthening of an alloy requires a combination of available phase diagram features for the alloy and specialized processing that includes a heat treatment into a single phase region followed by quenching into a two-phase region and a second heat treatment in the two-phase region (called aging) to bring out a fine dispersion of the second phase.

The typical phase diagram features required for an alloy to have the potential for precipitation strengthening (shown in Fig. 5.14*a*) include a single-phase region at the alloy composition from which cooling results in formation of a second phase. The steepness of the solubility boundary and alloy composition determine the amount of second phase that is available for strengthening. A special heat treatment process, as outlined in the process map of Fig. 5.14*b*, is required, because slow cooling of the alloy from the single-phase region into the two-phase region does not result in much strengthening. During slow cooling, the second phase often precipitates at the grain boundaries and forms large, widely spaced grains. If, however, the alloy is rapidly quenched from the single-phase region, the alloy will contain a higher amount of the alloying element than predicted by the equilibrium conditions shown on the phase diagram. Under these circumstances, the alloy is supersaturated, and as long as diffusion rates are slow, it can remain a single phase. If a mechanical test of hardness or strength is made on the alloy in this condition, it will be stronger than the pure alloy as a result of solid solution strengthening.

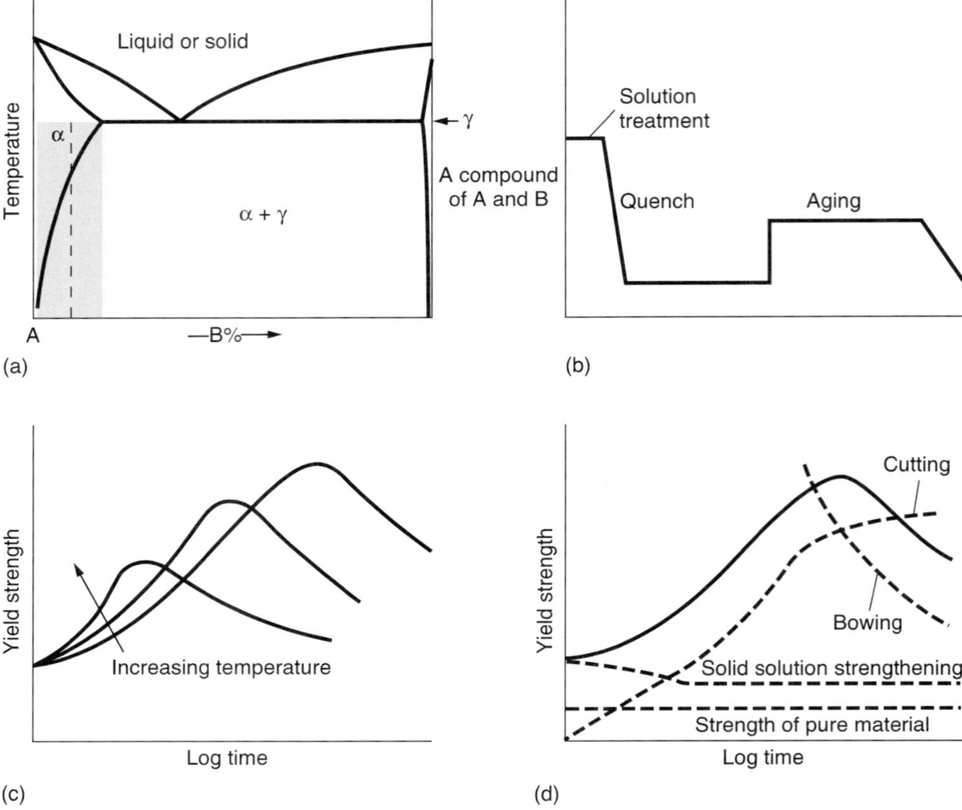

FIG. 5.14 (a) Typical phase diagram for the potential of precipitation strengthening with γ being an intermetallic of elements A and B. (b). A possible process map for the alloy designated with a dashed line in part (a). (c) The effect of aging temperature on the time to peak strength in precipitation strengthened alloys. (d) The contributions from different strengthening mechanisms given as dashed lines add up to the total strength as a function of aging time.

After quenching, the alloy is then aged to bring out a large number of distributed particles of the second phase. Although in some alloys (particularly some Al alloys) this aging occurs with the assistance of some quenched-in vacancies at room temperature, for most alloys the temperature and time of aging are designed to bring out well-dispersed precipitates by balancing a reasonable processing time with the desired final properties. Examples of the aging process versus time and temperature typical of a number of alloys are shown in Fig. 5.14*c*. Analysis of the aging process in these alloys most often includes the features depicted in Fig. 5.14*d*. In the first stage of heat treatment, the contribution to alloy strength of solid solution strengthening decreases as the first nuclei of the second-phase form. The initial particles of the second phase can form coherently with the initial alloy. This means that at least some planes of the second phase match in orientation, and nearly in spacing, those of the alloy matrix. Matching of some planes results in elastic stresses in addition to any derived from possible differences in the thermal expansion coefficients of the two alloys. As long as this additional elastic stress is present, it enhances the strengthening effect of the precipitates by inhibiting dislocation motion. As aging continues, the growth of precipitates has several implications. First, the increase in average size of precipitates means that some are eliminated at the expense of larger ones, thus increasing the interparticle spacing. If the particle spacing is increased sufficiently, the dislocations can just bow around the particles by the mechanisms discussed earlier. Second, the growth of the precipitates also leads to differential displacements between precipitate planes and the alloy matrix that cannot be supported by coherency. When the mismatch becomes too great, the precipitates can become incoherent with the planes of the matrix. This loss of coherency reduces the elastic stress field from lattice mismatch and thereby reduces the hardening.

EXAMPLE 5.4 *Strengthening of Al Alloys by the Formation of Guinier–Preston Zones*

The Al-Cu alloy system shown in Fig. 5.15 is the classic example of precipitation strengthening. At room temperature, the solubility of Cu in Al is approximately 0.1 weight percent. From this concentration to approximately 5 weight percent Cu, it is possible to process these alloys through a quench and age process to produce high-strength alloys. The strengthening of these alloys by $CuAl_2$ and the intermediate metastable particles leading up to $CuAl_2$ provide a convincing picture of elastic strain field effects on dislocation motion.

A typical alloy composition for high-strength Al-Cu alloys contains 4 weight percent Cu. By solution treating this alloy at approximately 550°C, a single-phase alloy containing 4 weight percent Cu should be obtained. Then, by rapidly quenching the alloy to room temperature, precipitation of $CuAl_2$ can be suppressed. If it was not rapidly cooled, large $CuAl_2$ particles would form preferentially at grain boundaries during slow cooling. By aging the quenched alloy between 100°C and 300°C, nucleation of a second phase can begin. In Al-Cu alloys, metastable phase precipitation precedes the formation of the equilibrium $CuAl_2$ phase. The first precipitates to form are very small metastable clusters of copper atoms that lie on {100} planes in the aluminum, as shown in Fig. 5.15*b*. The copper atoms cluster in these planes since <100> are the soft elastic directions in FCC metals. Typical dimensions of these plates are two atoms thick and 10 nm in diameter. Because copper atoms are smaller than aluminum atoms, the surrounding crystal is distorted. This distortion interacts with the strain field of dislocations to inhibit their motion. As these plates grow, the cluster morphology changes and the formation of additional intermediate phases precedes final formation of $CuAl_2$. Each of these intermediate phases is metastable and involves some elastic misfit. As the particles grow, the elastic strains between the particles and the matrix become so large that the planes no longer completely match. When only some of the planes match, we can say that the precipitates are semicoherent. While the particles are semicoherent, large elastic strains can still surround the particles. As the particles grow,

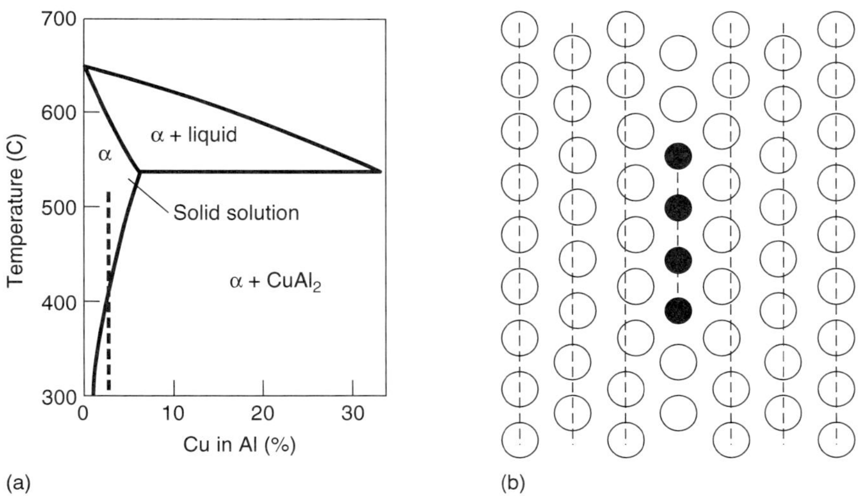

FIG. 5.15 (a) Phase diagram section for Al-Cu. (b) Mismatch between a copper platelike cluster (called Guinier–Preston zones) and the aluminum matrix shown by distortion of the planar position of the aluminum atoms.

the spacing between the particles decreases as long as the number of particles is increasing. Once the number of particles begins to stop increasing, large particles can grow at the expense of smaller particles. This point also will approximately coincide with the loss of coherency. Incoherent particles will not exert misfit strains and therefore will have a larger limited range of their effects. This stage of aging is called overaging and coincides with a reduction in hardness and strength.

During the time at which the particles are at least partially coherent, dislocations gliding through the alloy matrix can enter and cut through the precipitate, but this requires special relationships between the matrix and the precipitate. The resolved shear stress may also need to be higher than for matrix shear. If the precipitates have become incoherent and far enough apart, the process of bowing leaves dislocation loops around each precipitate as they pass. This leads to synergy between dislocation motion and the precipitates. ■

EXAMPLE 5.5 *Gamma Prime Ni_3Al in Ni-Base Superalloys*

In nickel-base superalloys, an ordered phase called gamma prime (γ') is used to inhibit plastic deformation. The high oxidation resistance of these alloys, which is derived from passive oxidation of Al and similar compounds, and high strength at high temperatures enable use of these materials in the turbines of jet engines. The γ' is present as precipitates that can be 70 volume percent of the alloy. These precipitates have an ordered structure with a cubic unit cell consisting of Al atoms on the corners and Ni atoms on the faces. Dislocations in the disordered phase move relatively easily. Motion of dislocations in the ordered phase is quite difficult. Each $\frac{a}{2}$<110> dislocation in the disordered phase must pair up with a second dislocation for plastic deformation to occur in the ordered phase, as shown in Fig. 5.16. The longer Burgers and complicated dislocation reactions result in a high critical resolved shear stress, as can be seen in the dislocation velocity versus shear stress data in Fig. 4.21. Unusually, the critical resolved shear stress in this and similar compounds apparently increases with increasing temperature. For this reason, and because of a resistance to diffusion-based high-temperature deformation mechanisms, many nickel-base superalloys have a yield strength that either increases or is almost unchanged with increasing temperature. ■

Dispersion Strengthening When rigid particles entrained within a metal alloy are not coherent, the strengthening is limited to that attained with the dislocation bowing mechanism with small enhancements from constraint and inhibition of grain growth resulting in strengthening from grain refinement. If the particles have a different thermal expansion coefficient than the matrix, there can also be a strengthening contribution from dislocations formed during cooling from elevated temperatures, but this strengthening contribution is often still less than that from coherency strains between a precipitate and matrix. As discussed in Chapter 6, composites with very fine particulates called dispersion-strengthened alloys provide this type of strengthening and are more stable for high-temperature applications. Larger-scale discontinuous composite reinforcements, platelets, and whiskers, with minimum dimensions of 1 μm, have limited effectiveness

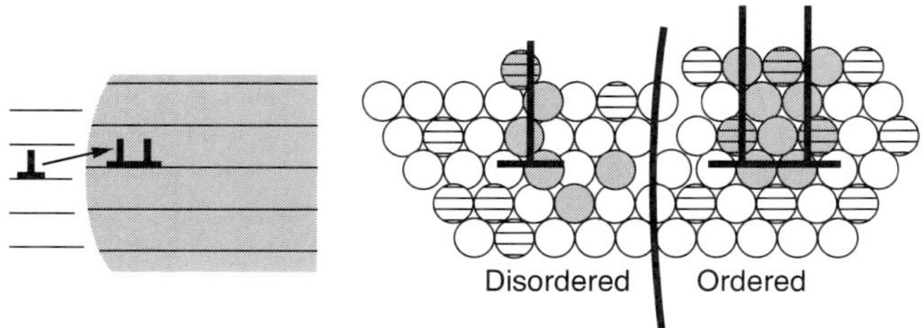

FIG. 5.16 Passage of a unit dislocation in disordered gamma into ordered gamma prime, resulting in a doubling of the length of a unit Burgers vector.

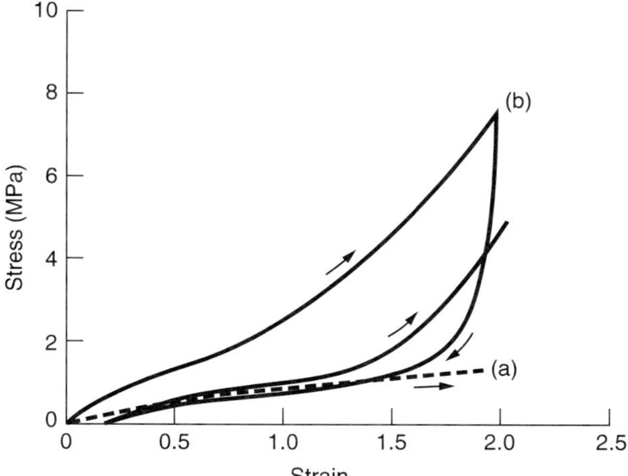

FIG. 5.17 Dispersion strengthening of rubber. (a) Natural rubber without carbon black powder. (b) Natural rubber with 50 percent carbon black powder. (McCrum, Buckley, and Bucknall, 1988, reprinted by permission of Oxford University Press).

in inhibiting dislocation motion since the spacing is large even for high-reinforcement additions. These reinforcements often reside at matrix grain boundaries because they are effective at pinning grain boundaries. These reinforcements do contribute to strengthening by constraint, multiplication of dislocations, and grain refinement, but are not as effective as precipitates distributed throughout a crystal. Filled polymers are another example of dispersion strengthening, as shown in Fig. 5.17. In many plastic parts, powdered metals or ceramics are added to raise the elastic stiffness and also to inhibit plastic deformation.

5.2.5 Grain Boundary (or Microstructural Refinement) Strengthening

Control of microstructural scale to produce desired properties is facilitated in most single-phase alloys by static recrystallization and grain growth following introduction of dislocations through cold work. Like the other strengthening mechanisms, justifications describing the fundamentals behind the strengthening are not as convincing as the empirical demonstration. The original expression for grain refinement strengthening is attributed to Hall and Petch with the expression

$$\sigma_{\text{yield}} = \sigma_{\text{ini}} + kd^{-h} \tag{5.16}$$

where $\sigma_{\text{yield}} = Y$, the tensile yield strength, σ_{ini} is the yield strength for a polycrystalline material with infinite grain size (or a hypothetical single crystal of random orientation), k is a proportionality constant, d is the average grain diameter, and h is an exponent of 1/2. Although h can vary from this value, by plotting Eq. 5.16 as σ_{yield} versus $d^{-1/2}$, a rather useful relationship description can be attained, as shown in Fig. 5.18 using expected values for Cu-30Zn brass and copper of commercial purity. Note that, unlike the strengthening expressions given in most cases above, Eq. 5.16 is expressed in terms of normal stresses because the shear stress expressions are not as meaningful if a polycrystal is under consideration. Examples of k-values assuming that $h = 1/2$ are given in Table 5.1. Values of σ_{ini} are not given, because they are strongly dependent on processing history (e.g., cold work and alloy content).

Attempts to represent the basis for this mode of strengthening usually follow one of several approaches. First, because it is not possible except under very special circumstances

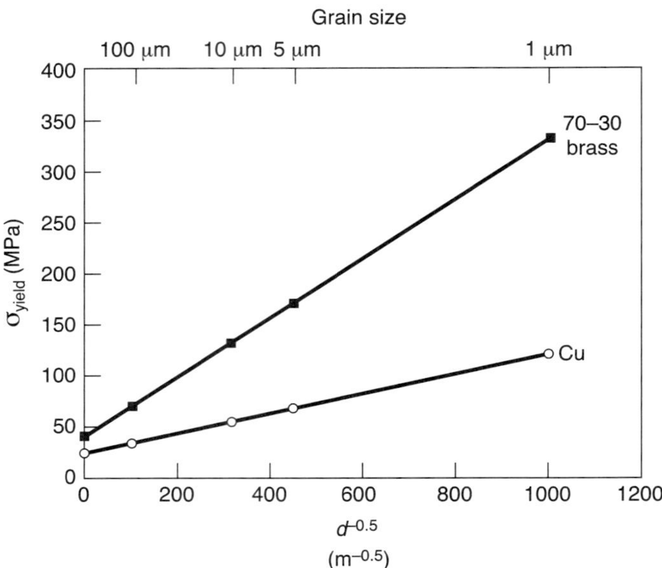

FIG. 5.18 Plot showing Hall–Petch behavior in brass and copper using typical values.

for a dislocation to cross a grain boundary, the region of material adjacent to the grain boundary is considered to offer constraint on each of the grains, as first mentioned in Section 5.1. Because of this constraint, multiple slip is ever more necessary as the relative interfacial area of each grain increases as each grain becomes smaller. For well-annealed materials, the dislocation density within the material does seem to scale inversely with grain size, which, if related to Eq. 5.12, yields Eq. 5.16.

Microstructural features other than grain boundaries, including second phases, can impart finer scales to microstructures. When extensive deformation has been induced in materials, dislocation networks form, further subdividing each grain. When these dislocation networks form, they both block dislocations and impart elastic strain fields throughout each individual crystal, interfering with dislocation motion. Thus, there is reason to believe that there is a significant synergy between grain size and dislocation multiplication. Although it is not possible to form exactly the same type of dislocation cell structure within

TABLE 5.1 Hall–Petch Parameters for Metals and Metal Alloys*

Material	***k*** **($MN/m^{3/2}$)**
Aluminum	0.07
Brass (Cu-30Zn)	0.3
Copper	0.1
Molybdenum	1.8
Silver	0.07
Tungsten	0.8

*Materials that have yield point discontinuities—e.g., low carbon steels—have not been included, because the magnitude of the upper yield stress can be changed by heat treatment. The values given are for well-annealed materials without significant alloying content, but because this data is strongly dependent on the history of the material, these data should be used with caution.

BIOGRAPHY

THE HALL–PETCH RELATION

Norman J. Petch and Eric O. Hall, both of the Cavendish Laboratory at Cambridge, are responsible for one of the most quoted relationships in materials science. Petch had a career that included research on steel crystallography, including the structure of cementite. Hall's other major contributions were in understanding yield point phenomena in steels, twinning, and martensitic transformations. (See Hall, 1951, and Petch, 1953.)

an unconstrained single crystal as inside a constrained grain within a polycrystal, the relationship between the two strengthening mechanisms might be given empirically as

$$\sigma_{\text{yield}} = \sigma_{\text{ini}} + \sigma_{\perp} + k'(\sigma_{\perp})d^{h} \tag{5.17}$$

where

$$\sigma_{\perp} = \kappa\mu\left(\frac{b\varepsilon}{x}\right)^{n}$$

A construction showing the combined effects of strain hardening and grain size on a hypothetical polycrystalline material is shown in Fig. 5.19.

Recent research into nanoscale microstructures of metals has suggested that at very small grain sizes the Hall–Petch relation breaks down with many nanocrystalline materials being softer than expected from data at larger grain sizes. Some data sets even suggest that the Hall–Petch relation may go through a maximum. Because the Hall–Petch relation was always an empirical correlation and has not been well described despite many attempts to develop a coherent theory, it is not surprising that there are limitations. Additionally, the challenges in producing dense, nanocrystalline materials with similar dislocation densities that are of sufficient size for reasonable tension and compression tests lead to most data coming from microscale or even nanoscale hardness tests. As discussed at the beginning of this chapter, the conditions of constraint in a hardness test make measurement of a yield stress quite difficult. Song, Guo, and Hu (1999) have compiled a great deal of the data showing a deviation from the Hall–Petch relation with nanoscale grains and have described one of the approaches for modeling the observed behavior.

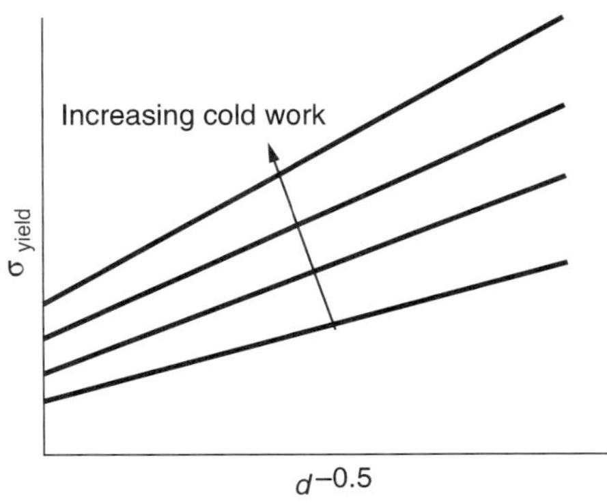

FIG. 5.19 The combined effects of strain hardening and grain size on strength.

5.3 ORIENTATION STRENGTHENING

Directional plasticity, like any form of anisotropy in either crystalline or molecular structured materials, is derived from a synthesis or processing history that causes a preferred orientation. The degree of plastic anisotropy is closely tied to the degree of orientation, the ability of shear deformation modes to operate at multiple orientations, and distinctions between the resistance to deformation at one orientation or another. The most profound examples of plastic anisotropy are found in highly aligned polymers (see Fig. 5.20) and single crystals of low-symmetry materials (review the discussion of the Schmid factor and texture in Chapter 3). Orientation softening or strengthening can occur in deformed metal polycrystals, but it usually coincides with other strengthening mechanisms.

Semicrystalline polymers, polymers that consist of 20-80 volume percent lamellar molecular crystals embedded in amorphous molecules, have numerous applications as consumer products (G'sell and Dahoun, 1994). For semicrystalline polymers, the crystal structures are usually orthorhombic or monoclinic. The crystal structures are fairly open and are about 20 percent higher in density than the amorphous phase. For polyethylene (PE), which consists of long carbon chains with each carbon bonded to two hydrogens, the likely crystal structure is orthorhombic, as shown in Fig. 5.21*a*. PE solidified from a melt forms a complex microstructure consisting of the spherulites shown in Fig. 5.21*b* and regions between the spherulites with more or less random molecules. Slip and twinning mechanisms have been described along and across the chains. The slip systems [001] (100) and [001] (010) are referred to as chain slip and [010] (100) and [100] (010) are referred to as transverse slip.[2] The thin lamellae of crystal are also free to bend or bow during deformation. Although semicrystalline polymers contain crystals, most plastic deformation occurs initially within the amorphous phase. At temperatures above the glass transition temperature, T_g, of the polymer, the amorphous phase shows rubbery behavior with no clear yield stress. For $T < T_g$, the secondary bonds between the polymer chains are sufficient to produce an elastic response with a definable yield stress.

The deformation response of these semicrystalline polymers is very sensitive to temperature, strain rate, and also self-heating during deformation. The deformation response is

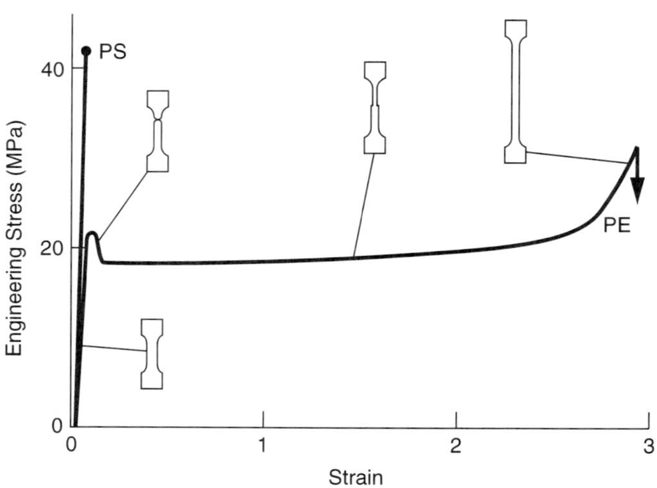

FIG. 5.20 The plastic deformation of polystyrene in tension shows formation of a neck early and then lengthening of this neck. Within the neck molecules become aligned, leading to orientation strengthening. (McCrum, Buckley, and Bucknall, 1988, reprinted by permission of Oxford University Press.)

[2]Note that the slip system indices are not given as families but rather specifically because the reference crystal structure is orthorhombic.

BIOGRAPHY

STEPHANIE KWOLEK (1923–)

Dr. Kwolek has received numerous awards for pioneering research that included the development of poly-p-phenyleneterephthalamide (the DuPont product Kevlar™) and other textile fibers. Besides controlling chemistry to make rigid polymer chains, stiff and strong fibers rely on orientation of the polymer chains.

also much more complex than can be readily accommodated by a von Mises criterion for deformation. Figure 5.22 shows stress-strain curves for tension and simple shear of PE and polyether etherketone (PEEK) plotted using effective true stress and effective true strain measured at a constant true strain rate and above their glass transition temperatures. The stress-strain responses in both deformation modes show very different hardening behaviors.

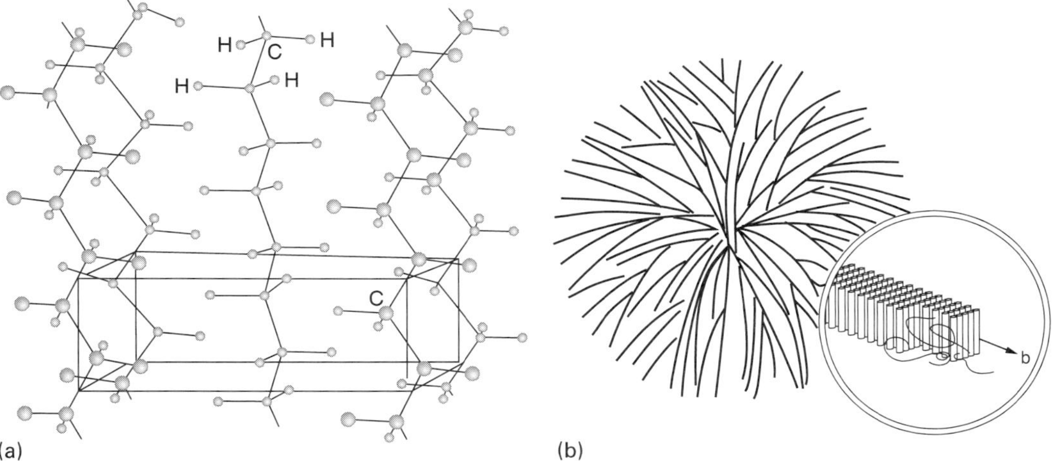

FIG. 5.21 (a) Chain configuration and the orthorhombic unit cell for polyethylene (G'sell and Dahoun, 1994, reprinted with permission of Elsevier Limited.) (b). Spherulites consist of folded chain lamellae (G'sell and Dahoun, 1994, reprinted with permission of Elsevier Limited.)

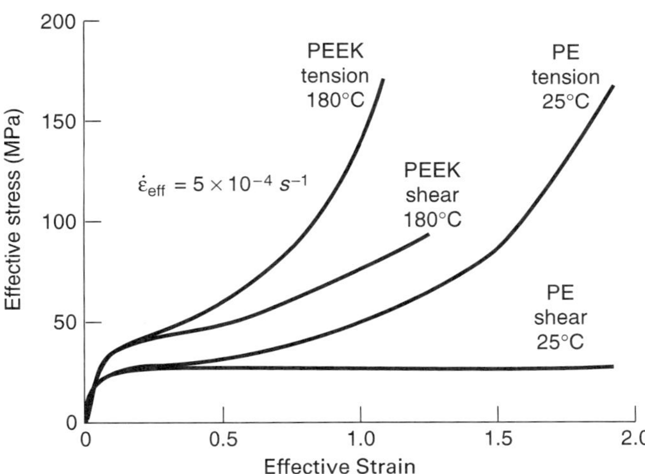

FIG. 5.22 Effective stress-strain behavior of polyethylene at 25°C and poly(etheretherketone) at 180°C under uniaxial tension and simple shear. The effective strain rate is $\dot{\varepsilon}_{eff} = 5 \times 10^{-4}\ s^{-1}$ in both modes (G'sell and Dahoun, 1994, reprinted with permission of Elsevier Limited).

In tensile deformation, both materials show a long plastic range with a hardening rate that accelerates with strain. This increasing rate of strengthening is a good indication of chain alignment producing orientation strengthening. Indications of chain orientation are shown by the pole figures in Fig. 5.23, which give an orientation normal to the chain lengths in both materials. The intensity of this preferred orientation increases with strain.

Another aspect of the deformation of PE and PEEK is that each material undergoes a decrease in density (or an increase in specific volume) with tensile deformation, but undergoes almost no density change in simple shear. This shows the effect of dilatant deformation from chain unraveling, lamella fragmentation, and a spreading of chain distances called crazing. Crazing is a dilatant mode of chain spreading observed as whitening at crack tips in some polymers.

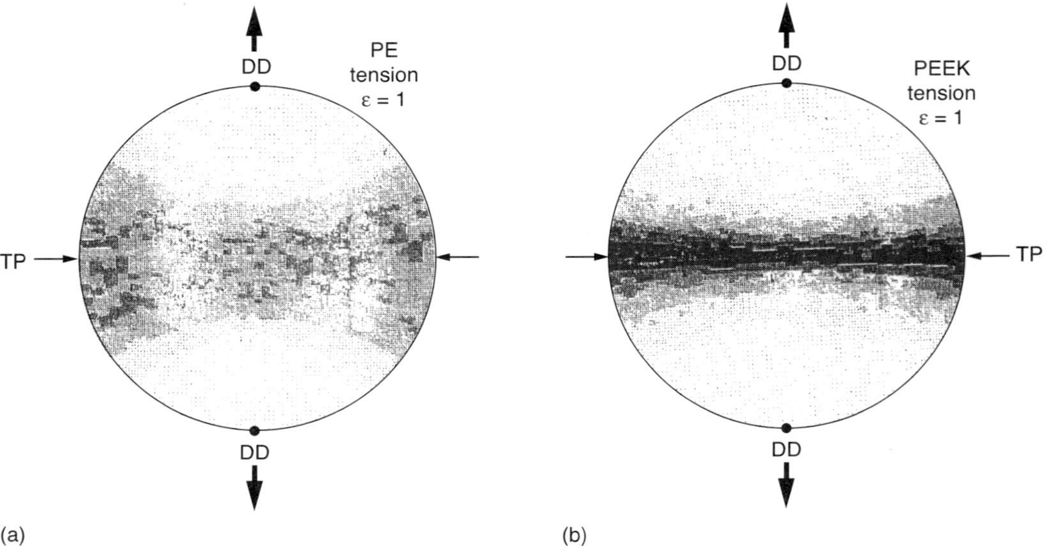

FIG. 5.23 Crystalline (200) pole figures of (a) PE and (b) PEEK after deformation in tension for $\varepsilon_{eff} = 1$. The darker the shading, the larger the number of 200 orientations aligned with that direction (G'sell and Dahoun, 1994, reprinted with permission of Elsevier Limited).

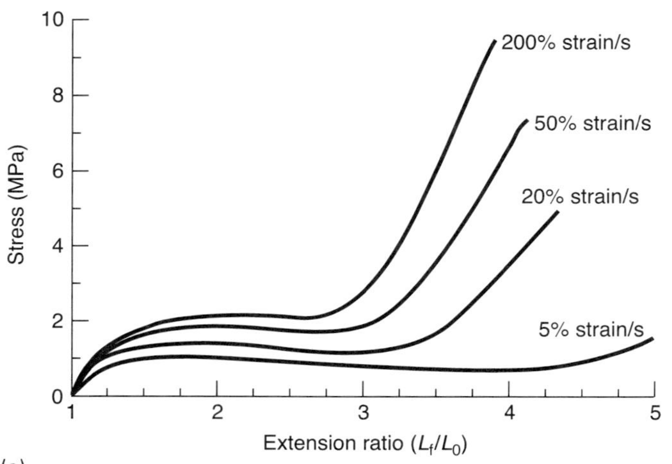

FIG. 5.24 (a) Stress-strain curves for PET specimens stretched biaxially in simultaneous mode at 100°C and at different strain rates (Chandran and Jabarin, I, 1993, Copyright ©This material is used with permission of John Wiley & Sons, Inc.).

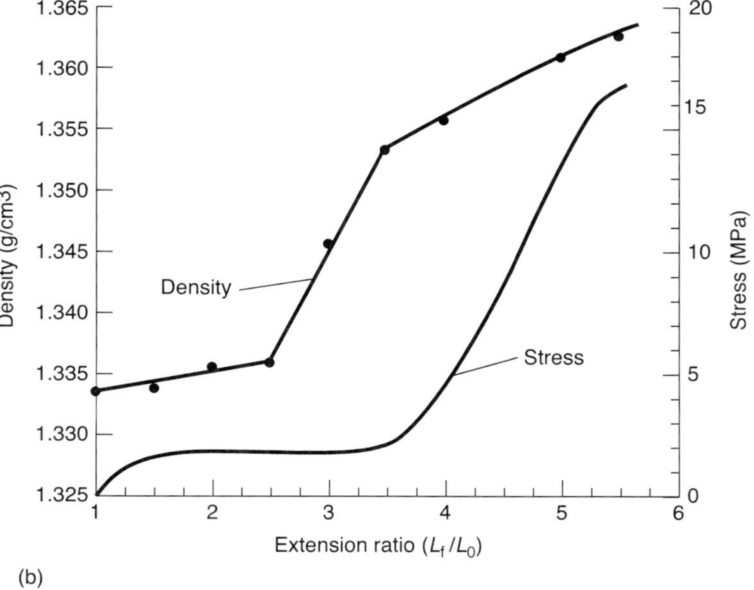

FIG. 5.24 (b) Comparison of the density vs. extension ratio curve with the stress measured during stretching of PET, for samples uniaxial stretched at 90°C and 50% strain per second (Chandran and Jabarin, II, 1993, Copyright © This material is used with permission of John Wiley & Sons, Inc.).

Another type of orientation strengthening is the strain-induced crystallization that occurs in polyethylene terephthalate (PET). PET, which finds applications as food and drink containers, recording tapes, and packaging films is not a very useful structural material in its amorphous form. The mechanical properties of stiffness and strength are too low. By carefully controlling orientation processes during deformation, PET can undergo crystallization that further strengthens the material. Transparent soda bottles (2-liter and smaller sizes) are produced by a stretch blow molding process that involves biaxial straining (like a balloon). Biaxial straining results in large increases in strength that are strain rate dependent, as shown in Fig. 5.24*a*. That crystallization coincides with plastic deformation has been verified by x-ray diffraction; however, the surprising *increase* in density with tensile deformation shown in Fig. 5.24*b* is also convincing evidence for crystallization. PET in this crystallized condition has very low CO_2 permeability, which enables it to store soda pop without loss of the fizz.

5.4 REFERENCES

R. F. Bunshah (Editor), *Techniques of Metals Research,* Volume V, 1971.

P. Chandran and S. Jabarin, *Advances in Polymer Technology,* Vol. 12[2], I 119–132, II 133–151, III 153–165, 1993.

T. H. Courtney, *Mechanical Behavior of Materials,* McGraw-Hill, 1990.

C. A. G'Sell and A. Dahoun, "Evolution of microstructure in semi-crystalline polymers under large plastic deformation," *Mater. Sci. Eng. A,* **175**, 183–199, 1994.

E. O. Hall, *Proc. Roy. Soc.*, **B64**, 474, 1951.

R. W. Hertzberg, *Deformation and Fracture Mechanics of Engineering Materials,* 4th Ed., Wiley, 1996.

D. Hull and D. J. Bacon, *Introduction to Dislocations,* 3rd Ed., Pergamon Press, 1984.

N. G. McCrum, C. P. Buckley, and C. B. Bucknall, *Principles of Polymer Engineering,* Oxford University Press, 1988.

D. McLean, *Mechanical Properties of Metals,* Wiley, 1962.

N. J. PETCH, *J. Iron Steel Inst.*, **174**, 25, 1953.

K. M. RALLS, T. H. COURTNEY, AND J. WULFF, *Introduction to Materials Science and Engineering,* Wiley, 1976.

G. SACHS AND J. WEERTS, *Z. Physik,* **62**, 473, 1930.

A. SEEGER, in *Dislocations and Mechanical Properties of Crystals,* edited by J. C. Fisher, W. G. Johnston, R. Thomson, and T. Vreeland, Jr., Wiley, New York, 243–329, 1957.

M. SHI, M. MCCORMICK, AND J. FEKETE, SAE Paper 980960, SAE Int. Cong. and Ex., Detroit, 1998.

C. S. SMITH AND R. W. VAN WAGNER, *1941 ASTM Proceedings,* **41**, 825–845, 1941.

H. W. SONG, S. R. GUO, AND Z. Q. HU, *Nanostructured Materials,* **11**, 203–210, 1999.

A. B. WATTS AND H. FORD, *Proc. Inst. Mech. Eng.,* **169**, 1141–1149, 1955.

5.5 PROBLEMS

A.5.1 Plot Eq. 5.7 as R versus τ for aluminum using $\alpha = 0.5$, $\mu = 25$ GPa, and $b = 0.286$ nm. What value of R is predicted for a critical resolved shear stress of 50 MPa?

A.5.2 How is κ in Eq. 5.11 related to β in Eq. 5.10?

A.5.3 Which type of dislocation should be most strongly influenced by a pressure (compressive hydrostatic stress), edge or screw?

A.5.4 When the hardness tests discussed in Chapter 1 are converted to units of stress (load over projected area), the magnitude of the stress is much greater than the yield stress for the same material. Why?

B.5.1 Explain why the strengthening effect of interstitial carbon is different in FCC and BCC iron alloys. In which type should interstitial carbon be more potent, and why?

B.5.2 Sketch and describe what happens to the dislocation loop after it reaches a half-circle. Why might the resistance to growth of the loop actually decrease as it extends beyond the half-circle? Consider the case for both a single half-loop and a continuous length of dislocation meeting several obstacles.

B.5.3 Assume that a thin sheet of a ductile material is attached to a substrate that has the same Young's modulus. Compare the condition for tensile yielding of the thin ductile sheet before and after it is attached to the substrate.

B.5.4 Place a rigid obstacle between the two pinning points of a Frank–Read source. Sketch the effects of the rigid obstacle on its operation.

B.5.5 Assume that a long fiber composite consists of aligned metal fibers with a true stress-strain behavior defined by $\sigma = 500\,\varepsilon^{0.7}$ and that the matrix deforms as $\sigma = 150\,\varepsilon^{0.2}$. Plot the expected composite flow behavior for a composite with equal volume fractions.

B.5.6 Plot true stress–true strain data for the high-friction, Fig. 5.3*a*, and low-friction, Fig. 5.3*b*, data using the Watts and Ford method to generate "friction-free" stress-strain curves.

C.5.1 Given that the energy of an edge dislocation in an isotropic material is greater than that of the screw dislocation, assume that dislocations cannot occur with mixed character in a hypothetical material. Sketch the expected shape and give the dimensions of such a "squared-off" loop if it pinned at two points for the following cases:

(a) an edge dislocation segment

(b) a screw segment

(c) an edge segment that meets the pinning points at a 45° angle.

C.5.2 Assuming that slip for a BCC crystal occurs on a slip system with the direction of slip in the [111] and the plane of slip given as $(\bar{2}11)$, sketch the path of rotation you would expect a crystal to take as it deforms in compression from an initial orientation of $[\bar{1}56]$. Do you expect orientation strengthening or softening to occur during deformation of this crystal?

C.5.3 A substitutional solute forms strongly covalent bands with the solvent metal. Would this change the strengthening interaction with edge and screw dislocations? Explain.

C.5.4 Second-phase particles often pin grain boundaries to inhibit grain growth. What type(s) of strengthening is (are) provided by the particles?

CHAPTER 6

HIGH TEMPERATURE AND RATE DEPENDENT DEFORMATION

TIME- OR RATE-DEPENDENT deformation of materials that occurs at high homologous temperatures can be expressed in a number of ways depending on the nature of the material and the perspectives of scientists and engineers evaluating a material. For those coming from fluid mechanics, the temperature and time dependence of deformation processes for many materials are simply examples of nonideal fluids, and expressions describing deformation can be readily produced. Materials scientists tend to envision temperature and time dependencies from the perspective of the fundamental steps required for shape change of a material that is composed of crystals or molecules. In many cases, the final expressions describing the shape change become quite similar. For most materials, the clearest definition of the nature of time-dependent deformation is observed in the mechanical processes called creep and stress relaxation.

6.1 CREEP

For creep, a material under a fixed tensile load deforms over time, as shown in Fig. 6.1. Immediately after the application of a load, most materials undergo (stage 1) nonlinear transient deformation with decreasing slope followed by the nearly linear region (stage 2) that is often termed steady-state creep. The creep rate in this region is then $\dot{\varepsilon}_{ss} = \frac{d\varepsilon}{dt}$ for the second region in Fig. 6.1. The last of the three stages often observed (stage 3) consists of the fracturing process that results in final failure across a specimen cross section. For materials with relatively stable chemistry and microstructure during creep, the time-dependent deformation usually increases with increasing temperature or stress, as shown in Fig. 6.2. Increasing stress level for the same temperature can produce similar effects. Note four important considerations regarding Fig. 6.2:

1. The strain at zero time is greater, because higher temperature or stress increases the instantaneous elastic strain.
2. The time to failure and the strain to failure decrease, because final rupturing mechanisms evolve more rapidly at higher temperatures or stresses.
3. With increasing temperatures and increasing stress, stage 2 and stage 1 may not appear at all, because the rupturing process may dominate the deformation versus time.
4. Environmental effects, including oxidation and stress corrosion, and microstructural changes, including grain growth and phase transformation, can change the deformation and failure mechanisms.

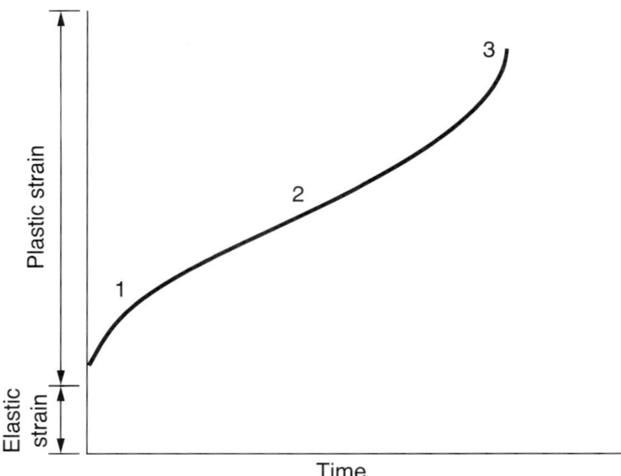

FIG. 6.1 Three stages of plastic strain accumulation for classical creep. (1) Transient or primary creep. (2) Steady-state creep. (3) Ternary or terminal creep.

Research investigations into fundamental creep mechanisms are often conducted under conditions of constant tensile stress to provide fixed values for deformation comparisons. For these experiments, the change in specimen cross section is compensated by a change in the applied load to maintain a constant stress as the specimen creeps. A constant stress test results in a less severe loading history for the specimen when compared to a test at constant load. For some applications, constant load may more accurately represent the actual conditions. Hence, engineering investigations are often conducted under constant load conditions. Initial trials on new materials and experiments designed to distinguish tensile damage from creep mechanisms are sometimes conducted in compressive loading.

The two most important parameters distinguished in creep testing are time to final failure of a material, or stress rupture life, and creep rate. The stress rupture life is readily applied to circumstances in which the part is expected to be used only for a short time—for example, in the engine components of a fighter aircraft, rocket booster, or missile. Mechanisms of deformation are not as important as the time to rupture, t_R. The stress rupture life is also useful in material development and quality control wherein the additional difficulties and costs associated with monitoring strain and strain rates over long time periods are not considered feasible. Extrapolation approaches are employed to predict the time to failure of a part based on tests conducted under more severe conditions. In this way, tests conducted at either higher stresses or higher temperatures, or both, are used to predict failure over longer periods of time. Creep of materials is strongly dependent on microstructure,

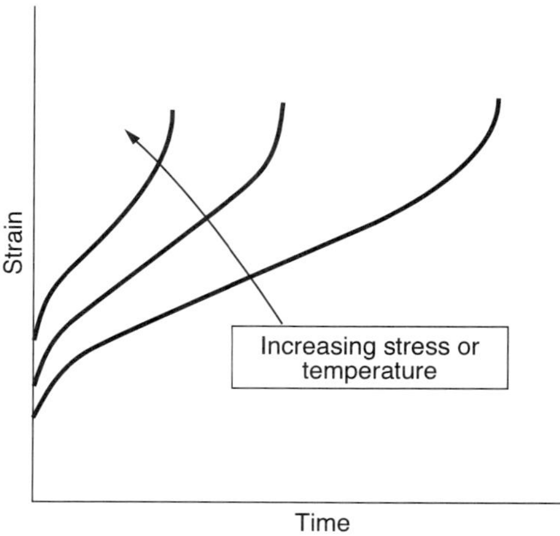

FIG. 6.2 Demonstration of how increasing stress or temperature changes classical creep behavior.

as shown in Fig. 3.24. Figure 3.24*a* shows that the viscosities of ceramic glasses and polymers vary continuously, whereas the viscosities of materials that readily crystallize—e.g., metals—change by many orders of magnitude at the melting temperature. Figure 3.24*b* was calculated using the expressions for diffusion creep in Section 6.4.3.

6.2 EXTRAPOLATION APPROACHES FOR FAILURE AND CREEP

Consider the challenge faced by a developer of a new material designed for long-term mechanical performance of 10 years or greater. If the material will be employed under conditions in which creep processes are anticipated, the engineers designing components with this material would like to know how the material will perform under the maximum loads and temperatures in the application. Then these engineers can design for safety factors appropriate to the application. The developer of the new material would appreciate sales of the material to offset development costs as soon as possible, and the design engineers would like to use the materials that best enhance their designs. So, how is this problem solved? Materials are often placed into service before the long-term tests are completed by testing materials for shorter times and at higher loads or temperatures and making very conservative assumptions. If concurrent long-term tests are conducted with the materials already in service, adjustments can be made in the predicted life and parts can be replaced before the original planned replacement dates. Unfortunately, if this strategy fails, component failure may result.

The second stage of steady-state creep labeled in Fig. 6.1 is the basis for many extrapolation approaches. As is shown later in this chapter, the strain rate for the steady-state creep regime can be modeled with a stress-dependent term and an activation energy term. The simplest form is

$$\dot{\varepsilon}_{ss} = A_{ss}\sigma^{n'} \exp\left(\frac{-Q_{\text{creep}}}{RT}\right) \tag{6.1}$$

where A_{ss} is a constant, σ is the applied stress, n' gives the stress dependence of the strain rate, Q_{creep} is the activation energy for the critical step in the deformation process, and T is the temperature. This relationship is applicable only for combinations of temperature and strain rate wherein steady-state creep is the major fraction of component life. Each of the deformation mechanisms dominates over a particular regime of stress and temperature. Consequently, the values for the constants in Eq. 6.1 apply only for conditions in which the particular mechanism makes the largest contribution to the total strain rate. Often the total strain rate may include contributions from several concurrent deformation mechanisms.

6.2.1 Larson–Miller Parameter

The Larson–Miller parameter (LMP) is an empirical expression using stress rupture data and some modifications of Eq. 6.1. By taking the logarithms of both sides,

$$\ln \dot{\varepsilon}_{ss} = -\frac{Q_{\text{creep}}}{RT} + f(\sigma) \tag{6.2}$$

is obtained wherein $f(\sigma)$ takes into account all other terms in Eq. 6.1. Stress rupture tests with time to failure, t_f, assumed to be inversely proportional to steady-state creep by $t_f = k/\dot{\varepsilon}$ can be related to Eq. 6.2 by

$$\ln t_f - \ln k + f(\sigma) = \frac{Q_{\text{creep}}}{RT} \tag{6.3}$$

By rearrangement and substitution for some terms and conversion of the logarithm to base 10, the normal form of the LMP is obtained as

$$\text{LMP} = T[\log_{10} t_f + C] \tag{6.4}$$

where the constant C is typically near 20 when we use temperature in Kelvin. The original expression employed the absolute Fahrenheit temperature, the Rankine temperature, as $T(^\circ\text{F}) + 460^\circ$. The method can also be applied using temperature in Kelvin.

EXAMPLE 6.1 *Larson–Miller Parameter Calculations Applied to a Steel Alloy*

An extensive set of stress rupture data was generated by Grant and Bucklin (1950). Using these tables, the data can be plotted as stress versus rupture times, as shown in Fig. 6.3*a*. The LMP constant C is adjusted to produce the best fit of data at a variety of stresses and temperatures. If we do so for these data, we can determine a master curve like that shown in Fig. 6.3*b*. How well the data fit to a single line is a function of the C-value chosen. The master curve enables an engineer to make an estimate for conditions other than those actually tested. For example, we might have an application that

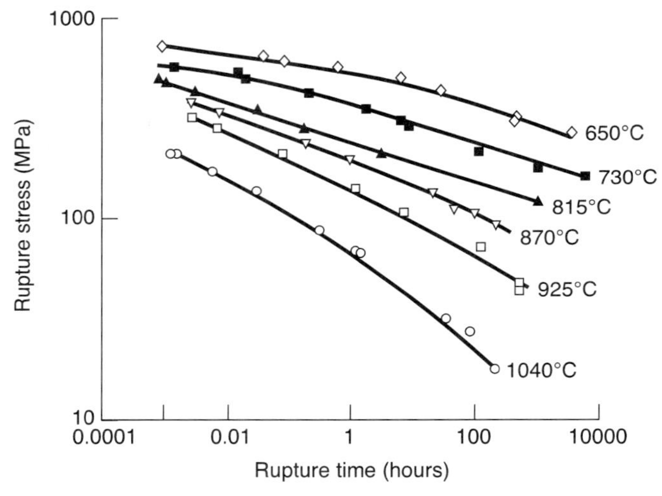

FIG. 6.3 (a) Stress fracture–time plots for an iron-base alloy (S590) at different temperatures (data from Grant and Bucklin, 1950, with permission of ASM Int.)

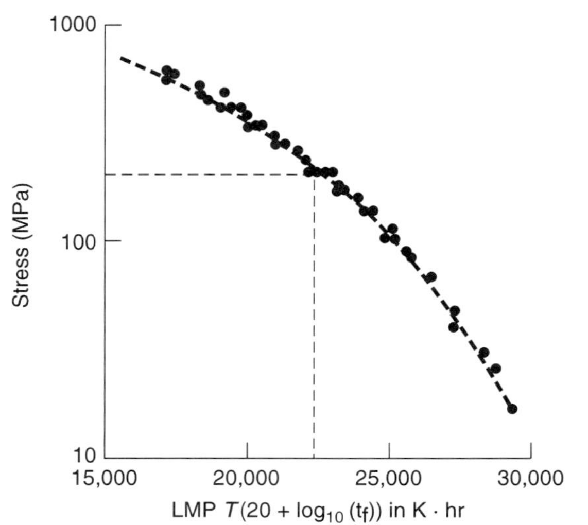

FIG. 6.3 (b) Larson–Miller master plot of the data in (a); the horizontal axis is the parameter $T(C + \log t_f)$. This diagram allows fracture times to be estimated for a variety of stress–temperature combinations).

requires a tensile stress of 100 MPa, a temperature of 630°C, and a lifetime of 100,000 hr, which is longer than the available data. We can use Fig. 6.3*b* to see if this particular alloy might be adequate to fulfill the required application in terms of stress rupture behavior. First, we calculate the LMP.

$$\text{LMP} = 900 \text{ K}[\log_{10}(100{,}000) + 20] = 22{,}500$$

Using this LMP, we can see that the estimated value is about 200 MPa on this log scale. This value would provide a safety factor of only 2 for the given application and would include the uncertainty present in any extrapolation. ■

6.2.2 Monkman–Grant Failure Criterion

Another empirical formulation that is applied to metals and ceramics—the Monkman–Grant failure criterion—employs a power law function of the steady-state creep rate with

$$t_f = A\dot{\varepsilon}_{ss}^{-\rho} \tag{6.5}$$

where A and ρ are material-dependent constants. As is evident in Fig. 6.4, the failure strain in ceramics can be inversely proportional to the steady-state creep rate. By reducing the creep rate, the time to failure can be increased. The exponent ρ is greater than 1 for materials showing this relationship between creep rate and time to failure. If ρ is equal to 1, the failure strain is independent of the strain rate.

6.2.3 Creep of Polymers

Creep of polymers is quite different from the creep of metals and ceramics described previously (see McCrum, Buckley, and Bucknall, 1988). Many polymers are viscoelastic, which means that although strain may increase with time under a given load, the deformation can, with time, reverse, after the load is removed, to nearly the original dimensions. This can be demonstrated by stretching transparent food storage bags to more than 0.05 to 0.1 plastic strain and then laying them on a table. The bags will continue to undergo visible relaxations for several minutes. At small strains ($\leq$1%), viscoelastic creep is nearly linear and can occur at temperatures above 77 K for most polymers. The temperature sensitivity of polymers is quite high. Empirical relations are often applied to extrapolate performance of polymers; however, physical aging processes in polymers that change the molecular structure and uncertainty over accuracy of the extrapolations limit their use in

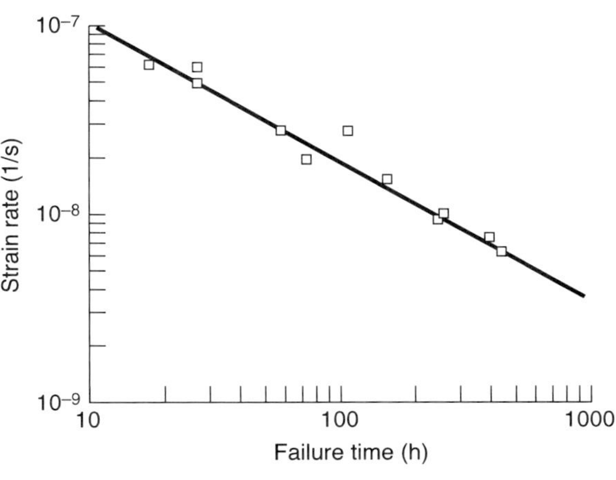

FIG. 6.4 Creep rupture of siliconized silicon carbide with a Monkman–Grant exponent of ρ = 1.45 (Wiederhorn and Hockey, 1991).

design. For polymers, compensation for temperature differences can be approximated by an Arrhenius expression giving the ratio of times

$$\frac{t_e}{t_o} = \exp\left[\frac{\Delta H_{relax}}{R}\left[\frac{1}{T_o} - \frac{1}{T_e}\right]\right] \tag{6.6}$$

where ΔH_{relax} is the activation enthalpy of relaxation, R is the gas constant, and t_e is the extrapolated time at the temperature T_e. T_e is less than the temperature T_o where the same condition (e.g., strain or failure) was reached in the time t_o. The enthalpy can be obtained from plots of the relaxation time versus inverse temperature. For most crystalline polymers and glassy polymers below the glass transition temperature, ΔH_{relax} is 100,000 to 200,000 J/mol.

Design with polymers often requires creep information obtained from manufacturers. Tensile creep curves up to several percent strain are often available. Figure 6.5*a* shows creep curves for a polyether sulphone (PES) material. Because polymers are very sensitive to the molecular weight distribution and processing history, it is important to use curves for identical materials. The data shown in Fig. 6.5*a* can be replotted as Fig. 6.5*b*, the stress-strain relation at fixed times (isochronous curves) or as Fig. 6.5*c*. The large variation in properties at 150°C is a hint that the properties of PES may also strongly depend on temperature. Failure data as a function of temperature in Fig. 6.5*d* confirm this. These data enable a design engineer to predict the stress and temperature conditions relative to an allowable amount of strain.

6.2.4 High-Temperature Deformation of Ceramic and Polymeric Glasses

The formation of glassy materials provides a wide range of properties for materials with the same chemistry. The size of molecules and their character—linear, branched, cross linked,

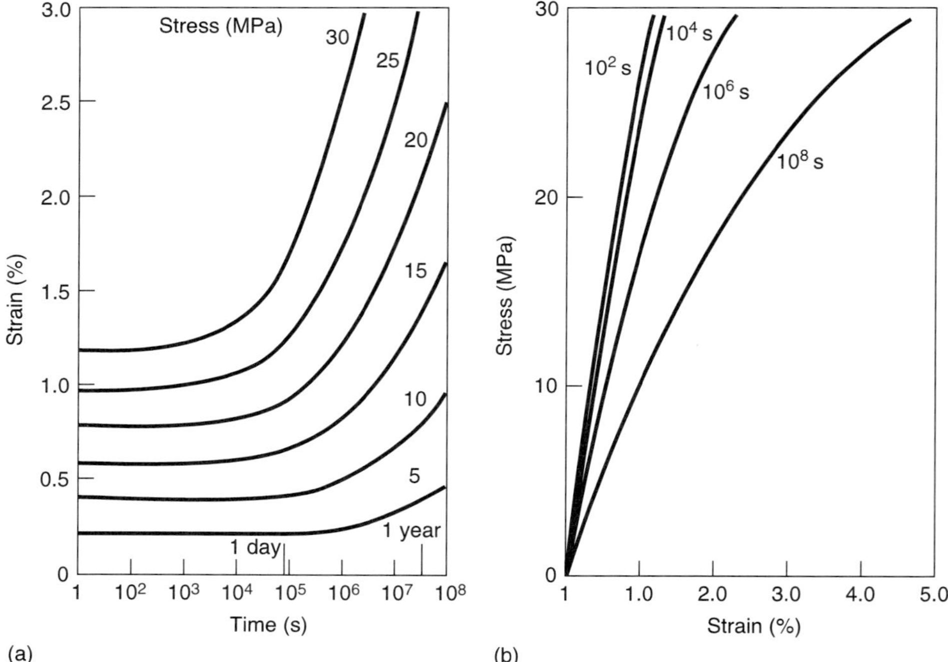

FIG. 6.5 (a) Creep curves for PES at 150°C. (b) Data in (a) replotted at fixed time.

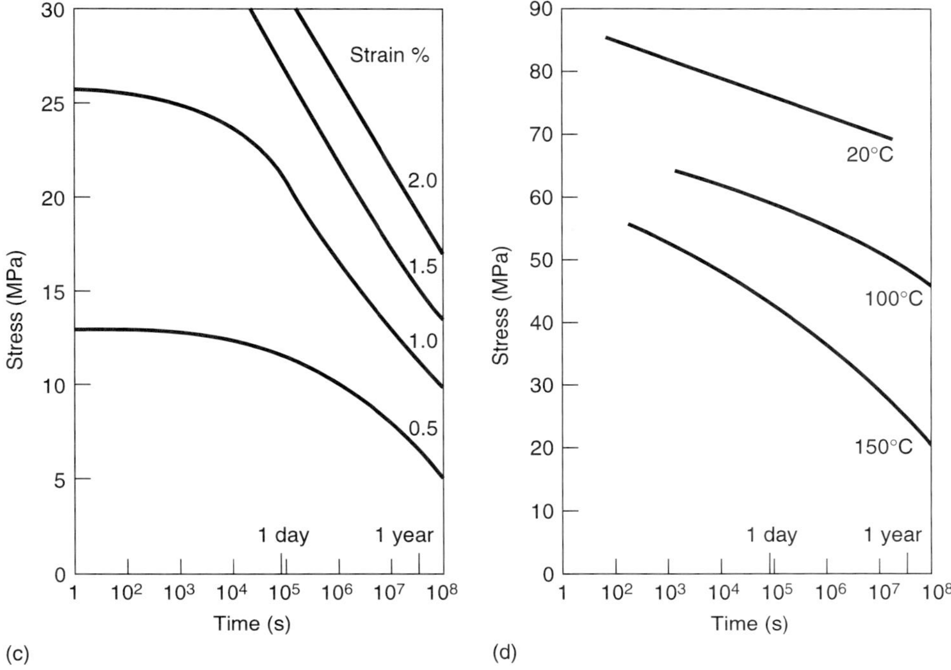

FIG. 6.5 (c) Data in (a) replotted at fixed total strain levels. (d) Temperature dependence of failure stress for PES (McCrum, Bucknall, and Buckley, 1988, used with permission of ICI Group).

or fully networked—determine the properties. These materials are often mixtures of very strong intramolecular bonds and relatively weak intermolecular bonds. For both ceramic and polymeric glasses, the density is also a good predictor of the expected mechanical response. The elastic and time-dependent responses both become stiffer or more rigid as the number of bonds per unit volume increases. Figure 6.6 shows the relationship between specific volume (or the inverse of density) and temperature for crystalline, partially crystalline, and noncrystalline glassy materials. Consistent with Fig. 3.14*a*, there is an abrupt transition in crystalline materials and a gradual transition in glasses. Many glass-forming materials can be at least partially crystallized by holding them just below the melting (or crystallization) temperature for long times. Some polymers and ceramics are specifically designed to

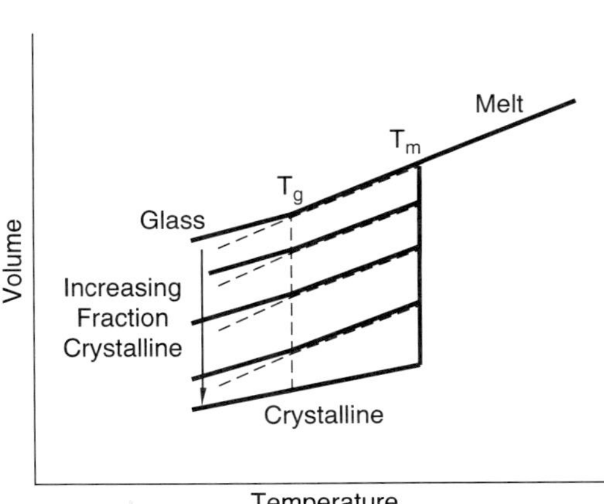

FIG. 6.6 Schematic figure showing volume versus temperature for a material that readily forms a glassy structure.

have a certain crystalline fraction. Polyethylene (PE) is an example wherein control of crystallinity determines the properties. In materials such as high-density polyethylene (HDPE), the high density comes about from having a high fraction of crystalline PE. The crystalline fraction strongly influences the elastic and plastic properties. Figure 6.6 shows the effects of partial crystallinity wherein the final material may consist of crystalline grains surrounded by a noncrystalline phase.

Glass-forming materials that are cooled below the melting temperature continue to undergo significant rearrangements of their structures until the glass transition temperature T_g is reached. The glass transition temperature is often given empirically as a fixed value of viscosity. As shown in Fig. 6.7, the cooling rate can affect the final volume of the glass and the T_g. Figure 6.7 suggests that the T_g is the maximum change in slope of specific volume versus temperature.

Glasses that are cooled faster have a higher T_g and also more "free volume" for intermolecular motion, and thus should deform more readily. The easiest possible form for a temperature-dependent relation would be based on an Arrhenius expression

$$\eta = \eta_o \exp\left(\frac{E_v}{RT}\right) \tag{6.7}$$

where η is the viscosity, η_o is a constant, and E_v is an activation energy for viscous flow. Because the molecular arrangements of a glass vary based on processing history, Eq. 6.7 is valid only for a very narrow temperature range.

The free volume available in a glass is a recognition that motion of part or all of one molecule past another is made easier if the molecules are not tightly packed. The free volume approach suggests an approximation for viscosity

$$\eta = \eta_o \exp\left(\frac{DV_c}{V_f}\right) \tag{6.8}$$

where D is a proportionality constant, V_c is the specific or molecular volume of the crystallized form, and V_f is the free volume. This relationship does not accommodate changes in networking or formation of macromolecules within the glass. Increased networking or an increase in molecular size (molecular weight) can result in higher viscosities.

The free volume consists of the fraction of free volume "frozen" in at the glass transition temperature, f_g, and the difference in the volume of the glass, V_g and V_c. Therefore we can write

$$V_f = (V_g - V_c) \tag{6.9}$$

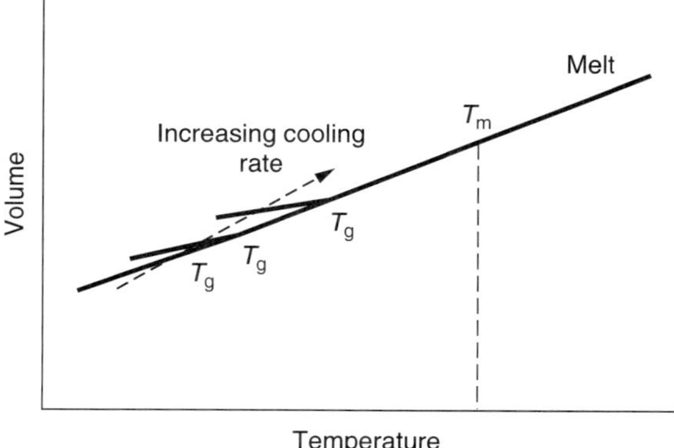

FIG. 6.7 Effects of cooling rate on the specific volume of a completely glassy material. This shows that completely noncrystalline materials can have a range of densities. The glass transition temperature increases with increasing cooling rate.

Differences in the glass and crystal volumes are related to the difference in specific volume between the glass and crystal at the glass transition temperature and the temperature dependence of the volumetric thermal expansion difference between the glass and crystal. Then we can write

$$\frac{V_c}{V_f} = \frac{1}{\Delta V_g + 3\Delta\alpha\left(T - T_g\right)} \tag{6.10}$$

where ΔV_g is the volume difference between the glass and crystal forms at the glass transition temperature and $3\Delta\alpha$ is three times the difference in the linear thermal expansion coefficients of the glass and crystal forms. The factor of 3 comes from the isotropic volume strain ($\varepsilon_{vol} = \varepsilon_1 + \varepsilon_2 + \varepsilon_3$) for volumetric expansion. The higher thermal expansion coefficient of the glass arises from the more open intermolecular arrangement.

6.3 STRESS RELAXATION

Stress relaxation (see Fig. 1.14) is a time-dependent response of a material to elastic straining. Because the bonds are stretched elastically, processes of rearrangement within the material can act to relieve elastic stress. The driving force for relaxation processes is simply the elastic energy per unit volume (elastic portion of the area under the stress-strain curve). Stress relaxation tests are conducted by stretching the material to a fixed elastic strain while it is in series with a load cell. The material undergoes stress relaxation as the elastic strain is replaced by plastic strain. The final component deforms at a rate that is driven by a constantly reducing stress. Following stress relaxation, the component has undergone plastic strain corresponding to the reduction in stress.

Stress relaxation is an important process in processing of engineering components. Rapid, nonuniform cooling of components from a high temperature can result in residual stresses. Residual stresses often result from nonuniform cooling or nonuniform deformation of components. Although all of the stresses within a component must balance, the elastic state of the material can vary with position. It is often favorable to induce a state of compressive stress near surfaces to enhance a material's resistance to crack initiation. Many heat treatments, mechanical working processes, and surface treatments are designed to produce this type of strengthening. Unfavorable residual stresses can be relieved by heating the material to a high temperature wherein creep mechanisms are sufficiently active to drive stress relaxation. Glass laboratory equipment or decorative articles made by glass blowers should always be heat treated to relieve internal stresses that might cause premature failure.

Stress relaxation also plays a part in the development of microscale stresses in composites and materials with noncubic crystal structures. As described in Chapter 2, the development of stresses as a result of differences in thermal expansion coefficients can result in stresses between different phases or differently oriented anisotropic grains. The starting point for calculations of residual stresses and internal microscale stresses owing to thermal expansion mismatch is usually a temperature at which the stress relaxation rate is sufficiently high that an assumption is made that the material is uniformly stress-free.

EXAMPLE 6.2 *Tailored Stresses in Glasses*

The resistance of glass to fracture can be improved by producing compressive residual stress on surfaces. This is usually preformed by rapidly cooling the glass surface through the use of air jets. Evidence of heterogeneous cooling is often observable in glasses by photoelastic fringes showing a spectrum of colors corresponding to the stress state of the glass. These fringes are usually visible if

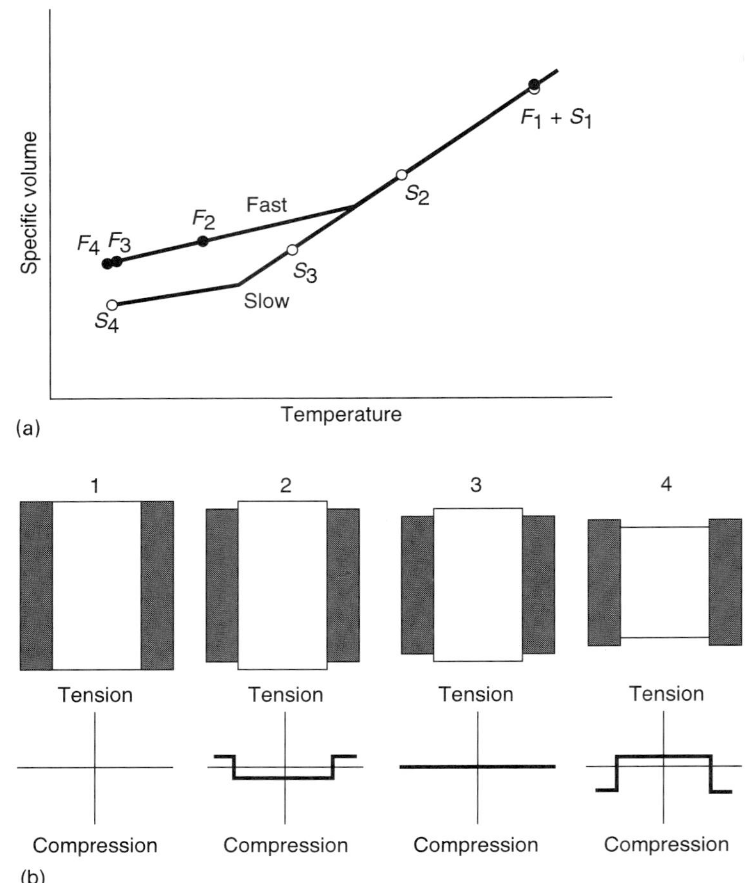

FIG. 6.8 (a) Specific volume versus temperature relationship for fast cooled surfaces (*F*) and a slowly cooled interior (*S*) corresponding to part (b). (b) The fast cooling exterior portion of this section from a glass plate subjected to cooling on both surfaces is shaded. The dimensional changes at sequential times corresponding to part (a) are shown with free displacements in the vertical direction (changes in thickness are not indicated). Schematic stress distributions are shown in the lower segments, with times 1 and 3 being essentially free of internal stresses.

Using residual stresses to improve the fracture resistance of ceramics is a powerful way to improve the performance of glasses and ceramics, but it includes a risk. If the internal tensile residual stresses are very high, deep scratches or damage that reach these internal tensile stresses can trigger dramatic and sometimes hazardous fractures.

---

you are wearing polarized sunglasses. The presence of compressive residual stress at a surface results in a suppression of crack propagation from surface flaws, which are the most likely initiators of fracture.

Figure 6.8 shows the process undergone by the fast cooled surface (F) and the slowly cooled interior (S) sections of the glass. The subscripts in Fig. 6.8*a* give the temperature and specific volume for each of the times indicated in Fig. 6.8*b*. At the start of the process (time 1), the exterior and interior sections are assumed to be stress-free. As the exterior is cooled rapidly, it would prefer to be smaller than the exterior, but the exterior and interior are mechanically coupled so that a stress distribution results with tension on the fast cooled surface and compression of the interior. The fast cooled material never relaxes to the smaller specific volume possible through slow cooling and remains elastic through the remainder of the cooling process. If the quench is too rapid at this point, the surface is susceptible to fracture. This shows that the quenching rate must be slow enough to avoid fracture at this point.

At time 3, the fast cooled (exterior) and slowly cooled (interior) sections have the same specific volume, which should relax nearly all stresses. This is assisted by the greater potential for relaxation of the slow cooling interior. By time 4, the slow cooling of the interior has resulted in a smaller specific volume than that of the exterior. This differential in specific volumes places the surface in the desired state of residual compression. ■

6.3.1 Mechanical Analogs for Creep and Stress Relaxation

The time-dependent mechanical responses of many partially crystalline and noncrystalline materials can be accomplished through simple mechanical analogs. The two most common elements for constructing mechanical analogs are springs and dashpots. A spring element is just that, a simple spring constant to describe elastic responses using Hooke's law for a uniaxial response, $\sigma = E\varepsilon$. A dashpot is the damping component in a shock absorber. The mechanical response of a dashpot is directly a viscous response where $\sigma = 3\eta\dot{\varepsilon}$. By assembling springs and dashpots in series (a Maxwell element) or in parallel (a Voigt element), different time-dependent mechanical responses can be modeled. The design of the final model is then related to the expected responses of the strong and weak bonding within the material.

For the Maxwell element, the spring and the dashpot are in series and undergo the same stress, and their strains are additive. The rate equation for the two elements in series is then

$$\dot{\varepsilon} = \frac{d\varepsilon}{dt} = \frac{1}{E}\frac{d\sigma}{dt} + \frac{\sigma}{3\eta} \tag{6.11}$$

The rate equation for the Voigt element requires the same strain, yielding

$$3\eta = \frac{d\varepsilon}{dt} = +E\varepsilon = \sigma \tag{6.12}$$

Table 6. 1 shows the separate responses to creep and stress relaxation for each element type.

6.3.2 "Jump" Measurements of Time-Dependent Responses

Evaluating the effects of stress and temperature on time-dependent deformation at first appears quite challenging. To recover a complete set of data incorporating the stress dependence and the temperature dependence of the deformation could require extensive testing. With some simplifying assumptions, it is often possible to recover the stress and temperature dependence of deformation for Eq. 6.1 on just one or a fairly small number of specimens. To recover the stress dependence of strain rate, we know that

$$\sigma^{n'} \propto \dot{\varepsilon}$$

where $n' = \frac{1}{m}$ and m is the strain rate sensitivity. If the data are plotted on a log–log plot of strain rate versus stress, the slope of this curve gives the stress exponent n' for individual deformation mechanisms for a number of materials, as shown in Fig. 6.9.

TABLE 6.1 Simple Spring and Dashpot Elements

	Voigt	Maxwell
General expression		
Mechanical model	Spring Dashpot	Spring Dashpot
Differential equation	$\frac{d\varepsilon}{dt} = \frac{1}{E}\frac{d\sigma}{dt} + \frac{\sigma}{3\eta}$	$3\eta\frac{d\varepsilon}{dt} + E\varepsilon = \sigma$
Stress relaxation	$\sigma = E\varepsilon e^{-Et/3\eta}$	$\sigma = E\varepsilon$
Creep deformation	$\varepsilon = \frac{\sigma}{E} + \frac{\sigma t}{3\eta}$	$\varepsilon = \frac{\sigma}{E}(1 - e^{-Et/3\eta})$

The stress exponent can also be recovered by performing a temporary "jump" up or down in strain rate, as shown in Fig. 1.15, or in stress level, as shown in Fig. 6.10. Assuming that the microstructure is essentially unchanged and there are no adiabatic heating effects, the respective change in stress or strain rate can be used to recover m or n'. The ratio of strain rates is given as

$$\frac{\dot{\varepsilon}_1}{\dot{\varepsilon}_2} = \left(\frac{\sigma_1}{\sigma_2}\right)^{n'} \tag{6.13a}$$

and the ratio of stresses is given by

$$\frac{\sigma_1}{\sigma_2} = \left(\frac{\dot{\varepsilon}_1}{\dot{\varepsilon}_2}\right)^{m} \tag{6.13b}$$

The temperature jump tests (see Fig. 6.11) enable recovery of the activation energy Q_{creep} in Eq. 6.1 for the active creep mechanism by taking a similar ratio of the two strain rates for differing temperatures, where

$$Q_{\text{creep}} = \frac{R\ln(\dot{\varepsilon}_1/\dot{\varepsilon}_2)}{1/T_1 - 1/T_2} \tag{6.14}$$

Assumptions included in this include the activity of a single mechanism and no change in the stress exponent. Similar formulations can be used to calculate n' and Q in tensile tests, as shown in Example 6.3.

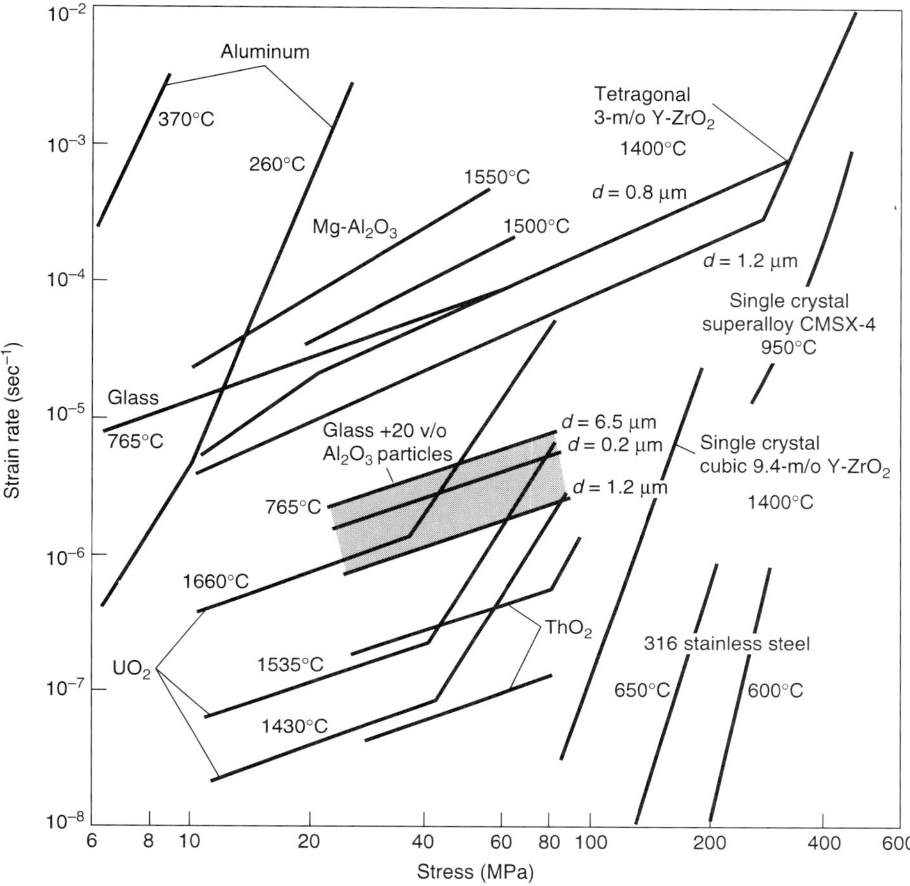

FIG. 6.9 Steady-state creep rate versus stress for UO_2 and ThO_2 with a grain size of 10 μm, Mg doped Al_2O_3 with a grain size of 2 μm, several yttrium oxide doped ZrO_2 materials (grain sizes indicated), 316 stainless steel, aluminum, a single-crystal superalloy, an aluminosilicate glass, and aluminosilicate glass reinforced with Al_2O_3 particles (particle sizes indicated). These materials show diffusion-based creep with an exponent approaching 1 at low stresses. At high stresses (regions with steeper slopes), power law creep controls deformation. The Mg-Al_2O_3 contains 300 ppm Mg, which acts to suppress grain growth. The tetragonal 3 mole percent Y-ZrO_2 alloys show a similar transition in deformation mechanism to the pure materials and demonstrate a power law behavior that is similar to that of the cubic single crystals doped with 9.4 mole percent Y-ZrO_2. The stainless steel and aluminum alloys show power law creep over the given ranges, although data for lower stresses would show a transition to diffusion creep mechanisms. The single-crystal superalloy CMSX-4 shows power law creep behavior with a very high resistance to creep deformation at high fractions of its melting temperature. The aluminosilicate glass, which has a glass transition temperature of 926°C, has a slope of 1, showing Newtonian response. The same glass containing various particle sizes of Al_2O_3 shows much lower creep rates and a similar Newtonian response. (Data for UO_2 and ThO_2 from Poteat and Yust, 1968; Mg doped Al_2O_3 from Kottada and Chokshi, 2000; Y-ZrO_2 from Chokshi, 2002; 316 stainless steel from Sasikala et al., 2000; aluminum from Sherby and Burke, 1968; single-crystal superalloy CMSX-4 from MacLachlan and Knowles, 2000; glass and Al_2O_3-reinforced glass from Sudhir and Chokshi, 2003.)

Making comparisons between materials for the type of data shown in Fig. 6.9 is difficult unless normalizations are applied relative to melting temperature and the elastic constants (see Mukherjee, Bird, and Dorn, 1968). Section 8.1 shows a strategy that provides such normalizations in what are called *deformation mechanism maps*.

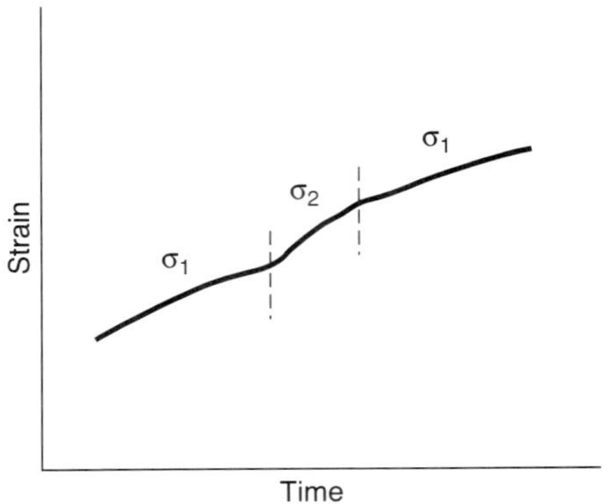

FIG. 6.10 The increase in strain rate corresponding to an increase in stress, $\sigma_2 > \sigma_1$.

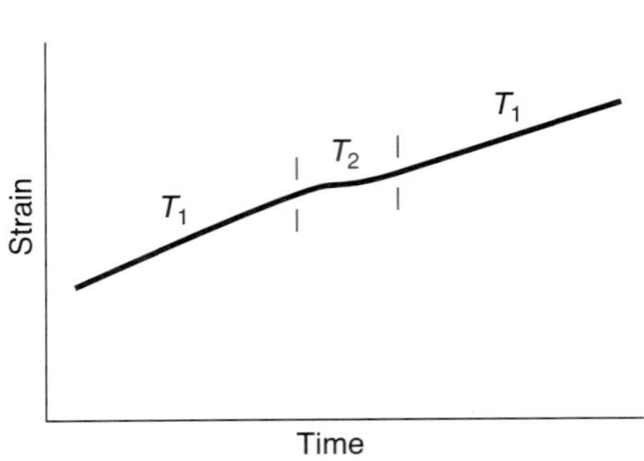

FIG. 6.11 The decrease in strain rate corresponding to an increase in temperature, $T_2 > T_1$.

EXAMPLE 6.3 *Stress Exponent and Activation Energy of Silicon Nitride with an Evolving Microstructure*

Mechanical test results on the ceramic material silicon nitride (Si_3N_4) demonstrate why it is important to recognize that changes in microstructure can alter the rate sensitivity and activation energy observed for a given material. As we will see in more detail in Section 6.4.5, it is possible to deform some materials to such large plastic strains the materials are considered "superplastic." If the microstructure changes through recrystallization or grain growth during the high-temperature deformation process, then the values of the "constants" in Eq. 6.1 may change. The effect of varying the displacement rate (strain rate jumps) is shown for a silicon nitride material by the true stress–true strain curve in Fig. 6.12*a*. This particular silicon nitride material has elongated grains that become aligned as they flow past one another as a result of the presence of a glassy grain-boundary phase that has a low viscosity at high temperatures. These grains have a hexagonal crystal structure, and dislocation motion is very difficult such that flow of the glassy grain-boundary phase is almost the only way plastic deformation can occur at significant strain rates. The alignment of the rodlike silicon nitride grains results in the orientation strengthening behavior discussed in Chapter 5 and is a very different mechanical response from that seen in a superplastic metal. Here the magnitude of the stress exponent $n' = 1/m$ increases with strain, as shown in Fig. 6.12*b*. The calculation used was

$$n' = \frac{\ln\left(\frac{CH_1}{CH_2}\right)}{\ln\left(\frac{s_1}{s_2}\right)}$$

where *CH* is the crosshead speed and *s* is the nominal stress. The subscript 1 represents the initial condition, and the subscript 2 is the jump value.

The effect of varying the temperature on the deformation of the same silicon nitride material enables recovery of the activation energy for the mechanism(s) that is (are) occurring to produce the observed deformation. An example of such a jump is shown in Fig. 6.13*a*. The dependence of the activation energy for a limited range of strains is shown in Fig. 6.13*b*. The activation energy was calculated using the expression

$$Q = \frac{Rn'(\ln s_2 - \ln s_1)}{\frac{1}{T_2} - \frac{1}{T_1}}$$

■

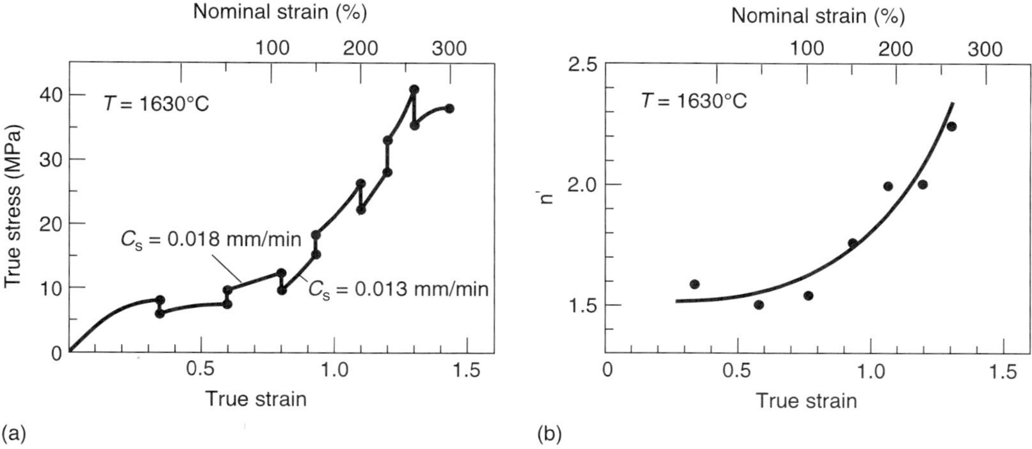

FIG. 6.12 (a) True stress–true strain behavior of a silicon nitride material with an evolving microstructure that results in elongated grains lining up parallel to the tensile axis. (b) Calculated values of n' measured in (a) (Kondo et al., 1999, used with permission of Elsevier Limited).

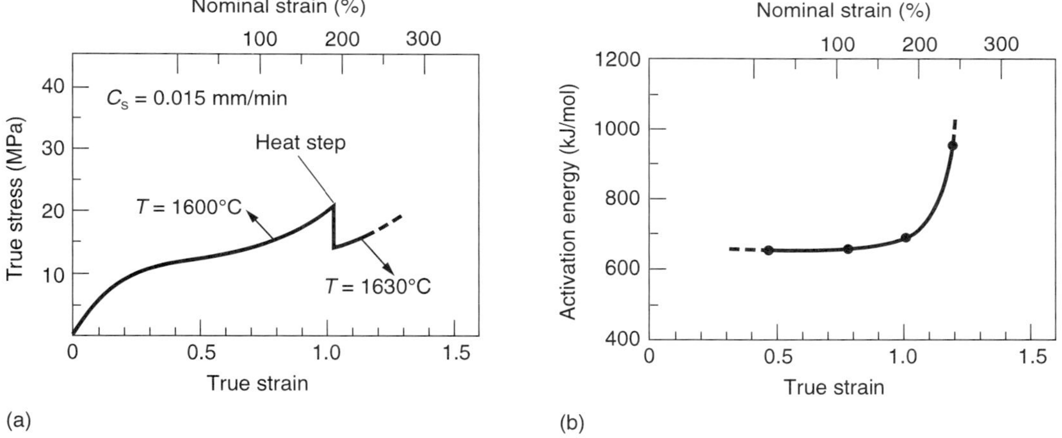

FIG. 6.13 (a) Reduction in the stress required for deformation of silicon nitride by a temperature jump of 30°C, which enables calculation of the activation energy. (b) Dependence of the activation energy for silicon nitride on the plastic strain that is accompanied by orientation strengthening (Kondo et al., 1999, used with permission of Elsevier Limited).

6.3.3 Adiabatic Heating

The temperature of a material increases during plastic deformation owing to the heat that comes from mechanical work. This is readily demonstrated by feeling the heat generated during rapid plastic deformation of a paper clip. Nearly all of the plastic work (area under the stress-strain curve) is released as heat. Part of the mechanical work is stored by rearrangement of the molecules in noncrystalline materials or by an increase in the internal strain energy through increased dislocation density in crystalline materials. Some work is also involved through changes in surface area by means of the strain energy. The total amount of work storage is usually less than 10 percent of the total mechanical energy. If we use the plastic work, the energy per unit volume is given as

$$w = \int_{o}^{\varepsilon_{eff}} \sigma_{eff} d\varepsilon_{eff} \tag{6.15}$$

If we ignore heat transfer to the surroundings, then the self-heating of a uniformly deformed specimen should be

$$\Delta T \approx \frac{f \int \sigma_{eff} d\varepsilon_{eff}}{\rho C} = \frac{f \overline{\sigma}_{eff} \varepsilon_{eff}}{\rho C} \tag{6.16}$$

where f is the fraction of stored energy, $\overline{\sigma}_{eff}$ is the average effective stress, ε_{eff} is the total effective strain, ρ is the density, and C is the mass heat capacity in J/(kg°C). Because heat transfer is dependent on time, deformation of materials at high rates can lead to substantial self-heating. Because an increased temperature results in a reduction in the flow stress, deformation instabilities become more likely at higher strain rates. This is particularly true for materials with poor heat transfer properties. Because heat transfer properties decrease with increasing temperature, the susceptibility of materials to localized deformation from self-heating is generally the greatest under hot working conditions. This synergy among self-heating, poor heat transfer, and temperature sensitivity of flow stress makes polymers particularly sensitive to localized deformation.

6.4 CREEP AND RELAXATION MECHANISMS IN CRYSTALLINE MATERIALS

For permanent deformation of a material to take place, a series of critical atomic scale steps must take place. In crystalline materials deforming by dislocation glide, stable kinks must form on dislocations and then the kinks must propagate to advance the dislocation by one Burgers vector. In crystalline materials deforming by dislocation climb, vacancies must move to or from jogs on dislocations to result in deformation strains. In noncrystalline polymers with large molecules, parts of the individual molecules must undergo relative rearrangements that result in an overall displacement. Each of these processes entails a critical step that depends at least in part on the thermal vibrations associated with bonding (see Kocks, Argon, and Ashby, 1975; Weertman, 1968, and Seeger, 1957). The energy that must be employed to overcome barriers determines both the temperature and strain rate dependence of the deformation. Since thermal energy is typically on the order of kT, the increased amplitude of thermal vibrations provides assistance in overcoming activation barriers.

6.4.1 Rate-Dependent Dislocation Glide in Crystalline Materials

The strain rate dependence of slip comes from the Orowan equation in the form

$$\dot{\gamma} = \rho_{\perp} b \overline{v} \tag{6.17}$$

and the rate expression

$$\dot{\gamma} = \dot{\gamma}_o \exp\left(\frac{-\Delta G(\tau)}{kT}\right) \tag{6.18}$$

where $\overline{v}$ is the average dislocation velocity and $\Delta G(\tau)$ is the stress-dependent activation energy for slip. The value of $\Delta G(\tau)$ is inversely dependent on the shear stress τ. The preexponential term in Eq. 6.18 is

$$\dot{\gamma}_o = \rho_\perp b\, s_\perp f_{o\perp} \tag{6.19}$$

where $s_\perp$ is the average distance swept out by a dislocation for every thermal fluctuation, $\rho_\perp$ is the mobile dislocation density, and $f_{o\perp}$ is the attempt or jump frequency. The local dislocation velocity $s_\perp \cdot f_{o_\perp}$ can be at most the speed of sound in the material. The dislocation jump frequency is a fraction of the vibration frequency $f_{\text{intrinsic}}$, scaled to the ratio of the Burgers vector and the dislocation segment length l by

$$f_{o_\perp} = \frac{b}{l}\left(f_{\text{intrinsic}}\right) \tag{6.20}$$

For low-temperature deformation of alloys, and particularly in pure BCC metals, the deformation is temperature-dependent. The extent of that temperature dependence is shown by the nearly fivefold change in yield stress for the stress-strain curves of high-purity BCC iron single crystals shown in Fig. 6.14*a*. The temperature dependence of deformation can be separated into an "athermal" component that scales with the temperature dependence of modulus τ_μ and a component that shows a stronger temperature dependence τ^* with

$$\tau = \tau_\mu + \tau^* \tag{6.21}$$

In BCC metals, the difficulty in moving screw dislocations provides most of the intrinsic rate dependence for low-temperature deformation. Yield stress data of the type in Fig. 6.14*a* can be normalized to the temperature-dependent shear modulus to give a relation similar to that shown in Fig. 6.14*b*.

The stress dependence for the activation energy of deformation can be expressed as

$$\Delta G(\tau) = \Delta G_{o,\gamma} - V^* \tau \tag{6.22}$$

where V^* is defined as the activation or critical deformed volume. The first term represents the modulus-dependent energy barrier to deformation

$$\Delta G_{o,\gamma} + \Delta G_o - V^* \tau_\mu \tag{6.23}$$

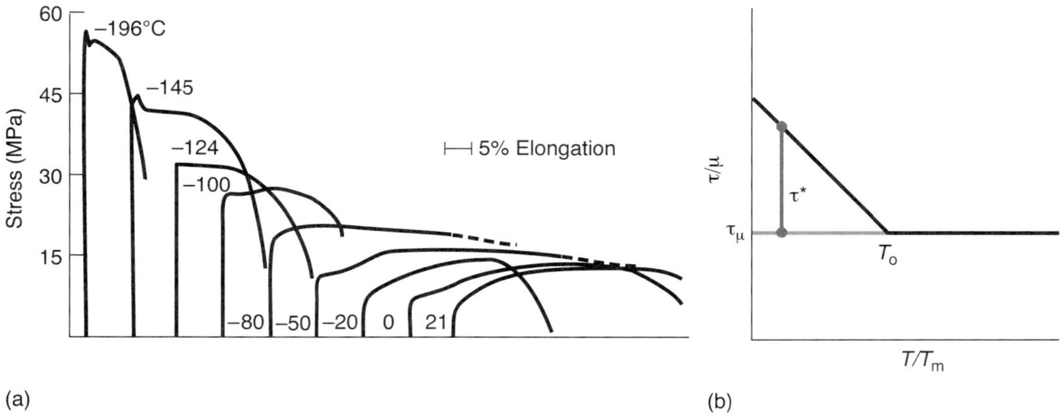

FIG. 6.14 (a) Series of tensile stress-strain curves for high-purity iron crystals with test temperatures given for each curve (after McLean, 1962). (b) Shear stress τ for deformation as a function of temperature normalized to the shear modulus. The terms τ_μ and τ^* indicate the magnitude of the athermal and thermal components of the shear stress. Below a critical temperature T_o, deformation is temperature- and rate-dependent.

which can be substituted into Eq. 6.22 as

$$\Delta G(\tau) = \Delta G_o - V^*(\tau - \tau_\mu)$$

yielding

$$\Delta G(\tau) = \Delta G_o - V^*\tau^* \tag{6.24}$$

The scaling of energy barriers associated with τ^* and τ_μ provides that the barriers associated with τ^* are more localized than those associated with τ_μ. Rearranging the expressions above yields

$$\tau = \tau_\mu + \tau^* \tag{6.25}$$

and substituting for τ^* leads to

$$\tau = \tau_\mu + \frac{\left[\Delta G_o - kT\ln\left(\frac{\dot{\gamma}_o}{\dot{\gamma}}\right)\right]}{V^*} \tag{6.26}$$

The activation barrier to slip at absolute zeron ΔG_o, is not dependent on strain rate. Below a critical temperature, slip becomes thermally activated and the critical temperature for this transition is strain-rate-dependent. The critical temperature can be given as

$$T_o = \frac{\Delta G_o}{k\ln\left(\frac{\dot{\gamma}_o}{\dot{\gamma}}\right)} \tag{6.27}$$

In this form, the critical temperature for the transition shown in Fig. 6.14 is strain-rate-dependent, as shown in Figure 6.15.

The strain rate sensitivity can be written as

$$m = \left(\frac{\partial \ln \tau}{\partial \ln \dot{\gamma}}\right)_{T,\gamma} \tag{6.28}$$

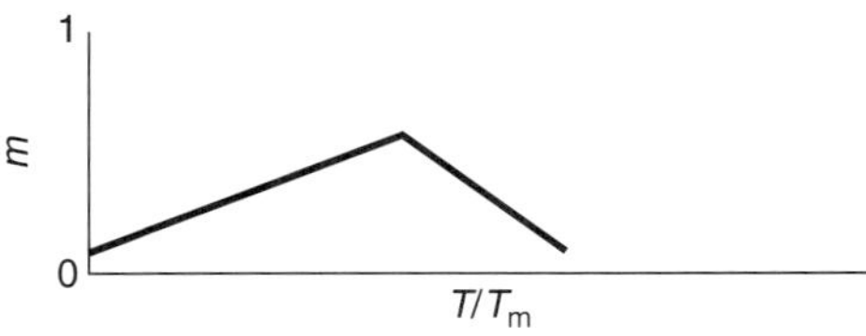

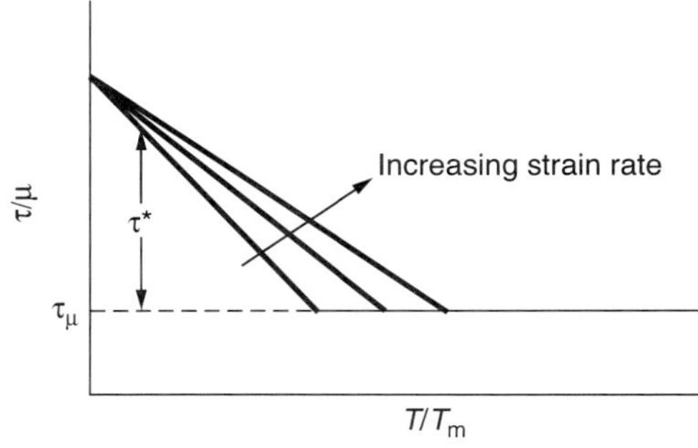

FIGURE 6.15 The theoretical change in flow behavior with increasing strain rate. The strain rate sensitivity, as inset schematically, is dependent on temperature.

with a fixed temperature T and strain history γ. Figure 6.15 shows that m is a maximum in the vicinity of the critical temperature for experiments conducted at two different strain rates.

An alternative and widely used description of the stress dependence for the slip activation energy (see Kocks, Argon, and Ashby, 1975) is

$$\Delta G(\tau) = \Delta G_o\left[1-\left(\tau/\tau_R\right)^p\right]^q \tag{6.29}$$

where τ_R is the stress required to overcome obstacles, $\mu b/l$, where l is obstacle spacing. Then, if p and q both equal 1, we get

$$\dot{\gamma} = \dot{\gamma}_o\left(\frac{\tau}{\tau_R}\right)^{\Delta G_o/kT} \tag{6.30}$$

which is equivalent to an expression for power law creep (see next section)

$$\dot{\gamma} \propto \tau^{n'} \tag{6.31}$$

For a range of obstacle spacings and obstacle geometries, the constants for dislocation glide fall into the ranges

$$\begin{gathered} 0 < p < 1 \\ 1 < q < 2 \end{gathered} \tag{6.32}$$

The obstacle strengthening strain rate expression can be given as

$$\dot{\gamma} = \dot{\gamma}_o \exp\left[\frac{\Delta G_o}{kT}\left(1-\frac{\tau}{\tau_R}\right)\right] \tag{6.33}$$

with

$$\dot{\gamma}_o = \alpha\left(\frac{\tau}{\mu}\right)^2 f_{o_\perp} \tag{6.34}$$

where α is a constant that depends on dislocation arrangements. For strong obstacles, the stress dependence of $\dot{\gamma}_o$ in Eq. 6.34 can be set as a constant from 10^5 to 10^6/s. Table 6.2 shows the relative values of ΔG_o and τ_R for different types of obstacles.

TABLE 6.2 Strengthening Components in Eq. 6.33 (after Frost and Ashby, 1982)

Obstacle strength	ΔG_o	τ_R	Example
Strong	$2\ \mu b^3$	$> \frac{\mu b}{l}$	Strong precipitates
Medium	$0.2\text{–}1\ \mu b^3$	$\approx \frac{\mu b}{l}$	Dislocations
Weak	$<0.2\ \mu b^3$	$<< \frac{\mu b}{l}$	Solution strengthening

EXAMPLE 6.4 *Slip at High Temperatures in Zirconia Ceramics*

The brittleness of most ceramic materials makes measurement of the flow stress for dislocations very difficult at low temperatures, because flaws tend to initiate failure before plastic deformation occurs. In these circumstances, the heavily constrained conditions of a hardness test can enable estimation of the flow stress even though it will always be exaggerated by the friction and constraint effects discussed at the beginning of Chapter 5. Figure 6.16*a* shows hardness measured in a special microhardness tester that allows the sample to be heated to high temperatures under vacuum. The specific measurements shown are for a number of different forms of zirconia (ZrO_2) ceramics. Zirconia can exist in many forms, depending on temperature, alloy content, and thermomechanical history. The cubic form has the fluorite crystal structure, and the tetragonal form is a distortion of the cubic structure with the c/a ratio slightly greater than 1. Each of the polycrystalline and single-crystal alloy materials tested shows a very strong dependence of hardness on temperature. Figure 6.16*b* shows highlighted slip markings around an indent in a single-crystal yttria-zirconia alloy indented on a {100} crystal face. These cubic single crystals contain 9.5 mole percent Y_2O_3, which should provide substantial solid solution strengthening for all temperatures. The Ce-TZP (tetragonal zirconia polycrystal) and Y-TZP materials have fairly small grain sizes and consist almost entirely of tetragonal grains. Both of these materials can undergo twin boundary motion and a phase transformation that causes shear and dilatant changes in the unit cell (tetragonal to monoclinic) that enable some permanent deformation to occur at low temperatures, as discussed in Chapter 3. This is most likely the reason that the polycrystalline forms are softer than the strongly solid solution strengthened cubic single crystal at temperatures below about 400°C. The Mg-PSZ (partially stabilized zirconia) material consists of a matrix of cubic grains containing tetragonal precipitates. This alloy benefits from solid solution strengthening and precipitation strengthening. The tetragonal precipitates can also undergo a shear and dilatant transformation from the tetragonal to the monoclinic form. This may help explain why the Mg-PSZ material is softer than the cubic yttria single crystal below about 200°C.

To demonstrate that slip is extraordinarily difficult in these materials, consider the stress-strain curves in Fig. 6.17. This figure shows compression data on single crystals of a tetragonal yttria-zirconia alloy tested along the <112> directions. This orientation should be quite soft, because the Schmid factor for the easy slip system <110>{001} is approximately 0.47. Because this is a tetragonal material, the <112> directions are not the same as the <121> directions, nor are the {001} planes the same as the

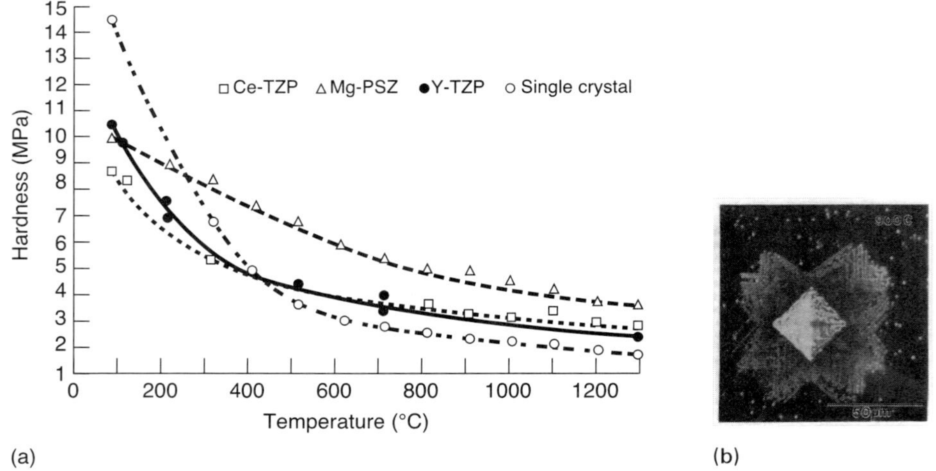

FIG. 6.16 (a) Hardness versus temperature for several zirconia materials (Tikare and Heuer, 1991). (b) Slip markings on the {100} face of a single-crystal yttria-zirconia material indented at 900°C. (Morscher, Pirouz, and Heuer, 1991). [Both (a) and (b) reprinted with permission of The American Ceramic Society, PO Box 6136, Westerville, Ohio 43086-6136. Copyright (1991). All rights reserved.]

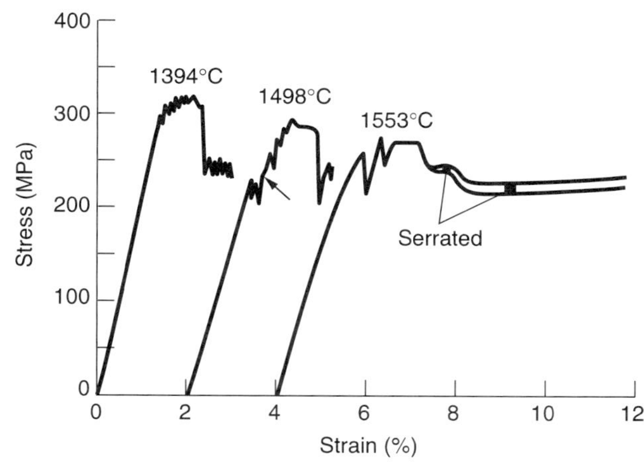

FIG. 6.17 Compressive stress-strain behavior of <112> oriented single crystals of a 3.4 weight percent Y_2O_3-ZrO_2 alloy. The deformation is very serrated, indicating difficulty in initiating dislocation multiplication and twinning. Also, because these are compression tests, the low plastic strains before failure suggest that tensile deformation might show little or no plasticity for similar strain rates (Muñoz, Wakai, and Dominguez-Rodriguez, 2001, with permission of Elsevier Limited).

{100} planes because of the reduced symmetry. The stress levels for deformation at these temperatures are comparable to room-temperature values for many high-strength polycrystalline metal alloys, but it should be remembered that pure zirconia melts at about 2800°C. Although transmission electron microscopy results confirm that some dislocation motion has occurred in the compression tests shown in Fig. 6.17, some twins are also present in the specimens. ■

EXAMPLE 6.5 *Anomalous Yield Behavior in Ni_3Al Intermetallics (γ')*

The nickel-aluminum intermetallic called γ' Ni_3Al has an unusual dependence of yield stress on temperature. Unlike most other materials, over a wide range of temperatures the yield stress of metal compounds with this structure increases to a maximum. This phenomenon is shown for two different γ' single-crystal orientations in Fig. 6.18. The yield stress increase continues over a wide range of temperatures until reaching a maximum between 800 and 1000°C. Precipitates with this structure are those found in nickel-base superalloys, suggesting one way in which superalloys are able to resist creep so well at high temperatures. The ordered structure of Ni_3Al, which is cubic with nickel atoms on the faces of a unit cell and aluminum atoms on the corners, constrains slip into dislocation pairs, as suggested by the TEM micrograph shown in Fig. 4.38. The dislocations that produce the deformation slip on the <101> {111} system, but the Burgers vector length of the dislocations $\frac{1}{2}$<101> only displaces the nickel atoms in the aluminum sites. This means that the order of the intermetallic will be locally interrupted (or out of phase). To be a unit dislocation, it must go from one aluminum atom to another or from one nickel atom to another. The two $\frac{1}{2}$<101> dislocations (Fig. 4.38) added together return the perfect structure. These dislocations also move differently for tensile and compressive deformation, and this difference is orientation-dependent, as can be seen in Fig. 6.18. The deformation behavior after the maximum becomes strongly influenced by the power law creep mechanisms discussed in the next section and slip on other slip systems is also observed. ■

6.4.2 Power Law Creep

For creep, the steady-state, stage 2 strain rate is often used as a basis for creep mechanisms. The strongly sloped portions of the creep curves in Fig. 6.9 show power law creep (PLC) for several materials. Between $0.3T_m$ and $0.6T_m$, metals and ceramics often show a strong dependence on strain rate, consistent with Eq. 6.30 and 6.31. In this intermediate temperature regime and at high stresses, deformation occurs by a combination of glide and diffusion. The power law exponent n′ in Eq. 6.31 can range from 3 to 10. The lower range of 3 is determined by the combination of a low value of ΔG_o in Eq. 6.33 and the preexponential factor in Eq. 6.34. For low ΔG_o, the applied stress in the exponential has a linear stress

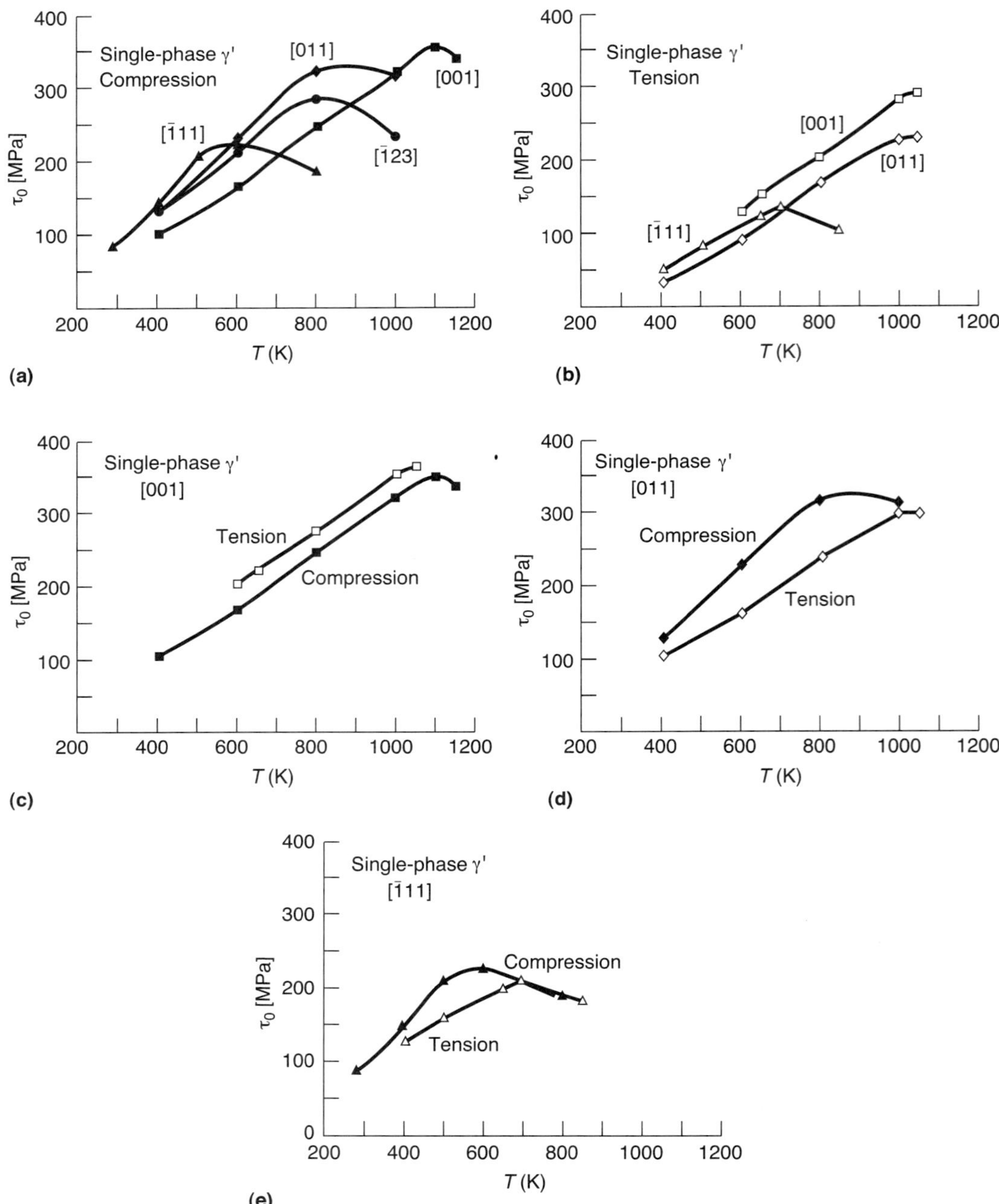

FIG. 6.18 Resolved shear stress for deformation of a γ′ material via the [1̄01](111) slip system as a function of temperature for (a) compression and (b) tension of multiple crystal orientations and a comparison of compression and tension for (c) [001] oriented crystals, (d) [011] oriented crystals, and (e) [1̄11] oriented crystals (Nitz and Nembach, 1997, with permission of Elsevier).

dependence. This, combined with a quadratic dependence of the applied stress in the pre-exponential term, leads to a cubic dependence of strain rate on stress. For the low ranges of $3 \leq n' \leq 4$, power law creep is limited by the slip processes of glide and cross slip.

At temperatures greater than $0.6T_m$, lattice-diffusion-controlled climb determines the rate of dislocation motion. The highest velocity at which a jog on an edge dislocation can climb is given by

$$V_{\text{jog}} \cong \frac{D_v \sigma_n \Omega}{bkT} \tag{6.35}$$

where D_v is the lattice diffusion coefficient, σ_n is the normal stress, and Ω is the atomic volume. Then the strain rate can be given by an expression of the form

$$\dot{\gamma}_{\text{PLC}} = A_{\text{PLC}}\left(\frac{D_v \mu b}{kT}\right)\left(\frac{\tau}{\mu}\right)^{n'} \tag{6.36}$$

where n' is equal to 3 and A_{PLC} is a dimensionless constant. Larger values of n' are not well explained unless cooperative dislocation motion and dislocation density mobility effects do not scale as described.

One other factor that is not considered in Eq. 6.36 is the effect of dislocations on diffusion. The distorted region near dislocation cores has a strong influence on diffusion. The volumetric strains about dislocation cores in edge dislocations result in an order-of-magnitude faster rate of diffusion along edge dislocations versus screw dislocations. When this core diffusion influences deformation, a modified form of this expression can be applied by substituting for D_v with

$$D_{\text{eff}} = D_v\left[1 + \frac{10(2w^2)}{b^2}\frac{D_c}{D_v}\right] \tag{6.37}$$

where $2w^2$ is the effective core diffusion area, w is similar to the boundary width defined for grain-boundary diffusion, and D_c is the core diffusion value. D_c typically scales with D_b, the grain-boundary diffusion coefficient defined for grain-boundary creep. For high temperatures, lattice diffusion dominates, but for low temperatures the core diffusion contribution in Eq. 6.37 plays a major role.

6.4.3 Diffusion Creep Mechanisms

Diffusion creep controlled by lattice diffusion was first described by Nabarro and Herring. Figure 6.19 shows the strong correlation between activation energies for creep versus those for self-diffusion (see also Chokshi, 2002). Nabarro–Herring creep is the dominant mechanism for high temperatures and low stresses in polycrystals with large grain sizes. Both authors modeled the process based on shear deformation and the relative effects of hydrostatic tension and hydrostatic compression. The basis of this creep model is that local stress alters the effective activation energy for vacancy formation by $\sigma\Omega$, where Ω is the atomic volume. The equilibrium vacancy concentration for tensile regions becomes

$$N_v(\text{tension}) = \exp\left(\frac{-Q_f}{kT}\right)\exp\left(\frac{\sigma\Omega}{kT}\right) \tag{6.38}$$

and

$$N_v(\text{compression}) = \exp\left(\frac{-Q_f}{kT}\right)\exp\left(\frac{-\sigma\Omega}{kT}\right) \tag{6.39}$$

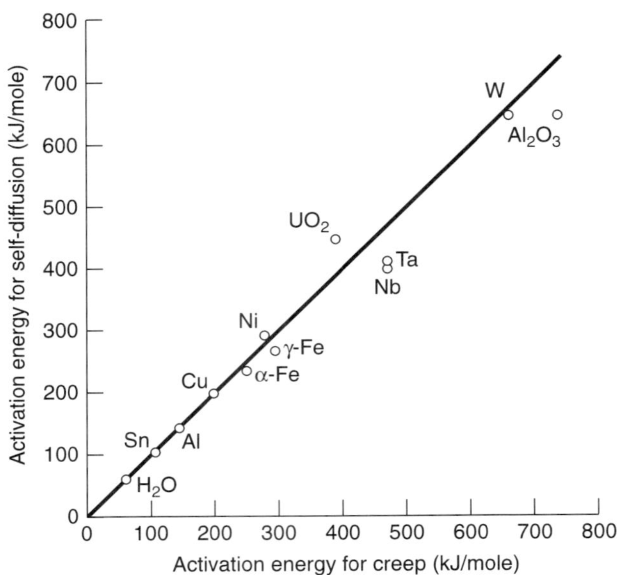

FIG. 6.19 Correlation between creep and self-diffusion activation energies (adapted from Weertman, 1968, used with permission of ASM International).

where Q_f is the vacancy formation energy. The different concentrations at regions within the crystal drive a net flow of vacancies from the tensile to the compressive regions (see Fig. 6.20). The vacancy flux J is given in vacancy volume/sec by

$$J = -\frac{\Omega}{b} D_m \left(\frac{\Delta N_m}{\partial x} \right) \tag{6.40}$$

where ∂x is proportional to grain size d and ΔN_m is given by the difference between the concentrations in the tensile and compressive regions. D_m is the diffusivity for vacancy motion, which depends on overcoming an activation energy for motion, Q_m, by

$$D_m = D_{o,m} \exp(-Q_m/kT) \tag{6.41}$$

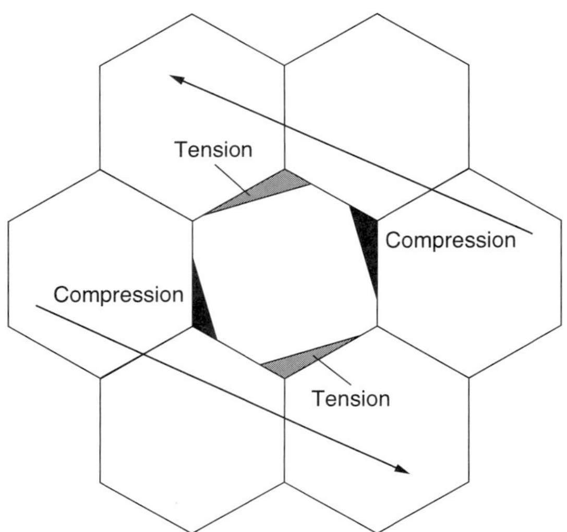

FIG. 6.20 Under an imposed shear stress (arrows), concentrations of net tensile or compressive stress can develop at grain junctions. Materials transport from compressive regions to tensile regions can relieve these stresses, producing deformation.

When this flux is related to the motion that must occur to produce deformation, the strain rate for Nabarro–Herring creep takes the form

$$\dot{\gamma}_{NH} = A_{NH}\left(\frac{D_V}{d^2}\right)\left(\frac{\sigma\Omega}{kT}\right) \tag{6.42}$$

The diffusivity for lattice diffusion includes both the separate activation energies for formation and motion of vacancies introduced above. Coble (1963) adapted the approach used by Nabarro and Herring to include the case wherein creep is driven principally by grain-boundary diffusion. For grain-boundary diffusion, the region through which diffusion takes place is within an effective width w. The size of this effective width must be scaled relative to the grain size. Thus, in addition to being based on the activation energy for grain-boundary diffusion, with

$$D_{GB} = D_{GB,o}\exp\left(\frac{-Q_{GB}}{kT}\right) \tag{6.43}$$

Coble creep is given as

$$\dot{\gamma}_c = A_c\left(\frac{D_{GB}w}{d^3}\right)\left(\frac{\sigma\Omega}{kT}\right) \tag{6.44}$$

For both diffusion-based creep mechanisms, the proportionality between the strain rate and stress is a linear Newtonian response, as shown by the data for several oxides, glasses, and glass composites in Fig. 6.9. At high temperatures, the contribution to creep from the Nabarro–Herring mechanism is greater than from a Coble mechanism, and at low temperatures the reverse is true. Example 6.6 shows Nabarro–Herring and Coble creep as a function of grain size and temperature.

EXAMPLE 6.6 *Diffusion Creep Mechanisms*

We can determine the temperature at which the two diffusion creep mechanisms make equal contributions to the total creep rate by setting their strain rates equal. Consider creep for germanium with the following data for Eq. 6.42 and 6.43:

$A_{NH} = 7$	$A_C = 50$
$D_{V,O} = 78.8 \times 10^{-4}$	$Q_V = 287$ kJ/mole
$D_{GB,O} = 10^{-17}$	$Q_{GB} = 172$ kJ/mole
$\Omega = 2.26 \times 10^{-29}\ \text{m}^3$	

BIOGRAPHY

ROBERT L. COBLE (1928–1992)

Coble specialized in understanding kinetic processes in ceramic materials and is responsible for the processing approaches used to produce many technical ceramic materials. He developed the Mg doping process that enabled "Lucalox," the transparent, polycrystalline aluminum oxide that would become a key part of the sodium-vapor lamp. His 1963 paper "Model for Boundary Diffusion-Controlled Creep in Polycrystalline Materials" (Coble, 1963) provides the basis for what we call Coble creep in this chapter.

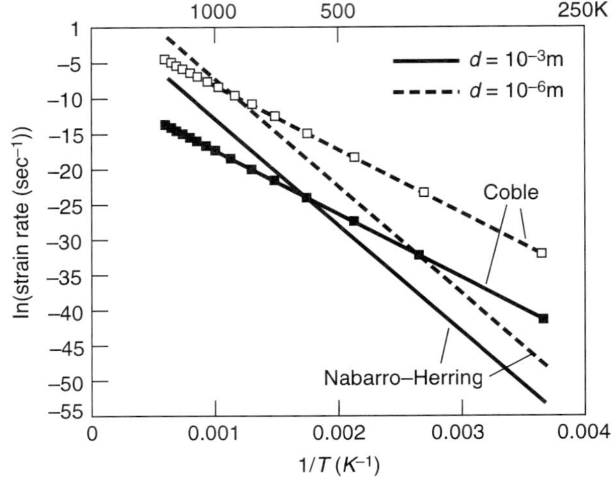

FIG. 6.21 Relationship between strain rates for Nabarro–Herring and Coble creep as a function of T for two grain sizes. The Nabarro–Herring creep curve is plotted without symbols, and the Coble creep curve is indicated by square symbols.

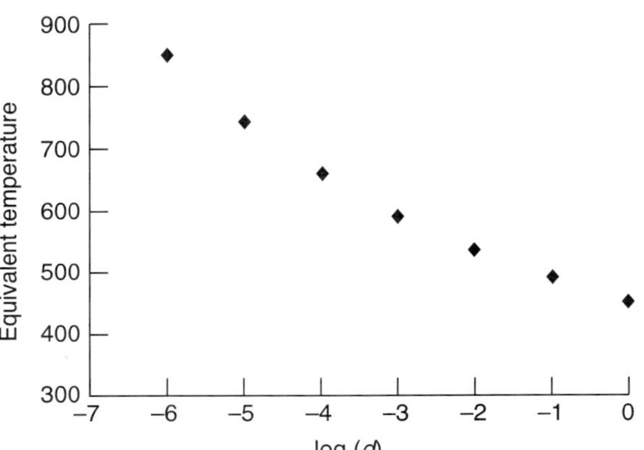

FIG. 6.22 Relationship between grain size and the temperature at which $\dot{\varepsilon}_{NH} = \dot{\varepsilon}_C$.

For a constant stress, the point at which the Nabarro–Herring and Coble creep rates are equal as a function of grain size and temperature can be obtained by a plot of the type shown in Fig. 6.21.

Figure 6.21 shows that decreasing grain size results in higher strain rates for both Nabarro–Herring and Coble creep. Also, the temperature at which $\dot{\varepsilon}_{NH} = \dot{\varepsilon}_C$ increases with increasing temperature. Or, in other words, Coble creep is dominant to higher temperatures with decreasing grain size (as long as grain growth is not occurring). This relation can be determined explicitly by setting $\dot{\varepsilon}_{NH} = \dot{\varepsilon}_C$ and solving for the two variables d and T. For the Germanium data, this yields

$$T(d) \cong \frac{13{,}800}{30 + \ln(d)}$$

This is plotted in Fig. 6.22. These relations between deformation rates for different mechanisms are employed in Chapter 8 to create deformation mechanism maps. ■

6.4.4 Grain-Boundary Sliding

To inhibit formation of internal voids, material must be transferred at the grain boundaries. Raj and Ashby (1971) suggested that the rate of grains sliding past one another under shear stress is determined by the grain-boundary shape. The sliding process consists of a relative

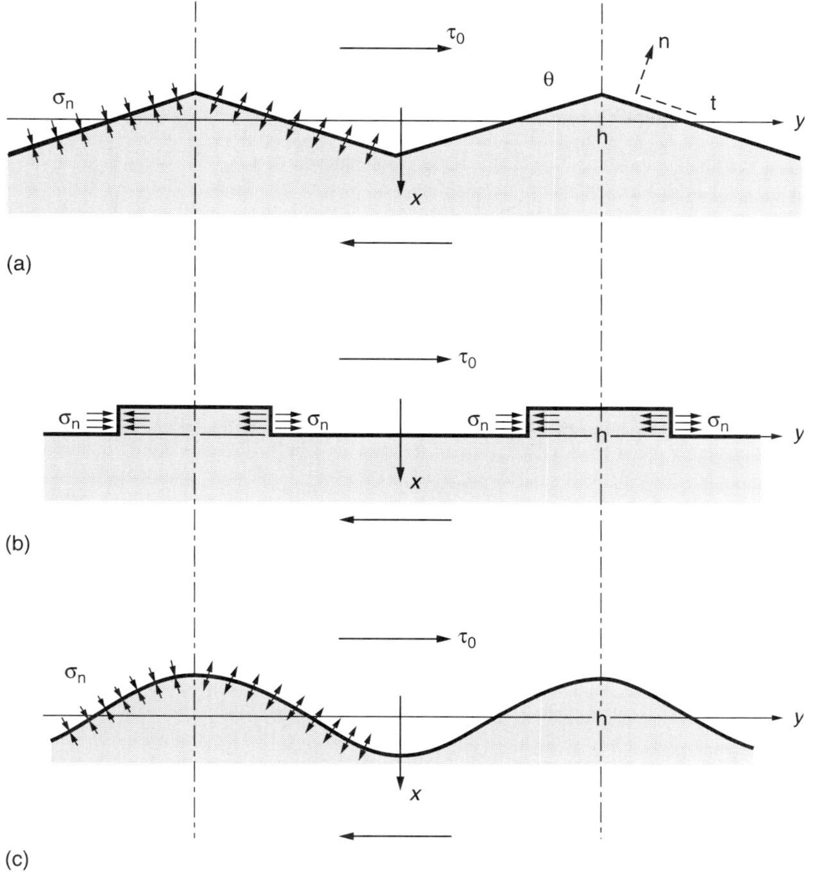

FIG. 6.23 The application of a shear stress τ_a results in normal stresses acting between grains with (a) sawtooth, (b) step, and (c) sinusoidal boundaries. The normal stresses between the grains at this plane are labeled σ_n (Raj and Ashby, 1971, used with permission).

displacement of the individual grains, but a pictorial image of grains skating past one another should be avoided. Grain-boundary sliding (GBS) experimental observations on Cu and Ag bicrystals have shown that specimens from the same bicrystal vary in creep resistance, grain-boundary migration changes creep resistance, hard precipitates slow GBS, and the activation energy for GBS is often equivalent to that for lattice diffusion.

The microscopic steps on boundaries between two crystals are shown in Fig. 6.23. A shear stress is applied across the boundary with an atomically smooth interface separated by a viscous fluid. For crystalline materials without a grain-boundary phase, the high transport rates near the boundary effectively provide such a condition, and many ceramics with a glassy grain-boundary phase possess such interfaces. For this case, the sliding rate between the two crystals is given by the effective viscosity of the grain-boundary phase. For rough interfaces, the sliding process must be accommodated by strains at the rough interface similar to those found in friction and wear processes.

The accommodating strains can be elastic or plastic. In the elastic case, once the stresses at the contact points are sufficiently high, the sliding process arrests. Plastic deformation at this interface can occur by dislocation motion at high stresses and low temperatures. At high temperatures, the local tensile and compressive stress gradients result in

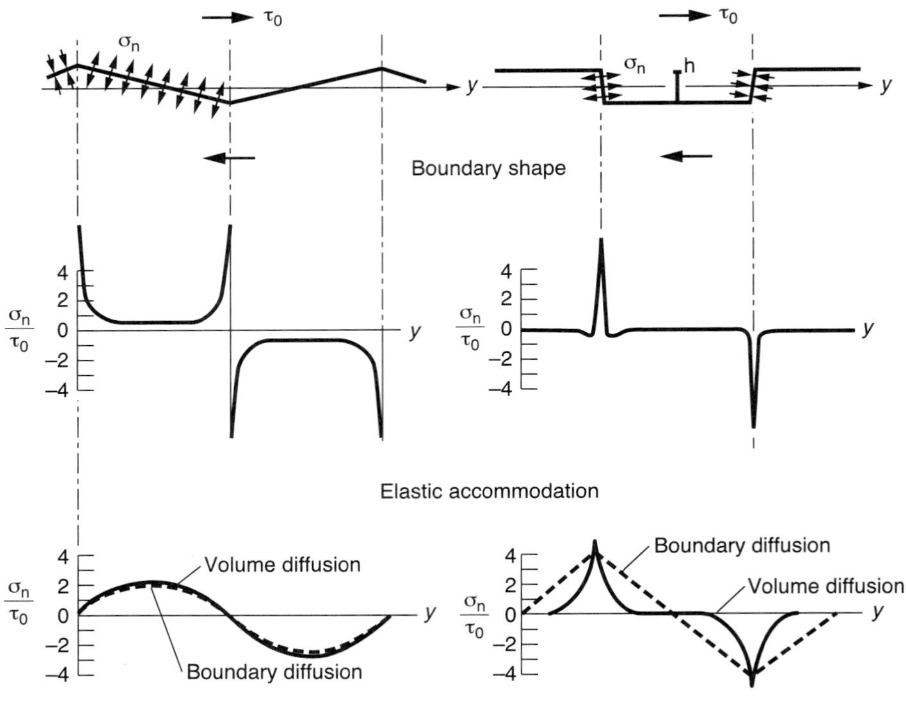

FIG. 6.24 Sliding of the sawtooth and step boundaries shown in the top set of diagrams results in the elastic stresses shown in the middle set of diagrams unless accommodated by diffusion. During steady-state grain boundary sliding, material flow by diffusion is driven to relieve these stresses (Raj and Ashby, 1971, used with permission).

diffusion creep through the grain boundary or the lattice. For the elastic case, the stress distribution normal to the grain boundary will have the periodicity shown in Fig. 6.24 when under applied shear stress. In elastic accommodation, the sliding between the two grains will arrest when the local interfacial stresses balance the applied shear stress. To relieve these stresses, diffusive transport can cause a net translation of material, resulting in a relative translation of the grains and changes in grain shape.

In some materials, the hydrostatic stresses can contribute to relative motion of the grains and can lead to failure by intergranular fracture. In these instances, the creep rate in tension can be attributed to nucleation and expansion of cavities (cavitation) at the grain boundaries (see Chen and Argon, 1981) or within glassy phases separating crystalline grains in some ceramic materials (see Luecke and Wiederhorn, 1995). The dilatant deformation process of cavitation is not available during compression, which is why there is a large difference between the tensile and compressive deformation results shown in Fig. 6.25.

6.4.5 Superplasticity

Superplastic behavior is a description of deformation wherein very large tensile strains can occur in polycrystalline materials without failure, as shown in Fig. 6.26*a*. Although often treated as a specific deformation mechanism, superplasticity can occur through operation of several mechanisms or even advantageous microstructures. Examples are shown in Example 6.3 and Fig. 6.26*b* and *c* for ceramic materials (see also Chen and Xue, 1990). The key

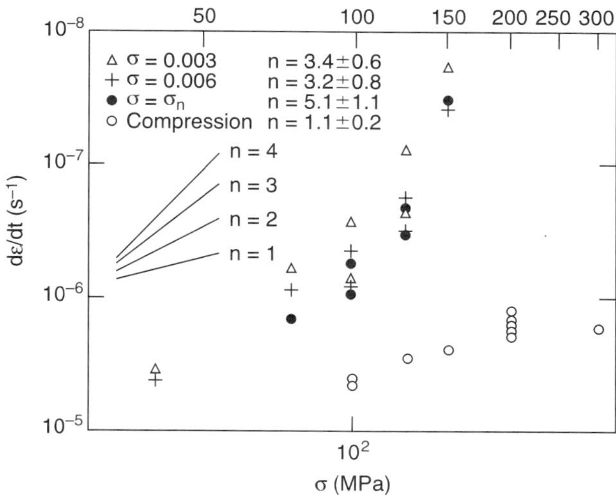

FIG. 6.25 Strain rate as a function of stress for tests at 1430°C on silicon nitride. The data points marked with circles are for compression. Cavitation in the grain boundaries results in faster creep in tension than in compression. Electron microscopy on specimens deformed in tension shows small bubbles within the glassy grain-boundary phase. (Luecke and Wiederhorn, 1995, reprinted with permission of The American Ceramic Society, PO Box 6136, Westerville, Ohio 43086-6136. Copyright, 1995. All rights reserved.) The silicon nitride material evaluated in this figure has a much smaller glassy grain-boundary phase content than the material discussed in Example 6.3. The failure strains for tensile creep of the materials shown here are also much lower than those in Example 6.3.

element in superplasticity is that the material must resist necking. If deformation proceeds without strain hardening and a high value of strain rate sensitivity, then the large strains shown for the materials in Fig. 6.26 are possible. Materials exhibiting this behavior are normally undergoing deformation at temperatures above half the melting temperature and strain rates between 10^{-4} and 10^{-2}/s. At these temperatures and strain rates, deformation can

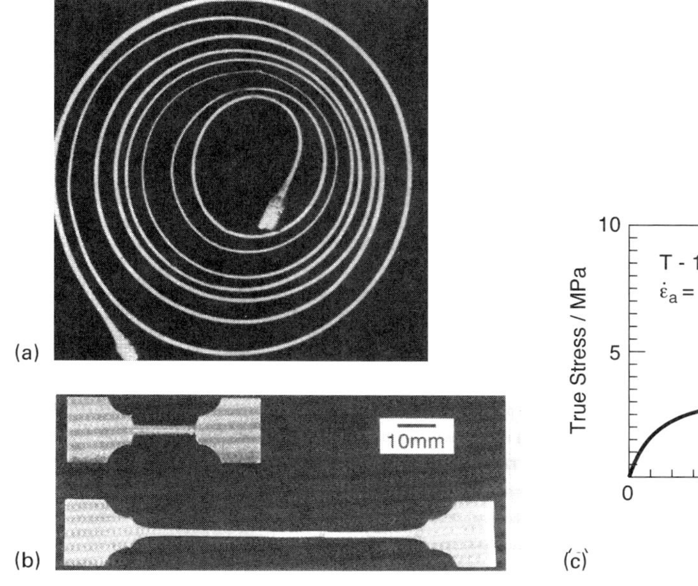

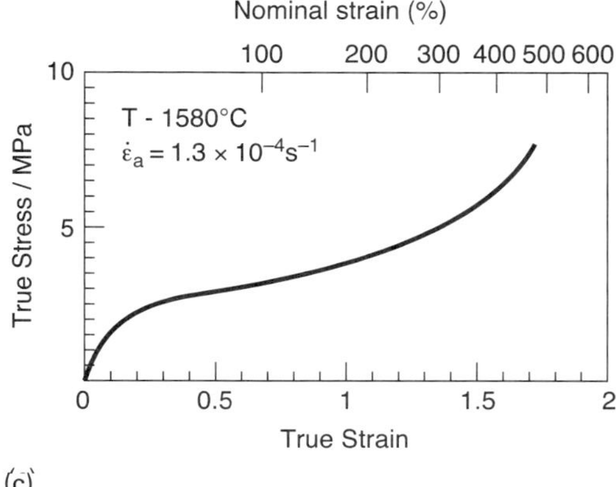

FIG. 6.26 (a) Superplastic extension of a Bi-Sn eutectic alloy (Pearson, 1934, with permission). (b) and (c) Superplastic silicon nitride with an initial very small grain size—a cylindrical grain diameter of approximately 0.1 μm and a grain length averaging about 0.3 μm. Just as the case in Example 6.3, some orientation strengthening is observed, but the stresses are much lower owing to a finer starting grain size (Kondo, Ohji, and Wakai, 1998, used with permission.)

proceed without strain hardening; however, at these temperatures, grain growth is expected. Without suppression of grain growth to maintain grain sizes less than ≈ 10 μm, superplasticity does not occur in most metal and ceramic systems. Although possible in single-phase alloys, two-phase alloys with nearly equal fractions and alloys with grain-boundary precipitates often have sufficiently slow grain growth rates to enable superplastic deformation. Materials with microstructures resulting from eutectic or eutectoid transformations also can exhibit superplastic behavior. Unlike conventional deformation, the grain shape normally does not correspondingly change during deformation. For true strains of 3 to 4, grain aspect ratios can remain less than 2.

We have already seen that strain rate sensitivity given as m relates stress to strain rate as

$$\sigma = C\dot{\varepsilon}^m$$

in Chapter 3. A high strain rate sensitivity generally translates to large elongations, as shown in Fig. 6.27. The correspondence among strain rate, stress, and the exaggerated values of m that result in superplasticity for a superplastic Mg-Al eutectic is shown in Fig. 6.28.

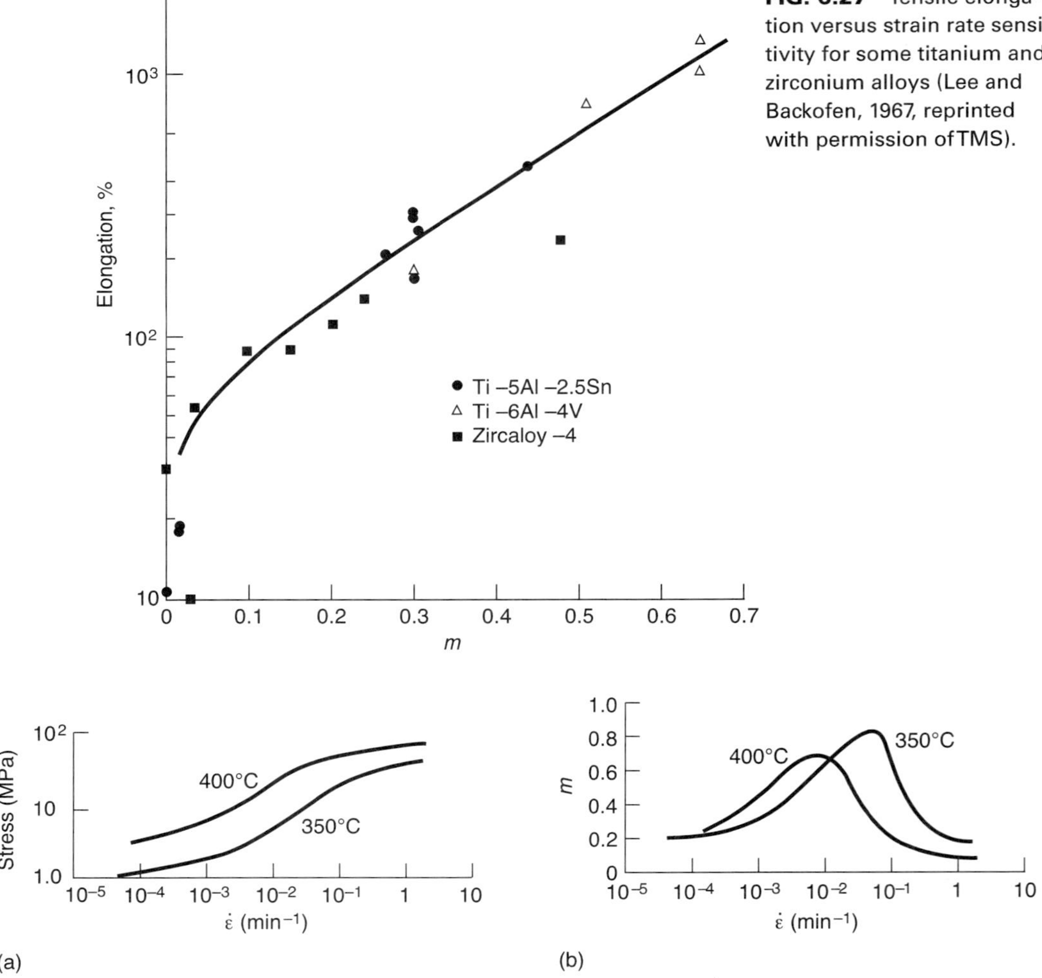

FIG. 6.27 Tensile elongation versus strain rate sensitivity for some titanium and zirconium alloys (Lee and Backofen, 1967, reprinted with permission of TMS).

FIG. 6.28 (a) Stress versus strain rate and (b) strain rate sensitivity m and strain rate relations for the superplastic Mg-Al eutectic alloy (d = 10.6 μm) (after Lee, 1969, used with permission of Elsevier).

As described in Chapter 3 for the relationship between the strain hardening exponent n and necking, the basis for suppression of necking in the presence of strain hardening is that the formation of a neck redistributes the deformation to the unnecked regions. The transient neck never grows unstably. For superplasticity with high values of m, a necked region might be anticipated to deform at a higher strain rate than the adjoining unnecked material owing to higher stresses. But the resistance to deformation is greater if the strain rate is higher. For $m = 1$, this should be the case. For small values of m, necking should occur.

6.4.6 Dynamic Recrystallization

At very high temperatures and the fairly rapid strain rates typical of hot working operations ($\dot{\varepsilon} - 10^{-1} \rightarrow 10^{2}$ sec^{-1}), deformation response is often manifested as an instability that is correlated with the formation of new grains through dynamic recrystallization. Conventional "static" recrystallization is accomplished by strain hardening a material to produce an increased dislocation density and nucleating new grains by heat treatment. If the dislocation density is sufficiently high, heat treatments can result in formation of new, essentially strain-free grains. This is in part a result of slow operation of diffusion-based recovery mechanisms for mutual annihilation of dislocations in heavily networked dislocation structures and in part a result of banding or cell formation within grains possessing very high dislocation densities. In dynamic recrystallization, the formation of new grains occurs while the material is undergoing plastic deformation.

Some metals do not dynamically recrystallize, but undergo dynamic recovery. During dynamic recovery, the dislocation density reaches a steady-state balance between strain hardening and dislocation annihilation. Materials with low stacking fault energies are likely to show dynamic recrystallization whereas materials with high stacking fault energies are more likely to show dynamic recovery. Many intermetallic and ceramic compounds also are strongly susceptible to dynamic recrystallization.

For both dynamic recovery and dynamic recrystallization, it is possible for deformation instabilities to occur during plastic working. Plastic instabilities are nearly always fatal to forming operations. Figure 6.29 shows the stress-strain response that is typical of dynamic recrystallization.

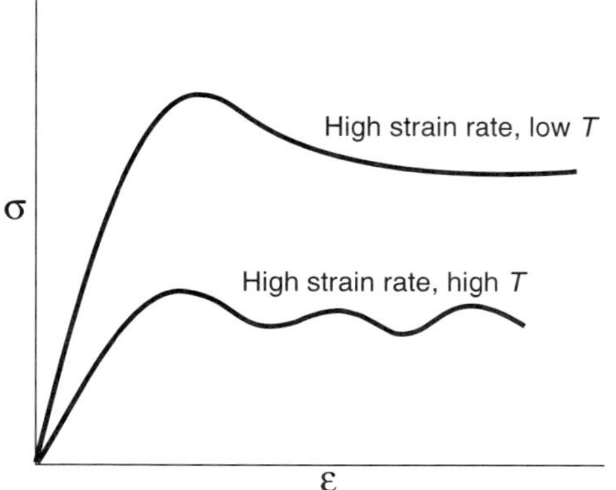

FIG. 6.29 Schematic of true stress–true strain behavior during dynamic recrystallization. The onset of recrystallization occurs shortly after the beginning of plastic deformation, which is the onset of recrystallization. A maximum stress is obtained for each strain rate. At low strain rates and high temperatures, the high-strain behavior is such that the stress oscillates about a steady-state value.

6.5 REFERENCES

I-W. CHEN AND A.S. ARGON, *Acta Metall.*, **29**, 1759–1768, 1981.

I-W. CHEN AND L. A. XUE, *J. Am. Ceram. Soc.*, **73**, 2585–2609, 1990.

A. CHOKSHI, *J. Euro. Ceram. Soc.* **22**, 2469–2478, 2002.

R. L. COBLE, *J. Appl. Phys.*, **34 [6]**, 1679–1682, 1963.

A. H. COTTRELL, *Dislocations and Plastic Flow in Crystals,* Oxford, 1953.

T. H. COURTNEY, *Mechanical Behavior of Materials,* McGraw-Hill, 1990.

H. J. FROST AND M. F. ASHBY, *Deformation Mechanism Maps,* Pergamon Press, 1982.

N. J. GRANT AND A.G. BUCKLIN, *Trans. ASM,* **42**, 720–751, 1950.

C. A. HANDWERKER, R. M. CANNON, AND R. H. FRENCH, *J. Am. Ceram. Soc.*, **77**, 293–298, 1994.

U. F. KOCKS, A. S. ARGON, AND M. F. ASHBY, *Thermodynamics and Kinetics of Slip, Progress in Materials Science,* Vol. 19, Pergamon Press, 1975.

N. KONDO, T. OHJI, AND F. WAKAI, *J. Ceram. Soc. Japan,* **106**, 1040–1042, 1998.

N. KONDO, Y. SUZUKI, T. OHJI, E. SATO, AND F. WAKAI, "Change in stress, stress sensitivity and activation energy during superplastic deformation of silicon nitride," *Mater. Sci. Eng.*, **A268**, 141–146, 1999.

F. S. KOTTADA AND A. H. CHOKSHI, *Acta Mater.*, **28**, 3905–3915, 2000.

D. LEE, *Acta Metal.*, **17**, 1057–1069, 1969.

D. LEE AND W. A. BACKOFEN, *TMS-AIME*, **239**, 1030–1040, 1967.

W. LUECKE AND S. WIEDERHORN, *J. Am. Ceram. Soc.*, **78 [8]**, 2085–2096, 1995.

D. W. MACHLACHLAN AND D. M. KNOWLES, *Met. Trans.*, **31A**, 1401–1411, 2000.

N. G. MCCRUM, C. P. BUCKLEY, AND C. B. BUCKNALL, *Principles of Polymer Engineering,* Oxford, 1988.

D. MCLEAN, *Mechanical Properties of Metals,* Wiley, 1962.

G. MORSCHER, P. PIROUZ, AND A. H. HEUER, *J. Am. Ceram. Soc.*, **74**, 491–500, 1991.

A. K. MUKHERJEE, J. E. BIRD, AND J. E. DORN, *Trans. ASM,* **62**, 155–179, 1968.

A. MUÑOZ, F. WAKAI, AND A. DOMINGUEZ-RODRIGUEZ, "High temperature plastic deformation of a tetragonal Y^2O^3-stabilized ZrO^2 single crystal," *Scripta Mat.* **44**, 2551–2555, 2001.

A. NITZ AND E. NEMBACH, "High-temperature compression and tensile tests of γ'– and γ' (50%)/(50%) γ-single crystals with <001>-, <123>-, <011>- and <111>-orientations," *Mater. Sci. Eng.*, **A239–240**, 164–168, 1997.

C. E. PEARSON, *J. Inst. Metals,* 111–123, 1934.

L. POTEAT AND C. YUST, in *Ceramic Microstructures,* edited by R. M. Fulrath and J. A. Pask, Wiley, 646–656, 1968.

R. RAJ AND M. F. ASHBY, *Met. Trans.*, **2**, 1113–1127, 1971.

G. SASIKALA ET AL., *Met Trans.*, **31A**, 1175–1185, 2000.

A. SEEGER, in *Dislocations and Mechanical Properties of Crystals,* edited by J. C. Fisher, W. G. Johnston, R. Thomson, and T. Vreeland, Jr., Wiley, New York, 243–329, 1957.

O. D. SHERBY AND P. M. BURKE, *Prog. Mater. Sci.*, **13**, 325–336, 1968.

B. SUDHIR AND A.H. CHOKSHI, *J. Non-Crys. Solids,* **316**, 167–173, 2003.

V. TIKARE AND A. H. HEUER, *J. Am. Ceram. Soc.*, **74**, 593–597, 1991.

J. WEERTMAN, *Trans. ASM,* **61**, 681–694, 1968.

S. M. WIEDERHORN AND B. J. HOCKEY, *Ceram. Int.*, **17**, 243–252, 1991.

6.6 PROBLEMS

A.6.1 When extrapolating creep data, what is the most important fundamental parameter related to deformation mechanisms that will make you certain that your deformation mechanism doesn't change?

A.6.2 Describe power law creep. Cite as many relevant variables as possible and show how they are related.

A.6.3 Sketch three-stage creep behavior for a material in (a) stress control and (b) load control. Assume that the starting applied stress is the same. Clearly label each section of the curves.

A.6.4 When a material is plastically deformed, why does it often become warm to the touch?

A.6.5 Sketch the formation of a kink on an edge dislocation wherein a stable (or critical-size) kink has a length of $4b$ and a kink height of $1b$. If we assume that the affected region is magnitude b in width, what is the activation volume?

A.6.6 Describe two different methods of finding m for a rate-sensitive material. Show how the data might be presented.

B.6.1 Plot ΔG in Eq. 6.29 for $q = 1, \frac{3}{2}, 2,$ and 3 for both $p = 1$ and $p = \frac{1}{2}$. Compare these curves and consider the effect on the barrier to slip in each case.

B.6.2 **(a)** Resistance to necking is a key to superplasticity in tension.

Assuming $\sigma = C\dot{\varepsilon}^m$

Derive $$\frac{dA}{dt} = -\frac{F}{Cm}A^{(m-1)/m}$$

with $\sigma = F/A$.

(b) Plot the ratio of dA/dt for a necked area versus the unnecked area of a tensile bar for $m = 0, 0.2, 0.5, 0.8$, and 1. Discuss your results.

B.6.3 Make a schematic plot of molar volume versus temperature for a given glass as a function of the macromolecule size using a range of curves spanning easy crystallization to difficult crystallization.

B.6.4 Two different types of mechanical testing equipment are commonly used: those that apply a force through a hydraulically actuated cylinder and those that apply a displacement by rotating screws that move the grips apart. Which type of equipment would be best for conducting a creep test, and which would be best for conducting a stress-relaxation test? Explain your answer.

B.6.5 A series of strain rate jumps (up and down) are given in Fig. 1.23 for polyurethane. Estimate the strain rate sensitivity of the polyurethane material.

B.6.6 Describe why the temperature rise for the same deformation of aluminum conducted at two different temperatures, say 100 and 200°C, might be different. Which temperature would give the greatest temperature rise for the same deformation?

B.6.7 Plot the activation energies for self-diffusion given in Fig. 6.19 versus melting temperature by estimating the values and looking up the corresponding melting temperatures. From this plot, estimate the activation energies for creep of Mo and Ag.

B.6.8 Plot the creep behavior of a Maxwell element that is placed under a load of 10 MPa for 1000 sec and then under no load for 1000 sec. The Maxwell element has the following properties:

(a) $E = 10$ GPa, $\eta = 5$ Pa/sec

(b) $E = 10$ GPa, $\eta = 100$ Pa/sec

(c) $E = 100$ GPa, $\eta = 5$ Pa/sec

(d) $E = 100$ GPa, $\eta = 100$ Pa/sec

B.6.9 Plot the creep behavior of a Voigt element that is placed under a load of 10 MPa for 1000 sec and then under no load for 1000 sec. The Voigt element has the following properties:

(a) $E = 10$ GPa, $\eta = 5$ Pa/sec

(b) $E = 10$ GPa, $\eta = 100$ Pa/sec

(c) $E = 100$ GPa, $\eta = 5$ Pa/sec

(d) $E = 100$ GPa, $\eta = 100$ Pa/sec

B.6.10 Plot the stress relaxation behavior of a Maxwell element that is stretched to a tensile strain of 0.1 and held at that length. The Maxwell element has the following properties:

(a) $E = 10$ GPa, $\eta = 5$ Pa/sec

(b) $E = 10$ GPa, $\eta = 100$ Pa/sec

(c) $E = 100$ GPa, $\eta = 5$ Pa/sec

(d) $E = 100$ GPa, $\eta = 100$ Pa/sec

B.6.11 Plot the stress relaxation behavior of a Voigt element that is stretched to a tensile strain of 0.1 and held at that length. The Voigt element has the following properties:

(a) $E = 10$ GPa, $\eta = 5$ Pa/sec

(b) $E = 10$ GPa, $\eta = 100$ Pa/sec

(c) $E = 100$ GPa, $\eta = 5$ Pa/sec

(d) $E = 100$ GPa, $\eta = 100$ Pa/sec

C.6.1 How can Eq. 6.36 be obtained from the Orowan equation and Eq. 6.35?

C.6.2 Why is a low stacking fault energy material likely to undergo dynamic recrystallization?

C.6.3 Estimate the maximum tensile residual stress for a rod of a material that consists of a fast cooled surface that remains glassy while the interior crystallizes. Assume that $\Delta V_g = 0.01$, $\alpha_{glass} = 10 \times 10^{-6}\ C^{-1}$, $\alpha_{crystal} = 5 \times 10^{-6}\ C^{-1}$, $(T - T_g) = 400$°C, and $E_{glass} \cong E_{crystal} = 150$ GPa.

C.6.4 Derive a relationship for strain rate in two Maxwell elements in parallel. Assume that the material constants (E and η) are identical in both Maxwell elements. Plot strain versus time and stress versus time for creep and stress relaxation of the two parallel elements, and compare these plots with the proportional behavior of a single Maxwell element.

C.6.5 Derive a relationship for strain rate in a Maxwell element and a Voigt element in series. Assume that the material constants (E and η) are identical. Plot strain versus time and stress versus time for creep and stress relation. Explain each transition.

C.6.6 If you found strain rate jump tests to give different values of strain rate sensitivity at different temperatures, what would you give as the possible reasons?

C.6.7 The data given in the following table are based on the original data given by Grant and Bucklin (1950, used with permission of ASM International), for alloy S-816. Convert the data to appropriate units and plot graphs similar to Fig. 6.3a and b, finding your best value for the constant C. How does this data set compare with that used in Fig. 6.3a and 6.3b?

Temperature (°F)	Rupture stress (ksi)	Time (hours)
1200	120	0.001
1200	100	2.1
1200	90	3.0
1200	80	20.9
1200	56	652
1350	90	0.092
1350	90	0.01
1350	80	0.12
1350	75	0.43
1350	70	1.0
1350	60	5.9
1350	50	13.0
1350	50	9.3
1350	40	249
1350	40	140
1350	35	363
1350	30	2019
1500	85	0.002
1500	80	0.006
1500	75	0.006
1500	70	0.02
1500	60	0.061
1500	45	0.37
1500	40	2.23
1500	34	7.9
1500	30	24.8
1500	25	112.0

CHAPTER 7

FRACTURE OF MATERIALS

THE PROCESS of fracture can be extremely complicated because it may be preceded by or may coincide with other deformation or failure mechanisms. Often the final fracture of a component occurs after prior deformation has rendered it unable to perform its function. Failures by fracture that result in injury or loss of life through structural damage of ships, airplanes, or bridges can be very dramatic, but failures by fracture are everyday occurrences in less critical applications. Some materials undergo large deformations before fracture occurs. For brittle materials (small strains to failure), surface or internal flaws may serve as fracture initiation sites and there may be a broad distribution of failure stresses for seemingly identical materials. In most cases, the local conditions at what becomes the final fracture surface were such that plastic deformation by shear mechanisms was not sufficient to support uniform deformation of the material, and mechanisms leading to rupture of the material are predominant. In other words, fracture or rupture intercedes before the stresses required for uniform deformation are reached.

EXAMPLE 7.1 *The Drama of Structural Failures*

The crisp report of a material fracturing during laboratory testing is routinely used to interest and intrigue visitors about the topics of fracture and failure. Failed specimens with interesting histories or stories are tools for fostering interest in the field of fractography. At the same time, the drama of failure also leads to ready speculation on how materials have "failed" us. During the unforgettable live view of the collapsing World Trade Center Towers (see Fig. 7.1), preparing a lecture for a class on temperature-dependent deformation mechanisms of materials was probably too relevant. Softening of elastic constants, changes in deformation mechanism, thermal expansion, and a reduction in yield stress were all hanging over a lecture to be given in 2 hours. None of the students wanted to cancel class, so we talked about deformation mechanisms in cubic metals in exquisite detail.

Following September 11, 2001, the media-political complex repeatedly enjoined evaluation of how we could prevent such failures of buildings in the future. Engineers led discussions of "preventability" or "lessons we could learn." Some responded as though we should design for structural damage by fuel-laden airplanes. In the litigation-oriented culture Americans have embraced, wherein no material or structure is allowed to fail without claims of preventable defects, the impressive engineering accomplishment of the World Trade Center Towers is that nearly all occupants of the buildings below the floors struck by the planes got out. The initial damage to portions of the buildings led to load transfer and concentration of stress in the intact portions of the damaged floors that, combined with structural performance degraded by high temperatures, led to final and devastating failure. Some hindsight measures may have increased the time to failure, but short of a miracle rescue, the final outcome was inevitable. That the two final failures were so similar seems to confirm this perspective. ■

FIG. 7.1 Photograph of the World Trade Center Towers from Summer 2000 (by K. J. Bowman).

Stress Concentration For a component, a change in specimen cross section or a fastening hole can serve as a source of stress concentration wherein a fracture may be initiated (for a comprehensive list, see Peterson, 1953). The presence of a stress concentration may lead to the growth of a crack from fatigue or creep damage or may come about in concert with oxidation or corrosion damage. Stress concentrations at cracks are shown in Fig. 7.2. In Fig. 7.2*a*, a crack in a piezoelectric material under load changes the electrical potential at the surface of the material. Applying a surface layer of liquid crystals that become polarized under field makes it possible to see the field concentration with polarized light. The kidney-shaped rings can be read like a topographic map of the stress level, with the highest stresses at the crack tip. Figure 7.2*b* shows similar contours of stress for a moving crack, but this time the contours result from changes in the optical characteristics of the transparent polymer when it is under load. Here, too, the contours can be used to describe the distribution of stresses.

One key step in understanding failures initiated by flaws within a material is the conceptual understanding of a stress concentration first developed by Inglis. Evaluation of a through circular hole subjected to a biaxial stress provides a clear indication that the resolution of forces and the balance of energy associated with a discontinuity provides a localized concentration of stresses. This localized concentration of stresses is often designated as

$$k_{conc} = \sigma_{max}/\sigma_{\infty} \tag{7.1}$$

where σ_{max} is the largest principal stress (found typically at the flaw surface) and σ_{∞} is the remote applied stress well away from the flaw.

Consider the plate with a round through-hole that undergoes biaxial tension shown in Fig. 7.3*a*. Since the interior of the hole is a free surface, all forces normal to the interior surface of the hole (excluding atmospheric pressure) must be equal to zero. Therefore, all forces resolved within the material must be carried by the material surrounding the hole. It is easy to expect that the circumferential stresses will be magnified near the hole surface.

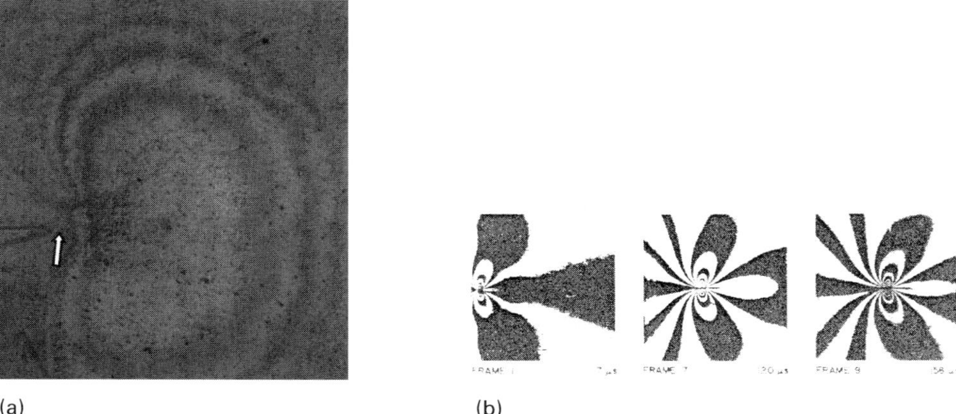

(a) (b)

FIG. 7.2 (a) Elastic stress field displayed using a thin film of planar nematic liquid crystal and observed with polarized light. The arrow indicates the crack tip of the material, and the loading direction is vertical on the page. The contours are visible because the application of stress to piezoelectric material causes changes in the local electric potential on the surface (provided by A. Kounga, D. Lupascu, and J. Rödel; see Lupascu et al., 2001). (b) "Snapshots" of a stress concentration in a photoelastic material for a moving crack (Anderson, 1995, used with permission of CRC Press). The contours represent constant stress level. The highest stresses are those nearest to the crack tip.

An inverse problem is commonly solved in basic mechanics or strength of materials classes for a thick-walled tube under pressure. In that case, the exterior of the tube is at zero stress and the interior is in compression. In both cases, the circumferential or hoop stress is magnified to twice the applied stress at the inner surface, as shown in Fig. 7.3*b*. For the thick-walled tube, the radial stress is equal to the pressure at the tube surface and the radial stress is zero at the exterior surface. See Appendix A for the derivation of the stress concentration in Fig. 7.3.

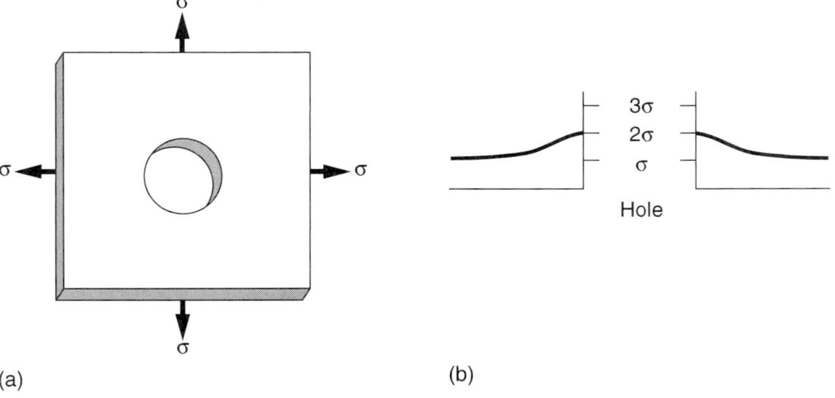

(a) (b)

FIG. 7.3 (a) A circular through-hole in a panel subjected to balanced biaxial tension. Note that to avoid end effects, the hole would need to be much smaller in comparison with the panel than pictured. (b) Stress concentration across the width (or the direction through the center of the hole). All points on the hole surface undergo a stress level equal to 2σ.

Although the derivation is considerably more involved, as the hole goes from a round hole to an elliptical hole with very sharp ends, the resolved stresses become most strongly concentrated at the smallest radius of curvature. The case for a single or uniaxial applied stress relative to the smallest radius of curvature is shown in Fig. 7.4*a*. The expression given by Inglis for a uniaxial stress applied to an elliptical hole is

$$k_{\text{conc}} = 1 + 2a/b \tag{7.2}$$

where a is the long axis and b is the short axis of the elliptical hole. The elastic solution for an infinitely sharp crack predicts an infinite stress. Since an infinite stress is not possible, it is clear that if the crack is not propagating and the part is under stress, the crack must have a finite curvature or the deformation is not purely elastic. For the circular hole in Fig. 7.4*b*, where $a = b$, k_{conc} equals 3 at the edges of the round hole that lie tangential to the applied stress direction. On the hole surfaces that lie perpendicular to the applied stress, the magnitude of the stresses is equal to the applied stress, but opposite in sign. Thus, for a tensile uniaxial stress, this part of the hole is under compression, but for a compressive uniaxial stress, tensile stresses are generated on the hole surface closest to the applied stress. This is one mechanism whereby flaws that lie nearly parallel to the applied compressive stresses can undergo crack propagation. Brittle materials that fail in compression often crack by splitting along the loading direction, and pores or other defects can provide the stress concentration that initiates the splitting.

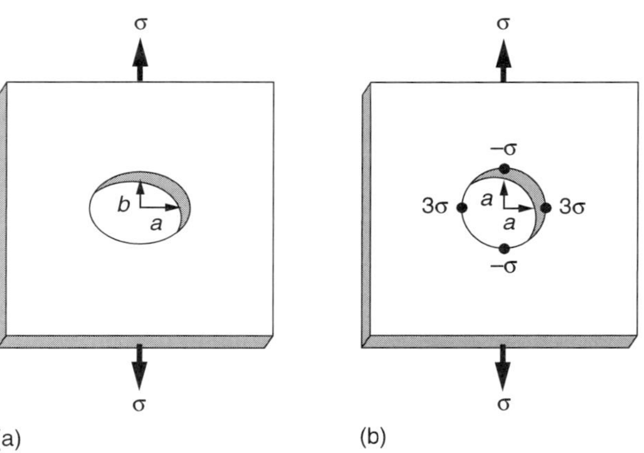

FIG. 7.4 (a) Uniaxial stress applied to a panel with an elliptical through-hole. Note that to avoid end effects the hole would need to be much smaller in comparison with the panel than pictured. (b) With $a = b$, the circular hole shows a stress concentration of extreme values of 3σ and $-\sigma$ in the positions shown.

BIOGRAPHY

A. A. GRIFFITH (1893–1963)

In 1920, Griffith wrote a paper that defined the field of fracture mechanics. In this paper, he defined a theory showing that removal of flaws increases the resistance of brittle materials to fracture. He performed experiments on glass that supported the theory that he had developed. He later moved into the area of engine design and is credited with the development of turbine jet engines. (See Griffith, 1968.)

7.1 STRESS DISTRIBUTIONS NEAR CRACK TIPS

The development of a singularity near a stress concentration is essential to fracture mechanics. The localization of the stress causes the material in that region to undergo damage that can lead to propagation of a crack through that region, thereby moving the damage zone further into the material. The activation of deformation mechanisms that toughen the material and increase resistance to further crack propagation is essential to arrest failure.

7.1.1 Griffith Approach to Linear Elastic Fracture Mechanics (LEFM)

The study of linear elastic fracture mechanics has a well-developed nomenclature and conventions. Most are applied in a relatively standard fashion across different science and engineering fields. The geometry shown in Fig. 7.5 demonstrates the relationship between the Cartesian and polar coordinates used to define the stress distribution near a crack.

The criterion developed by Griffith employed the approach that a central sharp elliptical flaw in a panel of a brittle material of infinite width and unit thickness can be loaded to a stress σ and then held at that stretched length, as suggested by Fig. 7.6. Without propagation of the crack, the material is then loaded as shown from zero stress and elongation to point A. At that point, the elastic strain energy stored in the material would then correspond to the area of 0AB. At that point, we hold the grips in place. If subsequently the crack grows by an increment da, the energy stored in the panel will decrease, because the ends of the panel are fixed and the panel is now more compliant. The new lower load is C and the new area under the curve is 0CB. The difference between these two areas corresponds to the elastic energy released when the crack advances from a to $a + da$, designated by 0AC. If this differential is sufficient to make the crack grow, then crack growth corresponds to

$$dE_t/dA = dU_{el}/dA + dW/dA \tag{7.3}$$

where E_t is the total energy, U_{el} is the elastic energy corresponding to 0AC, W is the energy level that must be reached for the crack to grow, and dA is the increase in crack surface area. If the total energy change is zero, then

$$-dU_{el}/dA = dW/dA \tag{7.4}$$

Griffith calculated that the magnitude of elastic energy is $(\pi\sigma^2a^2B)/E$ by using the analysis of Inglis with B as the thickness of the material and E as the Young's modulus. To derive the relationship between the external stresses and geometry, consider that the change

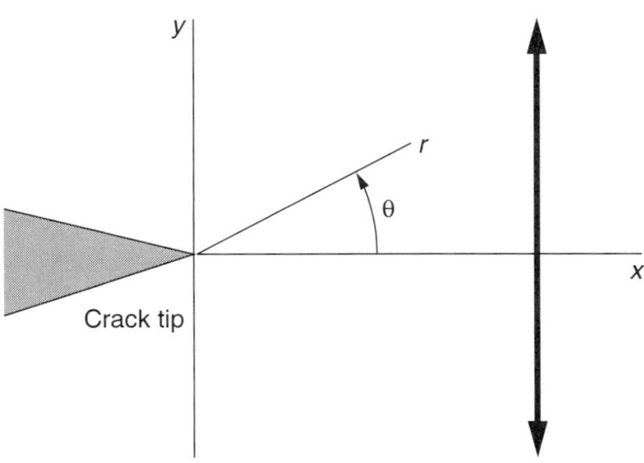

FIG. 7.5 Geometry and definition of terms for a crack tip. The crack tip is the gray triangle extending from left to right. The angle θ and the distance r represent cylindrical coordinates. The arrow indicates that the remote applied tensile stress is in the y-direction.

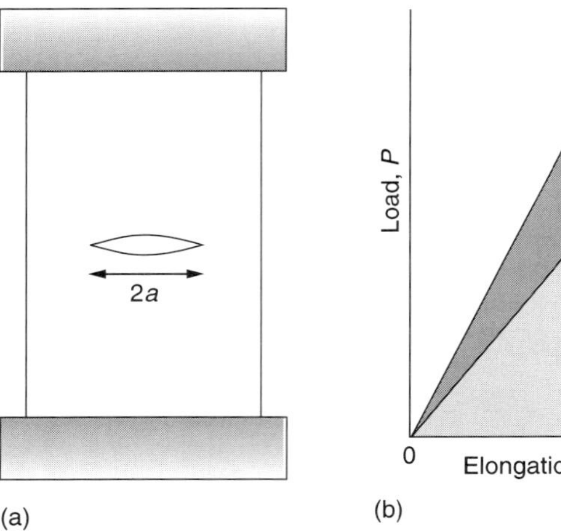

FIG. 7.6 (a) Panel containing a center flaw of length $2a$ stretched to a fixed elastic strain (the size of the flaw is exaggerated in comparison with the panel width). (b) Load versus elongation diagram showing the elastic unloading produced by extending the crack from length $2a$ to length $2(a + da)$.

in energy from crack propagation must also include a factor corresponding to the area of the crack and the surface energy of the material, γ_{surf}. For a center crack of half-length a and thickness B, the area of the crack is given by $2(2\pi aB)$ because both halves of the crack become new surfaces. Then the change in energy before and after crack propagation yields

$$dU_{el}/dA = \frac{\pi\sigma^2 a}{E} \tag{7.5}$$

and

$$dW/dA = 2\gamma_{surf}$$

The fracture stress σ_f can be obtained from these two relations to yield

$$\sigma_f = \sqrt{\frac{E\,2\gamma_{surf}}{\pi a}} \tag{7.6}$$

which for most brittle materials provides a reasonable estimate of the fracture strength of the material. Unfortunately, direct measurement of the surface energy is quite difficult unless the measurement takes place under conditions in which the material creeps. In addition, the occurrence of nonlinear deformation modes near a crack tip from plasticity, microcracking, phase transformation, and several other mechanisms changes the stress distribution and the energy released during fracture.

The crack-driving force expression dU/dA is often termed G, *the strain energy release rate*, after work by Irwin that sought a measure of the energy input for crack growth. The energy expended by crack propagation, dW/dA, is often also given as R, the crack growth resistance. Thereby, when $G \geq R$, crack propagation is possible.

If R has a constant value, then there is a critical value of G, which for simple tensile loading is given as

$$G_{Ic} = \frac{\pi\sigma_c^2 a}{E}$$

where σ_c is the fracture stress, the subscript I indicates a mode of tensile loading, and the subscript c stands for reaching the critical point at which fracture occurs. Using this result, the classic solution can be rearranged to yield the solution for plane strain conditions

$$K_{Ic} = \sqrt{EG_{Ic}} = \sigma_c\sqrt{\pi a} \tag{7.7}$$

To accommodate a variety of geometries, Eq. 7.7 can be written as

$$K_{Ic} = \beta\, \sigma_c \sqrt{\pi a} \tag{7.8}$$

which uses the dimensionless factor β to adjust for geometry. This factor β is a function of crack dimensions, specimen dimensions, and loading geometry that produces a fracture toughness in terms of the materials property K_{Ic}. A list of the plane strain fracture toughness for some materials at room temperature is given in Table 7.1. As we will see later, for ductile materials an inverse relation is observed between yield stress and K_{Ic}. For rolled or extruded materials, the fracture toughness is often anisotropic from the development of crystallographic texture and the alignment of defects during the forming operation. Also, as we will see later, the higher the toughness, the thicker the specimen must be to maintain plane strain conditions. For the ceramic materials, microstructure can have a strong influence on the fracture toughness that is measured. Toughening mechanisms that include elongated grains or transformation toughening in zirconia ceramics give the materials relatively broad ranges of possible toughness. In polymers, the molecular weight, branching, crystallization, and cross linking provide for a wide range of values, although the low Young's modulus of most polymers keeps the K_{Ic} values fairly low, with most of the changes taking place in G_{Ic}.

TABLE 7.1 Typical Ranges of Plane Strain Fracture Toughness and Yield Strength for Several Materials at Room Temperature

Material	K_{Ic} (MPa$\sqrt{m}$)	Y (MPa)
Al 2000 series	24–40	300–450
Al 7000 series	25–35	400–600
Ti-6A1-4V alloys	50–110	800–1100
4340 steel	55–105	1300–1700
Maraging steels	40–80	1400–2300
Alumina (Al_2O_3)	3–5	—
Boron carbide (BC)	4–6	—
Silicon nitride (Si_3N_4)	4–8	—
Silicon carbide (SiC)	2–5	—
Tetragonal zirconias (doped ZrO_2)	4–10	—
Epoxies	0.5–0.8	—
Borosilicate glass	0.5–1	
Polymethylmethacrylate (PMMA)	1–3	20–50
Polystyrene (PS)	1–2	30–80
Polycarbonate (PC)	2.5–3	60–70
Polyvinyl carbide (PVC)	2–3	40–50

BIOGRAPHY

GEORGE R. IRWIN (1907–1998)

Irwin joined the United States Naval Research Laboratory (NRL), where he was a research scientist and became supervisor of the Mechanics Division. He later became a professor at Lehigh University until 1972 and then at the University of Maryland. He led many of the accomplishments of ASTM's Committee E-24 on Fracture Testing and Committee E-08 on Fatigue and Fracture.

There is a slight difference between loading of the panel to a fixed displacement, as shown in Fig. 7.6, and loading with a fixed load. The energy increase from crack extension under a fixed load is nearly equal and opposite in sign to that for a panel with fixed ends.

Another quantity that becomes important in evaluation of specimens undergoing crack propagation is specimen compliance. Specimen compliance is simply the inverse slope of the load-extension curve.

$$\psi = \frac{\Delta u}{\sigma_\infty}$$

where Δu is the displacement and σ_∞ as the remote stress. Using the change in stored energy for crack advancement in the panel with fixed ends yields

$$G = -\left(\frac{1}{B}\right)\left(\frac{\partial U}{\partial a}\right) = \frac{-\Delta u}{2B}\left(\frac{\partial \sigma_\infty}{\partial a}\right)$$

Substitution of ψ leads to

$$G = \frac{\sigma_\infty^2}{2B}\frac{\partial \psi}{\partial a} \tag{7.9}$$

The same expression is obtained if one considers for the constant load case that there is an increase in energy of approximately the same magnitude. Additionally, as shown in several problems at the end of this chapter, the functional dependence of *G(a)* on crack length is determined by the loading and specimen geometry.

In most ductile materials, the resistance of a material to crack propagation is much greater than twice the surface energy, and may depend on the geometry of the crack and the size of the crack. Under these circumstances it is important to consider crack stability. If the resistance to crack propagation does not change with increasing a, then the relationship between G and R is as shown in Fig. 7.7*a*. In this figure, a is plotted on the horizontal axis and G and R are plotted relative to the vertical axis. For any initial crack size, when the stress is increased to a value at which G meets or exceeds the critical value of $R = G_c$, crack propagation will take place unstably. These relations are called resistance curves (*R-curves*).

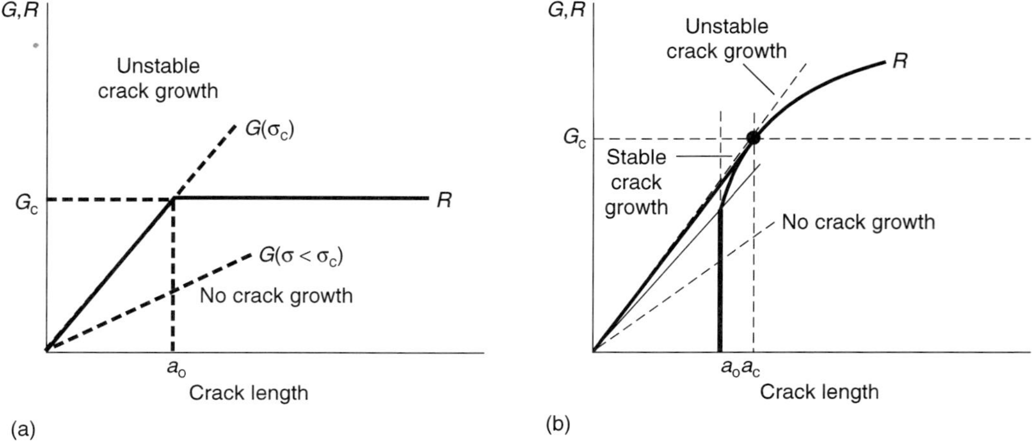

FIG. 7.7 Schematic driving *G* and *R*-curve diagrams. (a) Flat *R*-curve. (b) Rising *R*-curve.

Some materials have R-curve responses that show increases in crack growth resistance with crack size, as shown in Fig. 7.7*b*. For such materials, the crack grows and then stops when both

$$G = R$$

and

$$dG/da < dR/da \tag{7.10}$$

Thus, as the stress is increased, thereby increasing G, the crack can grow from its initial value of a_o to failure at a_c. For the crack to grow any longer than this critical crack length, propagation of the crack is unstable, because

$$dG/da > dR/da$$

Whereas different testing geometries involve different forms of G, a material with a crack-size-dependent R-curve cannot be assigned a single toughness value. This type of behavior is also described as a rising R-curve. A rising R-curve can result from transition from plane strain to plane stress fracture from formation of a zone of deformation near the crack tip called a *plastic zone*. Attributes in the microstructure that cause the R-curve to be a rising one are called toughening mechanisms.

Modes of Crack Propagation Although the examples discussed thus far focus on cracks with stresses that lie normal to the plane of the crack, which we have identified as Mode I, three distinct modes can be identified relative to a crack plane and the direction of crack propagation, as shown in Fig. 7.8. Every crack propagates by one or a combination of the modes. The critical stress intensities K_{II}, for shearing, and K_{III}, for tearing, usually represent deviations from Mode I propagation. As shown in Fig. 7.9*a*, crack growth tilted away from the Mode I crack growth direction is a mixture of Modes I and II. Crack growth twisted about the crack growth direction so that the crack plane is not orthogonal to the tensile loading direction is a mixture of Modes I and III, as shown in Fig. 7.9*b*. The stress fields surrounding through-cracks in an infinite plate under these different loading modes can be given as

$$\sigma_{ij}(\theta, r) = \frac{K_k}{\sqrt{2\pi r}} f_{ij,k}(\theta) \tag{7.11}$$

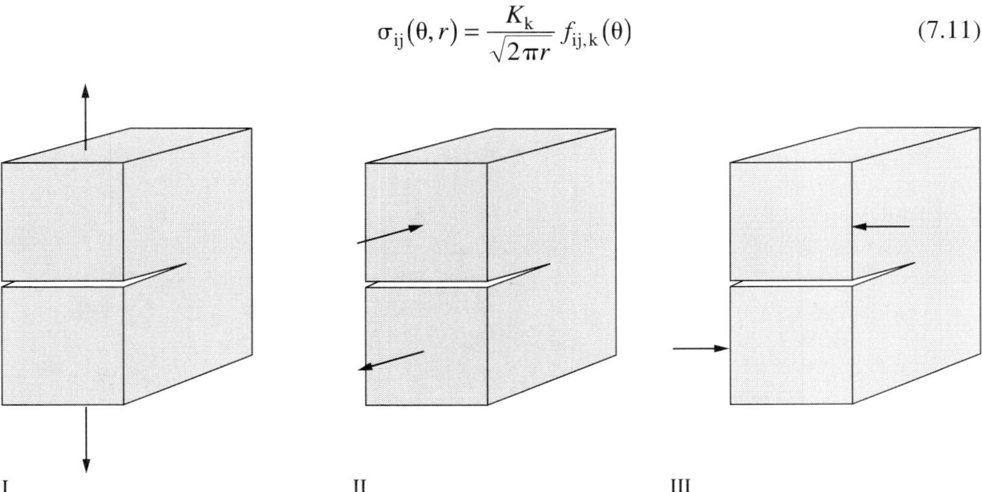

FIG. 7.8 Modes of crack propagation. Mode I, opening, is simple tension with the loading normal to the crack plane and the direction of crack growth. Mode II, shearing, is shearing across the crack plane and parallel or antiparallel to the crack growth direction. Mode III, tearing, is shearing across the crack plane and crack growth in an orthogonal direction.

and displacements at the crack tip can be defined as

$$u_i = \frac{K_k}{2\mu}\sqrt{\frac{r}{2\pi}}f_{i,k}(\theta) \tag{7.12}$$

where i and j define the terms of a stress tensor and the subscript k represents Mode I, II, or III.

For plane stress conditions at the surface of the loaded material with $\beta = 1$, the local stress distributions resulting from the stress concentration take the form of Eq. 7.11, as

$$\begin{aligned}\sigma_{11} &= \sigma\sqrt{\frac{a}{2r}}\cos\frac{\theta}{2}\left(1-\sin\frac{\theta}{2}\sin\frac{3\theta}{2}\right)\\ \sigma_{22} &= \sigma\sqrt{\frac{a}{2r}}\cos\frac{\theta}{2}\left(1+\sin\frac{\theta}{2}\sin\frac{3\theta}{2}\right)\\ \sigma_{12} &= \sigma\sqrt{\frac{a}{2r}}\sin\frac{\theta}{2}\cos\frac{\theta}{2}\cos\frac{3\theta}{2}\end{aligned} \tag{7.13}$$

These distributions give the stress level that is in addition to the remote applied stress. Each of these stress distributions applies only to fully elastic loading, which results in a stress singularity at the crack tip. The stress concentration shown in Fig. 7.2 would theoretically reach infinity at $r = 0$ in Eq. 7.13. The stresses are not infinite in real materials, because cracks are never infinitely sharp and also because most materials will undergo some nonlinear, nonelastic deformation at a fixed stress level. The nonlinear deformation can be either plasticity or some other mechanism (e.g., phase transformation or microcracking). The zone of nonlinear deformation is often called a plastic zone if plastic deformation is

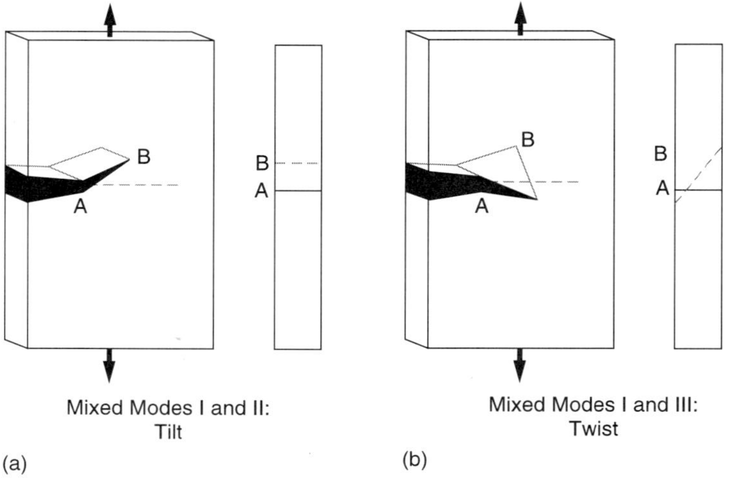

FIG. 7.9 (a) Tilting of the crack growth direction, showing a mixture of Modes I and II. (b) Twisting of the crack plane in a three-dimensional view and in projection, showing a mixture of Modes I and III. The letter A represents the end of Mode I crack propagation. The letter B shows a mixed-mode type of crack propagation. The crack paths shown would normally occur only if there were a variation in the microstructure that made Mode I propagation more difficult than mixed-mode propagation. Possible reasons for this deviation could include composite reinforcements or an easier crack path owing to either weak grain boundaries in the case of intergranular crack propagation or cleavage cracking on a favorable crack plane in the case of transgranular crack propagation.

TABLE 7.2 Geometric Factors *f* for Stress Distributions and Displacements

Stress	Mode I	Mode II	Mode III
σ_{11}	$\cos(\theta/2)[1 - \sin(\theta/2)\sin(3\theta/2)]$	$\sin(\theta/2)[2 + \cos(\theta/2)\cos(3\theta/2)]$	0
σ_{22}	$\cos(\theta/2)[1 + \sin(\theta/2)\sin(3\theta/2)]$	$\sin(\theta/2)\cos(\theta/2)\cos(3\theta/2)$	0
σ_{33}	0 for plane stress, $\nu(\sigma_{11}+\sigma_{22})$ for plane strain	0 for plane stress, $\nu(\sigma_{11}+\sigma_{22})$ for plane strain	0
σ_{23}, σ_{32}	$\cos(\theta/2)\sin(\theta/2)\cos(3\theta/2)$	$\cos(\theta/2)[1 - \sin(\theta/2)\sin(3\theta/2)]$	0
σ_{13}, σ_{31}	0	0	$\sigma_{13} = \sin(\theta/2)$ and $\sigma_{31} = \cos(\theta/2)$
u_1	$\cos(\theta/2)[\kappa - 1 + 2\sin^2(\theta/2)]$	$\sin(\theta/2)[\kappa + 1 + 2\cos^2(\theta/2)]$	0
u_2	$\sin(\theta/2)[\kappa + 1 - 2\cos^2(\theta/2)]$	$\cos(\theta/2)[\kappa - 1 - 2\sin^2(\theta/2)]$	0
u_3	0	0	$2\sin(\theta/2)$

For Table 7.2, $\kappa = (3 - \nu)/1 + \nu)$ for plane stress and $\kappa = 3 - 4\nu$ for plane strain, where ν is Poisson's ratio.

involved and a process zone for other mechanisms. Other mechanisms that can produce dissipation of energy near the crack tip and therefore enhanced toughness include all shear and dilatant mechanisms described in Chapter 3. In fact, the stresses shown above are only the first terms of a mathematical series that is used to solve the complex relations. If the crack size is large compared with the width or length of the part, then the other terms in the solution may become relevant. The values for each of the stress and displacement terms for all loading modes are given in Table 7.2 by the forms shown in Eq. 7.11 and 7.12.

EXAMPLE 7.2 *Elastic Distortions Near the Crack Tip in LEFM*

Plotting of contours is an important approach for representing the variance of a variable with a spatial reference frame. Using math software or a spreadsheet, it is possible to generate these contours as shown here. This example also imparts part of the meaning behind the contours constructed at stress concentrations and it provides a reminder of the relationship between Cartesian and polar coordinates. For a brittle material, the stress intensity is reported as

$$K = 10 \cdot 10^6 \cdot \mathrm{Pa} \cdot \sqrt{\pi \cdot 0.1 \cdot \mathrm{m}}$$

for a panel of material with an initial crack, as shown in Fig. 7.10. The normal stress distribution in the x-direction, σ_{xx}, can be expressed as functions of the polar coordinates of θ and r, and using the values of K for plane stress at the surface from Eq. 7.13. Because this state of stress in the x-y plane includes a shear component that varies with θ in a different manner than the normal stresses, the orientation of principal stress axes and the magnitude of the principal stresses must then change with θ.

The values of the principal stresses when there are no shear stress terms with z-components (i.e., $\sigma_{13} = \sigma_{31} = \sigma_{23} = \sigma_{32} = 0$) are given in Chapter 3 as

$$s_{\max,\min} = \frac{\sigma_{11}+\sigma_{22}}{2} \pm \sqrt{\left(\frac{\sigma_{11}-\sigma_{22}}{2}\right)^2 + \sigma_{12}^2}$$

Without reduction, direct application of this yields

$$s(\theta, r) = \frac{\sigma_{11}(\theta, r)+\sigma_{22}(\theta, r)}{2} \pm \sqrt{\left(\frac{\sigma_{11}(\theta, r)-\sigma_{22}(\theta, r)}{2}\right)^2 + \sigma_{12}(\theta, r)^2}$$

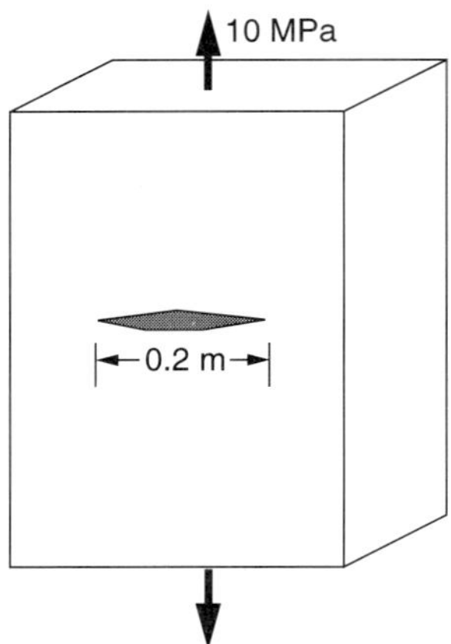

FIG. 7.10 Thick fracture specimen with a center crack of length 0.2 m and an applied tensile stress of 10 MPa. Assume that the material has fracture toughness and yield strength such that neither fracture nor yielding is relevant to the problem (in other words, ignore any plastic zone). The size of the crack is exaggerated in comparison with the specimen size.

where each term comes from Eq. 7.13. Now, the task is to find the contours determined in polar coordinates and then locate these same points on a Cartesian representation. One easy approach is first to establish an angular range for θ as 0 to 180° in increments of 10° using I as a range of integers from 1 to 19:

$$\text{Theta}_{\text{I}} = I \cdot \frac{\pi}{18} \cdot \frac{10 \cdot \pi}{18}$$

If values of r are chosen in a range of J from 1 to 20 corresponding to 0.1 to 2.0 mm using the equations above,

$$r_{\text{J}} = 0.0001 \cdot J \text{ (meters)}$$

We can establish a threshold for a particular principal stress contour using an "if" function

$$\text{Threshold}_{\text{I,J}} = \text{if } [(101 \times 10^{6} \cdot \text{Pa} - s(\theta_{\text{I}}, r_{\text{J}})) < 0 \cdot \text{Pa}, 1, 0]$$

to yield a stress threshold at 101 MPa. This function states that if $s(\theta,r)$ at any point is greater than 101 MPa, it will return a value of 1, and if it is less than 101 MPa, it will return a value of 0.

Then the matrix with rows in terms of θ and columns in terms of r appears as

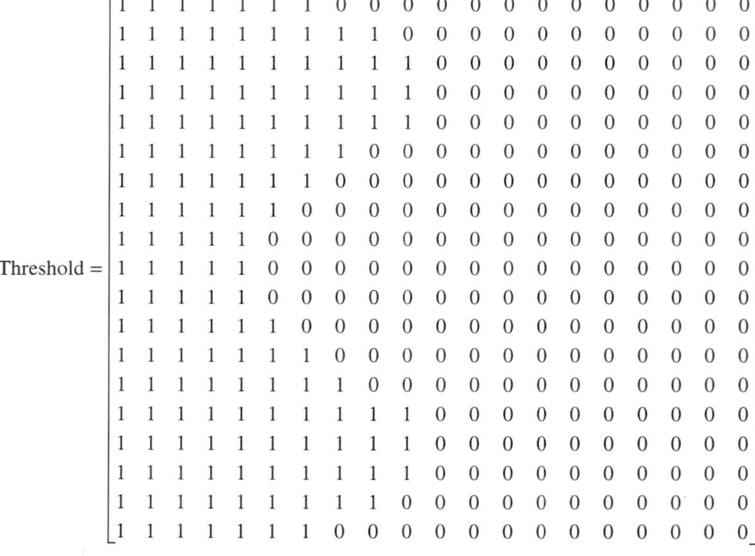

$$\text{Threshold} = \begin{bmatrix}
1 & 1 & 1 & 1 & 1 & 1 & 1 & 0 & 0 & 0 & 0 & 0 & 0 & 0 & 0 & 0 & 0 & 0 & 0 & 0 \\
1 & 1 & 1 & 1 & 1 & 1 & 1 & 1 & 1 & 0 & 0 & 0 & 0 & 0 & 0 & 0 & 0 & 0 & 0 & 0 \\
1 & 1 & 1 & 1 & 1 & 1 & 1 & 1 & 1 & 1 & 0 & 0 & 0 & 0 & 0 & 0 & 0 & 0 & 0 & 0 \\
1 & 1 & 1 & 1 & 1 & 1 & 1 & 1 & 1 & 1 & 0 & 0 & 0 & 0 & 0 & 0 & 0 & 0 & 0 & 0 \\
1 & 1 & 1 & 1 & 1 & 1 & 1 & 1 & 1 & 1 & 0 & 0 & 0 & 0 & 0 & 0 & 0 & 0 & 0 & 0 \\
1 & 1 & 1 & 1 & 1 & 1 & 1 & 1 & 0 & 0 & 0 & 0 & 0 & 0 & 0 & 0 & 0 & 0 & 0 & 0 \\
1 & 1 & 1 & 1 & 1 & 1 & 1 & 0 & 0 & 0 & 0 & 0 & 0 & 0 & 0 & 0 & 0 & 0 & 0 & 0 \\
1 & 1 & 1 & 1 & 1 & 1 & 0 & 0 & 0 & 0 & 0 & 0 & 0 & 0 & 0 & 0 & 0 & 0 & 0 & 0 \\
1 & 1 & 1 & 1 & 1 & 0 & 0 & 0 & 0 & 0 & 0 & 0 & 0 & 0 & 0 & 0 & 0 & 0 & 0 & 0 \\
1 & 1 & 1 & 1 & 1 & 0 & 0 & 0 & 0 & 0 & 0 & 0 & 0 & 0 & 0 & 0 & 0 & 0 & 0 & 0 \\
1 & 1 & 1 & 1 & 1 & 0 & 0 & 0 & 0 & 0 & 0 & 0 & 0 & 0 & 0 & 0 & 0 & 0 & 0 & 0 \\
1 & 1 & 1 & 1 & 1 & 1 & 0 & 0 & 0 & 0 & 0 & 0 & 0 & 0 & 0 & 0 & 0 & 0 & 0 & 0 \\
1 & 1 & 1 & 1 & 1 & 1 & 1 & 0 & 0 & 0 & 0 & 0 & 0 & 0 & 0 & 0 & 0 & 0 & 0 & 0 \\
1 & 1 & 1 & 1 & 1 & 1 & 1 & 1 & 0 & 0 & 0 & 0 & 0 & 0 & 0 & 0 & 0 & 0 & 0 & 0 \\
1 & 1 & 1 & 1 & 1 & 1 & 1 & 1 & 1 & 1 & 0 & 0 & 0 & 0 & 0 & 0 & 0 & 0 & 0 & 0 \\
1 & 1 & 1 & 1 & 1 & 1 & 1 & 1 & 1 & 1 & 0 & 0 & 0 & 0 & 0 & 0 & 0 & 0 & 0 & 0 \\
1 & 1 & 1 & 1 & 1 & 1 & 1 & 1 & 1 & 1 & 0 & 0 & 0 & 0 & 0 & 0 & 0 & 0 & 0 & 0 \\
1 & 1 & 1 & 1 & 1 & 1 & 1 & 1 & 1 & 0 & 0 & 0 & 0 & 0 & 0 & 0 & 0 & 0 & 0 & 0 \\
1 & 1 & 1 & 1 & 1 & 1 & 1 & 0 & 0 & 0 & 0 & 0 & 0 & 0 & 0 & 0 & 0 & 0 & 0 & 0
\end{bmatrix}$$

By counting from the right, the radial distance in increments of 0.1 mm at which the stress first falls below 101 MPa is indicated by a zero.

Now, the task is to translate this into a contour. By simply summing row-wise, the point approximating the contour position can be determined from the threshold matrix with the simple function

$$\text{Contour}_I = \sum_J \text{Threshold}_{I,}$$

which yields

$$\text{Contour} = \begin{bmatrix} 7 \\ 9 \\ 10 \\ 10 \\ 10 \\ 8 \\ 7 \\ 6 \\ 5 \\ 5 \\ 5 \\ 6 \\ 7 \\ 8 \\ 10 \\ 10 \\ 10 \\ 9 \\ 7 \end{bmatrix}$$

Translating the angular values and the limits set by the contour function back into Cartesian positioning, and smoothing the boundary between points in the matrix above and below the threshold, result in the plot given in Fig. 7.11.

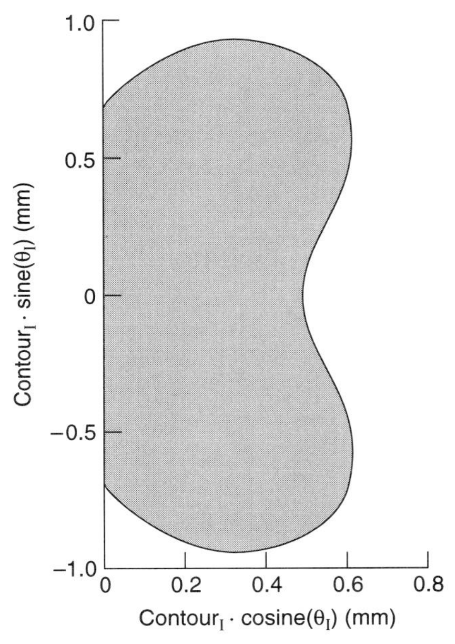

FIG. 7.11 Smoothed plot of a stress contour for 101 MPa starting near the right-hand tip of the crack shown in Fig. 7.10.

This rather explicit method of showing an individual contour hopefully provides a way to grasp the meaning behind the contours shown in Fig. 7.2. A similar strategy can be employed to calculate any of the stresses at a stress concentration.

Although this sample was solved numerically, we also could solve for a contour, as you will see in Fig. 7.12. The experimental contours shown in Fig. 7.2 can also be modeled numerically or analytically. Comparisons of experimental contours and contours from models for deformation or toughening mechanisms occurring near the crack tip can be used to verify how well the models describe the behavior taking place near the crack tip.

■

Loading on a crack in a structure often consists of multiple or mixed modes. The individual components are often treated as additive such that

$$\sigma_{ij}^{total} = \sigma_{ij}^{I} + \sigma_{ij}^{II} + \sigma_{ij}^{III} \tag{7.14}$$

Using the relations in Table 7.2, the normal stresses in the x- and y-directions, σ_{11} and σ_{22}, respectively, are equal when $\theta = 0$. Contours of stress for plane stress and plane strain conditions can be plotted using yield criteria as the effective stress level, as shown in Fig. 7.12. The contours in Fig. 7.12 define a zone within which we expect plastic deformation to occur, as discussed in the next section.

7.1.2 Plane Stress Versus Plane Strain Conditions

Consider once again the geometry in Fig. 7.5. Using Eq. 7.13 for the surface position at $\theta = 0$, we expect a state of balanced biaxial tension with $\sigma_{11} = \sigma_{22}$, $\sigma_{12} = 0$ at $(r, \theta = 0)$, and $\sigma_{33} = 0$ such that

$$\sigma_{ij} = \begin{pmatrix} \sigma_r & 0 & 0 \\ 0 & \sigma_r & 0 \\ 0 & 0 & 0 \end{pmatrix}$$

If this is true, when does yielding occur using a Tresca or maximum shear stress criterion for yielding? If we really have a state of plane stress at the surface, yielding should be possible whenever $\sigma_{r,\theta}$ reaches the yield stress. The Tresca criterion, as defined in Chapter 3, is given by

$$Y = s_{max} - s_{min}$$

where s_{max} is the largest principal stress and s_{min} is the smallest principal stress. Thus, whenever $\sigma_{11} = \sigma_{22} \geq Y$, plastic deformation should occur. The only position within a

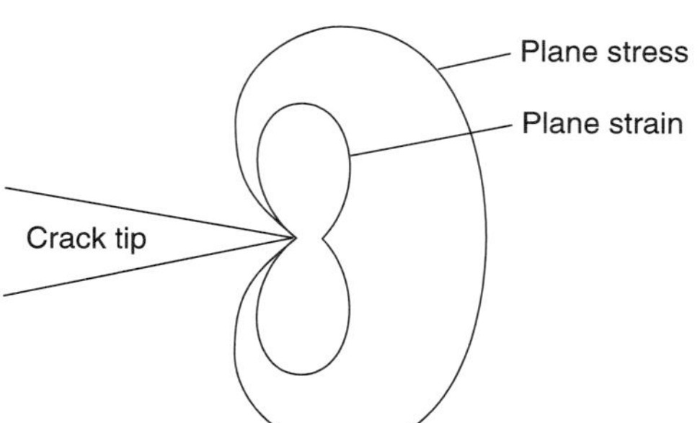

FIG. 7.12 Plastic zone shapes for plane stress and plane strain according to a von Mises yield criterion.

homogeneous specimen where $\sigma_{11} = \sigma_{22}$ and $\sigma_{33} = 0$ is at the free surface normal to the z-direction. Below the surface, the stress state becomes three-dimensional (or triaxial) owing to constraint (as discussed at the beginning of Chapter 5). This constraint inhibits plastic deformation by shear below the surface with the maximum constraint at the center of the specimen thickness.

Although plane stress at the center of the thickness can be reached completely only in an infinitely thin specimen, plane stress governs the response of the material when it is below a threshold thickness. This threshold is determined by geometry and material properties. Plane strain conditions are found below the surfaces of finite specimens, and the strain in the z-direction approaches a minimum as the center of the thickness is approached. Because all of the stresses are tensile, the difference between s_{max} and s_{min} will be smaller, leading to a smaller range over which we expect plastic deformation. The size of the plastic zone for the Tresca criterion is slightly larger for both plane stress and plane strain conditions, but very similar in shape to that shown in Fig. 7.12 for the von Mises criterion. This *plastic zone* (also called process zone for the nonlinear deformation processes occurring within it) varies in size from a maximum at the surface to a minimum at the center, as suggested by Fig. 7.13.

If a biaxial tensile stress is applied across the radial directions of a cylinder, a Poisson contraction takes place along the cylinder axis. Then, using Hooke's law, we can get the shortening of the cylinder. If we place this cylinder into a through-hole in a panel that has the same dimensions as the original dimensions of the cylinder, we must pull in tension to bring the cylinder surface back into line with the surface of the panel. The stress required to do so is $\sigma_z = \nu(\sigma_x + \sigma_y)$. This is a state of plane strain, because there is now no strain along the cylinder axis.

If we ignore any unloading of the stress concentration from the plastic deformation, the plastic zone for $\theta = 0$ can be estimated as

$$r^* = \frac{K_I^2}{2\pi Y^2} \tag{7.15}$$

which means that B/r^* determines whether plane strain or plane stress is dominant. From experiments on ductile metals, the criterion usually applied is

$$\frac{B}{(K_I/Y)^2} \geq 2.5 \tag{7.16}$$

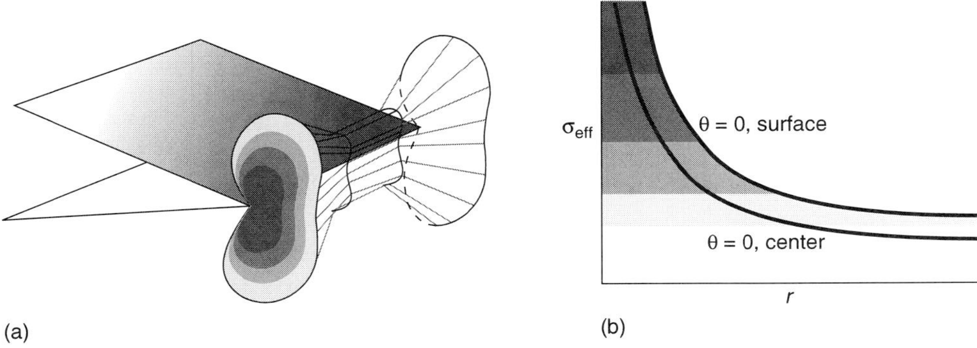

FIG. 7.13 Plastic zone size at the crack tip versus position, showing the variation in plastic zone size consistent with Fig. 7.12 (after Broek, 1982) with (a) showing a three-dimensional view and (b) showing the schematic levels of stress at θ = 0 as a function of the distance from the crack tip, *r*. The difference between the surface and the center is a function of specimen thickness.

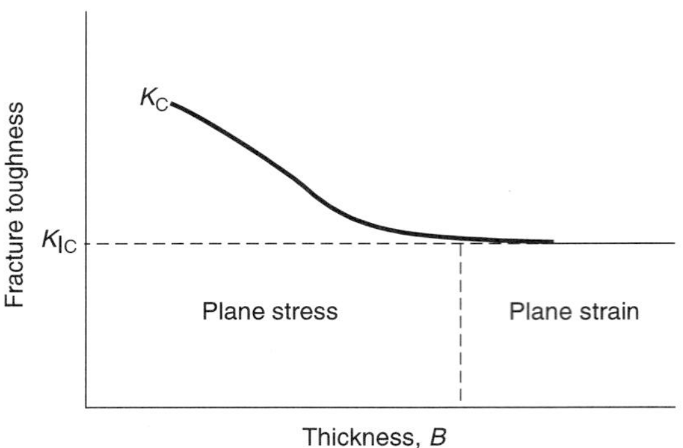

FIG. 7.14 Schematic diagram showing the thickness dependence of Mode I fracture _toughness.

for plane strain to be a good assumption. Below this ratio, plane stress conditions govern the response. Note that the stress intensity K_I appears in Eq. 7.15 and 7.16, rather than the critical plane strain stress intensity or fracture toughness K_{Ic}.

To get equivalent testing of different size parts, constraints must be equal. This suggests that failure should be analyzed only in plane strain (if full constraint is available) or in plates of equal thickness with through-thickness cracks. Thus, if the specimen is not thick, the toughness is a function of thickness—i.e., $K_I = f(B)$. The dependence of fracture toughness on thickness generally follows the pattern shown in Fig. 7.14.

The formation of the plastic zone in ductile materials adds to the work of fracture. Thus, the strain energy release rate G includes plastic deformation and the formation of new surface. The larger the plastic zone, the greater the material's resistance to fracture. Because decreasing yield stress leads to increasing plastic zone size, fracture toughness is inversely related to yield stress. At very low yield stresses, crack tip plasticity can also lead to blunting of the notch. This behavior also leads to the inverse dependence of fracture toughness on yield stress, as shown in Fig. 7.15.

In very ductile materials, the blunting of the notch is followed by large-scale yielding that can span the width of the material. Prediction of fracture in conditions approach-

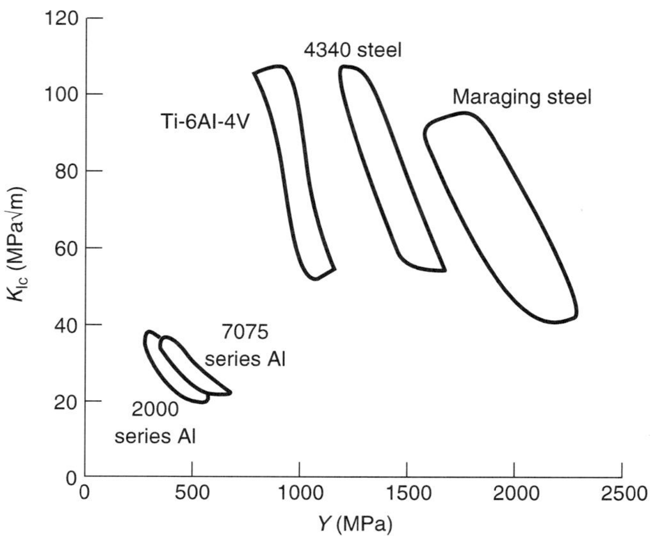

FIG. 7.15 Inverse relation between fracture toughness and yield stress based on data from Hertzberg (1996). Specific materials would be expected to have unique curves with a similar dependence of K_{Ic} on Y. Because decreasing Y results in an increased plastic zone size, the conditions for maintaining a valid plane strain condition may require thicker specimens. These data are given for demonstration purposes only; the possible ranges for each alloy type are much broader than shown.

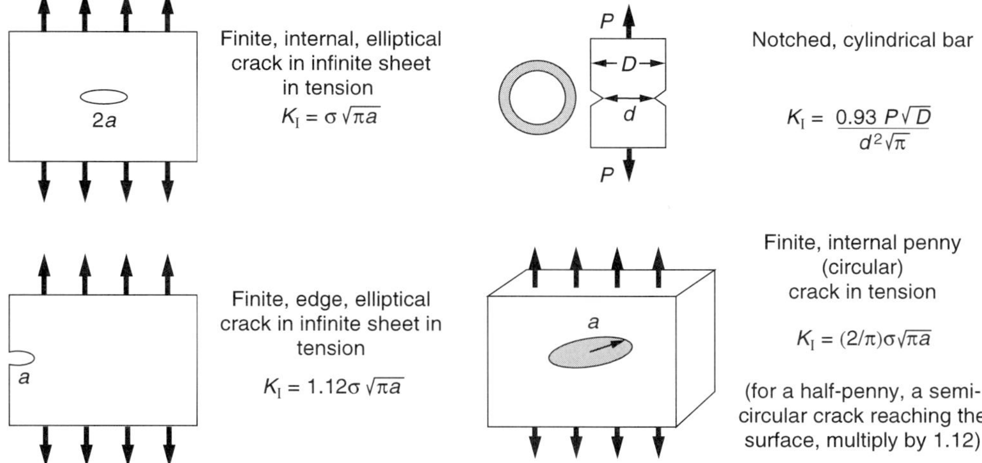

FIG. 7.16 Geometry and definition of terms for a crack tip, where P represents an applied force (after Broek, 1982).

ing large-scale yielding can be particularly difficult because it is strongly geometry-dependent and relies on a good understanding of strain hardening behavior. This area of investigation is called elastic-plastic fracture mechanics. One advantage of approaches that include the plastic response is that they require smaller test specimens than are required for the aforementioned linear elastic fracture mechanics. The best of these approaches employs the J-integral (where J_{Ic} is an elastic-plastic version of G_{Ic}), which enables a model for expected strain hardening behavior for the material in the plastic zone to be incorporated into a solution for the fracture toughness (see Anderson, 1995).

7.1.3 Practical Geometries and Superposition

For practical applications, the assumptions of infinite dimensions and central cracks could be quite limiting. To circumvent this problem, analytical and numerical calculations can be employed to obtain the magnitude of each value of K. Since K_I is the most severe crack propagation condition, it is very convenient to express different geometries in terms of K_I. Solutions for common geometries are shown in Fig. 7.16. When first considering complicated crack geometries, it is often advantageous to superimpose known stress fields in an additive way. Under these circumstances one can take simple solutions and generate the solution for a more complicated geometry using superposition (see Broek, 1982). A small crack at the edge of a round hole is a common example. To generate an expression by superposition for this problem, consider that the stress concentration at the edge of a round hole is simply three times the applied stress. Thus,

$$K_I \cong 3\sigma_{\text{applied}}\sqrt{\pi a}$$

if the size of the small flaw is enough smaller than the hole radius that it lies near the maximum stress concentration from the through-hole.

EXAMPLE 7.3 *Superposition for a Pressurized Internal Flaw*

Consider the hypothetical example of an internal flaw that is under pressure from containing a gas that is at a higher pressure than the atmosphere outside the specimen. For our example, the internal

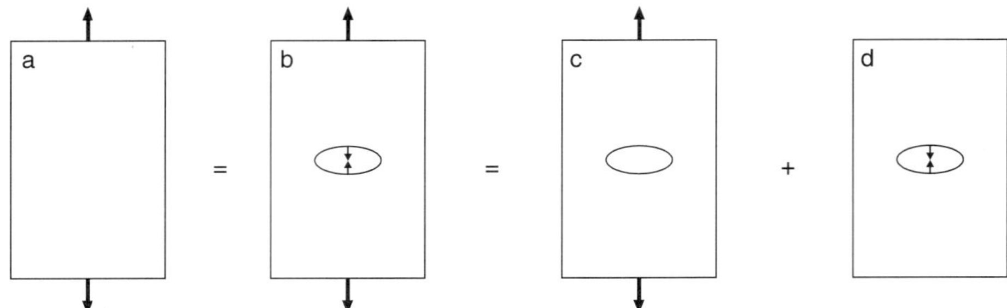

FIG. 7.17 Superposition for pressurized internal crack, where it is shown that a crack with an internal pressure pushing the crack open has the same stress intensity as would a tensile stress applied to the ends of the piece with the same initial flaw.

pressure is currently 5 atmospheres (0.5 MPa) above sea level pressure and the flaw size is $2a$ = 200 mm. We would like to see if the glass we are using will fail during expected heating that will raise the temperature from room temperature to 300°C. The K_{Ic} of the glass is almost unchanged over this temperature range, and we can neglect the small amount of thermal expansion that will marginally increase the flaw size. The glass being used has a fairly low fracture toughness of $K_{Ic} = 0.5$ MPa $\sqrt{\text{m}}$. We know that the pressure increase should be linear with temperature, so the internal pressure will increase to 1 MPa over sea level.

Figure 7.17 shows the superposition approach applied to a pressurized internal crack. Part (a) of this figure is a flawless plate that is under tension. We can treat this as having a stress intensity $K_I^a = 0$. Part (b) of this figure shows an internal flaw within a specimen loaded in tension on the ends and loaded with forces applied on the surfaces of the flaw. The forces are of a magnitude to exactly cancel the forces applied on the ends. Then these thought examples should have equal stress intensities $K_I^a = K_I^b = 0$. If we treat the elements of part (b) as separable, we can define parts (c) and (d). Then we can write that the superposition or sum of the stress intensities for these two cases is

$$K_I^b = 0 = K_I^c + K_I^d$$

We can write that $K_I^c = -K_I^d$. Since K_I^d is of course simply $K_I^c = \sigma\sqrt{\pi a}$, we can write $K_I^d = \sigma_p\sqrt{\pi a}$, where σ_p is a compressive (negative) pressure in the crack. For the given problem, K_I = 1 MPa $\sqrt{3.14 \cdot 0.1} = 0.56$ MPa $\sqrt{\text{m}}$.

This clearly should result in crack propagation, although extension of the crack will relieve the pressure and the crack should arrest without causing catastrophic failure. ■

EXAMPLE 7.4 *Superposition for a Cracked Panel Supported by a Fastener*

Superposition can also be used to evaluate the case shown in Fig. 7.18*a*. The loading in this example is similar to that found for a bolt or screw fastener that has led to crack propagation due to the supporting of a prior load. The mass of the panel hanging from the bolt is 80 kg, the width of the panel is 0.2 m, and the thickness is 0.01 m. The crack length is $2a$ = 0.02 m. The plexiglass (PMMA) has a fracture toughness of 1.2 MPa $\sqrt{\text{m}}$. For this example, P is a point force that is equal to the force resulting in the stress, σ, applied across the specimen area. We can then write

$$K_I^a = K_I^b + K_I^c - K_I^d$$

We already know that

$$K_I^b = \sigma\sqrt{\pi a}$$

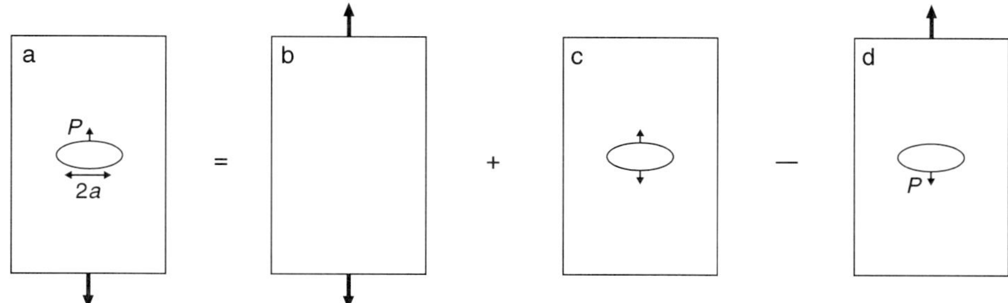

FIG. 7.18 Superposition for (a) point-loaded crack that is related to the sum of a (b) normal center-cracked panel and a (c) crack loaded with balanced central forces pulling it open, and (d) the negative of (a).

and we are given the following expression for the crack with central forces shown in Fig. 7.18*c*:

$$K_I^c = \frac{P}{B\sqrt{\pi a}}$$

Using the diagrams in Fig. 7.18, we can see the graphical relations among the different parts of the figure. For example, parts (a) and (d) are the same diagram switched by 180°. Applying the relations shown in Fig. 7.18, we can write

$$K_I^a = \frac{1}{2}\left(K_I^b + K_I^c\right) = K_I^d$$

which is simply

$$K_I^a = \frac{\sigma}{2}\sqrt{\pi a} + \frac{P}{2B\sqrt{\pi a}} \tag{7.17}$$

For the values given, the applied force is 9.8 m/sec^2 · 80 kg = 784 N, and the stress level is (784 N)/(0.2 m · 0.01 m) = 0.4 MPa.

Using the given crack length the resulting stress intensity is

$$\frac{0.4\text{ MPa}}{2}\sqrt{3.14\cdot 0.01\text{ m}} + \frac{784\text{ N}}{2\cdot 0.01\text{ m}\cdot\sqrt{3.14\cdot 0.01\text{ m}}} = 0.26\text{ MPa}\sqrt{\text{m}}$$

which is not even close to the expected fracture toughness. ■

EXAMPLE 7.5 *Thermal Mismatch Stresses in Ceramic Composites*

Consider an alumina (Al_2O_3) matrix composite containing silicon carbide (SiC) particles of average radius 10 μm. On cooling from the processing temperature, microscale cracks can form as a result of the difference in thermal expansion coefficients. This thermal expansion mismatch causes stresses between the two materials. The internal stresses vary in magnitude on a microscale within the composite:

$$\alpha_{SiC} = 4.3 \times 10^{-6}/\text{K}$$

versus

$$\alpha_{Al_2O_3} = 8.5 \times 10^{-6}/\text{K}$$

On cooling from high temperature to room temperature, we will assume that σ = 400 MPa for the stress applied by the platelets on the alumina matrix. We will use the solution for a penny crack under internal pressure as a conservative case (see Fig. 7.16) and the approach given for a through-crack in Fig. 7.17.

The penny crack stress intensity is given as

$$K_I = \frac{2}{\pi}\sigma\sqrt{\pi a}$$

which becomes

$$K_I = \frac{2}{\pi}\sigma_p\sqrt{\pi a}$$

for internal pressure. The resulting stress intensity is then approximately

$$\frac{2}{\pi}\sigma_p\sqrt{\pi a} = 1.43 \text{ MPa } \sqrt{\text{m}}$$

Of course, this is below the fracture toughness of the matrix ($K_{Ic} \approx$ 4–5 MPa $\sqrt{\text{m}}$ for aluminum oxide). It is possible, however, for this local stress intensity to interact with sintering flaws to result in some microcracks around the platelets. Sintering flaws may be more likely near the particles because they may affect sintering locally, leaving porosity adjacent to the reinforcements. The cracks from these SiC platelets will each be very small because propagation of the crack results in reduction of the stress intensity and the local elastic stiffness. ■

7.2 FRACTURE TOUGHNESS TESTING

Design of tests and test specimens for valid evaluation of fracture toughness can be very challenging. The design of such specimens in a cost-effective manner can be even more difficult. Standard tests for describing fracture behavior are developed for their reproducibility and simplicity. Standard tests—e.g., American Society for Testing and Materials (ASTM), Deutsches Institüt für Normung (German Standards: DIN), or Japanese Industrial Standard (JIS)—enable comparisons of materials under conditions that are considered fair indications of performance. The application of different tests to materials with large differences in mechanical properties should be done cautiously.

7.2.1 Fracture Toughness Testing in Brittle Materials

Bend tests are often employed to evaluate fracture toughness because of their ease of sample preparation and localization of stress. These tests are often conducted in three-point and four-point geometries with fairly complex equations for calculating the fracture toughness. Four fracture toughness techniques—Surface-Crack-in-Flexure (SCF), Single-Edge-Notched-Beam (SENB), Single-Edge-V-Notched-Beam (SEVNB), and Single-Edge-Precracked-Beam (SEPB)—shown in Fig. 7.19 are all employed to assess fracture toughness in brittle materials.

The Surface-Crack-in-Flexure (SCF) technique employs a Knoop or Vickers indenter to initiate a precrack on a bend bar surface. The bend bar is tested in three- or four-point bending (see Chapter 1) after polishing of the surface to remove residual stresses from the deformation zone produced by indentation. The fracture surface is examined using a microscope to determine the initial precrack dimensions. The fracture toughness (K_{Ic}) is calculated using the equations derived for this geometry.

The SENB and SEVNB techniques differ only in notch geometry. A single notch is machined across the bend bar surface for the SENB sample geometry using a diamond-impregnated saw. The size of the notch tip radius can influence the measured K_{Ic} with large notch tip radii (>50 μm), resulting in artificially high K_{Ic} values. A better technique employs

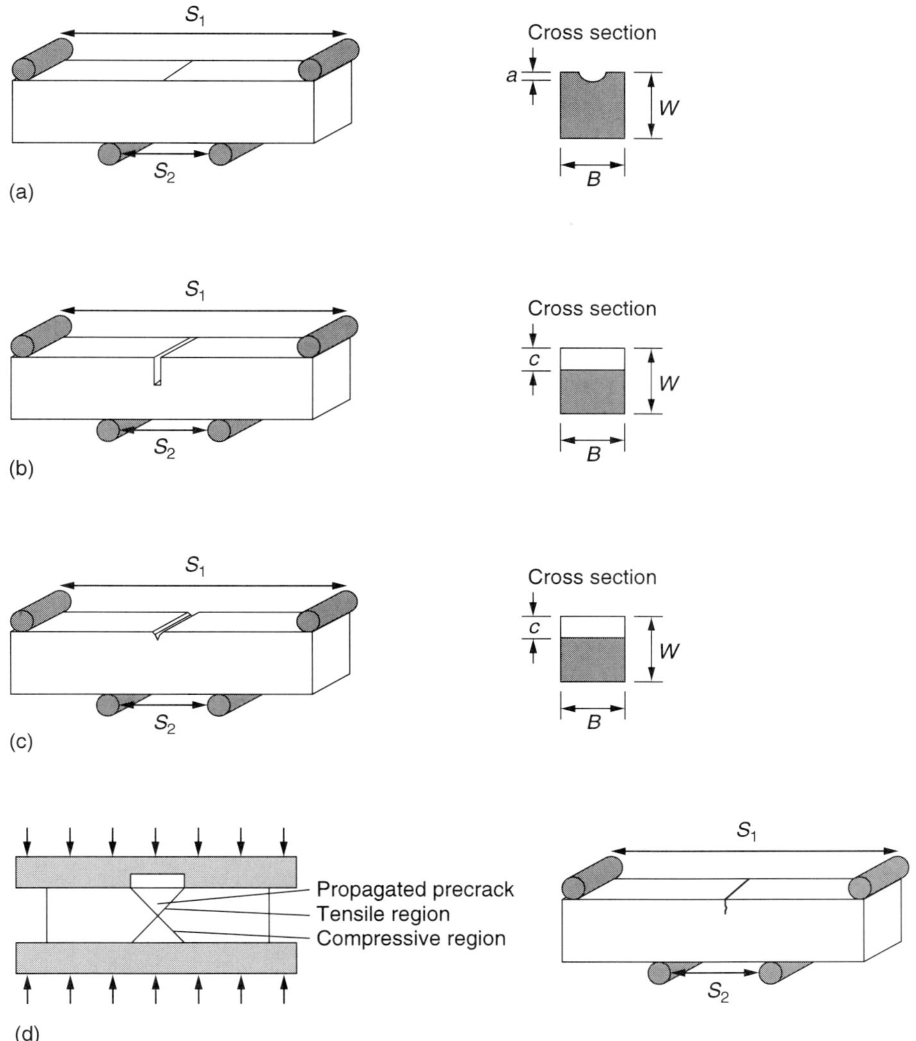

FIG. 7.19 Fracture toughness in bending: (a) Surface-Crack-in-Flexure, (b) Single-Edge-Notched-Beam, (c) Single-Edge-V-Notched-Beam, and (d) Single-Edge-Precracked-Beam (Moon, 2000, used with permission).

a V-notch machined into the sample using a razor blade. Using this technique, crack tip radii of > 15 μm for some material systems can be obtained.

The SEPB technique is similar to the SENB and SEVNB techniques and uses similar equations to calculate K_{Ic}. Unlike the relatively blunt notching of these tests, a sharp precrack is produced that is larger and more uniform than in the SCF technique. A starter crack is made using indentation or notching. The starter crack is then stably extended in a double-anvil loading fixture (see Fig. 7.19*d*). The SEPB technique employs a real crack with a sharpness similar to that of a "natural" flaw. Unfortunately, the precracks are typically long, < 400 μm. If the material shows an increased fracture resistance with crack length—i.e., *R*-curve behavior—then K_{Ic} values will be artificially high.

For the SENB, SEVNB, and SEPB specimens with $S_1/W = 8$, the geometric factor in Eq. 7.8 is

$$\beta = 1.93 - 2.47\left(\frac{c}{w}\right) + 12.97\left(\frac{c}{w}\right) + 14.53\left(\frac{c}{w}\right)^2 - 25.11\left(\frac{c}{w}\right)^3 + 58.8\left(\frac{c}{w}\right)^4 \quad (7.18)$$

for three-point bending and

$$\beta = 1.99 - 2.47\left(\frac{c}{w}\right) + 12.97\left(\frac{c}{w}\right)^2 - 23.17\left(\frac{c}{w}\right)^3 + 24.8\left(\frac{c}{w}\right)^4 \quad (7.19)$$

for four-point bending, where c is used as the crack or notch length.

Another popular fracture testing geometry is the double-cantilever beam (DCB), as shown in Fig. 7.20*a*. Because the elastic solution is fairly simple and a small specimen is possible, the DCB geometry is often applied to brittle materials. The fracture toughness from this test is given by

$$K_{\mathrm{Ic}} \cong \frac{\sqrt{12}Pc}{Bh^{3/2}} \quad (7.20)$$

Indentation is another technique used to approximate the K_{Ic} of brittle ceramics and ceramic glasses (see Lawn, 1993). Whenever a hardness indenter is used to make an impression in a brittle material, the deformation can be accompanied by cracking both during load application and after load removal. The cracks from Vickers indentation include *radial* cracks visible at the surface along the indenter diagonal (see Fig. 7.20*b*) and into the material and *lateral* cracks that form a bowl shape below the indentation. The radial cracks form during loading and continue to extend with unloading. Lateral cracks form on unloading. The radial crack extension is related to the magnitude of residual stress induced by the localized permanent deformation of the material in the indented region. Fig. 7.21*a* shows the radial and lateral cracks as viewed from the indentation direction, and Fig. 7.21*b* shows a side view of the radial crack formed in a glass. The radial cracks make the visible cross, and the shadow of interference shows the lateral cracks that lie nearly parallel to the surface.

By measuring the length of the surface crack in Fig. 7.21*b*, the fracture toughness can be estimated as

$$K_{\mathrm{Ic}} = 0.016\left(\frac{E}{H}\right)^{1/2}\left(\frac{P}{c^{3/2}}\right) \quad (7.21)$$

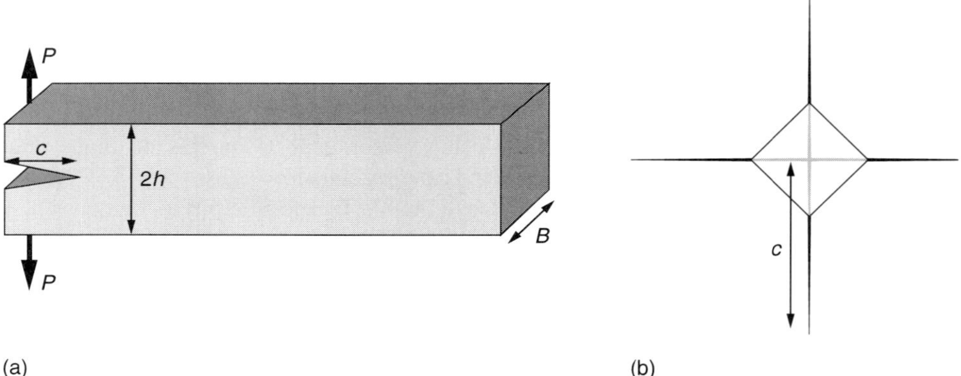

FIG. 7.20 (a) Double-cantilever beam fracture toughness geometry and (b) Vickers indentation toughness geometry, showing cracks measured to estimate fracture toughness.

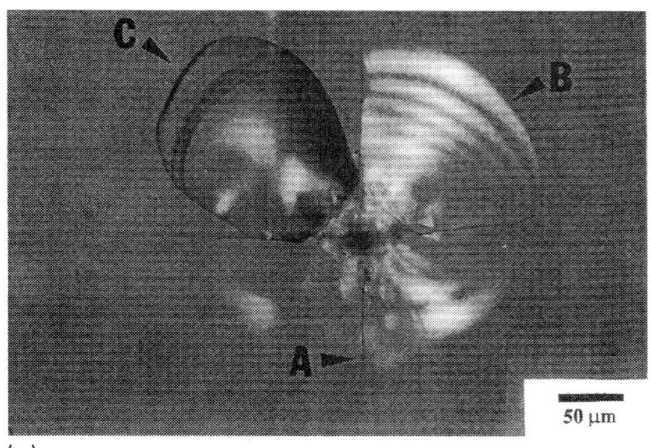

(a)

(b)

FIG. 7.21 (a) Cracking at Vickers indentation in glass. A is the radial crack surface marking, B shows the interference pattern from the lateral crack, and C shows the spalling of a lateral crack intersecting the surface. (b) View of radial crack formed from indentation of soda-lime glass. (Green, 1998, reprinted with permission of Cambridge University Press).

where E is Young's modulus, H is the hardness in units of stress, P is the force, and c is the crack length. If the material shows rising R-curve behavior, the K_{Ic} values measured in this technique can be artificially high.

7.2.2 Fracture Testing in Ductile Materials

For many ductile materials, the two greatest challenges in fracture testing are producing plane strain conditions and forming a sharp notch. The large thickness required for ductile materials to be tested in plane strain is often accommodated by the plastic fracture criterion called the J-integral or just testing the material in plane stress for the thickness of the structural component. This is particularly necessary for evaluation of sheet steel or aluminum products wherein the unique microstructural attributes achieved during rolling could not be readily reproduced in thicker materials. This can make comparisons of materials from different suppliers very difficult.

The formation of sharp notches in ductile materials is often a problem. Machining processes are usually inadequate for producing sufficiently sharp notches. Electrical discharge machining (also known as "spark" machining) is often effective in producing sharper notches than those produced by mechanical processes, but the most effective technique for producing flaws similar to natural ones is cyclic loading or fatigue of a machined notch. In any of these techniques, local damage of the material by heating or cold work should be considered in subsequent fracture analysis.

Fracture in ductile materials often includes failure by shear. For this reason, the fracture surface often shows a combination of features. The classic tensile failure mode of a ductile metal is necking. Soon after the formation of a notch, ductile voids can nucleate and grow in the center of the tensile specimen, producing a cup-and-cone fracture, as shown in Fig. 7.22. Once the intact cross section is reduced by the joining together (coalescence) of these voids, the final rapid failure occurs by the formation of ridges on the outside portion of the tensile fracture. This leads to final failure. The necking shown in Fig. 3.22 demonstrates the same flat fracture initiation in the center of the rectangular specimen's width with shearing at the edges. Figure 7.23 shows an example of the cavities that open in tensile and shear fracture. In notched fracture specimens, shear ridges or "lips" at the plane-stress-dominated regions form near the surface, as shown in Fig. 7.24. This leads to the qualitative assessment that the regions of more brittle fracture fail in a Mode I geometry and the regions of more

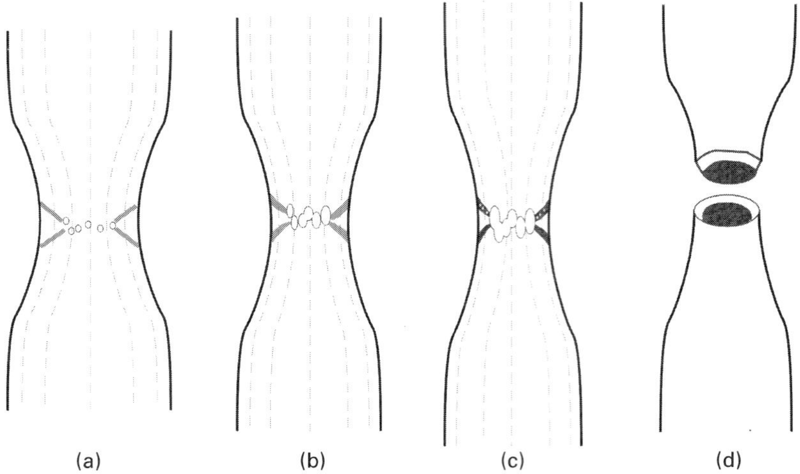

FIG. 7.22 Process of cup-and-cone fracture at a neck in a ductile material. (a) As the material necks down, a significant tensile triaxial stress state develops as a result of the outward radial components of stress. Voids nucleate at the center of the neck and shear bands form along the outer edge of the circumference. (b) The shear bands intensify as the voids grow, contact one another, and join together. (c) The voids grow into a single cavity in the center of the neck. At this point, voids begin to nucleate in the shear bands. (d) Final failure occurs when the material can no longer support the applied force and fracture results in the formation of a cup and cone on the respective ends of the failed tensile specimen. Large cavities are visible in the center of the specimen, and smaller cavities are visible within the shear ridges that are often called *shear lips*.

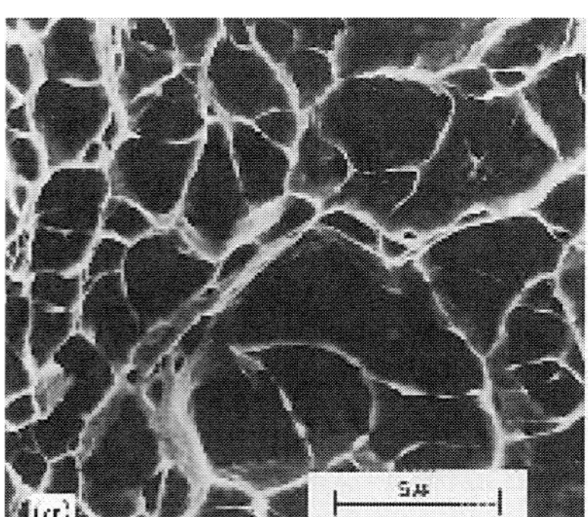

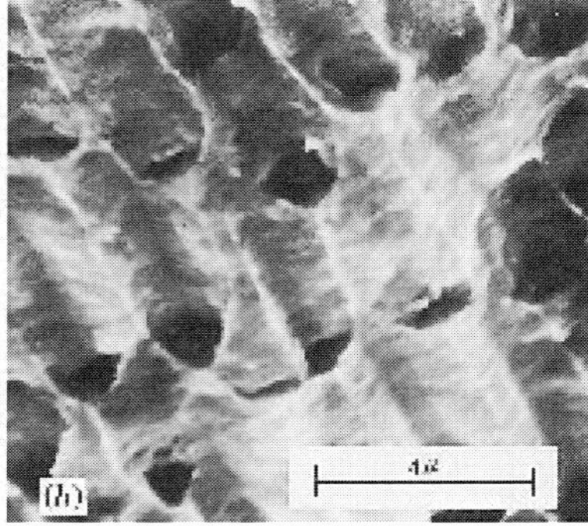

FIG. 7.23 Scanning electron micrographs showing the dimpled fracture surfaces resulting from (a) tensile loading of a ductile metal and (b) shear loading of a ductile metal. (Hertzberg, 1996, Wiley, reprinted with permission.)

ductile failure fracture by shearing with some apparent Mode II effects. This study of fracture surfaces to understand the mechanisms of failure is called fractography (for an excellent introduction to fractography, see Hull, 1999). The mechanisms of fracture in different material classes and their temperature dependence are discussed in Chapter 8.

7.2.3 Impact Fracture Tests: Charpy and Izod

Two types of dynamic impact tests are commonly used for screening ductile materials for their brittleness, the Izod and Charpy tests. Figure 7.25 shows a schematic of these gravity-driven testers, which swing a hammer into the notched materials. The loss in energy of the hammer from fracturing the specimen is the impact energy (also called "impact toughness"). These very rapid or dynamic tests are often used to compare materials for quality control or incremental process improvements. The uncontrolled strain rates make comparisons of different materials unwise.

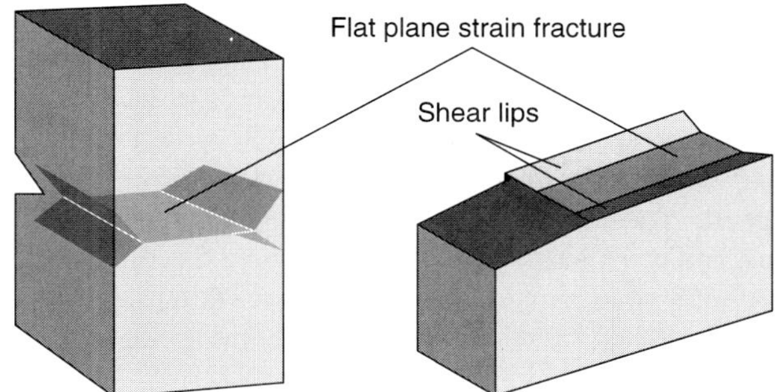

FIG. 7.24 Shear lip formation in notched ductile materials. The central region fails by essentially plane strain deformation, which appears more brittle. The central region also runs ahead of the shear lips that propagate through the deformation bands occurring by plane stress.

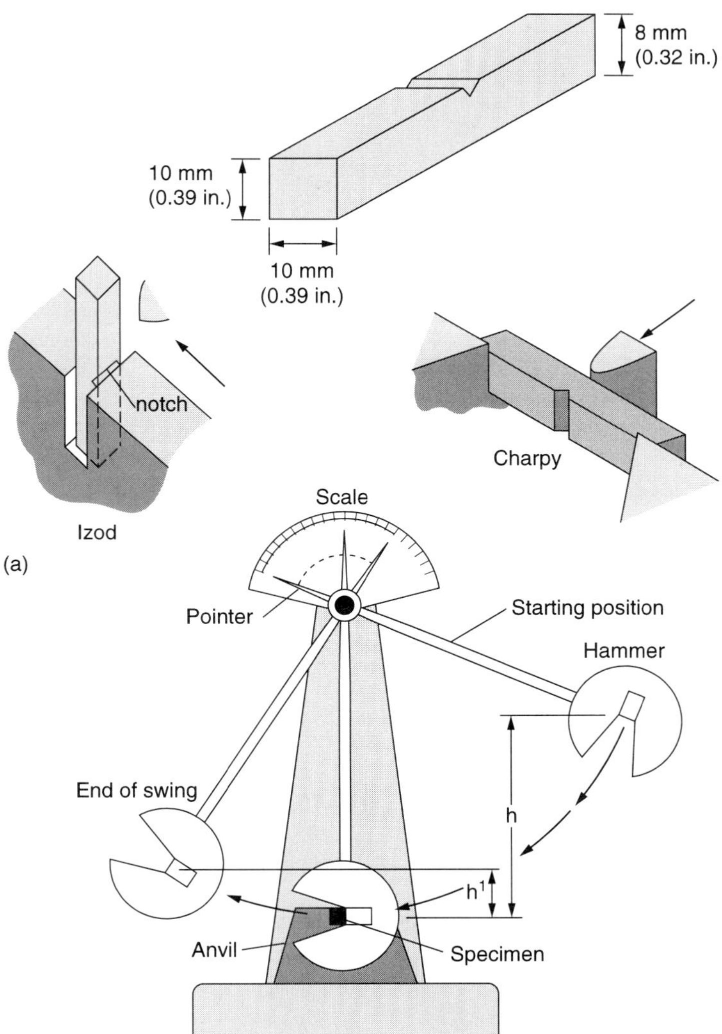

FIG. 7.25 (a) Charpy and Izod notched specimen and (b) the geometry for the Izod and Charpy tests wherein fracture is driven by the falling hammer (adapted from Hayden, Moffatt, and S. Wolff, 1965, Wiley, used with permission).

Figure 7.26*a* shows the temperature dependence of the Charpy energy for several materials. Only the steels, which have mostly BCC crystal structures, show a strong dependence of fracture energy and brittleness of the fracture on temperature. These dynamic tests are very commonly applied to steels wherein ferritic or BCC-structure steels can be screened for their brittleness by evaluating the effect of temperature on the energy expended to fracture identical specimens. The change in transition behavior can depend on heat treatment, impurities, inclusions, and mechanical processing history (e.g., cold work), but even the high-temperature (room temperature and above) upper level failure energies depend on yield strength, as shown in Fig. 7.26*b*. Lower transition temperatures are an indi-

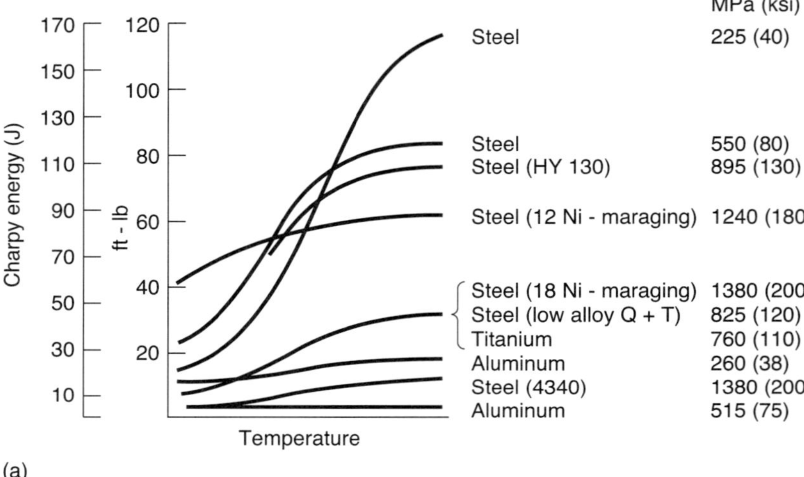

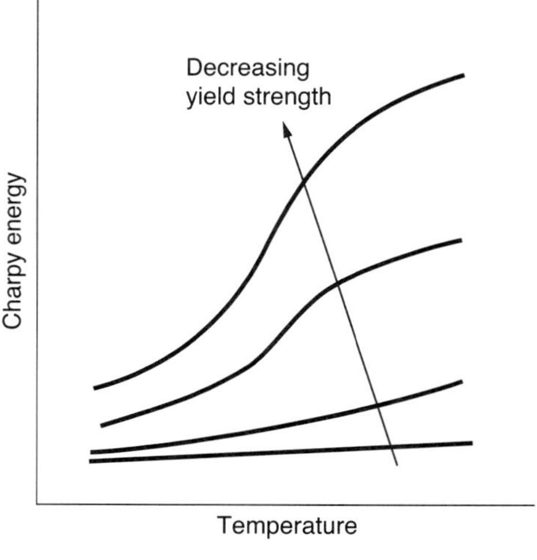

FIG. 7.26 (a) Temperature dependence of Charpy energy for various materials and related yield strengths (the range is typically from cryogenic to just above ambient temperatures) (adapted from Matthews, 1970, Copyright ASTM INTERNATIONAL, used with permission). (b) Effect of changes in yield strength from differences in grain size or alloying (for situations wherein yield strength is inversely proportional to ductility).

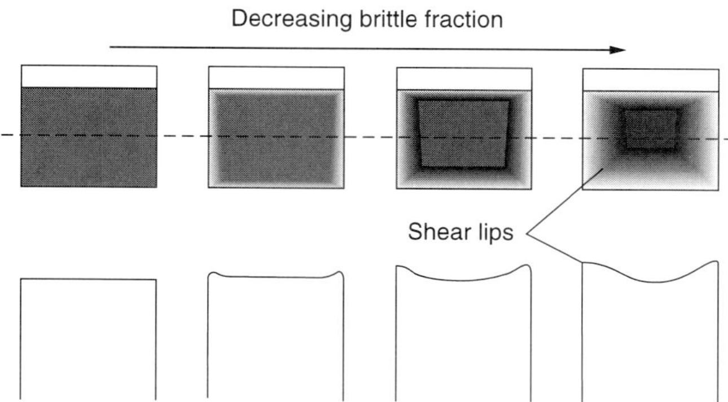

FIG. 7.27 Fracture surface appearance from Charpy tests. The light regions are the areas of shear failure, and the darker central regions are the fairly flat areas of brittle fracture.

cation of a reduced susceptibility to brittleness (or notch sensitivity) at high strain rates or low temperatures in fracture. In addition, the fracture surface can be evaluated for evidence of brittleness in the fracture, as shown schematically in Fig. 7.27. Completely brittle behavior leads to a nearly flat fracture surface with the crack propagating straight and parallel to the notch. As more ductility becomes evident, the shear lips on the edge of the specimen begin to grow and consist of a greater fraction of the fracture surface. For very tough specimens, deformation before fracture is extensive, resulting in substantial bending and a very jagged and distorted fracture surface.

7.3 FAILURE PROBABILITY AND WEIBULL STATISTICS

The analysis of fracture provided thus far in this chapter has been focused on the failure of a material containing a known crack or flaw. With recognition that the geometry of loading relative to the crack and the shape and location of a crack are important, it is not surprising that different brittle specimens of identical dimensions may demonstrate scatter in the stresses at which fracture takes place (see Creyke, Sainsbury, and Morrell, 1982, and Wachtman, 1996). Characteristics of the specific material, processing history of the material, and specimen preparation all contribute to flaw distribution. Consider the hypothetical specimen shown in Fig. 7.28*a*, with the largest flaws indicated. If the schematic two-dimensional specimen is loaded with a tensile stress vertically on the page, the flaw designated as flaw 1 is the flaw that is the most likely to initiate failure. If the specimen is loaded with a tensile stress horizontally across the page, the flaw that is most likely to initiate failure is the one labeled as flaw 2. Since flaw 1 is larger than flaw 2, it is also likely that the tensile stress for failure in the vertical direction is lower than that for failure in the horizontal direction. Not only is such a circumstance likely if any aspect of the processing was directional in nature, but also surface flaws introduced by machining damage can often cause such failures in brittle materials.

The dependence of brittle fracture on flaw size, and therefore on the critical flaw with an orientation favorable for crack propagation, means that there can also be a size dependence of the fracture process. This size dependence leads to effects of specimen size and stress distributions on the likelihood of fracture. If the piece of material shown in Fig. 7.28*a* is subdivided as shown in Fig. 7.28*b*, and all specimens are subjected to tensile loads along the vertical direction, the specimen that contains flaw 1 will fail at nearly the same stress as the larger specimen in Fig. 7.28*a*. On the other hand, each of the other specimens will sustain greater tensile loads because they have smaller flaws. If the average or mean

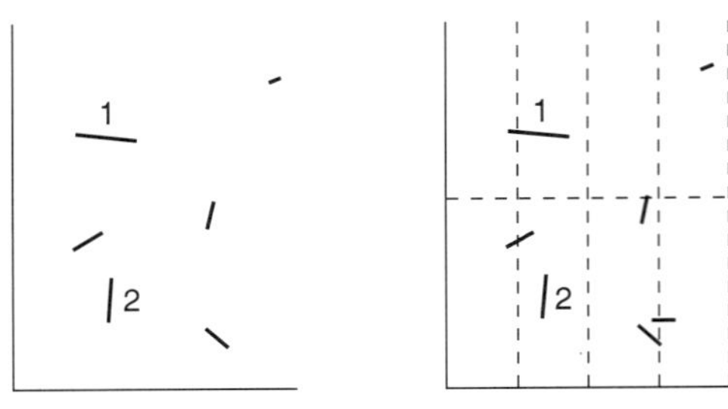

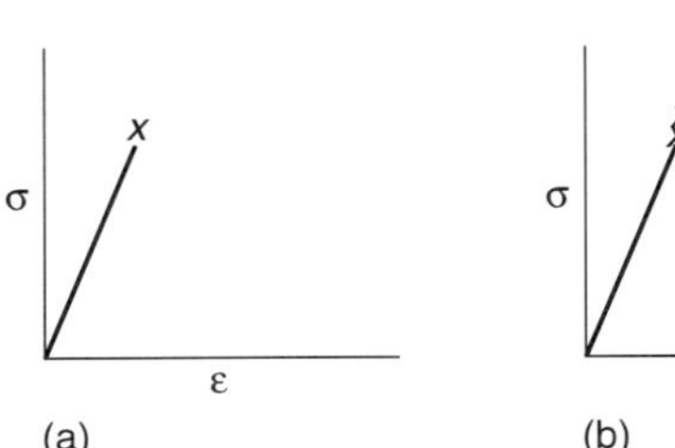

FIG. 7.28 (a) Schematic diagram of flaws in a large specimen and the brittle failure expected in a tensile test with stress applied vertically on the page and (b) the effect of dividing the larger specimen into smaller ones to yield a higher average failure stress.

failure stress is reported, then the average failure stress seems to increase as the size of the specimens is decreased.

The specimen size effect is also demonstrated when a specimen or component is subjected to a nonuniform stress distribution. If the highest tensile stress of a stress distribution does not coincide with the orientation or position of the largest flaw, then the material may survive larger stresses than might otherwise be anticipated.

Weibull is credited with development of a statistical function that describes fracture behavior in most materials relatively well. The Weibull function can be used to describe empirically the probability of failure with

$$P_f = 1 - \exp\left(-\left(\frac{\sigma_f - \sigma_{min}}{\sigma_o}\right)^m\right) \tag{7.22}$$

where σ_f is the variable fracture stress, σ_{min} is the minimum failure stress, σ_o is a constant with units of stress, and m is an exponent commonly called the Weibull modulus. A high value of m suggests a narrow distribution of failure stresses, and a low value suggests a broad distribution.

EXAMPLE 7.6 *Application of Weibull Statistics to Brittle Failure of a Ceramic*

Assume that we have conducted a large number of tensile tests on a brittle ceramic and the failure strengths of specimens that failed in the gage section (6.3 cm^3) of the specimen have been recorded and are given in Table 7.3.

If we plot these data as the Weibull parameter

$$\text{Weibull parameter} = \log\left(\log\left(\frac{1}{1 - P_f}\right)\right)$$

versus log (σ_f), we can get a straight line of slope m if the data fit the Weibull distribution well. To plot the Weibull parameter, we need to know P_f. If we have a sufficiently large number of specimens, we can rank-order the failure strengths as in Table 7.3. Then we can estimate that

TABLE 7.3 Failure Strengths of Brittle Specimens (MPa)

1.	187	11.	366	21.	422
2.	228	12.	371	22.	435
3.	243	13.	373	23.	442
4.	261	14.	381	24.	449
5.	289	15.	390	25.	458
6.	298	16.	395	26.	466
7.	311	17.	404	27.	482
8.	321	18.	406	28.	500
9.	343	19.	410	29.	508
10.	359	20.	419	30.	529

$$P_{\mathrm{f}} = \frac{n}{N+1}$$

if we have N test values and n represents the nth sample. A Weibull plot of the data in Table 7.3 is shown in Fig. 7.29. From this plot, the probability of failure at any stress can be predicted. For example, at a stress of 250 the logarithm to base 10 is 2.4. If we find this point in Fig. 7.29, we expect a failure probability of 10 percent from the right-hand axis.

Weibull predictions of failure probability should be made only on large data sets, $N \geq 20$. The Weibull distribution can be estimated from a normal distribution as

$$m \cong \frac{1.2\langle\sigma\rangle}{s_{\mathrm{d}}}$$

where $\langle\sigma\rangle$ is the mean of the distribution and s_{d} is the standard deviation. It is often more useful to base calculations on the mean stress to avoid finding σ_{o} from plotting of the Weibull distribution. Although a little more complex, the failure probability becomes

$$P_{\mathrm{f}} = 1 - \exp\left\{-\left(\Gamma\left(\frac{1}{m}+1\right)\right)^{m}\left(\frac{\sigma}{\langle\sigma\rangle}\right)^{m}\right\} \tag{7.23}$$

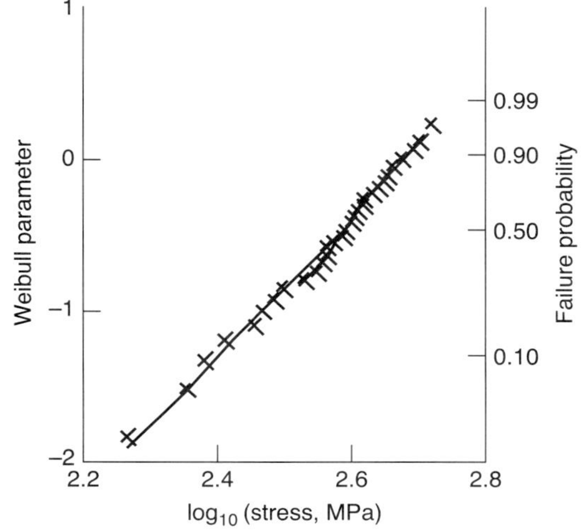

FIG. 7.29 Weibull plot of data in Table 7.3, showing that the slope gives the Weibull modulus. The right-hand axis gives the probability of failure as a function of the applied tensile stress level.

BIOGRAPHY

WALODDI WEIBULL (1887–1979)

Weibull's analysis was based on experimental data for performance of steels. Weibull was Research Director at the NKA Ball Bearing Company in Gothenburg, Sweden from 1916 to 1922 and was then a Professor of Mechanical Engineering at the Royal Institute of Technology, Stockholm. He was a member of four Royal Swedish Academies: Science, Engineering Science, Military Science, and Naval Science.

where Γ is the gamma function (which is found in tables on most advanced calculators and math software). The use of this expression to predict failure or safety in design is limited to (1) identical sizes and shapes, (2) identical loading conditions, and (3) identical processing history.

To adjust for different sizes and shapes or stress distributions, a parameter called the stress volume integral can be added to Eq. 7.23. The stress volume integral (SVI) and its application are discussed in Appendix B. One reason the size and stress distribution are important is that the Weibull approach relies on an assumption of a flaw distribution. Smaller specimens loaded to the same stress have a smaller probability of failure because they are less likely to contain a critical flaw. Stress distributions are important because the maximum stress in a component may or may not coincide with the critical flaw. In bending, for example, the maximum stress is at the surface (see Chapter 1). Machining flaws on the tensile surface of a bend specimen may lead to an exaggerated likelihood of fracture. On the other hand, if the surface is well prepared, the lower average stresses relative to the maximum at the surface and the half of the specimen under compression can greatly reduce the probability of failure. It is for these reasons that design using Weibull statistics must be done with great care. ■

7.4 MECHANISMS FOR TOUGHNESS ENHANCEMENT OF BRITTLE MATERIALS

The *R*-curve for ideally brittle materials is given as a constant because the resistance to crack extension is an intrinsic material property independent of crack size. For most ceramic materials at less than 0.7 T_m, crack tip plasticity is not evident. A rise in *R*-curves of some ceramic materials results from other mechanisms producing resistance to crack extension. These toughening mechanisms can improve the Weibull modulus, making brittle materials more flaw-tolerant. Rising *R*-curve behavior in ceramics results from events occurring behind the crack tip, termed "wake effects."

Three mechanisms of increasing the resistance to crack extension in ceramic materials are crack deflection, crack bridging, and phase transformations. Toughening from crack deflection results from changes in mode from Mode I to contributions from Modes II and III (Fig. 7.30*a*). For crack bridging, toughening occurs by grains or "ligaments" that bridge the small crack opening (Fig. 7.30*b*). The resistance increases as the crack extends and accumulates more bridging points. The grain "pull-out friction" keeps the bridged grains in place until the grains either pull out or fracture. As we will see in Example 7.7, this increase results in a rising *R*-curve.

Toughening from phase transformations occurring by the high stresses in front of the crack tip induces the material to undergo a stress-initiated phase transformation. The nonlinear deformation occurring via phase transformation within the process zone shields the crack tip from the magnitude of stress it would undergo in a purely linear elastic response. Substantial toughening also results from the superimposed compressive force, or "wake

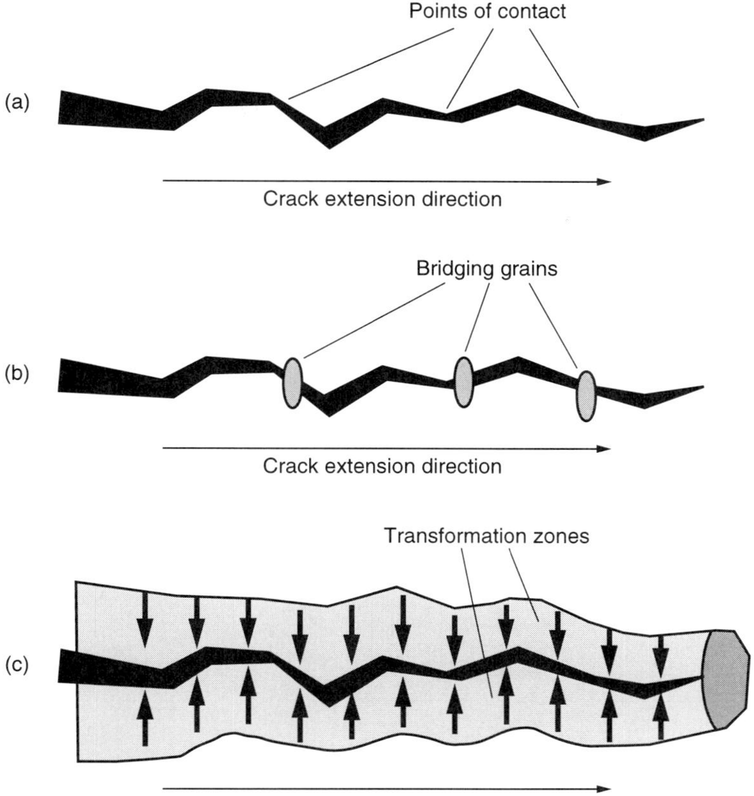

FIG. 7.30 Schematic crack configurations for three toughening mechanisms in ceramics: (a) crack deflection, (b) crack bridging, and (c) phase transformation (Moon, 2000, used with permission).

effect," caused by the transformed material acting along the entire crack length (Fig. 7.30*c*). A similar wake effect can occur from distributed microcracking around the crack tip. Each of these deformation processes results in a process zone. Increasing the size of the process zone results in an increasing wake effect and fracture toughness. As with plasticity, the lower the critical stress (for plasticity, the yield stress) for nonlinear deformation, the larger the process zone and the greater the resulting toughness.

For monolithic ceramics with rising *R*-curves, the maximum (plateau) fracture toughness can occur at crack lengths many times those of natural flaws. When a constant load is maintained during crack growth, the slope of the *R*-curve is often insufficient for stable crack growth.

EXAMPLE 7.7 *R-Curves in Oxide Ceramics*

The toughening of ceramic materials is dependent on the microstructure of the ceramic or composite. The effectiveness of a particular microstructure in creating an increasing resistance to crack propagation, which is called a rising *R*-curve, depends on the operation of the toughening mechanisms behind the crack. Moon and coworkers (2001, 2002) have observed the propagation of SEVNB cracks in a microscope and documented the effects of changes in microstructure on the resistance to further crack propagation. Fig. 7.31*a* shows an *R*-curve for a material consisting of 80 volume percent aluminum oxide and 20 volume percent of a zirconia alloy. For this material, no evidence of transformation toughening was observed, and the R-curve behavior shown occurs most likely from the development of some bridging in the material. The modest increase in the crack growth resistance is typical of this material. An increased grain size or other microstructural changes can increase the toughening and the steepness of the R-curve. An example of this is shown in Fig. 7.31*b*, where a gradient layer consisting a larger grain size that begins abruptly at about 210 μm causes a rapid increase in the R-curve. A combination of residual stress and microstructure is believed to cause this large rise in the resistance to crack growth. ■

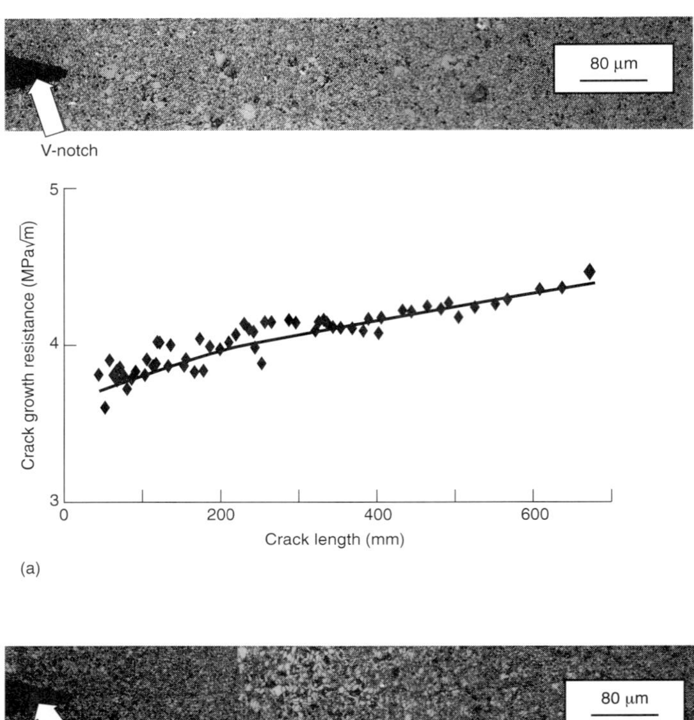

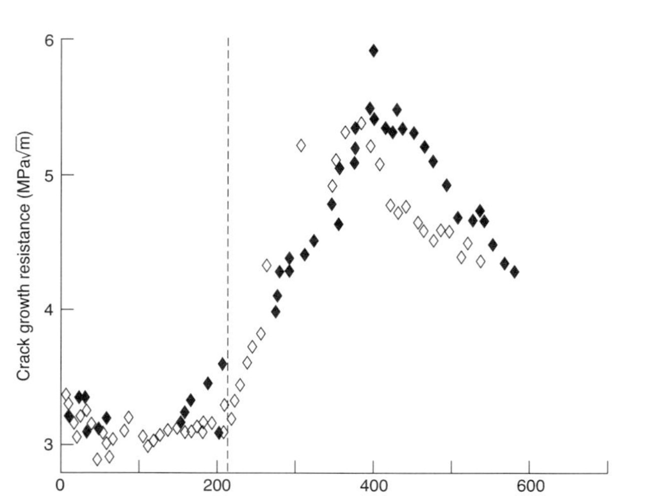

FIG. 7.31 (a) R-curve for 80 Al_2O_3–20 ZrO_2 with the crack growth resistance, R, plotted versus crack extension. (b) R-curve for material with nominally the same composition but manufactured with an abrupt increase in grain size at a crack length of about 210 μm. The effect of the larger grain size is to promote rising R-curve behavior. The grain size in this material is graded and gradually decreases over the remainder of the measurement. (Moon, 2000, used with permission; see also Moon, 2001, 2002)

7.5 APPENDIX A: DERIVATION OF THE STRESS CONCENTRATION AT A THROUGH-HOLE

The derivation of the stress concentration in a thick sheet of material containing a cylindrical hole (Fig. 7.3) under biaxial tension can be accomplished using the following assumptions.

First, we can be certain that the number of variables matches the number of available equations by

Variables	***Equations***
σ_r, σ_0, σ_z	3 equations - Hooke's law
ε_r, ε_0, ε_z = const.	1 - force equilibrium
	1 - strain compatibility
5 variables	5

We can apply Hooke's law in the form

$$E\varepsilon_r = \sigma_r - \nu(\sigma_\theta + \sigma_z) \tag{7.A.1}$$

$$E\varepsilon_\theta = \sigma_\theta - \nu(\sigma_r + \sigma_z) \tag{7.A.2}$$

$$E\varepsilon_z = \sigma_z - \nu(\sigma_r + \sigma_\theta) \tag{7.A.3}$$

Because we know that $\varepsilon_z \approx 0$ in thick materials, we only need to consider the radial and tangential stresses:

$$\frac{\sigma_\theta}{E} = \frac{\nu\varepsilon_r + (1-\nu)\varepsilon_\theta}{(1+\nu)(1-2\nu)} \tag{7.A.4}$$

$$\frac{\sigma_r}{E} = \frac{\nu\varepsilon_\theta + (1-\nu)\varepsilon_r}{(1+\nu)(1-2\nu)} \tag{7.A.5}$$

and the corresponding strains:

$$\varepsilon_r = \frac{\partial u}{\partial r}, \qquad \varepsilon_\theta = \frac{u}{r} \tag{7.A.6}$$

Then we can substitute Eq. 7.A.6 into Eq. 7.A.4 and 7.A.5 to yield:

$$\frac{\sigma_\theta}{E} = \frac{\nu\left(\frac{\partial u}{\partial r}\right) + (1-\nu)\frac{u}{r}}{(1+\nu)(1-2\nu)} \tag{7.A.7}$$

and

$$\frac{\sigma_r}{E} = \frac{\nu\left(\frac{u}{r}\right) + (1-\nu)\left(\frac{\partial u}{\partial r}\right)}{(1+\nu)(1-2\nu)} \tag{7.A.8}$$

We can then use these stresses as a summation of forces in the radial direction. We can list each of these forces as

Inside force:	$(\sigma_r)\ (r)\ (\partial\theta)$ per unit thickness
Outside force:	$(\sigma_r + \partial\sigma_r)\ (r + \partial_r)\ (\partial\theta)$
Side faces:	$\sigma\theta\ (\partial r)$ (but not in r direction)
Radial component on side faces:	$\sigma\theta\ (\partial r) \sin\frac{\partial\theta}{2}$

Using these forces, we can write the sum as

$$\Sigma F_R = (\sigma_r + \partial\sigma_r)(r + \partial r)(\partial\theta) - (\sigma_r)(r)(\partial\theta) - 2(\sigma_\theta)(\partial r)\left(\sin\frac{\partial\theta}{2}\right) = 0$$

which can be reduced to

$$\Sigma F_R = (\sigma_r)(\partial r) + (r)(\partial\sigma_r) - \sigma_\theta(\partial r) = 0 \tag{7.A.9}$$

We can next consider the compatibility of strain that is necessary for describing a solid finite object subjected to distortion:

$$\frac{\partial^2 u}{\partial r^2} + \frac{1}{r}\frac{\partial u}{\partial r} - \frac{u}{r^2} = 0 \tag{7.A.10}$$

If we integrate twice, we can write a function of the form

$$u = Ar + \frac{B}{r} \tag{7.A.11}$$

where A and B are constants of integration. Using this result, Eq. 7.A.7 and 7.A.8, and appropriate boundary conditions, we have

$$\sigma_{rr} = \sigma_{\text{applied}}\,(1 - a^2/r^2) \tag{7.A.12}$$

$$\sigma_{\theta\theta} = \sigma_{\text{applied}}\,(1 + a^2/r^2) \tag{7.A.13}$$

$$\sigma_{zz} = 2\nu\sigma_{\text{applied}} \tag{7.A.14}$$

Thus, for $a = r$, $\sigma_{\theta\theta}$ is twice the applied stress and $\sigma_{rr} = 0$.

7.6 APPENDIX B: STRESS-VOLUME INTEGRAL APPROACH FOR WEIBULL STATISTICS

The stress-volume integral is an approach that enables adjustment of the probability of failure for circumstances in which there is a distribution or gradient of stresses in a test specimen or in a structure (see Creyke, Sainsbury, and Morrell, 1982). It also enables conversion of data from test specimens to structures of different sizes. The expression

$$P_f = 1 - \exp\left\{-\left(\Gamma\left(\frac{1}{m}+1\right)\right)^m \left(\frac{\sigma_{\text{nom}}}{\langle\sigma\rangle_{\text{ref}}}\right)^m \frac{V}{V_{\text{ref}}}\, SVI(V)\right\} \tag{7.B.1}$$

enables compensation for volume and stress distribution. The term V is the volume of the part under consideration, and V_{ref} is the volume of the test specimens that gave the average failure stresses. The nominal stress, σ_{nom}, is typically the maximum stress defined in the specimen, and the average reference stress, $\langle\sigma\rangle_{\text{ref}}$, is the mean of the reference volume in uniform tension. The term $SVI(V)$ is called the stress-volume integral, which is a dimensionless quantity. The common definition of the stress-volume integral is

$$SVI(V) = \int_v \left(\frac{\sigma(V)}{\sigma_{\text{nom}}}\right)^m \frac{\partial V}{V} \tag{7.B.2}$$

wherein the stress distribution is given as the function $\sigma(V)$, which is then normalized using σ_{nom} and V.

Consider the difference in stress distribution for a three-point bend specimen versus a tensile specimen. We can use the bending equations from Chapter 1 to derive

$$\sigma_{\text{nom}} = \frac{3Pl}{2wh^2}$$

and the stresses vary through the volume of each symmetric half of the bar as

$$\sigma(x, y) = \frac{4xy}{lh}\sigma_{\text{nom}} \tag{7.B.3}$$

where x is the distance along the bar from the outer loading points and y is the distance from the neutral axis toward the tensile surface. We will use a range of x from 0 to l/2 and a range of y from 0 to h/2, because we can ignore the compressively loaded portions of the specimen. The stress-volume integral is then

$$\begin{aligned} SVI(V_{\text{bend}}) &= \int_v \left(\frac{\sigma(V)}{\sigma_{\text{nom}}}\right)^m \frac{\partial V}{V} \\ &= \int_0^{l/2}\int_0^{h/2} \left(\frac{4xy}{lh}\right)^m \cdot 2\frac{\partial x \partial y \cdot b}{lhb} = \frac{1}{2(m+1)^2} \end{aligned} \tag{7.B.4}$$

If we rewrite Eq. 7.B.1 for this case of bending, we get

$$P_f = 1 - \exp\left\{-\left(\Gamma\left(\frac{1}{m}+1\right)\right)^m \left(\frac{\sigma_{\text{bend}}}{\langle\sigma\rangle_{\text{ref}}}\right)^m \frac{V_{\text{bend}}}{V_{\text{ref}}} SVI(V_{\text{bend}})\right\}$$

from which we can write

$$\frac{\langle\sigma\rangle_{\text{ref}}}{\langle\sigma\rangle_{\text{bend}}} = \left\{\frac{V_{\text{bend}} SVI(V_{\text{bend}})}{V_{\text{ref}} SVI(V_{\text{ref}})}\right\}^{1/m} \tag{7.B.5}$$

We can use the result from Eq. 7.B.4 with Eq. 7.B.5 for the case when $V_{\text{ref}} = V_{\text{bend}}$ to derive

$$\frac{\langle\sigma\rangle_{\text{ref}}}{\langle\sigma\rangle_{\text{bend}}} = \left\{\frac{1}{2(m+1)^2}\right\}^{1/m} \tag{7.B.6}$$

which equals 0.794 for $m = 10$. This result shows that the average stress for fracture in tension is lower than that for bending. The reason for this is that the volume under the higher levels of stress is smaller in the three-point bend specimen than in a tensile specimen (see Fig. 1.3.3).

We can also use Eq. 7.B.5 for equivalent stress geometries and different volumes to demonstrate that the average stress for failure increases as volume decreases such that

$$\frac{\langle\sigma\rangle_1}{\langle\sigma\rangle_2} = \left(\frac{V_2}{V_1}\right)^{1/m} \tag{7.B.7}$$

For $m = 10$ and $V_2 = 10V_1$, the stress ratio is 1.26, showing that the smaller samples are apparently 26% stronger than the larger specimens (assuming that loading rate and surface finish are comparable).

7.7 REFERENCES

T. L. ANDERSON, *Fracture Mechanics: Fundamentals and Applications*, CRC Press, Boca Raton, 1995.

D. BROEK, *Elementary Engineering Fracture Mechanics*, 3rd Ed., Martinus Nijhoff, Boston, 1982.

W. E. CREYKE, I. E. J. SAINSBURY, AND R. MORRELL, *Design with Non-ductile Materials*, Applied Science Publishers, London, 1982.

DAVID GREEN, *An Introduction to the Mechanical Properties of Ceramics*, Cambridge University Press, 1998.

A. A. GRIFFITH, *Philos. Trans. R. Soc.* London, **Ser A 221**, 163, 1920 (reprinted with commentary in Trans. ASM, 61, 871, 1968).

H.W. HAYDEN, W.G. MOFFATT AND J. WULFF, *The Structure and Properties of Materials, Vol. III, Mechanical Behavior*, John Wiley & Sons, New York (1965).

R. HERTZBERG, *Deformation and Fracture Mechanics of Materials*, 4th Ed., Wiley, 1996.

D. HULL, *Fractography*, Cambridge University Press, 1999.

B. LAWN, *Fracture of Brittle Solids*, Cambridge University Press, 1993.

D. C. LUPASCU, S. L. DOS SANTOS E LUCATO, J. RÖDEL, M. KREUZER, AND C. S. LYNCH, *App. Phys. Letters*, **78**, 2554–2556, 2001.

W. MATTHEWS, *Impact Testing of Metals*, ASTM STP 466, 1970.

R. J. MOON, Ph. D. Thesis, Purdue University, 2000.

R. J. MOON, K. J. BOWMAN, K. P. TRUMBLE, AND J. RÖDEL, *Acta Mater.*, **49**, 995–1003, 2001.

R. J. MOON, M. HOFFMAN, J. HILDEN, K. J. BOWMAN, K. P. TRUMBLE, AND J. RÖDEL, *Engineering Fracture Mechanics*, **69**, 1647–1665, 2002.

P. C. PARIS AND G. C. SIH, ASTM STP 381, 1965.

R. E. PETERSON, *Stress Concentration Design Factors*, Wiley, NY, 1953.

H. TADA, P. C. PARIS, AND G. R. IRWIN, *Stress Analysis of Cracks Handbook*, ASM, 2000.

J. WACHTMAN, *Mechanical Properties of Ceramics*, Wiley, 1996.

7.8 PROBLEMS

A.7.1 What are possible units for fracture toughness?

(a) $N \cdot m^{-3/2}$

(b) $J \cdot m^{-1/2}$

(c) $MPa \sqrt{m}$

(d) ksi $\sqrt{in.}$

(e) All of the above

A.7.2 Give the mode(s) of fracture expected for each described geometry given below. Consider both *initiation* and *propagation*. Describe the stress distribution near the crack tip in each case. Sketches may help.

(a) Brutus the torque-wrench-less technician has again broken another bolt by twisting off the head. (b) Mischievous Madge has popped your birthday balloon with her steak knife. (c) California has finally slipped into the ocean. All locations west of the San Andreas Fault have sunk 100 feet.

A.7.3 Which of the stress intensity expressions in Fig. 7.16 could be applied as an approximation for a screw that has broken by being overtightened?

A.7.4 Explain the difference in shear dilatant fracture that occurs across a fracture surface in a ductile material.

B.7.1 After increasing the yield strength of an alloy by strain hardening, it was recognized that the fracture toughness K_{Ic} decreased. Why?

B.7.2 A new high-toughness glass ceramic has a fracture toughness of 6.5 MPa $\sqrt{m}$. For a double-cantilever beam specimen, estimate a_c assuming that there is no R-curve behavior. Assume that the starting flaw size c = 4 mm, thickness B = 20 mm, and h = 20 mm.

B.7.3 One of the assumptions commonly made in Weibull analysis is that failures do not occur in compression. This leads to an assumption that the minimum failure stress used in Weibull analysis is zero. Thus, a designer might reason that the probability of failure in all compressive loads will be zero, even for a material with a low Weibull modulus. Is it always okay to assume zero probability of failure in compression using Eq. 7.22? Explain your answer.

B.7.4 Figure 7.6 shows the energy relationships for propagation of a crack under constant displacement conditions. Show with a sketch crack propagation under constant load conditions in this type of plot. Discuss the difference in energy of these two approaches (you can use geometry to solve this).

B.7.5 A 2-m-wide panel has a circular through-hole with a radius of 1 cm and an elliptical hole with radii of a = 1 cm and b = 0.5 cm. Assuming that they are far enough apart not to interact, which will produce the greater stress concentration?

B.7.6 A 10-mm-thick panel has a crack of length $2a$ = 100 μm oriented normal to the loading direction. Calculate the plastic zone size predicted by Eq. 7.15 and determine if the criterion given in Eq. 7.16 is fulfilled for each of the following conditions:

(a) σ = 10 MPa, Y = 100 MPa

(b) σ = 25 MPa, Y = 50 MPa

(c) σ = 50 MPa, Y = 250 MPa

(d) σ = 1 MPa, Y = 20 MPa

B.7.7 Take a half-sheet of paper and fold it on its diagonal. Use a pair of scissors to cut a notch perpendicular to the fold. Open the fold, and you have an angled notch that should result in a mixed-mode geometry. Grasp the ends of the paper and pull it in tension until it fails. What combination of modes did the initial notch represent? By what mode(s) did the final fracture propagate? Explain this result.

C.7.1 Write the Mode I stress distributions (Table 7.2 or above) in Cartesian coordinates.

C.7.2 Write the Mode II stress distributions in Table 7.2. in Cartesian coordinates.

C.7.3 Write the Mode III stress distributions in Table 7.2 in Cartesian coordinates.

C.7.4 Write the Mode I displacement distributions in Table 7.2 in Cartesian coordinates.

C.7.5 Write the Mode II displacement distributions in Table 7.2 in Cartesian coordinates.

C.7.6 Write the Mode III displacement distributions in Table 7.2 in Cartesian coordinates.

C.7.7 If the formation of a process zone is based only on dilation (volume increase), what would you expect to happen to the process zone size in a thick specimen (how would it compare with Fig. 7.12)?

C.7.8 Determine the relative shapes of the plane stress contours for the plastic zone using the Tresca criterion. This will require use of a spreadsheet, a program, or math software.

C.7.9 A volcano has erupted on your small coastal island and you know it is urgent to move all of your belongings. The capacity of the steel bridge that connects you to the coast is about that of your truck even before you load it. Unfortunately, you have noticed a crack freshly initiated by an earthquake associated with the volcano. The crack is in a primary load-bearing span of the bridge. It looks bad. Helicopters and boats are not an option. What could you do to reduce the risk of the bridge collapsing when you drive across it? You have very little time, but a good set of tools (and safety glasses).

C.7.10 Toughening of brittle materials has been accomplished by transformation toughening. The toughening is accomplished through the effects of the transformation strains—the differences in unit cell dimensions before and after transformation. Discuss the effectiveness of the following transformation strains in producing toughening in both plane stress and plane strain conditions. (Also see Chapter 3.)

(a) $$\varepsilon_T = \begin{vmatrix} 0.04 & 0 & 0 \\ 0 & -0.04 & 0 \\ 0 & 0 & 0 \end{vmatrix}$$

(b) $$\varepsilon_T = \begin{vmatrix} -0.04 & 0 & 0 \\ 0 & -0.04 & 0 \\ 0 & 0 & 0.01 \end{vmatrix}$$

(c) $$\varepsilon_T = \begin{vmatrix} 0.01 & 0.04 & 0 \\ 0.04 & 0.01 & 0 \\ 0 & 0 & 0.02 \end{vmatrix}$$

C.7.11 Find the stress contour for 101 MPa for plane strain conditions using the data in Example 7.2 and the values of f given in Table 7.2. Compare your result with that given in Example 7.2.

C.7.12 Use the von Mises criterion to determine the value of r at which the elastic stress solution will reach a yield stress of 2 MPa at $\theta = 0$ for the data in Example 7.2. Give solutions for both plane stress and plane strain conditions.

C.7.13 The fracture surface of a failed bolt shows that a small half-penny crack has formed at the bottom of one of the threads. First make a sketch of the fracture surface, and then use superposition to calculate the failure load given the following data:

The thread: $D = 5$ mm, $d = 4$ mm, $K_{Ic} = 18$ MPa $\sqrt{\text{m}}$

The half penny: $a = 0.08$ mm

C.7.14 Pete the glass blower spits on a small notch he has made in glass to make a cleaner break before applying a bending load. What is so special about Pete's saliva that makes the glass fracture easier? (It works!)

C.7.15 For an infinite thin plate with a center crack with a length of $2a = 8$ cm and a stress of 10 MPa: (a) Calculate the principal stresses at

$(r,\theta) = 1$ μm, 30°
100 μ, 30°
1 μm, 45°
100 μm, 45°
1 μm, 60°
100 μm, 60°
1 μm, 90°
100 μm, 90°

(b) Which points will exceed the Tresca yield criterion assuming that $Y = 200$ MPa?

(c) Show the "orientations" of the principal stresses.

CHAPTER 8

MAPPING STRATEGIES FOR UNDERSTANDING MECHANICAL PROPERTIES

STRATEGIES FOR MAPPING properties or performance have become widely applied to engineering of materials. These maps range from maps that identify safe and unsafe conditions to those that identify expected lifetimes across a complex set of variables. Three types of maps relevant to mechanical behavior are briefly described here, beginning with deformation mechanism maps, followed by fracture mechanism maps, and then mechanical design maps, which assemble elements based on multiple properties. We have already seen some examples of mapping in earlier chapters, with yield criteria being the most prominent example. The development of databases for enabling material selection based on multiple properties is a recent development, and hopefully their accessibility and use will broaden with time. For yield criteria, plotting of the principal stresses as Cartesian coordinates provides a surface that is useful for both design to prevent failure and for design of a forming operation. The von Mises yield criterion is the foundation for the first example in this chapter.

8.1 DEFORMATION MECHANISM MAPS

Deformation mechanism maps (DMMs) have been established as a useful way to document deformation mechanisms by Professor Michael Ashby of Cambridge University and his coworkers (see Ashby, 1972, and Frost and Ashby, 1982). DMMs provide a concise framework for condensation of existing information on the deformation rates of a material as a function of microstructural parameters, stress, and temperature. Their primary utility is as an early step in materials selection and as a tool in the early stages of design. In addition to providing a good starting point for "back of the envelope" analysis, they also enable comparisons of different materials.

Figure 8.1*a* shows the shear deformation mechanisms that are relevant to most crystalline materials with their positions given relative to effective shear stress τ_{eff} (see Eq. 3.20), normalized by the shear modulus, μ versus the homologous temperature. Each region signifies the dominance of a specific deformation mechanism for that combination of stress and temperature for an assumed constant microstructure (e.g., constant grain size d) using data derived from databases of experimental information. Because deformation mechanisms are assumed to coexist, the dominant mechanism is the one making the largest contribution to the total plastic strain rate. The boundaries between the deformation mechanisms shown in Fig. 8.1*a* represent the positions at which the plastic strain rates from two mechanisms are equal. At each point on the map, the effective shear strain rate (given by Eq. 3.22) of the specimen is the sum of the strain rate components from each of the deformation mechanisms.

For the examples shown in this text, the equations used are based on those introduced for rate-dependent deformation in Chapter 6 expressed in terms of the effective shear strain rate. For some of the maps, a deformation mechanism called the ideal shear strength is shown. This horizontal line occurs near the top of the DMMs and represents the theoretical strength of a material in the absence of defects. To find the value of the total effective shear strain rate, $\dot{\gamma}_{Total}$, for any combination of material, grain size, temperature, and stress level, we sum the shear strain rate contributions from each of the expected deformation mechanisms described in Chapter 6, e.g., $\dot{\gamma}_{PLC}$, $\dot{\gamma}_{NH}$, etc., with the recognition that for some values of shear stress and temperature the contribution of a given mechanism may be insignificant. Once the values have been determined, then contours of their values can be placed on the DMM, as shown in Fig. 8.1*b*.

The following important considerations should be kept in mind.

1. Any mappings of data are only as good as the data and the equations used to construct them. Only order-of-magnitude analysis is appropriate without turning to more detailed investigations on behavior of the specific material in question.
2. The maps represent a steady-state condition of the material in terms of microstructure and phase composition. Any changes that could be expected, particularly at high

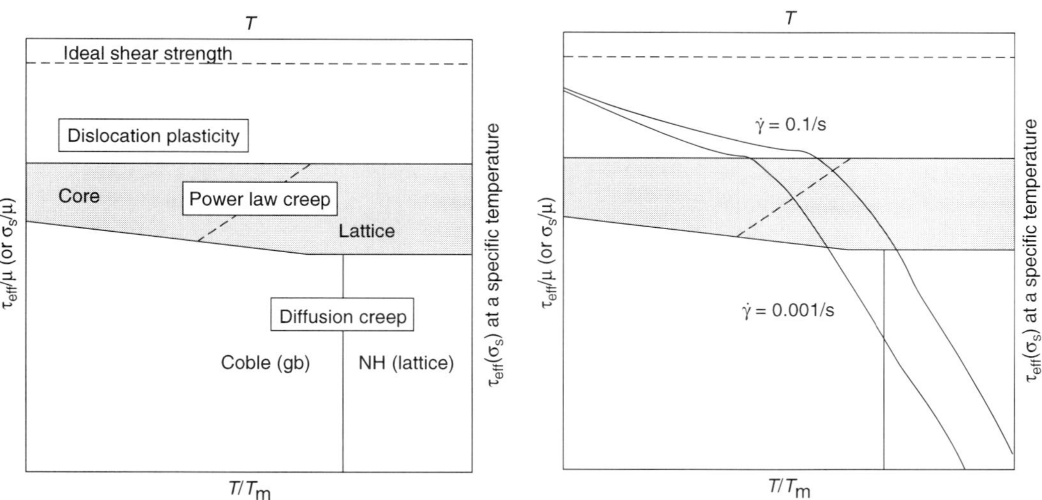

FIG. 8.1 (a) Sketch of a deformation mechanism map (DMM) with normalized effective shear stress and temperature. The individual regions shown represent plasticity based on dislocation motion above the solid horizontal line and the ideal or theoretical shear strength above the dashed line. Power law creep is shown with separate regions wherein dislocation core diffusion and lattice diffusion are dominant and two diffusion creep regions, one in which grain-boundary diffusion is dominant via Coble creep (gb) and one in which lattice diffusion (lattice) is dominant via Nabarro–Herring creep. (b) Two contours of strain rate over a complete range of temperatures and stresses for a hypothetical crystalline material. To use a map in a design problem, the relevant ranges of stresses and temperatures must be selected, and then the plastic strain rate can be estimated.

temperatures, will result in behaviors that differ from those predicted by the deformation mechanism maps. For example, grain growth at high temperatures could strongly affect any predictions of long-term performance by changing the mechanisms and strain rates.

3. Other issues, such as compatibility between materials, oxidation, and corrosion, must be dealt with separately, and the regions given on maps for special phenomena (e.g., dynamic recrystallization and phase transformation) are typically based on experimental observations.

Because deformation behavior occurs primarily through shear transport of material, the effective shear stress and effective shear strain rate for plastic deformation are used in expressing deformation mechanism maps. Using the effective shear strain rate, the principal strain rates can be recovered by expressing the Levy–von Mises relations in Eq. 3.23 as rate-dependent. By combining the expressions for shear and principal values of stress and strain and assuming that strains directly correspond to strain rates, we can write a relationship between any one of the principal strain rates and the plastic resistance to deformation. For the principal strain rate $\dot{\varepsilon}_1$, this relation is given as

$$\frac{\dot{\varepsilon}_1}{\left[s_1 - \frac{1}{2}(s_2 + s_3)\right]} = \frac{2}{9}\frac{d\dot{\gamma}_{\text{eff}}}{d\sigma_{\text{eff}}} \tag{8.1}$$

With Eq. 8.1 and Eq. 3.23 providing a way to calculate all three of the principal strain rates from any given set of applied principal stresses, the related effective stress and the effective shear strain rate can be recovered from a DMM. To recover the stress level for a given temperature, it is necessary to know the temperature dependence of the shear modulus. Table 8.1 gives the estimated temperature dependence of the shear modulus (see also Fig. 1.25) for several materials using the expression

$$\mu = \mu_o\left[1 - \frac{(T - T_{\text{ref}})}{T_{\text{m}}}\left(\frac{d\mu}{dT}\right)^*\right] \tag{8.2}$$

where T_{ref} is 300 K (or the melting temperature of materials that melt below room temperature), and the reference temperature shear modulus, μ_o, and the temperature coefficient of the modulus are given as a dimensionless quantity, $(d\mu/dT)$.

It is also possible to plot the data in the DMMs shown here with axes given in terms of strain rate or even grain size, but here we will focus on just one way to represent DMMs.

TABLE 8.1 Linear Approximations of Temperature Dependence of Shear Modulus

Material	T_{ref} (K)	T_m (K)	μ_o (GPa)	$(d\mu/dT)$*
Iron (alpha)	300	1810	65	1.3
Iron (gamma)	300	1810	81	0.91
Nickel	300	1726	80	0.64
Tungsten	300	3683	160	0.38
Zinc	300	693	49	0.50
Silicon	300	1687	64	0.08
Sodium chloride	300	1070	15	0.80
Al_2O_3	300	2320	155	0.35
Ice	273	273	2.9	0.35

*See Frost and Ashby, 1982, or Simmons and Wang, 1971.

EXAMPLE 8.1 *Deformation Mechanisms in Ice*

Consider uniaxial loading of ice at –50°C and a compressive stress of 10 MPa. We use Eq. 3.20:

$$\tau_{\text{eff}} = \sqrt{\frac{(10)^2 + (10)^2}{6}} = 5.8 \text{ MPa}$$

To get the actual value of the stress ratio on the y-axis in Fig. 8.2, we need the shear modulus of ice at that temperature. We can see that at the solidification temperature the shear modulus of ice is about 1 MPa/3 × $10^{-3} \approx$ 3 GPa. At a temperature of –50°C, we expect the modulus to be higher, so we can expect that it might range from 3.1 to 3.3 GPa. This estimate fits pretty well with the data in Table 8.1. The corresponding ranges are shown on the DMM with resulting shear stress ratios of 1.5 × 10^{-3} to 1 × 10^{-3}. The expected effective shear strain rate is then $\dot{\gamma}_{eff} \approx 10^{-7}/s$. Then, using Eq. 8.1 and the Levy–von Mises relations, we can calculate the principal strain rates as follows:

$$\frac{\dot{\varepsilon}_1}{-10 - \frac{1}{2}(0+0)} = \frac{\dot{\varepsilon}_1}{-10} = \frac{2}{9}\left(\frac{10^{-7}}{5.8}\right) \qquad \varepsilon_1 = -3.8 \times 10^{-8}/s$$

$$\frac{\dot{\varepsilon}_2}{0 - \frac{1}{2}(-10+0)} = \frac{\dot{\varepsilon}_2}{5} = \frac{2}{9}\left(\frac{10^{-7}}{5.8}\right) \qquad \dot{\varepsilon}_2 = \dot{\varepsilon}_3 = 1.9 \times 10^{-8}/s$$

■

Research on ice is conducted by scientists and engineers who study motion of the polar ice caps and glaciers. High pressures and high homologous temperatures can lead to significant creep rates, and dynamic recrystallization is often part of the deformation seen in glaciers.

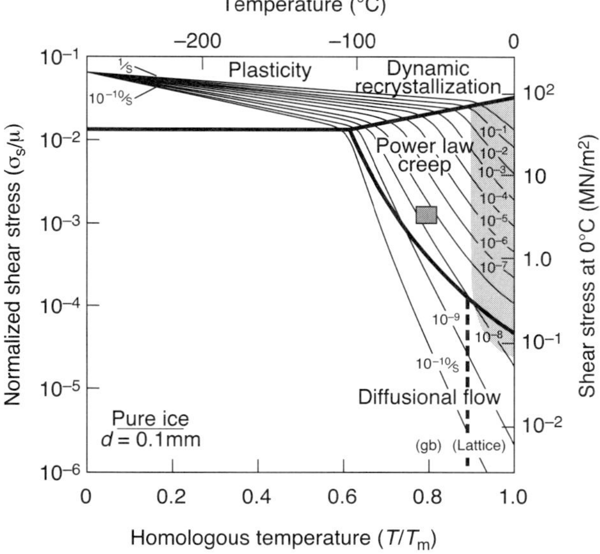

FIG. 8.2 Plot of normalized stress level versus homologous temperature of ice with a 0.1-mm grain size. Ice shows a very high resistance to deformation just below its melting temperature. The highly directional character of its bonding imparts this high strength relative to its elastic constants. The small gray square inset represents the range of deformation conditions discussed in Example 8.1 (Frost and Ashby, *Deformation Mechanism Maps,* Pergamon Press, New York, 1982, with permission).

The positions of some boundaries in DMMs depend strongly on microstructure, with dislocation density affecting glide and climb and grain size having a significant effect on diffusion creep mechanisms. The effects of grain size and dislocation density on nickel with commercial purity are shown in Fig. 8.3. In Fig. 8.3, increasing grain size in a work hardened material results in strong decreases in the diffusional creep rates, expanding the regime of stresses and temperatures over which power law creep is the dominant mechanism. In the first of these materials, shown in Fig. 8.3*a*, the small (1 mm) grain size results in a diffusional flow regime in which boundary diffusion (Coble creep) is the dominant diffusion-based mechanism nearly to the melting temperature. Because Eq. 6.33 is used for dislocation glide, the obstacle spacing resulting from prior work hardening has an effect. For this and all subsequent DMMs, where appropriate, the obstacle spacing, *l*, is given in the caption. Figure 8.3*c* shows that a distinctly separate range of boundary diffusion can be separated from lattice diffusion (Nabarro–Herring creep) at a grain size of 1 mm. The position of this transition between dominant regions for Coble creep and Nabarro–Herring creep is determined using the same procedure as that shown in Example 6.5. Figure 8.3*d*,

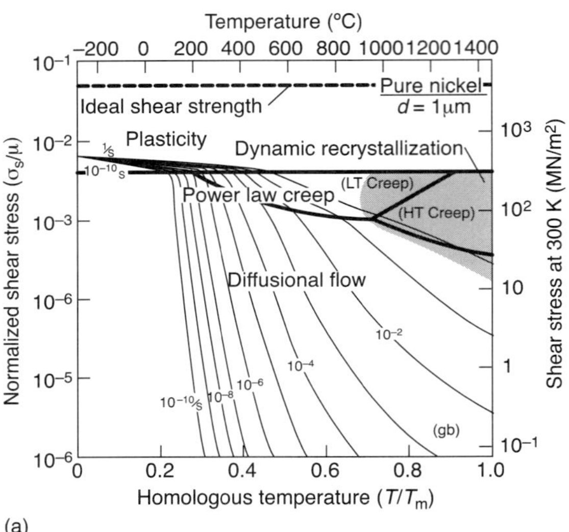

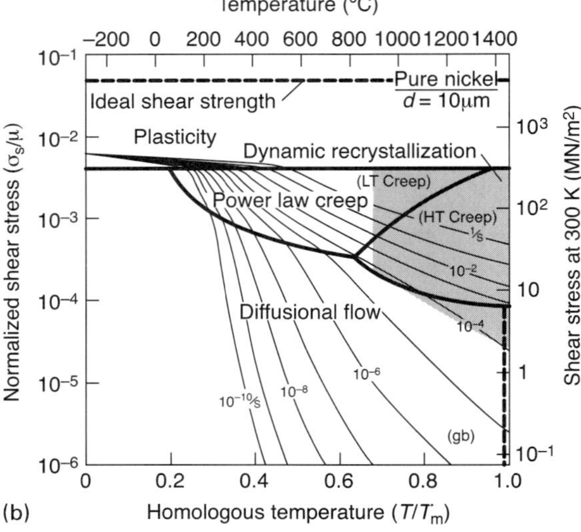

FIG. 8.3 DMMs for pure nickel with grain sizes of (a) 1 μm and (b) 10 μm with a work hardened obstacle spacing of $l = 4 \times 10^{-8}$ m (Frost and Ashby, *Deformation Mechanism Maps,* Pergamon Press, New York, 1982, with permission).

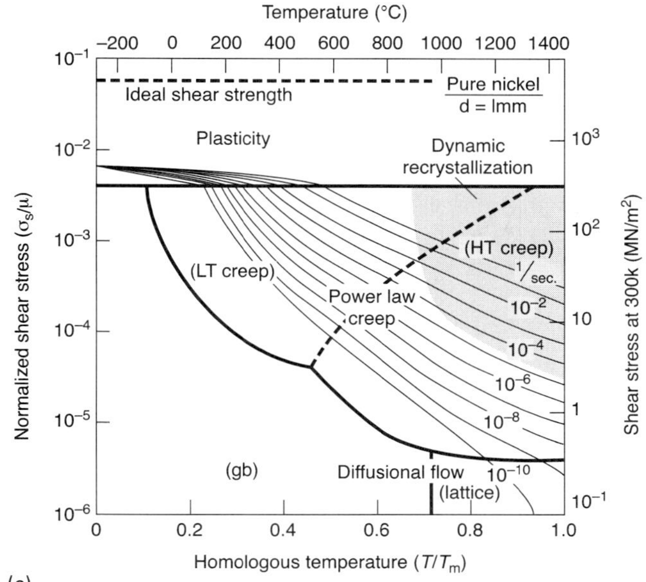

(c)

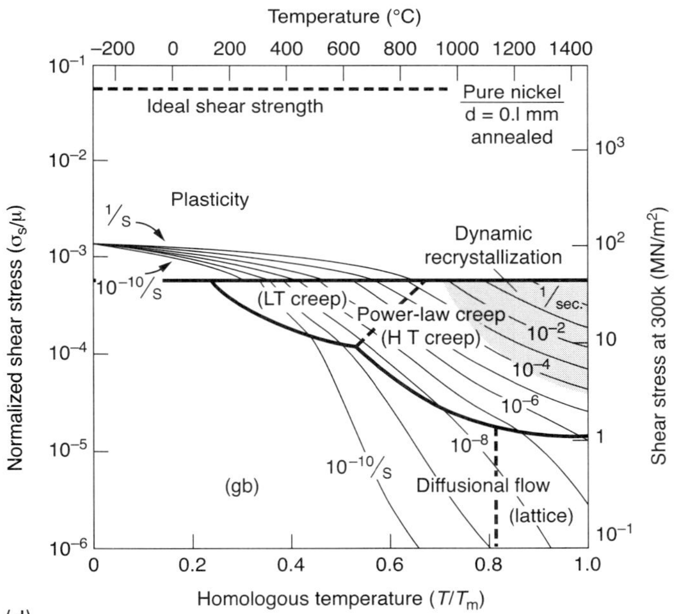

(d)

FIG. 8.3 (cont.) DMMs for pure nickel with grain sizes of (c) 1 mm with a work hardened obstacle spacing of $l = 4 \times 10^{-8}$ m and (d) 0.1 mm that has been annealed to an obstacle spacing of 2×10^{-7} m (Frost and Ashby, *Deformation Mechanism Maps,* Pergamon Press, New York, 1982, with permission).

which would have diffusion creep contributions between those in Fig. 8.3*b* and Fig. 8.3*c*, has a much larger dislocation obstacle spacing because of thermal relaxation processes—viz., annealing-driven reduction of dislocation density.

The DMMs are indicative of the structures of individual materials and show consistent characteristics across classes of similar materials. The pure FCC metals have similar DMMs, with the most significant differences attributable to differences in stacking fault energy. For this reason, it is often possible to make a rough estimate of expected deformation properties by scaling the upper x-axis and the right-hand y-axis with the proportional properties corresponding to a material for which a map is not available. For example, one can rescale the top axis of the DMM for the melting temperature of the material for which a map is not available and then recalculate the room-temperature values for the stress levels given on the right-hand y-axis. Clearly, such a strategy would be fairly successful for the refractory BCC metals tungsten and niobium shown in Fig. 8.4. Although the details of the dislocation plasticity would be different for these two metals (as discussed in Chapters 4 and 6), the creep-related mechanisms, dynamic recrystallization, and the resulting effective shear strain rate contours are fairly consistent. Like all pure BCC metals, in the temperature range 0.1 to 0.2 T_m, there is high strain rate sensitivity and a steep rise in the shear strain rate contours with decreasing temperature.

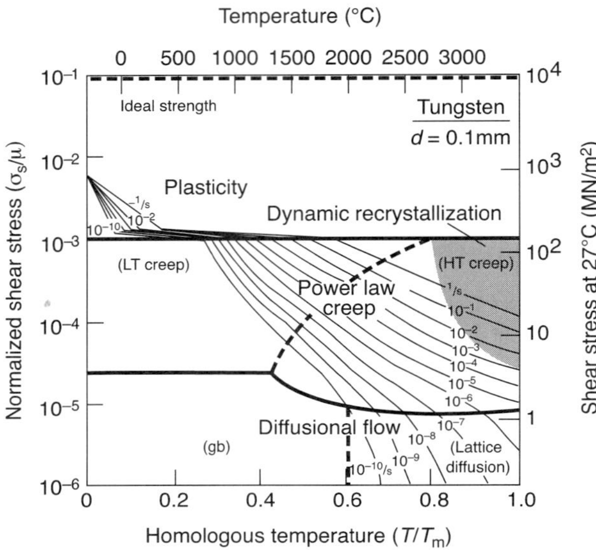

(a)

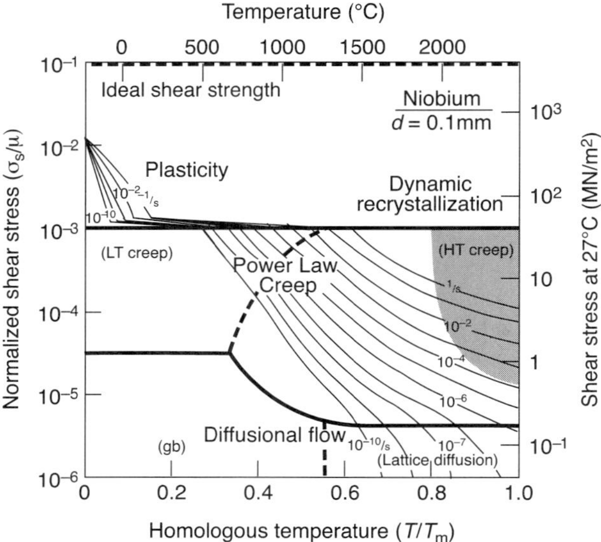

(b)

FIG. 8.4 DMMs for (a) tungsten and (b) niobium with the same grain size. Although there are some differences in the homologous temperature dependence of the dislocation plasticity, the positions of the regions for different creep mechanisms and the related effective shear strain rate contours are very similar (Frost and Ashby, *Deformation Mechanism Maps,* Pergamon Press, New York, 1982, with permission).

The axis-rescaling approach can provide an initial guide for properties of a pure material for which there is insufficient data to construct a map. The example given below for zirconium dioxide includes information from maps based on materials with similar structures, uranium dioxide and thorium dioxide (Frost and Ashby, 1982).

DMMs can also be constructed for alloy systems, with the effects of the alloying being readily evident in how they are manifested on the maps. Figure 8.5 shows the effects of different modifications, including alloying (Fig. 8.5*a*) and particulate composition (Fig. 8.5*b*). The effects of solid solution strengthening are apparent when Fig. 8.5*a* is compared with the diagrams in Fig. 8.3. The effects of incoherent obstacles are evident in Fig. 8.5*b* and 8.5*c*. The figures show the remarkable engineering of suppressed creep deformation accomplished in conventional nickel-based superalloys. Most of these alloys benefit from a combination of solid solution and precipitation strengthening, with MAR M-200 containing a majority of the ordered intermetallic compound called gamma prime, Ni_3Al. The excellent creep and stress-rupture properties of MAR M-200 are displayed in the DMM shown in Fig. 8.5*c*.

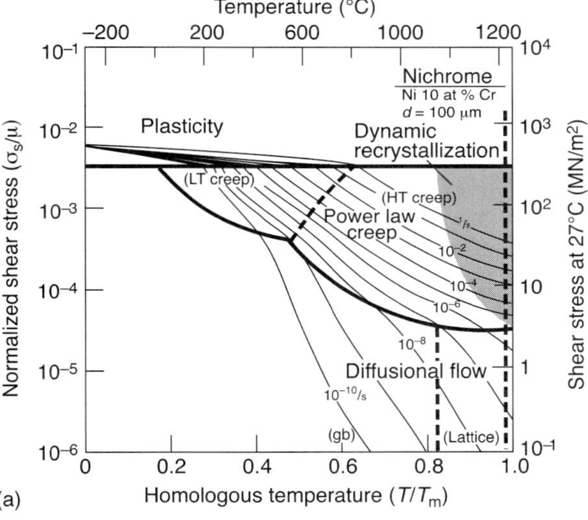

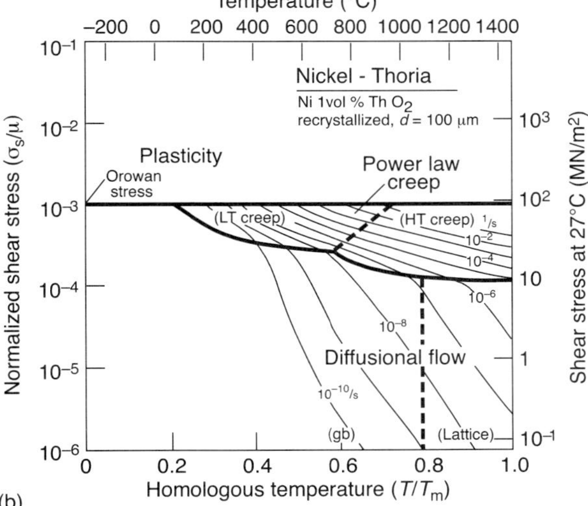

FIG. 8.5 (a) DMM for a nickel–10% chromium alloy with a grain size of 100 μm. The melting point of pure nickel is used for this map. (b) DMM of nickel reinforced with a particulate of thorium dioxide particles at 1 volume percent. The grain size of the nickel is 100 μm. (Frost and Ashby, *Deformation Mechanism Maps,* Pergamon Press, New York, 1982, with permission).

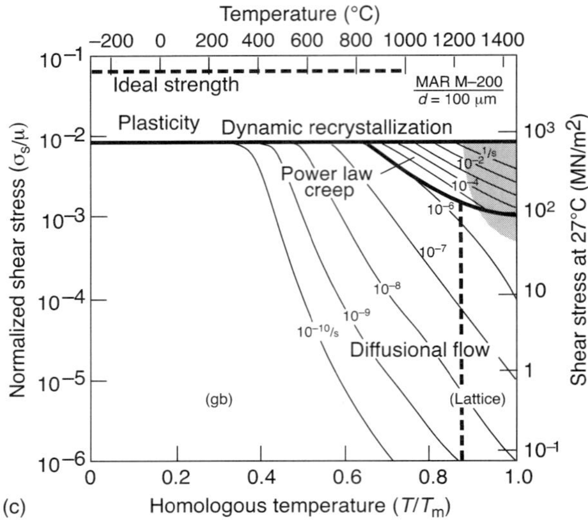

FIG. 8.5 (cont.) (c) DMM for the nickel-based superalloy MAR M-200 in the as-cast condition with a grain size of 100 μm. (Frost and Ashby, *Deformation Mechanism Maps,* Pergamon Press, New York, 1982, with permission).

DMMs can also allow us to depict readily the effects of other important materials phenomena and in many cases help in clarifying the effects of phase changes. Shown in Fig. 8.6 is the DMM for pure iron with a grain size of 0.1 mm. The allotropy endemic to pure iron is readily shown by the phase transitions between the ferrite (α–BCC), austenite (γ–FCC), and ledeburite (δ–BCC) phases. The close-packed structure of FCC iron leads to a slow rate of creep compared with the BCC phases. The higher rates of diffusion with the BCC phases of iron result in substantial shifts in the strain rate contours when crossing in either direction (up or down in temperature) from FCC iron. Just as for tungsten and niobium, iron has a very high strain sensitivity at ~0.15 T_m. The critical resolved shear stress for slip in this regime of temperatures rises quickly with decreasing temperature owing to the lattice resistance to screw dislocation motion.

DMMs can be constructed for almost any material, including ceramics and noncrystalline materials. All that is required is a set of rate-dependent deformation equations. Examples of DMMs for zinc, silicon, and aluminum oxide are shown in Fig. 8.7.

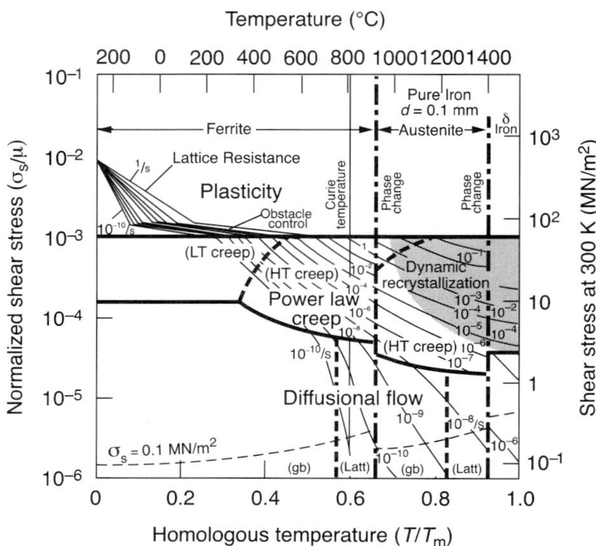

FIG. 8.6 DMM for pure iron with a grain size of 100 μm. The effects of the phase transformations are clearly evident, with the respective regions showing creep behavior that would be expected for each of the two BCC regions and the one FCC region. (Frost and Ashby, *Deformation Mechanism Maps,* Pergamon Press, New York, 1982, with permission).

For pure zinc, dislocation plasticity shows almost no dependence on strain rates although the stresses to produce plasticity in a polycrystal are fairly high owing to the limited number of slip systems. In contrast, the regimes over which power law creep is the dominant deformation mechanism are quite large. Recrystallization also occurs over a relatively wide range of stresses and temperatures.

For polycrystalline silicon, it is clear in Fig. 8.7*b* that the highly covalent diamond cubic crystal structure does not readily enable slip at low stresses and also that diffusion contributions to deformation are very sluggish. In fact, many of the features of the DMM for silicon are similar to the example for ice shown in Fig. 8.3. In ice, the highly directional bonding between individual water molecules and the associated difficulty of diffusion of the water molecules through the ice result in a material that is substantially resistant to plastic deformation even near its melting temperature. Although Si is an element, the directional-

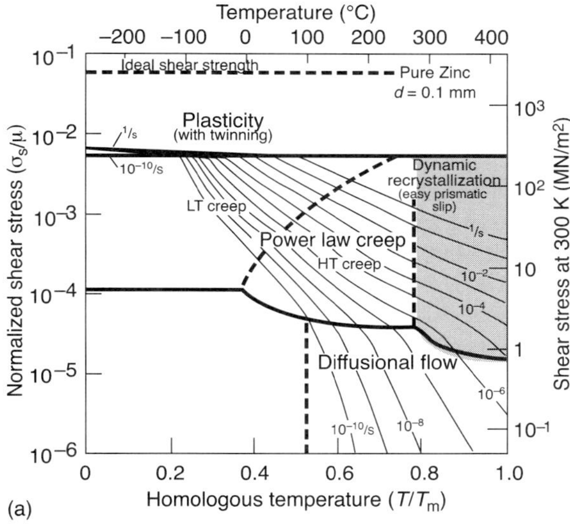

FIG. 8.7 DMMs for (a) the close-packed hexagonal metal zinc with a grain size of 0.1 mm, (b) the diamond cubic material silicon with a grain size of 100 μm and the strain rate contours given in logarithms of base 10. (Frost and Ashby, *Deformation Mechanism Maps,* Pergamon Press, New York, 1982, with permission).

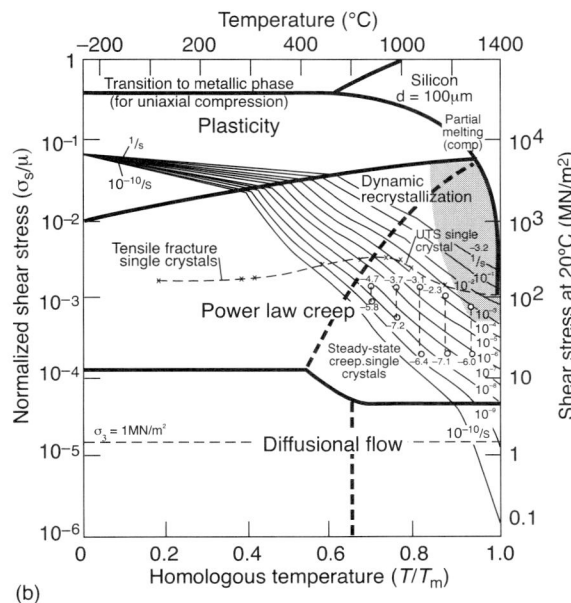

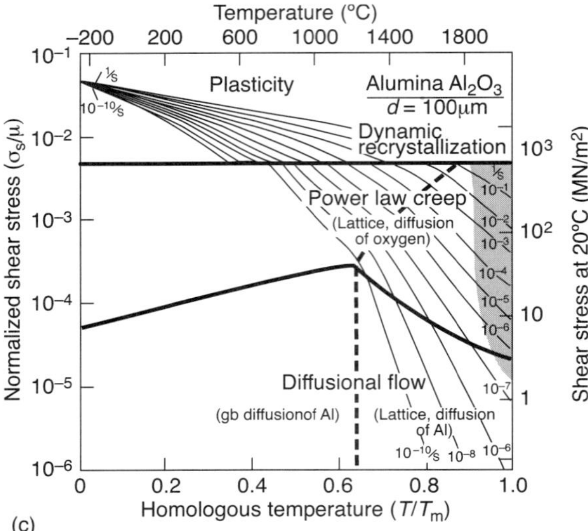

FIG. 8.7 (cont.) (c) the oxide ceramic aluminum oxide with a grain size of 100 μm. (Frost and Ashby, *Deformation Mechanism Maps,* Pergamon Press, New York, 1982, with permission).

ity of its bonding similarly leads to a material that is very brittle up to nearly its melting point. The special section at the top of the Si DMM discussing transformation under compressive stresses is a special phenomenon seen in some materials in which compressive stresses can result in a phase transformation. In contrast to the temperature-based phase transformations shown on Fig. 8.6 for pure iron, this is a phase transformation that is generated by very large compressive hydrostatic stresses.

The DMM for aluminum oxide (Al_2O_3) in Fig. 8.7*c* shows the effects of the strong bonding in this ionic ceramic material with a rhombohedral crystal structure. The strain rate contours occur at high stress values until quite high fractions of the melting temperature are reached. Because Al_2O_3 is an ionic compound, diffusion rates depend on the diffusivity of the slowest species. Further investigation of conventional aluminum oxide materials might reveal substantially different behavior than that shown, depending on the impurities present in the material. Like many ceramic materials, the high-temperature performance of aluminum oxide is strongly dependent on the presence of impurities that can segregate to the grain boundaries and produce a high susceptibility to creep.

As mentioned previously, DMMs can be constructed for materials for which the database is limited by using isostructural materials as the basis. Figure 8.8 shows a DMM calculated from a combination of property data on the alloy yttria-doped zirconia and the expected values from the DMMs for thoria (ThO_2) and urania (UO_2) (Frost and Ashby, 1982). Details of the data used are given below in Example 8.2.

EXAMPLE 8.2 *Generation of Deformation Mechanism Maps for a Yttria–Zirconia Alloy (Duan, Elsner, Oppong, and Vest, 1993)*

The mapping of a material can be accomplished by using existing maps with the same structures (isostructural). For zirconia, this is a bit complicated, because the pure material is monoclinic at room temperature and transforms to tetragonal and then cubic phases with increases in temperature. Control of these phase transformations with alloying is the principle behind transformation toughening. Alloying additions that stabilize the high-temperature phases can produce a variety of materials with interesting properties. At sufficiently high additions of yttrium oxide (Y_2O_3), the high-temperature cubic phase exists at all temperatures below the melting temperature. Mapping of such a material requires data that document the effects of the alloying on the mechanical properties. Because zirco-

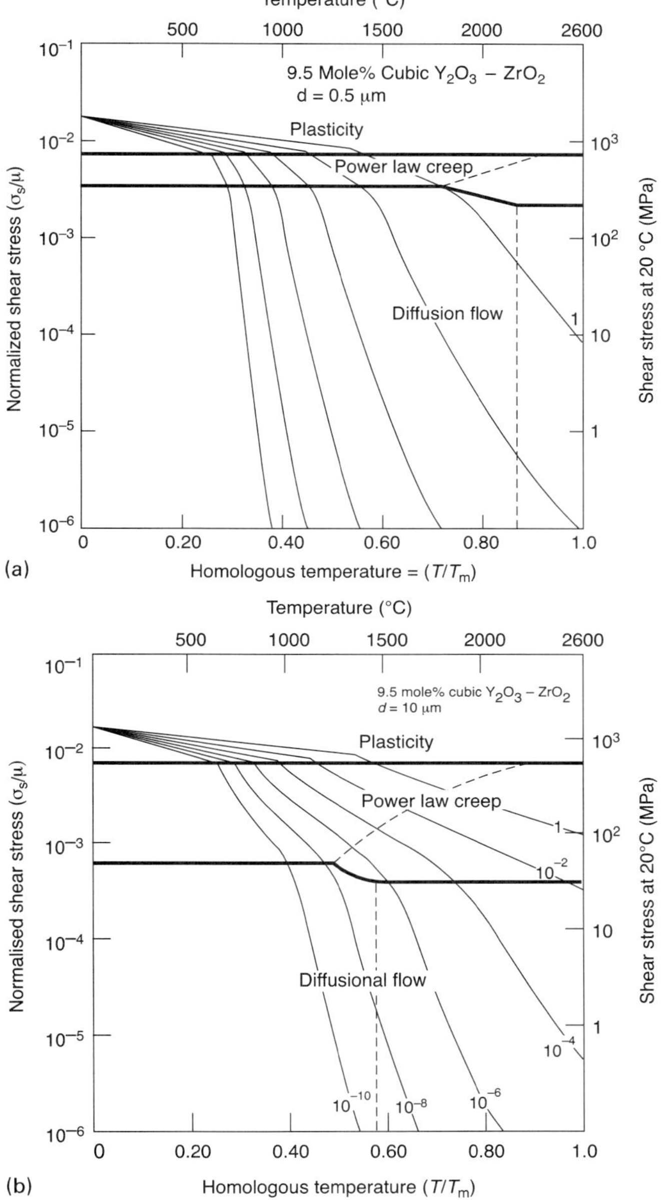

FIG. 8.8 DMMs for grain sizes of (a) 0.5 μm and (b) 10 μm for a ceramic alloy consisting of yttria and zirconia. The composition was chosen so that considerations of phase transitions that would be present at other compositions would not be necessary. (DMM constructed by Purdue MSE 555 students Duan, Elsner, Oppong, and Vest, 1993)

nia alloys have been extensively investigated, this example of an estimated DMM can be constructed as described here. The data used for the generation of the zirconia maps are given in Table 8.2 along with the references and relevant equations from Chapter 6.

The temperature dependence of the shear modulus comes from data by Kandil et al. (1984), who reported shear moduli for yttria-containing crystals at 293 K and at 973 K. For each temperature, the shear modulus is linearly extrapolated to the composition 9.5 mole percent yttria. These values suggest a dimensionless temperature-dependent coefficient ($d\mu/dT$) of 0.72 if we assume that the change in shear modulus with temperature is essentially linear. The activation energy for plasticity reported by Dominguez-Rodriguez et al. (1986) for the alloy 9.4 mole percent Y_2O_3–ZrO_2 is nearly three times larger than the activation energies for lattice resistance controlled plasticity used by Frost and Ashby (1982) for ThO_2 and UO_2, so this should have a significant influence on the DMM of

Y_2O_3–ZrO_2. Because no data are available for core diffusion values, a core diffusion constant of D_c about equal to D_b was employed (Frost and Ashby, 1982). It was assumed that $w \approx b$ and $a_c \approx 2w^2$ to estimate the core diffusion values. The value used for the dimensionless Dorn constant A is based on that reported by Oishi et al. (1990). ■

TABLE 8.2 DMM Data for Yttria–Zirconia

Burgers' vector magnitude:		
b (Å)	3.62	Martinez-Fernandez et al., 1990
Melting temperature:		
T_M (K)	3032	Cheong et al., 1990
Shear modulus:		
μ (GPa)	93–0.021T	Kandil et al., 1984
Nabarro–Herring creep:		
D_V (m^2/s)	$2.7 \times 10^{-5}\exp[-4.38\ eV/kT]$	Oishi et al., 1989; Eq. 6.42
Coble creep:		
wD_{GB} (m^3/s)	$1.2 \times 10^{-14}\exp[-3.04\ eV/kT]$	Oishi et al., 1989; Eq. 6.43
Core diffusion:		
$10(a_c)D_c$ (m^4/s)	$8.7 \times 10^{-24}\exp[-3.04\ eV/kT]$	Oishi et al., 1989, and Frost and Ashby, 1982; Eq. 6.37
Stress exponent:		
n'	4.5	Martinez-Fernandez et al., 1990; Eq. 6.36
Dorn constant:		
A	1937	Oishi et al., 1989; Eq. 6.36
Plasticity activation energy:		
ΔG_o (eV)	6.8	Dominguez-Rodriguez et al., 1986; Eq. 6.33
Athermal flow strength:		
$\hat{t}_p$ (MPa)	0.018	Frost and Ashby, 1982; Eq. 6.33
Plasticity preexponential:		
$\dot{\gamma}_p$ (s^{-1})	10^{11}	Frost and Ashby, 1982; Eq. 6.38

8.2 FRACTURE MECHANISM MAPS

The development of fracture mechanism maps (FMMs) by Ashby and coworkers follows closely the recipe for DMM, but is focused on modes of fracture and the combinations of stress and temperature that produce them. Tensile stress normalized to the elastic modulus is used in the FMMs because tensile stresses drive the fracture process. Many modes of

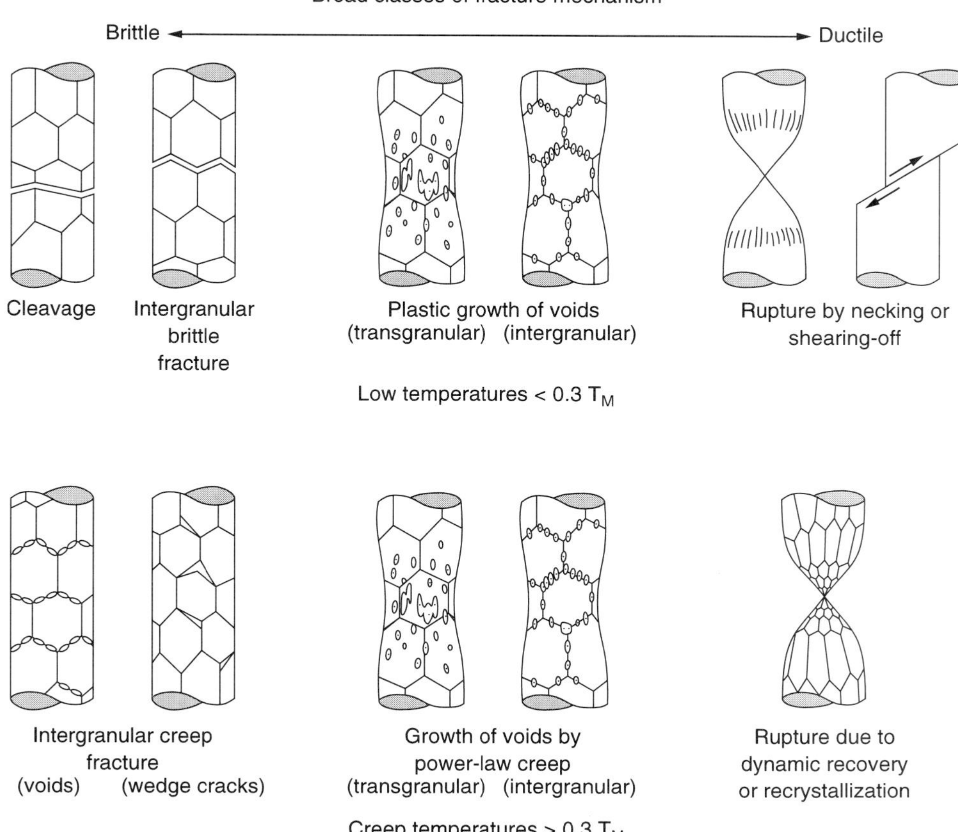

FIG. 8.9 Simplified classification of fracture mechanisms for use on fracture mechanism maps (Ashby, Gandhi, and Taplin, 1979, "Fracture-mechanism maps and their construction for f.c.c. metals and alloys," *Acta Metallurgica*, 27, 699–729, reprinted with permission of Elsevier Science). The upper row is for temperatures below $0.3T_m$ in which there is a small temperature or time dependence. The lower row is for conditions above $0.3T_m$ in which creep may occur. These temperature ranges apply specifically to FCC metals and alloys whereas strongly ionic or covalent materials may undergo brittle failure up to higher temperatures.

fracture are often modeled for the regimes over which they occur so that models for failure by these different mechanisms can be included. FMMs are mostly determined by mapping out the observations from fractographic evaluations of material failure. Figure 8.9 shows the mechanisms defined by Gandhi and Ashby (1979) for fracture. Each of the mechanisms can be associated with models for that particular type of failure. Figure 8.10 shows FMMs for pure iron (Fig. 8.10*a*) and pure nickel (Fig. 8.10*b*). The gray shading along many of the lines represents overlap between the mechanisms.

The regions of fracture behavior in FMMs are based on classifications of the fracture mode. The FMM for iron in Fig. 8.10*a* shows the effects of the same phase transitions depicted in Fig. 8.6 for a DMM. For example, notice that high stresses and low temperatures are associated with regions labeled with cleavage failure or brittle intergranular failure (BIF) followed by a number value. As the numbers increase, from 1 to 3 in the

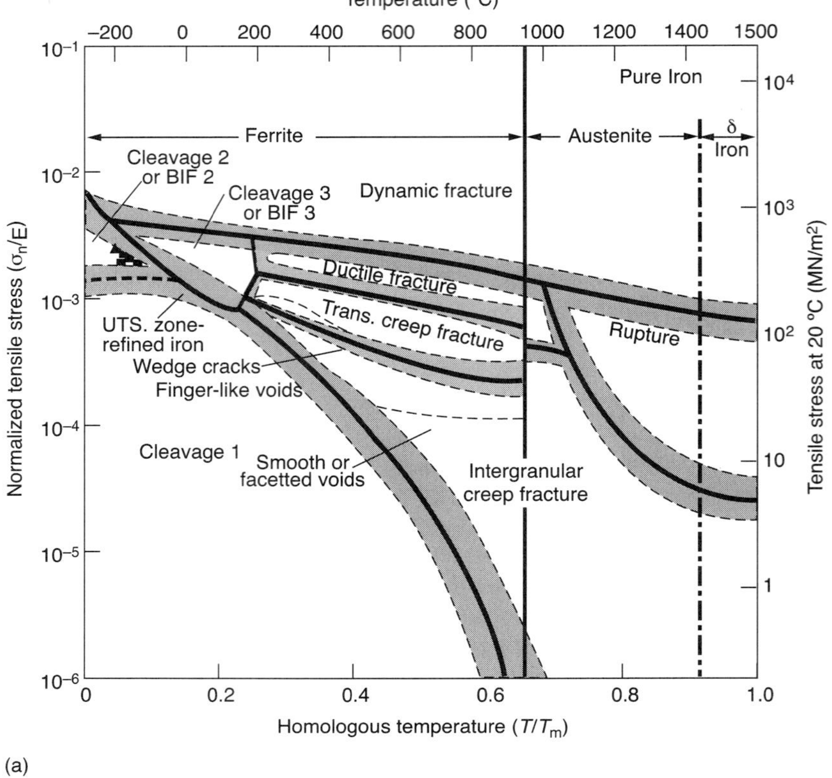

FIG. 8.10 (a) Fracture mechanism map for pure iron. Mechanisms of fracture shown include cleavage, ductile fracture, transgranular creep fracture, intergranular creep fracture, and ductile rupture (Fields, Weerasooriya, and Ashby, 1980, with permission of Metallurgical Transactions).

classification scheme of Ashby and coworkers, the amount of plastic deformation preceding brittle failure increases from (1) no microscale plasticity, to (2) microscale plasticity, to (3) macroscopic plasticity. For cleavage failure, the fractures are mainly transgranular and their occurrence depends on grain size. The presence of impurities that segregate to grain boundaries and thereby weaken them can increase the likelihood of intergranular fracture. These regions occur in BCC metals but are essentially absent from the maps of FCC metals (see Fig. 8.10*b* and 8.10*c*) because they have no regime of temperatures over which they demonstrate brittle fracture. Or, in other words, they do not demonstrate the behavior that is often described as a "brittle-to-ductile transition." Figure 8.10*c* shows the effects of alloying on the fracture behavior and the superposition of the time-to-failure on the maps. These often can be correlated with time-dependent deformation and failure mechanisms to describe regimes over which particular types of fracture are expected.

The transition in regimes of fracture behavior follows a trend consistent with the bonding strength. Figure 8.11 shows FMMs for compounds with ionic and covalent bonding. Materials with these strong bonding types display brittle behavior for almost any type of tensile loading.

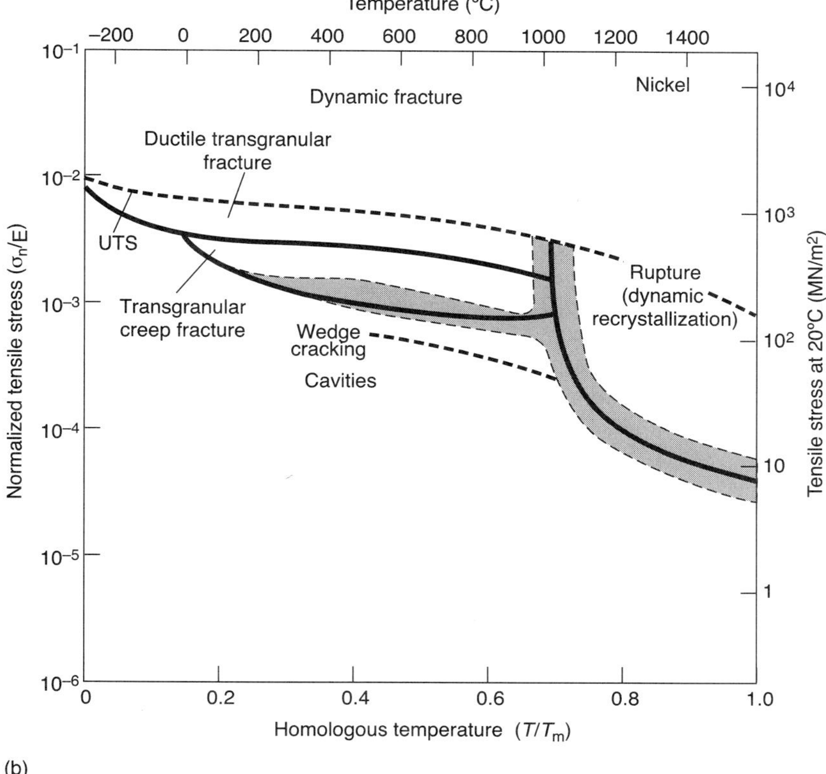

FIG. 8.10 (continued) (b) Fracture mechanism map for pure nickel showing extensive regions of ductile failure at even low temperatures. Individual specimens are indicated by each symbol, and the $\log_{10}$ of the time to failure in seconds is indicated. Intergranular failure is indicated by solid symbols and a mixed-mode failure by shaded symbols (Ashby, Gandhi, and Taplin, 1979, "Fracture-mechanism maps and their construction for f.c.c. metals and alloys," *Acta Metallurgica*, 27, 699–729, reprinted with permission of Elsevier Science).

For sodium chloride (NaCl) (Fig. 8.11*a*), the availability of some limited slip systems allows for some plasticity before brittle failure, but not enough to demonstrate ductility. The yield stress for slip on the easy slip system, <110>{110}, is plotted on this map, but these slip systems do not produce enough independent slip systems for deformation of polycrystalline NaCl. Review of Table 3.A.2 shows that only two of these slip systems will be independent. At temperatures greater than 200°C, plastic deformation of NaCl in compression is possible as other slip systems become viable, but tensile deformation is still quite difficult. Careful examination of table salt shows that the often cuboidal particles were created by cleavage failure on {001} planes.

For alumina (Al_2O_3) and silicon nitride (Si_3N_4), only high temperatures and substantially compressive stresses enable anything resembling plastic deformation. As shown here, tensile stresses invariably result in relatively brittle failure in these strongly bonded materials. For the FMM of Al_2O_3 shown in Fig. 8.11*b*, the possibility of slip on basal planes and twinning are the only viable slip mechanisms below $0.5T_m$, and then this would be possible only in a hardness test or under constrained compressive loading. Fracture occurs

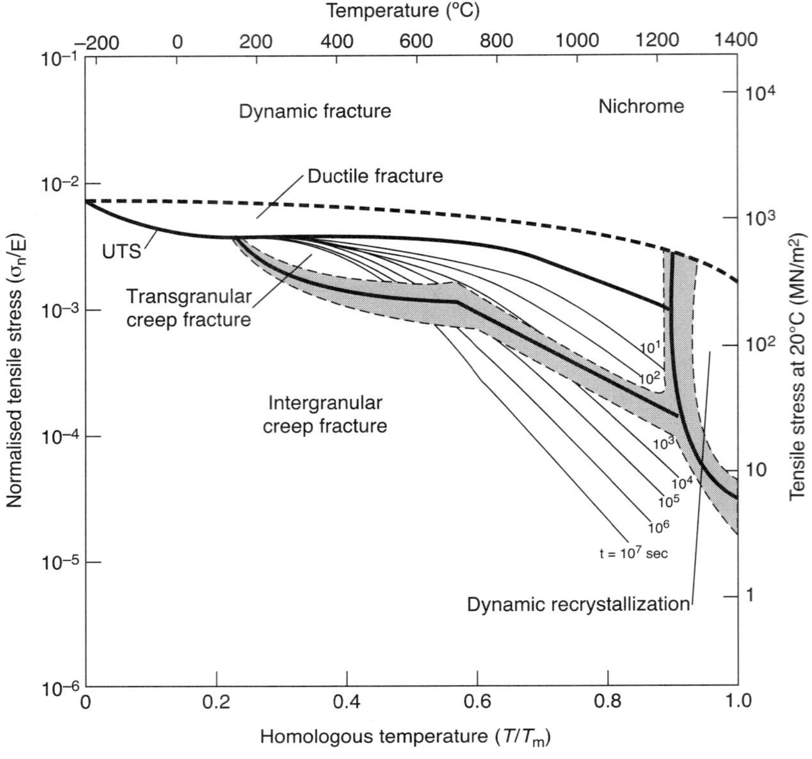

(c)

FIG. 8.10 (continued) (c) Fracture mechanism map for the alloy Nichrome showing the effects of alloying. The contours superimposed on the FMM represent approximately equal times to failure (Ashby, Gandhi, and Taplin, 1979, "Fracture-mechanism maps and their construction for f.c.c. metals and alloys," *Acta Metallurgica*, 27, 699–729, reprinted with permission of Elsevier Science). Data are cited in the original publications.

principally without plasticity, and mostly by cleavage (transgranular fracture) at low temperatures. However, it is possible to get significant intergranular fracture in materials containing significant fractions of a glassy grain-boundary phase.

The Si_3N_4 data in Fig. 8.11*c* demonstrate that the highly directional covalent bonding of this ceramic suppresses the operation of any plastic deformation mechanisms until quite high temperatures are reached. Although contributions of plasticity are evident in materials deformed above 1000°C, these materials usually show plasticity that is greatly assisted by glassy grain-boundary phases. Because Si_3N_4 materials are available in a wider array of alloys, phase contents, and microstructures, the ranges given here should only be considered an example and definitely should not be used for design purposes.

Figure 8.12 summarizes the trend in FMMs that stronger bonding decreases the size of the ductile zones and expands the regimes of brittle failure. For ceramic materials, the effects of grain-boundary phases on the observed deformation behavior must be considered, because they can greatly alter the response. For the FMM of Si_3N_4 shown in Fig. 8.11, the presence of a grain boundary phase can greatly alter the response. The crystallization of the typically glassy grain-boundary phase can slow the creep contribution to deformation and affect the failure mechanisms.

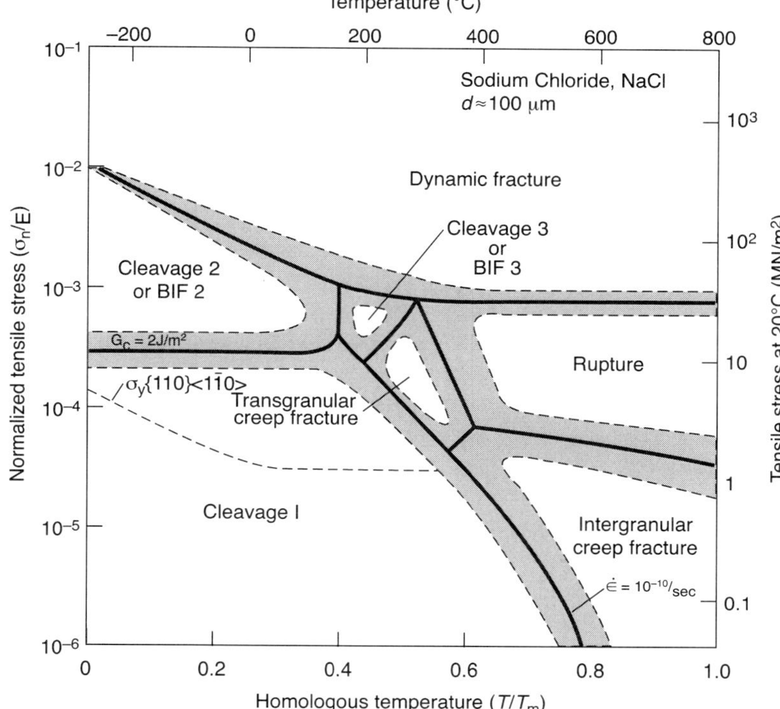

(a)

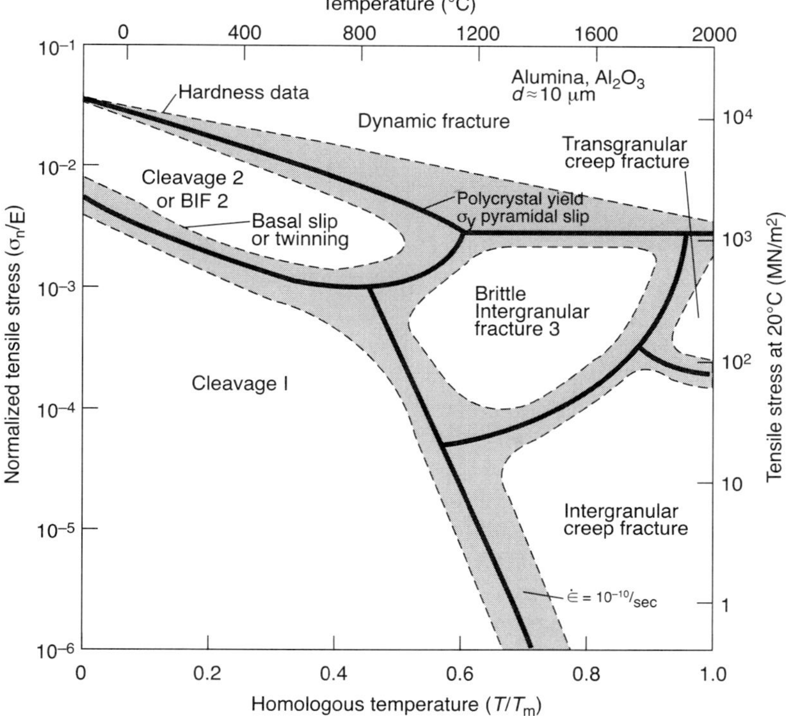

(b)

FIG. 8.11 FMMs for (a) NaCl, and (b) Al_2O_3, showing the regimes of fracture behavior. (Gandhi and Ashby, 1979, "Fracture-mechanism maps for materials which cleave: F.C.C., B.C.C. and H.C.P metals and ceramics," *Acta Metallurgica*, 27, 1565–1602, reprinted with permission of Elsevier Science).

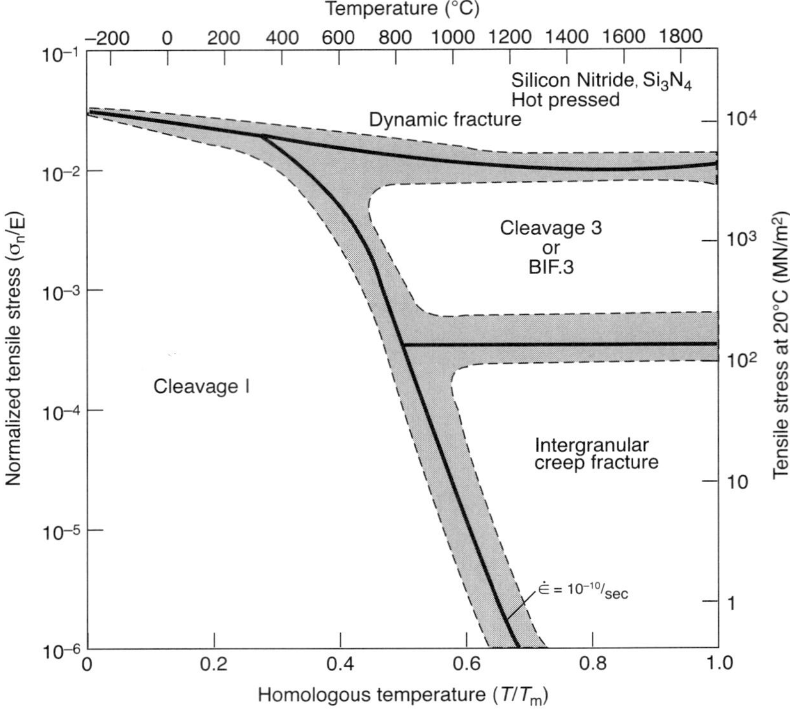

FIG. 8.11 (continued) FMMs for (c) Si_3N_4, showing the regimes of fracture behavior. (Gandhi and Ashby, 1979, "Fracture-mechanism maps for materials which cleave: F.C.C., B.C.C. and H.C.P metals and ceramics," *Acta Metallurgica*, 27, 1565–1602, reprinted with permission of Elsevier Science).

8.3 MECHANICAL DESIGN MAPS

Mechanical design maps (MDMs) are a design tool developed by Ashby (1992, 1999) to enable optimization of materials selection based on multiple properties. An example of this type of map can be derived from Fig. 1.25 to 1.27, which show the relations between specific pairs of properties for a range of materials. Combinations of properties for which figures of merit can be established enable the plotting of lines showing trends for specific property relationships and defining regions over which a material can fulfill the needs of an application safely. The mechanical design principles can be expressed in terms of properties and also in terms of shapes, which enable design engineers to consider optimizing the application for performance including the best combination of material and shape. This latter concept has been extraordinarily difficult to make normal practice in some industries because most attempts to employ new materials in existing structural applications focus on replacement in designs made for properties of the original materials. On the other hand, the rapid pace of advances in electronic materials has resulted partly from necessary strong coupling of the material to its function.

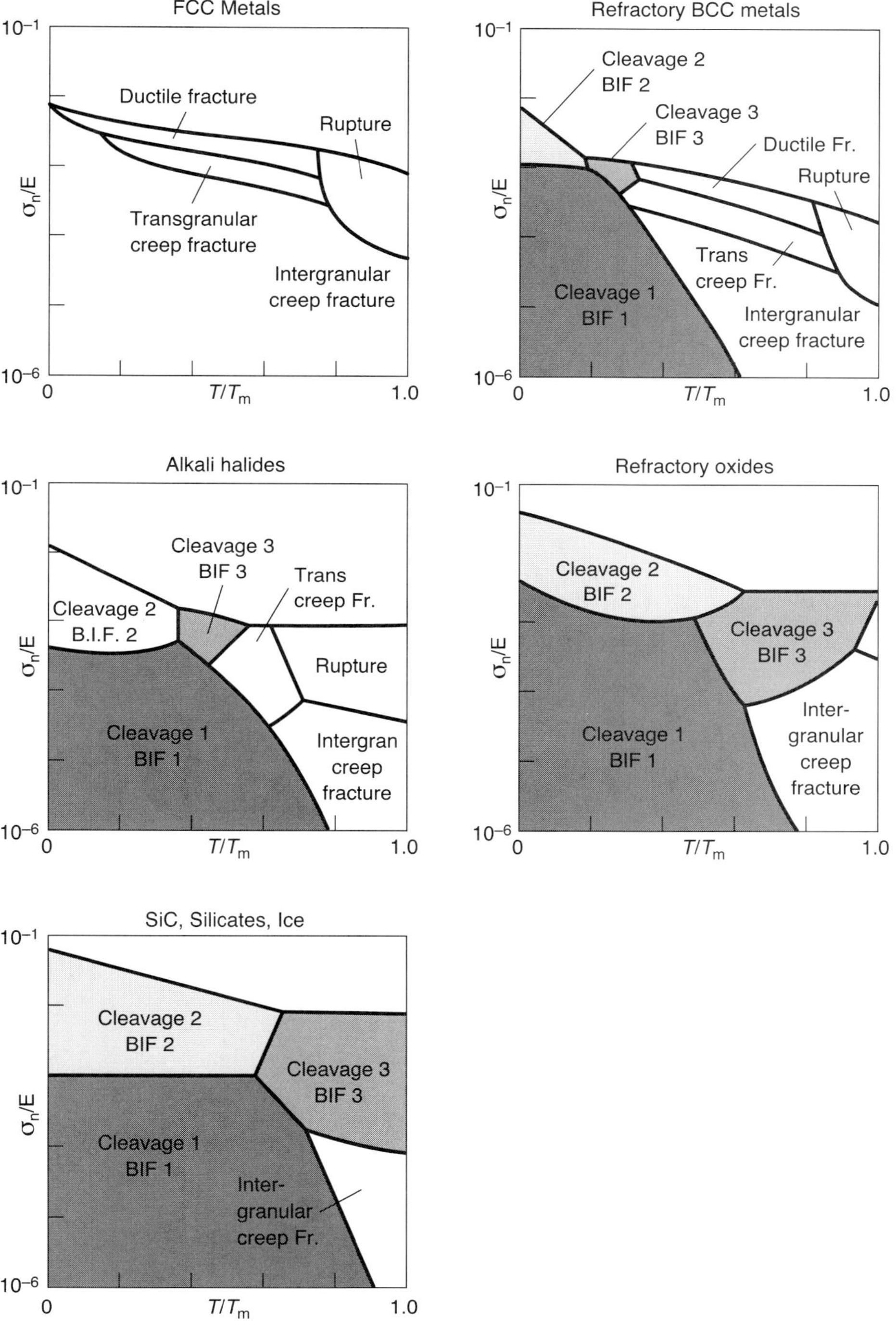

FIG. 8.12 Progressive shift of fracture mechanism fields from metallic to ionic to covalent or directional bonding (from Gandhi and Ashby, 1979, "Fracture-mechanism maps for materials which cleave: F.C.C., B.C.C. and H.C.P metals and ceramics," *Acta Metallurgica*, 27, 1565–1602, reprinted with permission of Elsevier Science).

EXAMPLE 8.3 *The Design of Pressure Vessels*

Safety in pressure vessels can be best accomplished by allowing the material to either fulfill one of two criteria: (1) *yield before breaking* or (2) *leak before breaking* (after Ashby, 1992). In the first condition, overall yielding can be observed before failure, enabling intervention prior to catastrophic failure. This condition is useful for small and compact pressure vessels but not for larger pressure vessels, which are often tied to larger structures. If crack propagation is stable to lengths greater than the wall thickness of the pressure vessel, then a leak can be detectable prior to catastrophic failure. In this case, the material must be chosen in terms of a combination of the yield strength Y and the fracture toughness K_{Ic}. For this example, the simplest geometries are considered and material properties are treated as uniform. In a real pressure vessel, varying geometry and varied processing history (e.g., welds) would require much more detail, and if the contained substance were hazardous, the second criteria might not be an option.

For a spherical thin-wall pressure vessel, the maximum stress tangential to the surface is given by

$$\sigma_\theta = \frac{Pr}{2t} \tag{8.3}$$

where P is the internal pressure, r is the radius, and t is the wall thickness. The wall thickness must be chosen so that $\sigma_\theta < Y$ at the operation pressure. The greatest degree of security is obtained by considering the effects of cracks and making certain that the crack will not propagate even if the stress reaches the general yield stress. Using Eq. 7.8 with the yield stress as the stress level, we can design to

$$a \le \frac{\beta^2}{\pi}\left[\frac{K_{Ic}}{Y}\right]^2$$

In this relation, the critical element is the ratio of the plane strain fracture toughness to the yield stress. That gives our first criterion as

$$\mathrm{Crit}_1 = \frac{K_{\mathrm{Ic}}}{Y} \tag{8.4}$$

We can also consider the second criterion. For leaking to occur, a stable crack must be just large enough to cross the wall thickness. Because crack length is always defined in terms of twice the distance from the center of the crack, $2a$, the stress level as a function of thickness gives us a critical relation wherein

$$\sigma_\theta \le \frac{\beta K_{\mathrm{Ic}}}{\sqrt{\pi \frac{t}{2}}}$$

If we set the stress level to the yield stress Y and use the expression for the tangential stress given in Eq. 8.3 then we can write

$$\beta^2 \frac{\pi P r}{4} = \left[\frac{K_{\mathrm{Ic}}{}^2}{Y}\right]$$

This leads to the maximum pressure safely borne when it leaks before failing catastrophically as

$$\mathrm{Crit}_2 = \left[\frac{K_{\mathrm{Ic}}{}^2}{Y}\right] \tag{8.5}$$

If we consider only Eq. 8.3, we might imagine that the material that enables a very thin wall would be the best to fulfill the *leak before breaking* condition. This would lead to the selection of the dimensions based on

$$t \geq \frac{Pr}{2Y}$$

and a condition that depends only on the yield strength

$$\text{Crit}_3 = Y$$

These design criteria are shown in Fig. 8.13 with examples given. In this figure, the strength is given as $\sigma_f = Y$. The conditions Crit_1, Crit_2, and Crit_3 are figures of merit for different criteria. ■

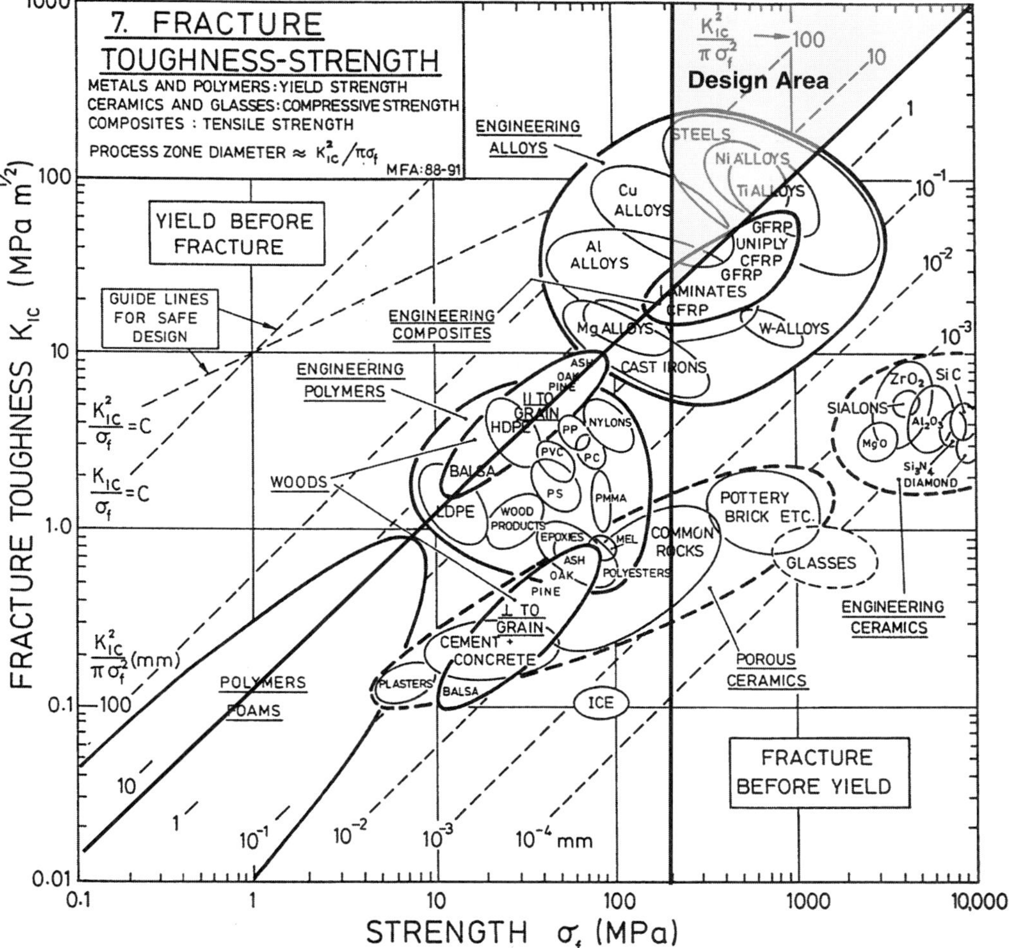

FIG. 8.13 Mechanical design map for fracture toughness versus strength that explores different criteria for failure. Crit_1 and Crit_3 are shown overlaid for a design with $Y = 200$ MPa and $K_{Ic} = 20$ MPa$\sqrt{\text{m}}$. (Ashby, 1992, "Materials Process and Selection Charts," Chart 7, reprinted with permission from Elsevier).

8.4 REFERENCES

M. F. Ashby, *Acta. Met.*, **20**, 887–905, 1972.

M. F. Ashby, C. Gandhi, and D. M. R. Taplin, *Acta Metall*, **27**, 699–729, 1979.

M. F. Ashby, *Materials Selection in Mechanical Design*, Pergamon Press, 1992.

M. F. Ashby, *Materials Selection in Mechanical Design*, 2nd Ed., Butterworth-Heinemann, 1999.

D. S. Cheong, A. Dominguez-Rodriguez, and A. H. Heuer, *Phil. Mag. A*, **63**, 377, 1990.

A. Dominguez-Rodriguez, K. P. D. Lagerløf, and A. H. Heuer, *J. Am. Ceram. Soc.*, **69**, 281, 1986.

Z. Duan, J. Elsner, E. Oppong, and S. Vest, Purdue MSE 555 Project, Keith J. Bowman, Instructor, 1993.

R. Fields, T. Weerasooriya, and M. F. Ashby, *Metallurgical Trans. A*, **11A**, 333–347, 1980.

H. J. Frost and M. F. Ashby, *Deformation Mechanism Maps*, Pergamon Press, New York, 1982.

C. Gandhi and M. F. Ashby, *Acta Metall*, **27**, 1565–1602, 1979.

H. M. Kandil, J. D. Greiner, and J. F. Smith, *J. Am. Ceram. Soc.*, **67**, 341, 1984.

J. Martinez-Fernandez, M. Jimenez-Melendo, and A. Dominguez-Rodriguez, *J. Am. Ceram. Soc.*, **73**, 2452, 1990.

Y. Oishi, K. Ando, and Y. Sakka, in *Advances in Ceramics*, Vol. 7, *Additives and Interfaces in Electronic Ceramics*, edited by M. F. Yang and A. H. Heuer, American Ceramic Society, Westerville, Ohio, 1989, pp. 208–219.

G. Simmons and H. Wang, *Single Crystal Elastic Constants*, MIT Press, 1971.

8.5 PROBLEMS

A.8.1 What is the smallest grain size of nickel that would allow for a strain rate less than 10^{-8}/sec at a normalized shear stress level of 10^{-4} and a temperature of 800°C?

A.8.2 According to the DMMs given in this chapter, which of the following materials will have a normalized shear stress of 10^{-3} resulting in a strain rate of less than 10^{-3}/sec at 0.6 of the melting temperature?

(a) 0.1 mm grain size ice
(b) 1 μm grain size Ni
(c) 10 μm grain size Ni
(d) 1 mm grain size Ni
(e) 0.1 mm grain size W
(f) 0.1 mm grain size Nb
(g) Nichrome
(h) Nickel–thoria
(i) MAR M-200
(j) 0.1 mm Fe
(k) 100 μm Si
(l) 2 μm ZrO_2

A.8.3 Which of the materials listed in Problem A.8.2 is the strongest at 0.2 of its melting temperature (compared with its own shear modulus)?

A.8.4 Which of the materials listed in Problem A.8.2 is the strongest at 0.9 of its melting temperature (compared with its own shear modulus)?

A.8.5 Which of the materials listed in Problem A.8.2 is the strongest at room temperature?

A.8.6 Which of the materials (other than ice) listed in Problem A.8.2 is the weakest at room temperature?

A.8.7 At a temperature of 0.6 T_m and a tensile stress level of 10^{-3} E, write the expected fracture mode for each of the following materials.

(a) Iron
(b) Nickel
(c) Nichrome
(d) Sodium chloride
(e) Alumina
(f) Silicon nitride

B.8.1 Which of the materials (other than ice) listed in Problem A.8.2 has the highest shear modulus at 20°C?

B.8.2 According to the DMM for Al_2O_3, what is its shear modulus at 20°C?

B.8.3 Lithium fluoride has the same rock salt crystal structure as NaCl. How would you construct a quick estimated DMM for this material?

B.8.4 Find the temperature and the magnitude of the effective shear stress for PLC at 0.6 of the melting temperature for the DMM of each of the following materials.

(a) Nickel
(b) Zinc
(c) Tungsten
(d) Silicon
(e) Al_2O_3

B.8.5 The actual creep rate is typically much slower than predicted by DMMs. Why?

C.8.1 Construct an estimated plot of the steady-state tensile creep rate versus temperature for pure iron that is loaded to a tensile stress of 10 MPa. The required temperature range is 600 to 1000°C. Describe the transitions that occur across this temperature range.

C.8.2 Compare the expected lifetime to failure of Nichrome and MAR M-2000 loaded in balanced biaxial tension to 10 MPa at 1000°C assuming steady-state values on the DMM given in this chapter. Assume that failure occurs when any one of the tensile strains reaches 2 percent.

C.8.3 Construct an estimated DMM for Purdunium. Purdunium has the same crystal structure as silicon, but it has a melting temperature of 320 K, making it essentially useless for most any application.

C.8.4 For the DMM and FMM of Nichrome, consider the applied stress level of $10^{-8}\,\sigma_f/E$ at 0.8 of the melting temperature. Find the expected time to failure on the FMM and then determine the expected tensile strain at which failure is predicted to occur from the DMM.

C.8.5 Using the stress tensor values (MPa) given below, predict the expected principal strain rates for Nickel at 800°C with grain sizes of 1 μm and 1 mm and give the dominant deformation mechanism. To do so, you will need to find the effective shear stress, take an effective shear strain rate from the DMMs, and then use the Levy–von Mises equations to predict the principal strain rate ratios.

$$\begin{vmatrix} 10 & -2 & 0 \\ -2 & 5 & 0 \\ 0 & 0 & 3 \end{vmatrix}$$

C.8.6 Using the stress tensor values (MPa) given below, predict the expected principal strain rates for nickel and Nichrome at 1000°C with a grain size of 100 μm and give the expected dominant deformation mechanism. To do so, you will need to find the effective shear stress, take an effective shear strain rate from the DMMs, and then use the Levy–von Mises equations to predict the principal strain rate ratios.

$$\begin{vmatrix} 0 & 0 & 0 \\ 0 & 10 & 12 \\ 0 & 12 & -10 \end{vmatrix}$$

CHAPTER 9

DEGRADATION PROCESSES

THE RECOGNITION that the mechanical performance of materials can change with time even below the conditions at which we expect recognizable plastic strains to occur has often come about from experience. By relying on handbook quotations of material performance for static loading in short-term tests, many engineers risk such an education when they ignore the effects of cyclic loading or the interaction of environment with a material under load. For many years the intrusion of failure from unknown causes under apparently unchanged operation conditions inspired suggestions that something about the material had changed. Although there are many mechanisms and groups of mechanisms that can cause material degradation over time, the leading mechanical causes for materials to fail to perform their desired functions after long-term operation are *fatigue*, *frictional wear*, and *oxidation/corrosion* or combinations of these three phenomena. The signatory combinations include the phenomena called *corrosion fatigue* for combined corrosion and fatigue, *stress corrosion cracking* for combined static loading and corrosive environments, and *fretting* for a combination of fatigue and frictional wear. The first two of these synergistic phenomena require an extensive foundation in the electrochemistry of corrosive processes, and such detailed explanations of oxidation and corrosion processes are beyond the scope of this text. Another combination degradation process is known as creep fatigue, which is a combination of high-temperature creep deformation under cyclic mechanical loading.

The operation of materials under repeated loading can lead to failure that occurs even though the material has already undergone loads of the same level without apparent damage. This apparent weakening of the material is called fatigue, with the literal meaning that the material has tired from the prior loadings.

Cyclic loading occurs in reciprocating engines, bridges loaded with traffic passing over them, and surfaces of aircraft that are loaded by fluctuations in the loads from air passing over them. Failures can occur by the propagation of final cracks, as discussed in Chapter 7, or by increasing compliance of the component from the growth of *subcritical* cracks—that is, cracks that have not reached the length at which they would propagate to final failure in a single tensile load cycle. Strategies for preventing fatigue failure in many ways guide the approach to testing the fatigue response of materials. One strategy is to avoid the processes that initiate fatigue cracks entirely, and another is to accept that cracks may initiate in the material but that they will propagate in a predictable fashion to sizes at which conventional inspection processes are readily able to detect them before catastrophic failure occurs. Within each of these strategies, we must also consider the nature of the loading for the application and the nature of the final failures that are likely to occur.

One of the more dramatic examples of fatigue failure came about from the window design of the deHavilland Comet. Three unexplained crashes within 2 years after its 1952 introduction as the first commercial passenger jet had hurt the prospects for jet-powered passenger planes. On recovery of the needed evidence from the third crash of the 36-passenger jet, it was clear that crack initiation at the fairly large windows could lead to complete failure of the fuselage by explosive decompression of the pressurized cabin (see Stephens, Fatemi, Stephens, and Fuchs, 2000).

9.1 CYCLIC FATIGUE OF MATERIALS

The cyclic loading of engineering components and the resulting failure can occur in any mode of loading, leaving us with the general concern of how transferable fatigue tests conducted in simple geometries are to more complex loading histories (see Suresh, 1999, and Stephens, Fatemi, Stephens, and Fuchs, 2000). Fatigue can occur in loadings in which fatigue damage results in accelerating stresses, as in the case of crack propagation under constant loads, or fatigue can occur in situations in which distributed damage results in dimensional changes of the component that result in the reduction of stresses with increased cycling. The loadings in fatigue can come about singly or in combinations of direct mechanical loads and indirect thermal gradient loads. Generally, the magnitude of the loads in terms of stress scales with the amount of fatigue damage, but this depends strongly on how close the value of the effective stress is to the yielding condition. For loads above the initial yield strength, fatigue damage is rapidly accumulated and strongly affected by the difference between the maximum and minimum loads. For loads generally below the yield strength, the heterogeneity of the effective stress distribution (e.g., at stress concentrations at phase boundaries or pores) and the extremes of tensile stresses relative to minimum or even compressive excursions are important in predicting fatigue response.

9.1.1 Cyclic Plasticity in Ductile Materials

Extensive scientific investigations of fatigue have been accomplished on ductile materials with research into the cyclic plasticity behavior of single crystals and polycrystals. Cyclic testing of specimens similar in shape to a standard tensile specimen has enabled recognition of how dislocation structures evolve to produce damage leading to the formation of cracks that can propagate to failure. The fatigue properties of high-purity single crystals may appear to have little direct importance for fatigue of polycrystalline engineering alloys, but the mechanisms involved are also observed in engineering failures. In addition, the cyclic stress-strain behavior of single crystals shows characteristics that help us to understand polycrystals.

Several strategies for cyclic plasticity have been employed, including testing to constant stress levels, constant total stress levels, and constant plastic strain levels. As shown in Fig. 9.1, constant plastic strain levels in initially soft, well-annealed single crystals produce increasing stress levels as the material hardens. Figure 9.1*a* also shows the definitions of the maximum and minimum stress levels, σ_{max} and σ_{min}, and the maximum and minimum plastic strain levels, $\varepsilon_{pl,max}$ and $\varepsilon_{pl,min}$. In this figure, the numbers adjacent to the maximum stresses are the numbers of cycles. The curve of data in Fig. 9.1*b* gives the maximum stress on the tensile side for each of the hysteresis loops designated in Fig. 9.1*a*. When these stresses are plotted against the accumulated plastic strain (the sum of the tensile and compressive plastic strains) for each cycle, we can obtain what is called a cyclic hardening

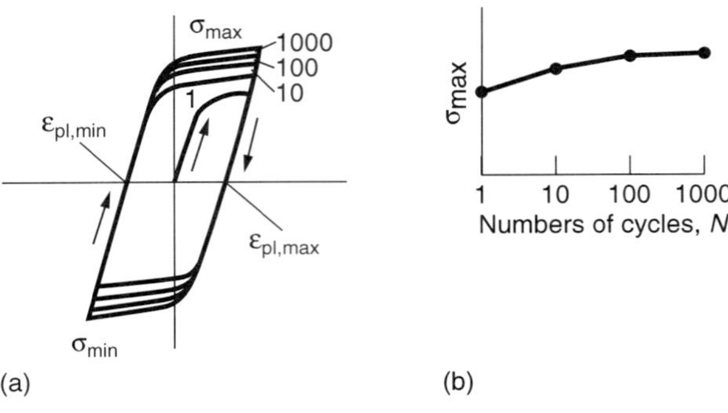

FIG. 9.1 (a) Schematic hysteresis loops for cyclic hardening of a ductile metal with the defined values for maximum and minimum stresses and strains. (b) A cyclic hardening curve for a fixed plastic strain amplitude plotted in terms of the number of cycles. To get the cyclic hardening rate versus accumulated strain, we multiply N by $4\varepsilon_{pl}$, which is the total forward and reverse plastic strain per hysteresis loop. By the time 1000 cycles have taken place, the example shown has begun to show saturation behavior.

curve. When this type of experiment is conducted on a single crystal, the values of cyclic tensile and compressive stresses and strains can be converted to shear stresses and strains based on the observed slip systems, just as is often done for tensile or compressive testing of single crystals. For this reason, a great deal of the data on cyclic stress-strain behavior is plotted with shear values on the axes.

The cyclic behavior of a given material typically proceeds to a saturation stress level wherein there is little change with further cycling unless a fracture initiates. Although soft, annealed materials typically harden, work hardened materials can soften down to the saturation stress. The progressive change of the hysteresis loops is also very different if the deformation is controlled by stress level or total strain level. Figure 9.2 shows the effects of cyclic hardening, saturation, and cyclic softening for high-purity copper. Because copper is an FCC metal with a relatively high stacking fault energy (see Table 4.2) that enables cross slip, it is possible for the dislocation substructures within the material to reform, producing about the same dislocation substructure in softening as in hardening. In materials with a low stacking fault energy, the restriction of slip to specific planes does not allow for sufficient changes in the dislocation substructure.

Once cyclic saturation has been reached in a single crystal, the continued cycling can result in the formation of surface offsets or steps that roughen the surface and can lead to fracture initiation. These surface steps typically follow specific planes of slip in the crystal and are largest at the point at which the active Burgers vector points out of the crystal, as shown in Fig. 9.3. If the distribution of cyclic slip is occurring on a small number of planes rather than being distributed widely, fracture is more likely to initiate sooner owing to the formation of larger slip steps at the surfaces. With a low stacking fault energy it is possible to produce observable slip steps in materials that have been loaded only to small plastic strain levels but over a repeated number of cycles. In materials with low stacking fault energies, slip tends to be localized on just a few slip planes. Figure 9.3 shows the relationship between stacking fault energy and the type of slip steps expected, if they are visible. The same type of process occurs in polycrystals, with cracks often being nucleated at or near surfaces and also at grain boundaries. In addition, the higher rates of hardening and multiple slip present in cyclic plasticity of polycrystals result in much higher stress levels. The growth of the slip steps typically occurs after saturation has been reached. The saturation level is also important because the dislocation substructures that develop in the work hardened material become quite stable even while the surfaces of the crystals may continue to increase in the level and prevalence of slip steps.

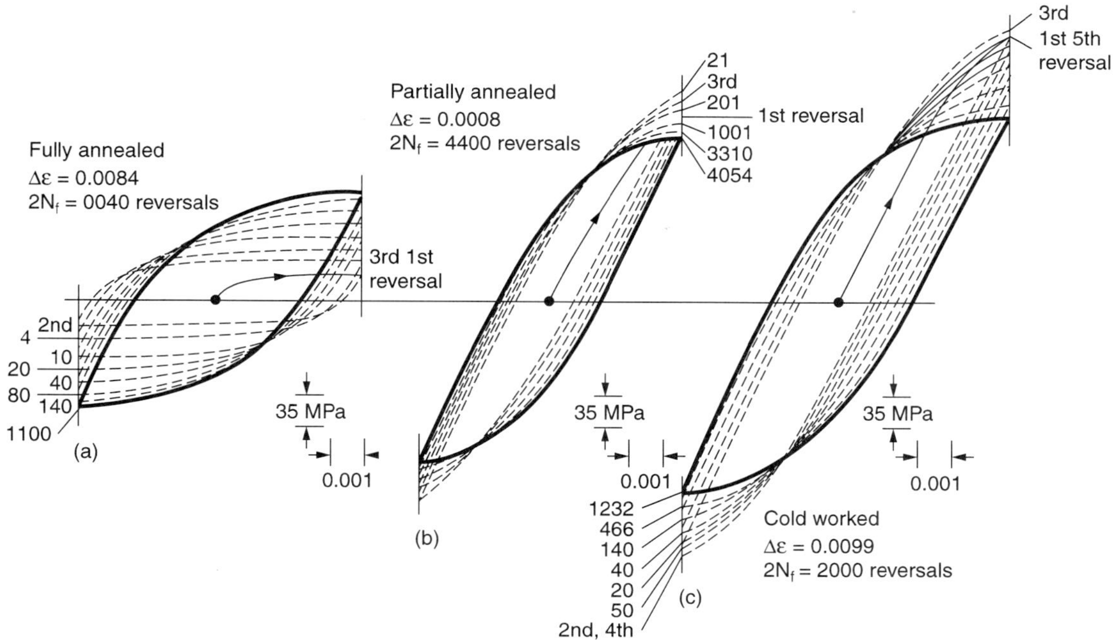

FIG. 9.2 (a) Cyclic hardening, (b) saturation, and (c) softening as functions of initial work hardened state for high-purity copper tested in total strain control (Morrow, 1965, Copyright ASTM INTERNATIONAL, used with permission).

The saturation stress can be plotted against the plastic strain amplitude for plastic strain controlled tests to give what is called a cyclic stress-strain curve for the material. This can be constructed from plastic strain controlled hysteresis loops in materials hardened to saturation, as shown in Fig. 9.4.

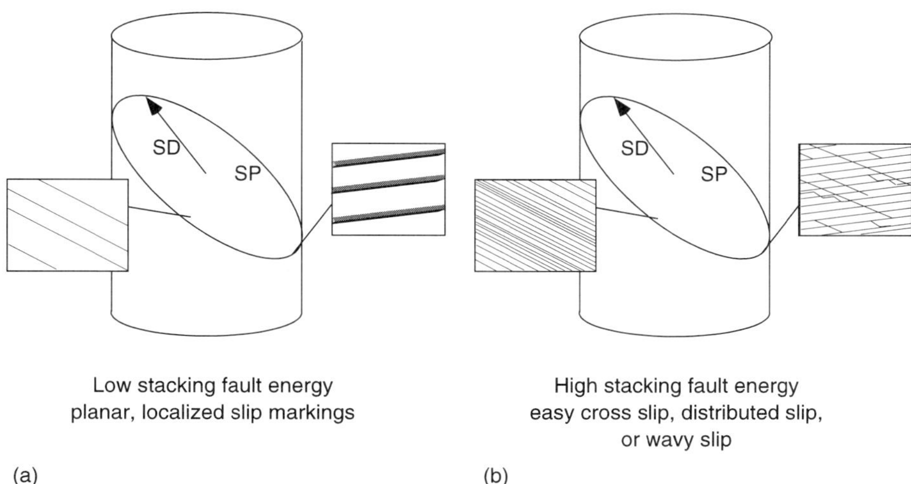

FIG. 9.3 The stacking fault energy normally correlates with the formation of slip steps at the surfaces of single crystals. (a) Localized slip in low stacking fault energy materials that show concentrated planar slip. (b) Distributed or wavy slip in high stacking fault energy FCC metals and most BCC metals.

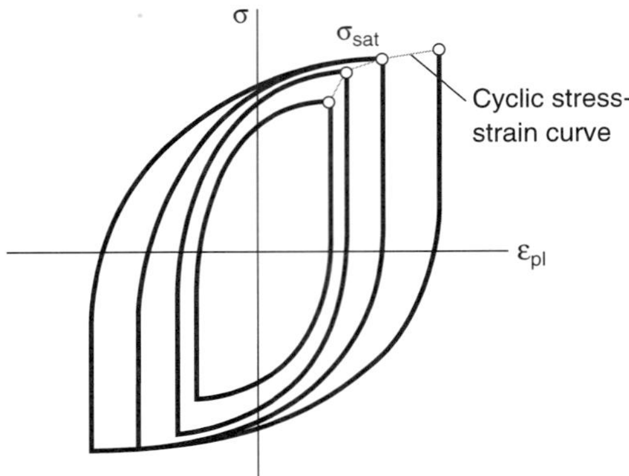

FIG. 9.4 Schematic hysteresis loops following saturation at different values of controlled plastic strain amplitude. Here, only plastic strain is plotted, so there is no tilt from the effects of elastic strain on elastic loading. At the top right corner of each saturation loop is the maximum value of the stress, often called the saturation stress σ_{sat}, which is plotted against the plastic strain value at that point to generate the cyclic stress-strain curve.

Cyclic stress-strain curves for nickel at two temperatures are shown in Fig. 9.5*a*. In this figure, the stress and strain values are given in terms of shear after calculating the orientation relationship between the expected single operative slip system and the specimen axis. Most FCC metals show a three-stage behavior with a middle plateau region persisting across one or two orders of magnitude in plastic strain level. Figure 9.5*a* also enables the effects of temperature to be considered. As mentioned above, the type of dislocation structures observed in metals following cyclic deformation depend on the ability of the material to cross slip. Figures 9.5*b* and 9.5*c* show such a comparison. Materials that undergo sig-

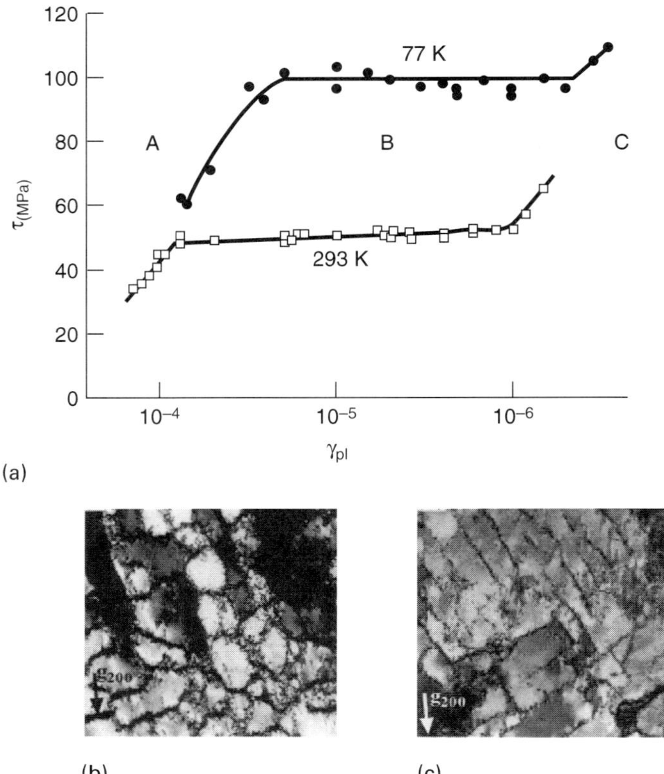

FIG. 9.5 (a) Cyclic shear stress–shear strain curves for 99.99% nickel single crystals tested at 77 K and 293 K. The crystals were oriented to produce single slip on a <110>{111} slip system, and the shear stress and shear strain values were calculated using the orientation relationship to the fatigue specimen axis (after Hollmann, Bretschneider, and Holste, 2000, Wiley, used with permission). (b) Dislocation cell structure produced at an accumulated plastic strain of 1.25% in Mo-containing 316 austenitic stainless steel that has a relatively high stacking fault energy. (c) Clearly planar dislocation activity in the same alloy fatigued as in part (b) doped with nitrogen to reduce the stacking fault energy (Murayama et al., 1999, Elsevier, used with permission, picture width ~2μm).

nificant cross slip owing to the presence of nondissociated dislocation cores often form the cell structures shown in Fig. 9.5*b*. In this electron micrograph, the dark regions form a cellular network consisting of entangled dislocations. These local cell wall regions have very high dislocation densities. When the stacking fault energy is lower, dislocation structures typically remain more confined to a plane and cell structures are not readily formed, as shown in Fig. 9.5*c*.

The development of cyclic stress-strain curves for ductile metals enables the exploration of these materials at high dislocation densities that otherwise would be attainable only in a large strain forming operation. The dislocation structures that develop in these materials and the stress levels required to sustain the plastic hysteresis behavior are indicative of substantial cold work with a large number of entangled, immobile dislocations. As discussed in Chapters 4 and 6, the temperature dependence of the flow stresses and strain rate sensitivity in BCC metals are much greater than in FCC metals. This can also be seen pictorially in the deformation mechanism maps shown in Chapter 8, wherein the increase in the stress level for plasticity rises dramatically below $\sim 0.15T_m$. Figure 9.6 demonstrates these differences via cyclic stress-curves for FCC and BCC single crystals. The data shown here are for crystals with center stereographic triangle orientations (see Chapter 3) and with data normalized to the room-temperature Young's modulus of each material. The data are given in normal stresses because slip with a single Burgers vector, but on multiple slip planes, is often observed in BCC single crystals.

The basis for the strong temperature dependence of the cyclic stress-strain curves for BCC metals is the large difference in the mobilities of the motion of edge versus screw dislocations at low temperatures. We first saw this in Fig. 6.14*a* for tensile tests on iron crystals. At very small plastic strain amplitudes, the work hardening rates and thus the saturation stresses for BCC metals are quite low as long as there are sufficient edge dislocations to maintain the strain levels required to produce a particular level of plastic hysteresis. At these so-called *microstrain* levels, screw dislocation motion may contribute very little to the plastic deformation. Once macrostrain plastic strain levels are reached and screw dislocation motion is required, the hardening rates are very high because of the difficult and complex asymmetric motion of the screw dislocations at low homologous temperatures. The differential in the microstrain and macrostrain levels increases with decreasing homologous temperature, as shown in Fig. 9.6.

In BCC metals, the availability of multiple slip systems at cyclic microstrain levels leads to a greater number of edge dislocations that can help produce motion at relatively low strain hardening rates. Figure 9.7 shows that cyclic hardening curves at microstrain levels occur at higher stress levels because the low symmetry of the specimen orientation leads to fewer equivalently stressed slip systems. At macrostrain levels, the trend is reversed—that is, decreasing symmetry leads to lower saturation stresses, owing to the requirement that screw dislocations with different Burgers vectors cross each other when multiple slip systems are operating.

Another implication of low temperatures is that asymmetry in the motion of screw dislocations becomes more apparent. Similar to the W data shown in Fig. 4.37*b*, the data for cyclic saturation stresses at room temperature ($\sim 0.08T_m$) indicate significant asymmetry at macrostrain levels, as shown by the asymmetric hysteresis loops in Fig. 9.8. The asymmetry is clearly dependent on orientation and may be tied to how orientation affects the motion of dissociated screw dislocations. This asymmetry of slip can lead to changes in specimen shape that accumulate with increasing numbers of cycles. Initially, cylindrical crystals loaded to equal levels in tension-compression can develop ellipsoidal cross sections as a result of the plastic asymmetry of deformation even though the tensile and compressive plastic strains are kept equal.

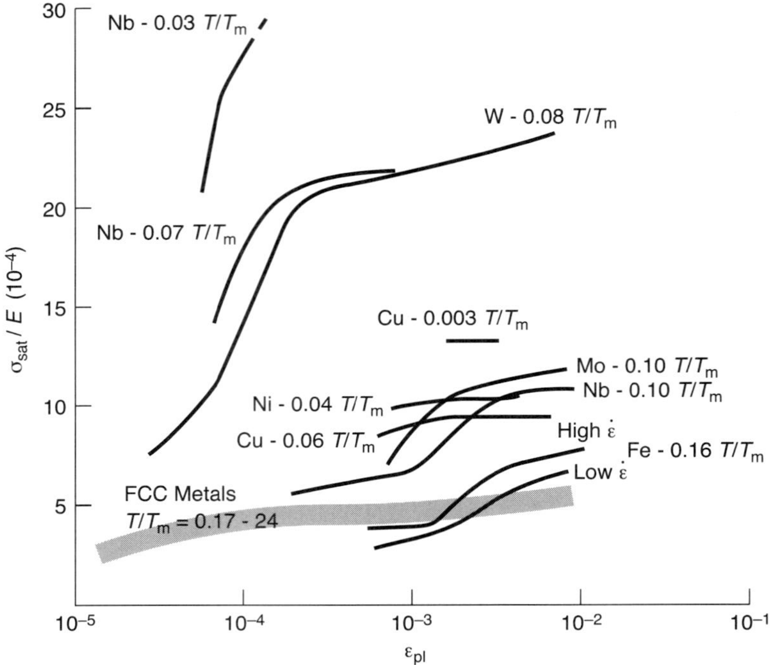

FIG. 9.6 Cyclic stress-strain curves for FCC and BCC metals with saturation stresses normalized to the room-temperature Young's modulus of each material. The FCC metals are represented by gray curves and the BCC metals are represented by black curves. The examples plotted here are for near center positions in the standard stereographic triangle of high-purity crystals. The effects of strain rate and temperature, given as the homologous temperature, are also observed. The BCC metals show a much greater dependence on plastic strain amplitude with decreasing temperature than do the FCC metals. (Bowman and Gibala, 1986, and Bowman, 1987)

In metal alloys, the stress levels are correspondingly higher with higher rates of strain hardening to the saturation level. In noncubic metals and cubic materials with limited slip systems (e.g., intermetallics), the orientation dependence of cyclic stress-strain behavior is often quite strong. In polycrystals of these slip-limited materials, hardening rates are high and cyclic plasticity levels may be possible without failure only at small plastic strain amplitudes.

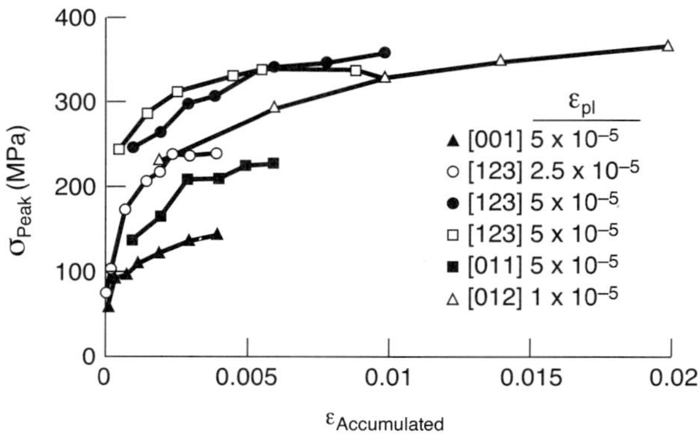

FIG. 9.7 Microstrain-level cyclic hardening curves (tensile peak stresses versus accumulated strains) for the single-crystal W tested at room temperature at the indicated plastic strain amplitudes. The orientations can be compared with the stereographic triangle shown in Fig. 4.37*a*. Decreasing symmetry of the crystal orientation results in higher saturation stress levels (Bowman, 1987).

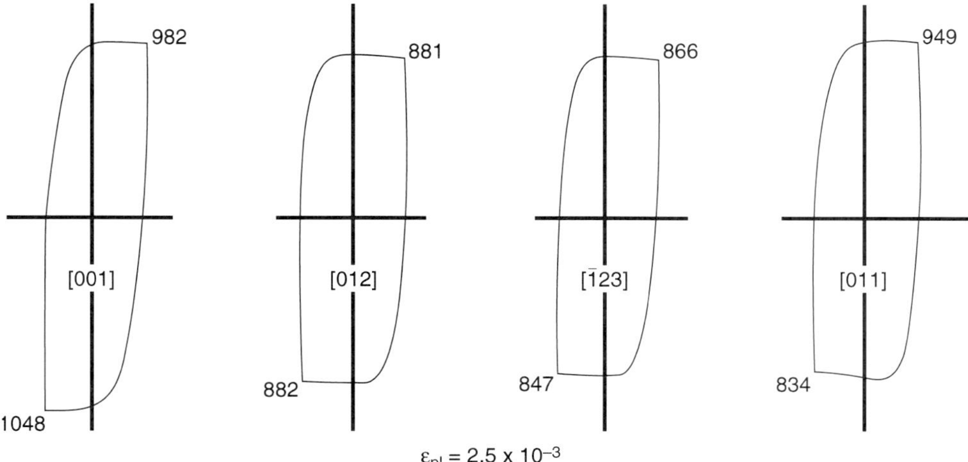

FIG. 9.8 Saturation hysteresis loops at the macrostrain level for the BCC metal W cycled to saturation at room temperature (~$0.08T_m$). The single crystal orientations used can be compared with the definition of ψ in the stereographic triangle shown in Fig. 4.37*a*. The cyclic saturation stresses (in MPa) are indicated for the tension and compression portions of the curves, suggesting that the asymmetry depends on ψ. (Bowman, 1987; see also Bowman and Gibala, 1992)

9.1.2 Cyclic Deformation in Nonmetals

In brittle materials, including brittle polymers and brittle ceramics, the absence of plasticity often precludes significant fatigue sensitivity in the absence of flaws that grow to form cracks. The lack of a nonlinear deformation mode resembling plasticity often results in a substantially elastic response for most ceramic materials up to the point at which they fail from internal flaws. Efforts to toughen ceramic materials that result in some limited degree of plasticity may increase the stress at which cracks initiate, but may lead to more rapid propagation rates once crack growth begins. This is discussed further in Section 9.2.2.

In polymers with significant ductility, fatigue failure can be initiated by damage that accumulates within the polymer. Because polymers often have strongly rate-dependent elastic and plastic properties, the fatigue responses of polymers can be affected by the loading frequency. The cyclic loading responses of polymers are often adversely impacted by adiabatic heating. Increasing temperature increases the amount of plastic deformation and the associated heating. To make matters worse, thermal transport properties of polymers, which are already low compared with those of most metals, also are reduced by increased temperature. These additive effects can cause localized weakening and failure of the cyclically loaded polymer, as shown in Fig. 9.9.

For polymers that form the dilatant damage called crazing (in transparent polymers it is observed as the formation of white creases in the material), damage occurs mostly on the tensile portion of loading. This type of dilatant behavior is shown in the yield behavior depicted in Fig. 3.5*b*. Figure 9.10*a* shows a schematic hysteresis loop, and Fig. 9.10*b* shows the successive tensile and compressive peak stresses for a polycarbonate (PC) tested at 77 K. Under these conditions, most damage occurs from the hydrostatic mean stress during the tensile part of the loading cycle.

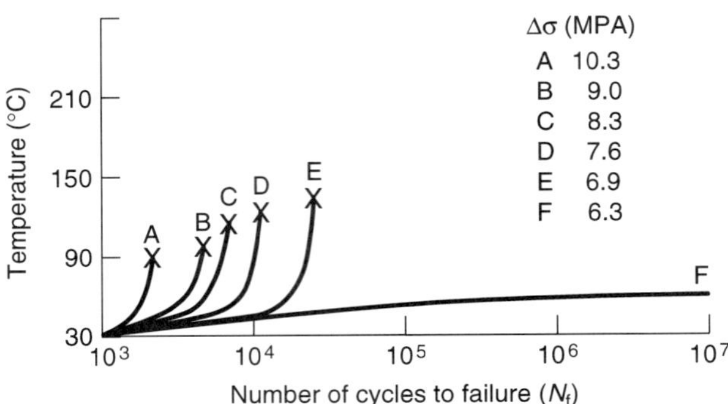

FIG. 9.9 Effect of stress magnitude on temperature rise in PTFE under stress-controlled cyclic loading. The "x" marks indicate specimen failure from localized heating (Riddell, Koo, and O'Toole, 1967, with permission of the Society of Plastics Engineers).

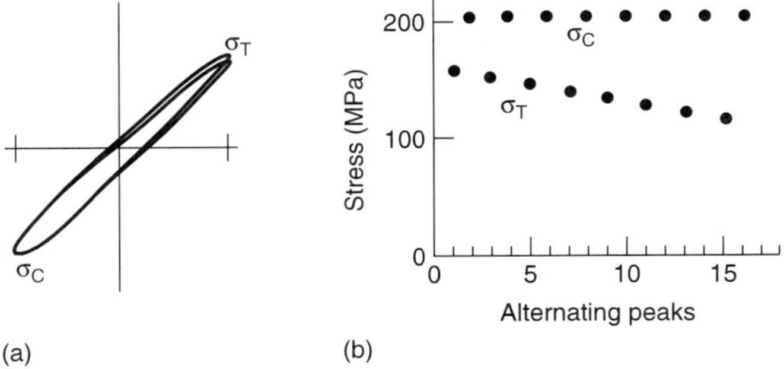

FIG. 9.10 (a) Schematic figure showing cyclic loading behavior of polycarbonate (PC) loaded at 77 K, demonstrating how crazing during tensile loading at low temperatures causes asymmetric deformation via softening of the tensile portion of the hysteresis loop. (b) The data are the peak stresses for alternating tension–compression with a controlled strain level of 0.08 in tension and compression. The horizontal-axis provides the sequence of peak stresses in tension and compression (after Rabinowitz and Beardmore, 1974).

9.2 ENGINEERING FATIGUE ANALYSIS

9.2.1 Stress-Driven Fatigue Failure Analysis

The simplest approaches to evaluating the fatigue life of a material are to load it to a fixed set of stresses, ignore the strains that it undergoes, and count the number of cycles to failure, N. Then the important parameters become the frequency, the tensile and compressive extremes of the loading cycle, the mode of mechanical loading (e.g., push-pull, bending, or rotating bending), the surface conditions of the material, and the environment to which the material is exposed. Although this is a simple technique, it can have limited value in predicting the fatigue performance of a component in service conditions different from those in the test. Service conditions can be readily emulated in a simple laboratory test stand for applications such as rotating bending of axle components. Changes can then be made in the material or the component design to prevent the type of failure shown in Fig. 1.15. Because the failure under these conditions initiates at the surface where the maximum stresses are reached, approaches used to improve the resistance to failure in rotating bending include hardening of the surfaces and inducing compressive residual stresses at the surface through shot peening. These tests are conducted under accelerated conditions, enabling comparisons of axles under different processing conditions. A schematic illustration showing rotating bending is given in Fig. 9.11.

The maximum stress (at the surface) during each rotation cycle for the loading can be plotted along with the number of cycles to failure to give final failure curves of the type shown in Fig. 9.12*a*. This type of curve is often called an *S–N* curve, as originally conceived by Wöhler. The only information that is missing is the actual position of the crack initiation curve, as shown in Fig. 9.12*b*. If we do not know when initiation occurred, we may be able to evaluate the fracture surfaces to determine if the crack initiated at a preexisting scratch or if cracks propagated from multiple initiation sites. Studying high-magnification images of fracture surfaces such as that shown in Fig. 1.15 is often critical to this type of assessment. Nonetheless, we may not be able to determine the cycle range over which propagation took place. This type of determination is sometimes possible if the ridges on the fracture surface that form on the loading and unloading cycles are resolvable, but oxidation and contact of surfaces in moving parts often make this quite difficult for service failures.

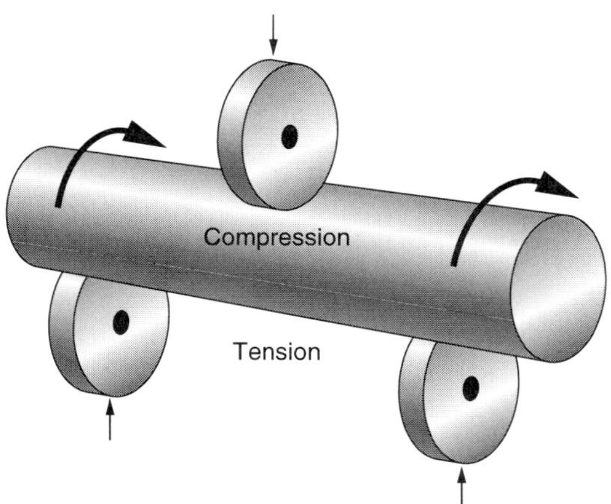

FIG. 9.11 Schematic illustration showing the rotating bending that an axle undergoes during operation. The disks contacting the surface represent loading points, which usually are rollers that also sustain the rotation. Rotating bending can occur in three-point loading, four-point loading, or even more complex geometries depending on the application and the shape of the axle.

BIOGRAPHY

AUGUST WÖHLER (1819–1914)

Wöhler is considered the first practitioner of systematic fatigue testing and designed the first stress versus life (S–N) diagrams. From 1847 to 1889, he directed Imperial Railroads in Strasbourg. In 1870, he suggested that the difference between the maximum and minimum stresses is more important to the occurrence of fatigue failures than is the maximum stress. (See Stephens, Fatemi, Stephens, and Fuchs, 2000.)

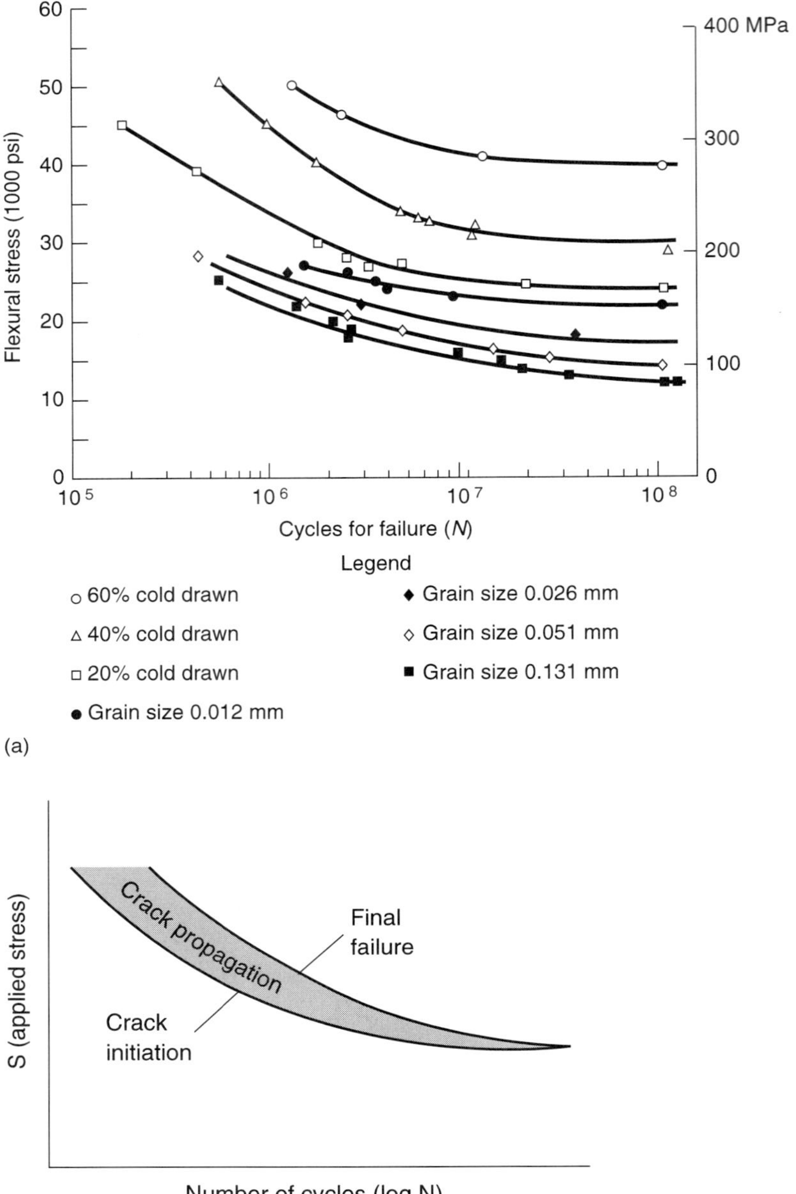

FIG. 9.12 (a) S–N curves for a series of 70Cu–30Zn brass materials subjected to different histories in fully reversed loading. The effects of cold work and grain size reduction are shown. The points on the curves represent the numbers of cycles to failure for the indicated stress levels. The points at 10^8 cycles have errors indicating a "run-out" condition wherein the test was stopped without failure. (Sinclair and Craig, 1952, used with permission of ASM). (b) Stress levels producing failure in the given number of cycles. Often labeled an S–N curve, the number of cycles to failure, N_f, on the horizontal axis depends on the applied stress levels on the vertical axis.

EXAMPLE 9.1 *Fractography of Fatigue Failures*

The study of fracture surfaces is often part of providing expert testimony for consultants in legal cases on possible fatigue failures. Careful preservation and preparation of fracture surfaces (see Hull, 1999, and ASM, 1987) combined with reasoned study enable the forensic evaluator to explain the history of the failure. Figure 9.13 shows schematic diagrams of four possible fatigue failure surfaces from rod-shaped parts that have failed in fatigue. In many fatigue failures, the rubbing together of the already fractured surfaces causes the concentric arcs of the crack front to be readily resolved on the fracture surface. Because oxidation or corrosion often attacks the fracture surface while fatigue is ongoing, the associated discoloration can be a ready marker of the fracture process, although extreme attack may actually obscure the fracture history. The initiation points are indicated in Fig. 9.13 by the dark shaded arcs originating at the part surfaces. In Fig. 9.13*a*, only one source of failure is apparent, suggesting a failure in bending, wherein the fracture initiation site is most likely the most tensile portion of the fracture, or uniaxial tension–tension or tension–compression loading. For bending, since the failure initiates on the surface, it is likely that the loading is not fully reversed. That is, it is possible that the loading cycle may include a deflection in only one direction. The actual loading information is, of course, critical to this assessment.

The diagram in Fig. 9.13*b* shows two fracture initiations that occurred approximately at the same time. Because they lie across from one another along a line through the center of the part, it is likely that the loading is in bending and is fully reversed such that propagation occurs alternately on the tensile loading of each surface.

Figure 9.13*c* shows two initiation sites that originated at different times. The difference in initiation could be a result of differences in material or surface condition at these two locations or of the eccentricity of the loading.

The diagram in Fig. 9.13*d* shows multiple initiation sites occurring along the surface, which is often observed in rotating bending (see Fig. 9.11). The final failure proceeds in an annular fashion, which is characteristic of rotating bending. Careful observation of Fig. 1.15 shows a fracture surface that apparently initiated at the surface and then proceeded to final failure in an annular pattern. ■

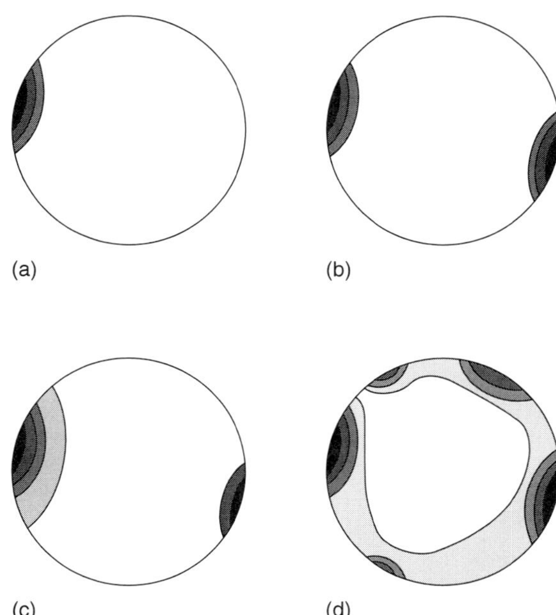

FIG. 9.13 Hypothetical fatigue failure surfaces for different conditions. (a) A single initiation site. (b) Two failure sites that initiated approximately at the same time. (c) Two failure sites that initiated at different times. (d) Multiple initiation sites followed by joining together of the initiation sites in a manner typically seen in rotating bending. The final failure process is not depicted, but its nature depends on the ductility and the type of loading.

Like any other failure process, the fatigue lifetimes of materials are strongly dependent on the monotonic deformation properties. For ductile metals, the yield strength and hardness can strongly influence failures in cyclic loading. In Fig. 9.12*a*, the higher yield strength materials have the greater resistance to failure in fatigue.

In each curve shown in Fig. 9.12*a*, there is an apparent stress level at which it might be anticipated that the material could be cycled indefinitely. Although not all materials in all conditions show this behavior, this *endurance limit* is often reported. Also, it should be recognized that the value typically given on the curve represents the average number of cycles to failure and there may have been a distribution of failures across a wide range of *N*.

Figure 9.14 demonstrates that the variables increase in number when axial loading of a material is considered. This figure shows the effects of what is called the mean fatigue stress, $\sigma_{fat,m}$, which is the average of the maximum tensile and compressive stresses in a given cycle:

$$\sigma_{fat,m} = (\sigma_{max} + \sigma_{min})/2 \tag{9.1}$$

This is a very different mean stress from the one first discussed in Chapter 2, which was the average hydrostatic stress for a material under multiaxial loading. As the maximum tensile stress becomes more tensile in nature, the likelihood of crack formation and propagation increases. This is clearly demonstrated by the data in Fig. 9.14, wherein the most compressive of the mean stresses results in the highest lifetimes for each stress level. If we define an endurance limit, it increases with the mean fatigue stress. The effect of the average stresses in addition to the stress amplitude,

$$\sigma_a = (\sigma_{max} - \sigma_{min})/2 \tag{9.2}$$

suggests that we need a way to model the effects of mean stress. The mean stress can also be described in terms of the loading ratio,

$$R = \frac{\sigma_{min}}{\sigma_{max}} \tag{9.3}$$

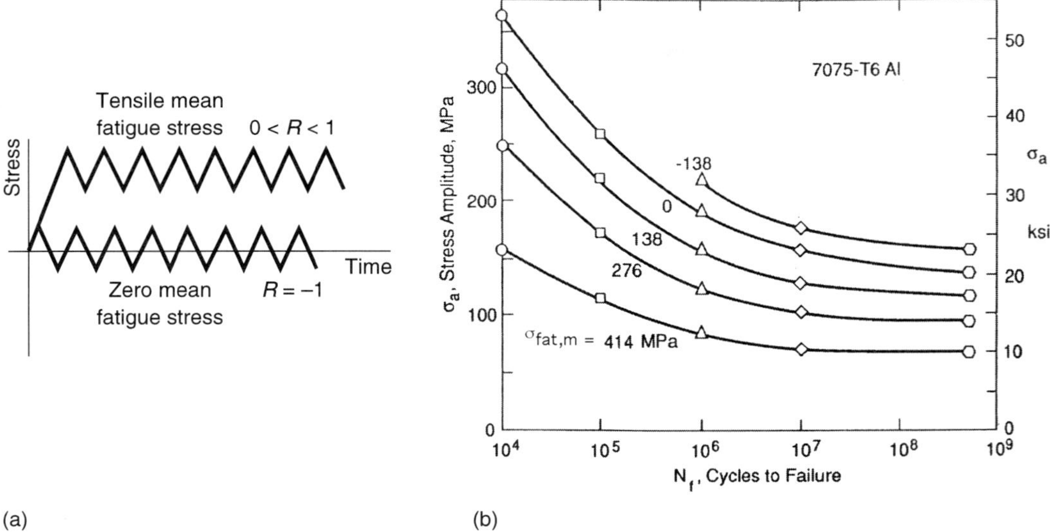

FIG. 9.14 (a) Different mean stresses in cyclic loading. (b) Tension–compression S–N curves for various mean stress levels in unnotched specimens of a high-strength Al alloy. (MECHANICAL BEHAVIOR OF MATERIALS, 1/E by DOWLING, © Reprinted by permission of Pearson Education, Inc., Upper Saddle River, NJ, data from Howell and Miller, 1955.)

For $R = -1$, the magnitudes of the tensile and compressive stresses are equal.

Because of the variety of parameters involved, several different approaches have evolved to represent the effects of stress amplitude and mean stress in combination. Models attributed to Gerber (1874), Goodman (1899), and Soderberg (1939) fulfill this purpose (see Dowling, 1993). These models allow construction of the constant life diagrams shown in Fig. 9.15*b*. The relations are written as follows:

$$\text{Gerber:} \qquad \sigma_a = \sigma_{fat}\left\{1 - \left(\frac{\sigma_{fat,m}}{\sigma_f}\right)^2\right\} \tag{9.4}$$

$$\text{Goodman:} \qquad \sigma_a = \sigma_{fat}\left\{1 - \left(\frac{\sigma_{fat,m}}{\sigma_f}\right)\right\} \tag{9.5}$$

$$\text{Soderberg:} \qquad \sigma_a = \sigma_{fat}\left\{1 - \left(\frac{\sigma_{fat,m}}{Y}\right)\right\} \tag{9.6}$$

In Eq. 9.4 to 9.6, the value of σ_a represents the stress amplitude at the endurance limit for any mean stress level, σ_{fat} is the endurance limit for a mean stress of zero in fully reversed loading, σ_f is the tensile fracture stress, and Y is the yield strength. The usual assessment of these different models is that the Gerber expression is the best choice for ductile alloys, the Goodman expression applies most often to relatively brittle metals, and the Soderberg model is simply the most conservative of the models.

Thus far, the behavior described here for fatigue has been focused on materials for which we assume that the surface preparation was ideal: specimens with smoothly polished surfaces absent of residual stresses and notches. Because fatigue fractures are often initiated at surfaces, the condition of the surface can be as important as any other characteristic of the material. Figure 9.16 shows the effects of notching the surfaces of specimens tested in rotating bending. The effects of the stress concentrations (see Chapter 7) are lower fatigue lifetimes for every stress level and a corresponding lower endurance limit.

9.2.2 Fatigue Crack Propagation

The intentional propagation and observation of existing cracks in cyclic loading defines a relatively young but very large field of mechanical property investigations. The development of materials as a result of fatigue crack propagation (FCP) research has led to an understanding of crack propagation that has enabled designers to design structures in which the lifetimes of cracks of a certain size can be predicted. This allows inspection programs

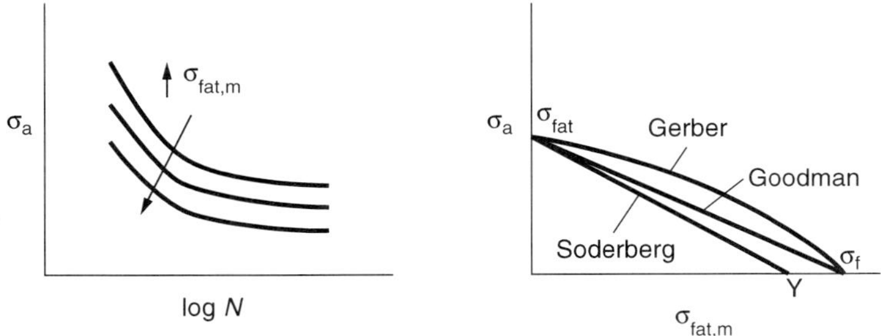

FIG. 9.15 Three models representing the mean stress effects on fatigue shown in part (a) are demonstrated in the constant life diagrams shown in part (b).

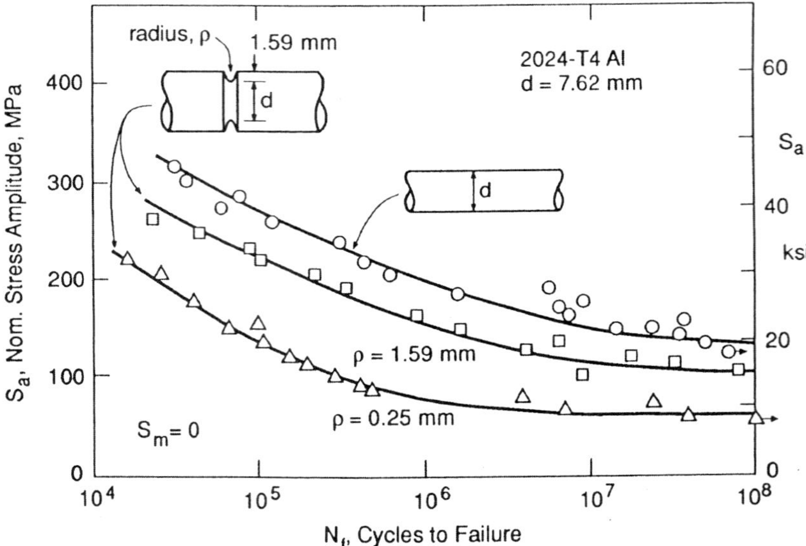

FIG. 9.16 Effects of notches with stress concentrations $k_{conc} = 1.6$ and 3.1 on fatigue in rotating bending for an aluminum alloy. (MECHANICAL BEHAVIOR OF MATERIALS, 1/E by DOWLING, © Reprinted by permission of Pearson Education, Inc., Upper Saddle River, NJ, data from MacGregor and Grossman, 1952)

that can detect cracks before they reach a size that could propagate unstably. Acceptance of the predictability of fatigue crack growth is even found among the designers of such critical components as aircraft wings and helicopter rotors. Studies of fatigue crack propagation are typically conducted in a Mode I geometry, and the testing is conducted with control of parameters such as frequency, loading spectrum, and loading level.

Testing geometries for FCP studies are often material-dependent and are usually similar to those used in fracture toughness testing. Direct Mode I fracture geometries are the simplest to evaluate, because one of the most important parts of the testing is the ability to document the increase in crack length, *da*, versus the number of cycles, *N*. If we consider the simple geometry shown in Fig. 9.17*a*, it is easy to see that crack growth is expected to occur during the tensile loading portion of the loading cycle. In fact, it is often possible to observe evidence of the crack propagation by using a microscope on the fracture surfaces, but gages that track crack growth and changes in compliance with crack propagation are also employed. Plastic or process zone shape changes that have occurred at the crack tip may be resolvable as microscale ridges on the fracture surface, as suggested schematically in Fig. 9.17*b*. In investigations of service failures, this can be a tool in estimating how many cycles were required between crack initiation and propagation. Similar to monotonic fracture processes, the fractures can be intergranular or transgranular in nature or some combination of the two cracking paths.

FCP testing is typically conducted in tension–tension loading wherein the maximum and minimum applied stresses are both tensile ($R > 1$ in Eq. 9.3). But it is the difference in the applied stresses, $\Delta\sigma = \sigma_{max} - \sigma_{min}$, that provides the driving force for crack propagation. By inserting $\Delta\sigma$ into Eq. 7.8 as the stress term, we can write the cyclic driving force for crack propagation as

$$\Delta K = \beta \Delta\sigma \sqrt{\pi a} \tag{9.7}$$

This effect of ΔK on crack propagation is shown in Fig. 9.18. In order to assess the effects of ΔK on crack propagation in a laboratory study, testing is often performed by reducing the applied stress to maintain a constant ΔK level through what is called "load shedding." Under this type of testing, it has been observed that a simple empirical expression suggested by Paris in 1961 can be fitted to the steady-state FCP behavior of so many materials that it is often called the Paris law,

$$\frac{da}{dN} = C\Delta K^m \tag{9.8}$$

where C is a fitting parameter with units that depend on the magnitude of the exponent m.[1] A log–log plot of FCP data can consist of three distinct regimes, as shown in Fig. 9.19. In addition to the linear behavior, the threshold level for the start of FCP and the catastrophic failure regime are identified in this figure.

Data supporting Paris law behavior for a wide range of materials are given in Fig. 9.20 to 9.22. In the data for cubic metals shown in Fig. 9.20, a fairly wide range of steady-state crack growth rates is observed, with one possible trend being the expected yield stresses of the materials.

In FCP behavior of engineering plastics, the strongest trend observed is that the resistance to crack growth increases with increases in fracture toughness. This is borne out by the data shown in Fig. 9.21 (which is given with crack propagation rates in millimeters per cycle). The materials that show FCP at very low values of ΔK and high rates of crack

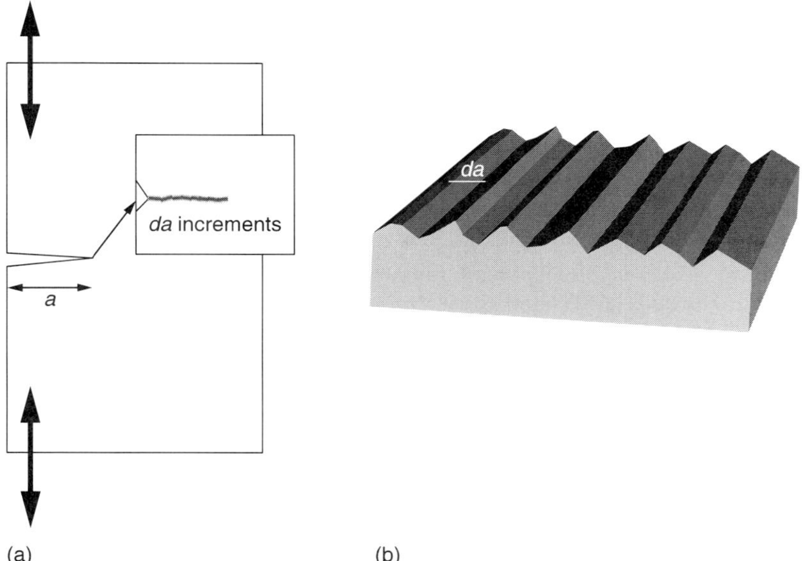

FIG. 9.17 (a) Simple testing geometry for Mode I FCP. Test specimen and testing system may require the capability for reversal to compressive stresses if this is required in the testing protocol. (b) Schematic figure showing microscopic ridges or *striations* that are often observed on fracture surfaces following failures. The ridges often form during the excursions to the minimum stress as the deformed plastic or process zone causes the fracture surfaces to come back into contact. Under laboratory conditions, constant ΔK amplitudes may result in uniform spacing, whereas in service failures, stress-controlled failures typically show increasing spacing between the striations and strain-controlled failures typically show decreasing striation spacing.

[1]This m is not related to strain hardening or strain rate sensitivity.

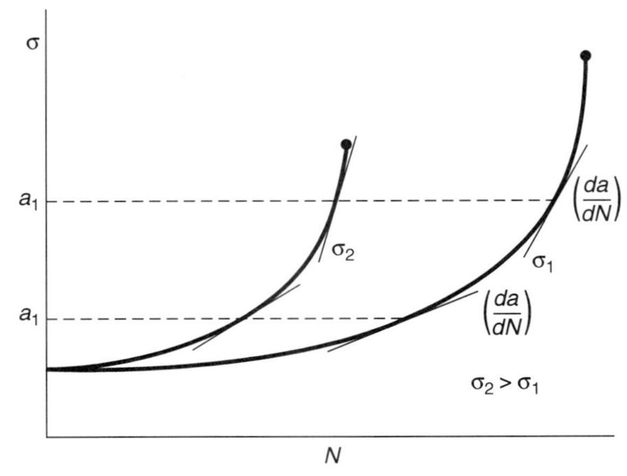

FIG. 9.18 Effects of repeated cyclic tensile loading on crack propagation in loading. Higher ΔK values from higher applied stress and larger crack length result in higher crack growth rates, da/dN.

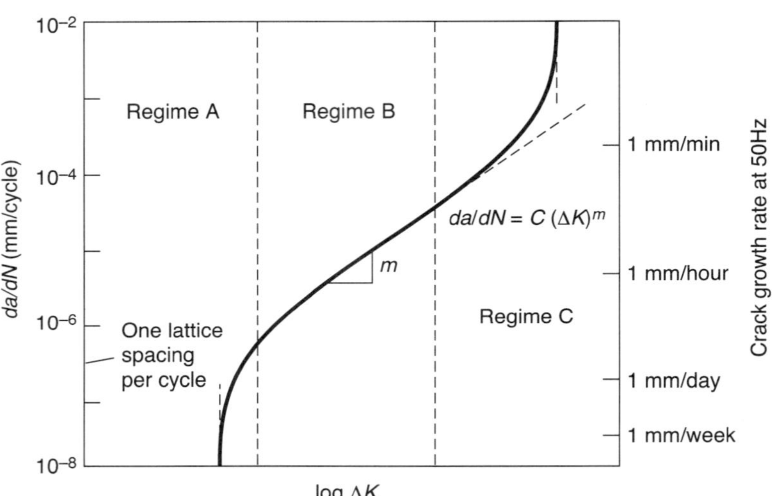

FIG. 9.19 Diagram showing the typical approach for plotting FCP data as da/dN vs. ΔK and the steady-state regime over which the Paris law is fitted. Three regimes are designated. In regime A, the threshold stress level at which FCP can be defined, the crack growth rates are at or less than one crystal lattice per cycle. Regime B is the range of crack growth rates over which the Paris law is defined. Regime C is the range in which crack growth rates accelerate to failure. (Suresh, 1999, reprinted with permission of Cambridge University Press)

BIOGRAPHY

PAUL C. PARIS (1930–)

In addition to being responsible for the most widely used expression for describing fatigue crack propagation, Paris has continued to oversee the development of education and research into fatigue crack propagation. He first conceived of relating the fatigue crack growth rate to the stress intensity factor range in 1957 while a faculty associate at the Boeing Company. The final paper was not published until 1961, because reviewers from several major publications objected to the use of the elastic-based stress intensity in a fatigue relation for ductile materials. According to Paris, the methodology he introduced means that, "Instead of analyzing structural fatigue as if the structure were perfect, you have to consider that there is always the possibility of a flaw in the structure that will grow." (see Paris, 1998)

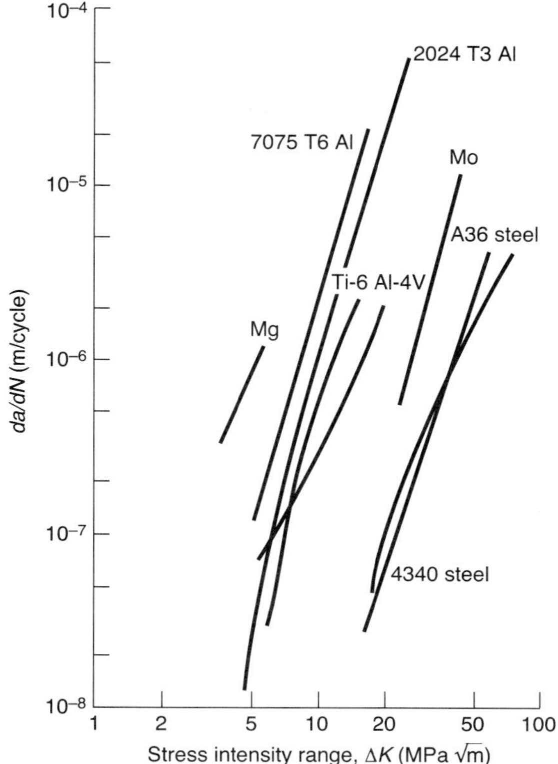

FIG. 9.20 FCP for metal alloys, demonstrating correspondence to the Paris law with $R = 0$ and da/dN given in meters per cycle. (Data for Mo, Mg, and 7075 Al; Paris, 1964. All other data from compilation by Ramsamooj and Shugar, 2001)

propagation are some of most brittle polymers available, with both PMMA and epoxies typically possessing fracture toughnesses of less than 1 MPa$\sqrt{\text{m}}$. By comparison, the fracture toughness values of PC and nylon are normally greater than 2 MPa $\sqrt{\text{m}}$.

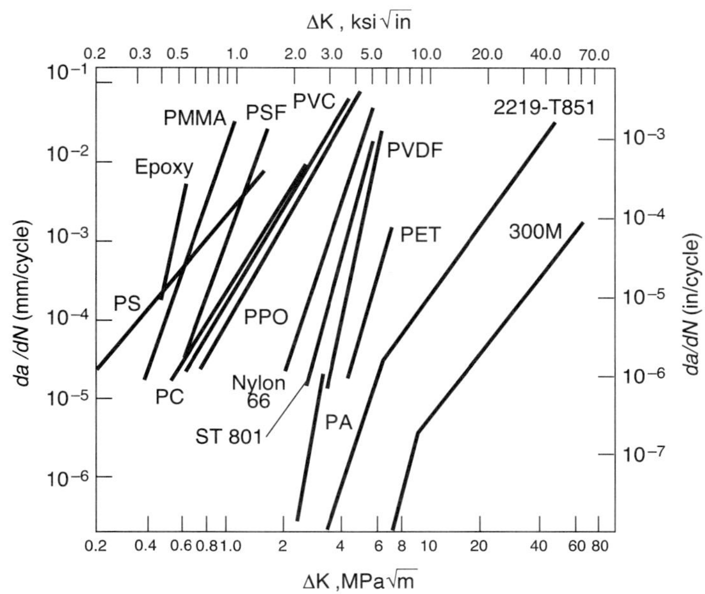

FIG. 9.21 FCP for several noncrystalline and semicrystalline polymers in comparison with aluminum and steel (Hertzberg, 1996, Wiley, used with permission of John Wiley & Sons, Inc.).

Similar to polymers, the trend for FCP in ceramics also apparently depends strongly on the toughness level, as shown in Fig. 9.22 (see Dauskardt and Ritchie, 1991, and Jacobs and Chen, 1995). Any of the mechanisms that enhance the toughening of ceramics apparently offer some advantages in terms of suppression of FCP to higher threshold levels even if they may not have the same positive impact once crack propagation levels are reached. One disconcerting impression from Fig. 9.22 is the very high m values for ceramics compared with metals. Ceramics also are susceptible to what is called *static fatigue*, which is a type of corrosion-assisted crack propagation that is called *stress corrosion*. Ceramics under tensile load can undergo crack propagation that is accelerated by attack of the material in the region of stress concentration by the environment. Certainly any stress corrosion effects will also accelerate the crack propagation in a material, as demonstrated for a steel alloy in Fig. 9.23. In Fig. 9.23, slower frequencies allow for more attack at the crack tip and higher rates of crack growth per cycle. In addition, another factor in crack growth, for loading ratios in which the crack faces are allowed to come in contact (which is the topic of the next section), is mechanical wear at the crack face.

EXAMPLE 9.2 *Fatigue Crack Propagation in Cast Iron*

Using the Paris law, let us consider the number of cycles to failure for a cast iron component subjected to repeated loading. For most cast irons, the value of m is about 4, and we will use a value of C of 2×10^{-8} (MPa $\sqrt{\text{m}}$)$^{-4}$. Our inspection procedure can only discover cracks > 1 cm in the relatively large cross section of 0.5 m. If the maximum tensile stress to which the part is loaded from zero stress is 25 MPa on each cycle ($R = 0$), and the cast iron has a fracture toughness of 15 MPa $\sqrt{\text{m}}$, for how many cycles will it last?

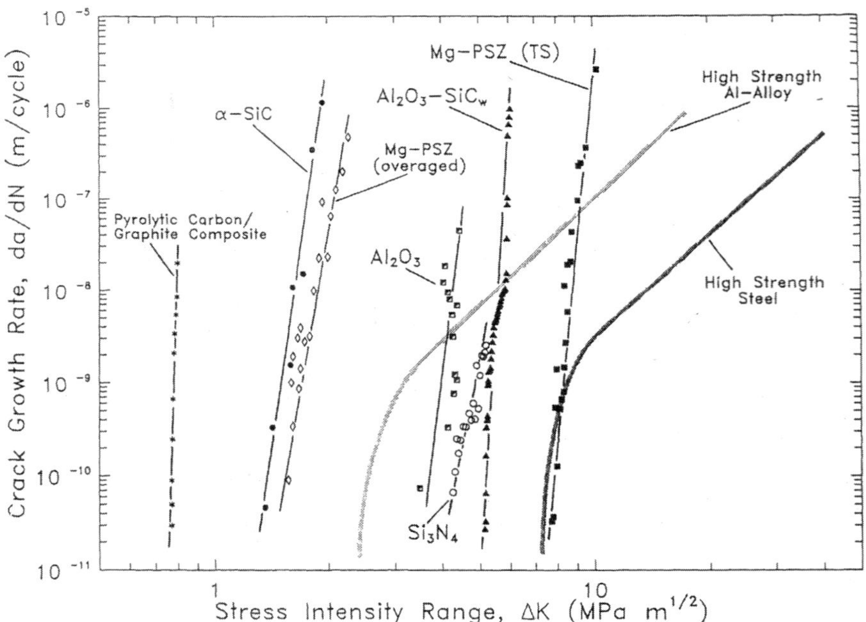

FIG. 9.22 FCP for transformation-toughened, whisker-reinforced, and single-phase ceramics and composites in comparison with data for aluminum and steel. The steady-state cyclic response of ceramics shows a much higher crack growth rate than do the metal alloys (Dauskardt and Ritchie, 1991, reprinted with permission of The American Ceramic Society, PO Box 6136, Westerville, Ohio 43086-6136. Copyright [1991]. All rights reserved. See also Steffen, Dauskardt, and Ritchie, 1991.)

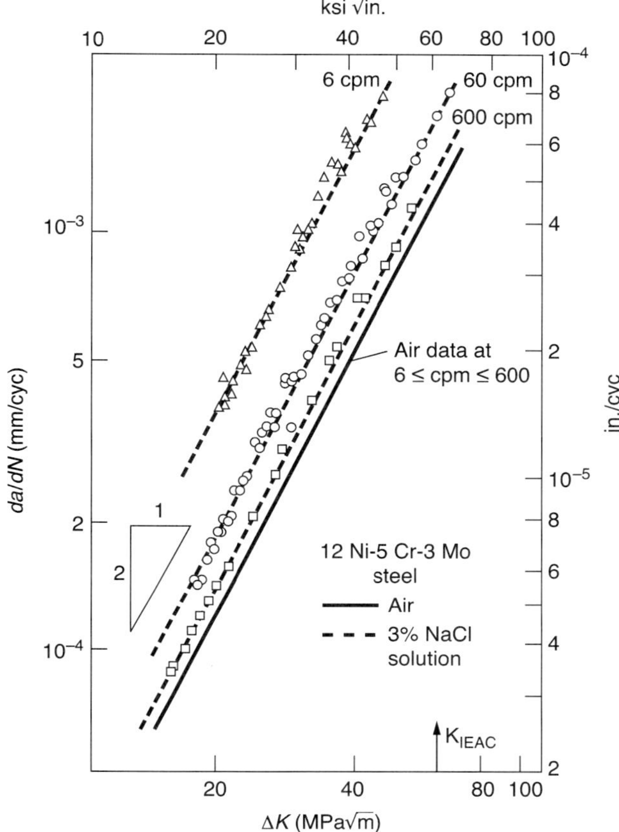

FIG. 9.23 Demonstration of corrosion fatigue by exposure to a 3% NaCl aqueous solution of a molybdenum-containing steel (Imhof and Barsoum, 1973, Copyright ASTM INTERNATIONAL, reprinted with permission). The effect of test frequency, given as cycles per minute (cpm), is that lower cycling rates in a saline solution result in more time for chemical attack and increased crack propagation per cycle.

First, we should check whether or not it will fail on the first cycle. Assuming that $\beta = 1$,

$$K = \sigma\sqrt{\pi a} = 25\sqrt{\pi 0.01}\ \text{MPa}\ \sqrt{\text{m}}$$
$$= 4.43 < 15\,\text{MPa}\ \sqrt{\text{m}}$$

So, we should be safe for one cycle. If we reverse the calculation, we can see that failure is predicted at 11 cm.

To calculate the fatigue life, we use

$$\Delta K = \Delta\sigma\sqrt{\pi a}$$

Because the remote stress level is constant, extensive crack propagation should result in an increasing ΔK value. We can write

$$\frac{da}{dN} = C\Delta\sigma^4\pi^2 a^2$$

and then

$$dN = \frac{1}{\left(C\Delta\sigma^4\pi^2\right)}\frac{da}{a^2}$$

Integration leads to

$$N = \frac{1}{C\sigma^4\pi^2}\left[\frac{1}{a_{\text{initial}}} - \frac{1}{a_{\text{final}}}\right]$$

If we use 1 cm for the initial flaw length, we expect that failure will occur at the final crack length of 11 cm calculated above—that is, if we allow for no safety factor (not a good idea). Using these values, we get a prediction of only 1200 cycles to failure. Remember that these are often order-of-magnitude estimates if we apply them to an application, so failure could occur much earlier. ■

9.3 WEAR, FRICTION, AND LUBRICATION

Friction and associated processes of wear often lead to failure of materials to perform well in components containing moving parts. At the same time, efficient wear processes are critically important to the material removal and polishing processes enabling preparation of components with high tolerances. The field of study incorporating friction, wear, and lubrication is usually called tribology. The character of surfaces, including roughness, residual stresses, machining damage, surface phases (oxides), and chemistry differences, all influence friction and wear processes (see Stachowiak and Batchelor, 2001, and Rabinowicz, 1995). The complexity of the topic and the wide variety of engineering and science disciplines involved in the study of tribology make it difficult to apply broadly, or even assess, concepts. Added challenges have arisen in recent years as smaller-scale mechanical devices (e.g., hard drives) have led to a need to understand friction and lubrication with lubricants applied in thicknesses on the order of a single molecule (see Khurshudov and Waltman, 2001). Computer hard drives operate with the device that reads and writes data flying on a thin aerodynamic cushion of air only nanometers above the magnetic media surface. The disk may be moving by at speeds of more than 7000 revolutions per minute. Startup, shutdown, contact with dust particles, and shock loading can all lead to contact between the head and the disk surface. Hard carbon coatings protect the magnetic material that stores the data, and very thin lubricants are used to limit the friction when the head comes in contact with the disk surface.

The true area of contact, the relative values of shear and normal loads, and the environment affect a material's performance in wear applications. The volume of material removed in a wear application or test is generally proportional to the load and sliding distance. Of all the mechanical properties, hardness has the greatest effect on wear. This is the reason that materials designed for material removal processes are some of the hardest materials available (e.g., diamond). Hardness is most often inversely proportional to wear as long as the friction coefficient does not change significantly.

The Archard model for adhesive wear is often employed as a general rule of thumb for wear. We can express the frictional wear rate W_f as

$$W_f = \frac{\text{Volume removed}}{\text{Sliding distance}}$$

Then for low loads, expressed as a pressure P, the frictional wear rate is

$$W_f = K_{Ar}AP \tag{9.9}$$

where K_{Ar} is the Archard constant and A is the nominal contact area. The Archard constant defined here is inversely proportional to the hardness defined in units of stress. Although originally based on an adhesive model of friction and wear, the frictional wear rate given above is reduced if the zone of plastic deformation near the surface is reduced by using a harder material.

9.3.1 Contact of Surfaces

The topography of all surfaces includes some roughness that affects the true area of contact between two mated surfaces. As the contact stress is increased, the contact area increases as

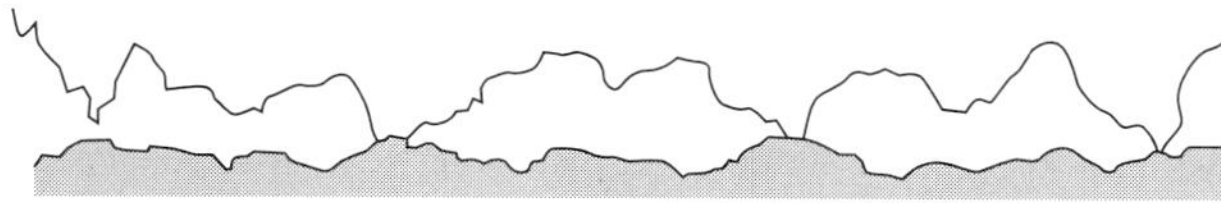

FIG. 9.24 Asperities in contact between two different surfaces. The upper surface has a higher degree of roughness than the lower one.

elastic strain occurs at the contact points of the surface, as shown in Figure 9.24. Furthermore, additional asperities become contact sites. At high enough contact stresses, plastic deformation of these contact points will occur. The effective contact area is often given as

$$A_c \propto \frac{\sigma_{\text{contact}}}{Y}$$

where σ_{contact} is the compressive stress defined using the nominal area and Y is the yield stress.

For dry (without lubricant) sliding conditions, wear rates at steady state are not strongly influenced by surface finish, although the initial wear rates *are* affected by surface finish. As a hard material slides across a soft material, the rate of material removal from the soft material is increased for rougher surfaces. Because surface roughness changes during continued sliding, the initial surface roughness often becomes of modest importance to long-term wear processes.

9.3.2 Friction

Friction is the resistance to sliding of one body over another. The amount of friction between the surfaces undergoing relative motion can be affected by

1. Contact stress
2. Surface topography
3. Elasticity and plasticity of the materials
4. Temperature and lubrication
5. Humidity
6. Sliding speed
7. Vibration

And Leonardo da Vinci defined two laws of friction:

1. The friction force is proportional to the load.
2. The frictional properties are independent of the apparent contact area.

Both of these laws hold pretty well, with the exception that the first does not apply very well to polymers at or above their glass transition temperatures because of the very low yield stresses and the rate dependence of their properties. Because the surfaces in contact may also change over the time of contact, it is difficult to know a great deal about expected values of friction without testing the materials in simulated service conditions.

The coefficient of friction μ_F is often defined as the ratio of interfacial shear stress (or force) required for motion to contact stress (or force)

$$\tau = \mu_F \sigma_n \tag{9.10}$$

where τ is shear stress and σ_n is the normal stress. This proportional relationship can be used as a rule of thumb. Friction coefficients are often dependent on load as well as on sliding velocity and temperature.

For static friction, there is a critical shear stress at which deformation proceeds. For sliding or kinetic friction, the value of μ_F is lower and modestly dependent on sliding speed. Rolling friction, for bearings, rollers, or wheels, is dependent on their geometry, elastic stiffness, and rolling speed. In most cases, the rolling friction coefficient is much less than those for static or sliding friction. Both static and sliding friction are strongly influenced by lubrication, with both liquid and solid lubricants providing dramatic reductions in friction. An example of this effect is shown in the metal-on-metal friction coefficients shown in Fig. 9.25. Table 9.1 lists values of μ_F for a number of materials (see Rabinowicz, 1995).

The overall transition from static to sliding friction can be expressed in the elastic–plastic terms of yielding. The initial small displacement includes elastic deflection of asperities and the displacements required to reach asperity-to-asperity contact that resists sliding. To initiate sliding, a critical load level is required. In the absence of creep processes, this behavior is not time-dependent. Until sliding takes place, the stress resisting the overall motion of the object is equal to the applied force, and the component will undergo proportionate elastic displacements. When sliding occurs, it is apparent that the force resisting motion is less than the force producing it.

We can explore a simple model that is most applicable to ductile metals. If we assume that yielding can take place at the contact points shown in Fig. 9.24, then the relative shear strength of soft metal interface material, τ_i, and the shear strength of the material in contact with it, τ_o, will determine the coefficient of friction. If the two materials have the same shear yield strength, then the coefficient of friction will become very large. A simple expression that satisfies this is as follows (see Hutchings, 1992).

$$\mu_F = \frac{1}{2\left(\frac{\tau_o}{\tau_i} - 1\right)^{1/2}} \tag{9.11}$$

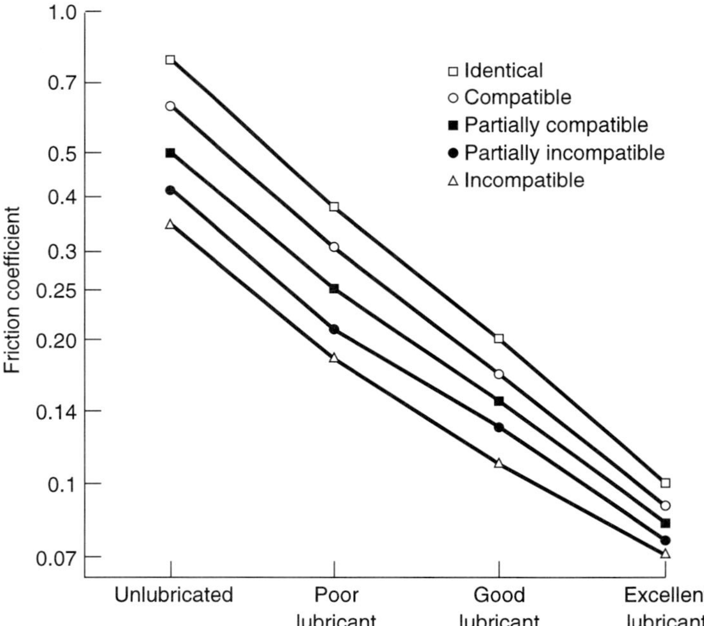

FIG. 9.25 Metal-on-metal friction coefficients and the empirical effects of lubrication (Figure 4.57, p. 119, Rabinowicz, 1995; Wiley, used with permission of John Wiley & Sons, Inc.).

TABLE 9.1 Coefficients of Friction for Sliding (Room Temperature)*

Materials and conditions	Coefficient of friction, μ_F
Rubber on harder materials	0.5–1
Leather on wood	0.3–0.5
Wood on wood (dry)	0.2–0.5
Metal on metals (dry)	0.15–0.2
Greased smooth surface	0.05–0.08
Teflon on harder materials	0.05–0.1
Graphite on other materials	0.08–0.15
Hard materials on ice at −20°C	0.2–0.3
MoS_2 on other materials	0.2–0.1
Hard materials on ice at 0°C	0.01–0.1
Hydrodynamic film lubrication	0.001–0.01
Rolling contact of hard materials	0.001 or less

* All examples include an assumption of atmospheric conditions.

This relation is consistent with the observation that soft metal layers are successful in reducing the coefficient of friction in bearing applications. It also readily provides an understanding of why liquid lubricants, which by definition have almost no ability to sustain a shear stress, can be so successful. If the interface material has about half the strength of the contact material, the coefficient of friction is 0.3.

At high loading levels in solid–solid interfaces, sliding friction proceeds by plowing of the surface asperities. Subsequent evolution of friction with sliding distance often exhibits one of the two three-stage behaviors shown in Figure 9.26 (see Suh, 1986). These can be summarized as follows.

1. For most metals, adhesion does not begin right away, because the surfaces are normally contaminated with oxides or other compounds that are strongly bonded to their own substrates. With lubrication, the first stage of frictional motion may remain until or unless the lubricant degrades.
2. Without lubrication or after the properties of the lubricant have been degraded, the coefficient of friction will rise once wear of the original surfaces brings uncontaminated materials into contact. A steep increase in friction coefficient occurs as a result of adhesion and the formation of wear particles from fractured asperities. Strain hardening of the surface material and fracture from cyclic fatigue also add to the wear debris.

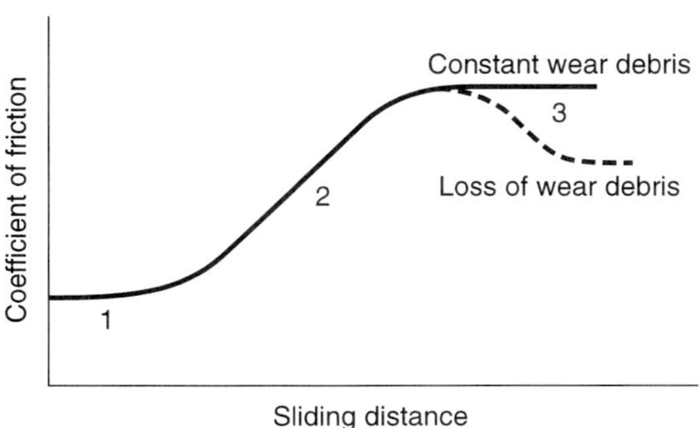

FIG. 9.26 Friction evolution with sliding distance (after Suh, 1986).

3. (a) If the surface roughness does not change significantly, the friction level can continue almost indefinitely, or (b) if there is interfacial transfer and smoothing of the surfaces, it is also possible for the coefficient of friction to decrease as a result of a reduction in wear particle formation.

9.3.3 Lubrication

Lubrication, and the resulting reduction of friction coefficients, are most often associated with fluids. In that context, lubrication becomes more of a fluids problem. Solid lubricants are less well known but are very important for applications wherein contamination by a liquid (e.g., in vacuum) is unacceptable. Mechanisms of solid lubrication also help explain the nature of friction. Two primary methods of solid lubrication are applied: soft films and layered-structure materials.

When a soft, adherent film of material is applied to a relatively hard substrate, the combination of material properties can lead to low coefficients of friction. If the soft layer is sufficiently thin, the stiffness and high strength of the substrate keep the contact area small. The low shear strength of the soft film provides a low resistance to asperity deformation. If the soft film is too thick, this synergistic benefit is lost.

Highly anisotropic, layered-structure materials with weak interlayer bonding—e.g., graphite, MoS_2, and polytetrafluoroethylene (PTFE)—lubricate by the sliding of strongly bonded layers. The bases for their effectiveness as lubricants are as follows.

1. Graphite: The principle of graphite lubrication was originally expressed by Bragg in 1928. Weak bonding between basal planes in these hexagonal crystals leads to separation between basal planes and the ability of these planes to slide over one another with very little resistance. The low surface energy of basal planes may in part result from adsorbed hydroxides for graphite, because graphite has a larger friction coefficient under vacuum.
2. Molybdenum disulfide (MoS_2): MoS_2 is a planar crystal similar in nature to graphite. MoS_2 can be readily applied by vacuum deposition techniques. It has often been used in space applications.
3. Teflon (PTFE): The structure of PTFE is functionally similar to those of graphite and MoS_2 relative to wear, although the actual crystal structure and size scale are different. PTFE is normally semicrystalline. The crystals are normally platelike and are separated by weakly bonded noncrystalline regions. These layered materials lubricate hard materials by transfer of oriented layer fragments to form an interfacial zone of easy sliding.

9.3.4 Wear

Defining wear as a quantity is fairly difficult, although clearly it is an important aspect of performance for moving parts and is critical to machining processes. We can define wear in terms of the volume V of material removed as

$$V = kLx/p \tag{9.12}$$

where k is the dimensionless quantity called the wear coefficient, which is related to the probability of forming a wear particle. The other terms in Eq. 9.12 are the load L, the sliding distance x, and the hardness p of the material being worn away. This simple expression is often used as a comparison of how materials might wear against one another, and the wear coefficient is often cited for material combinations. As might be expected, the wear coefficient is related to the coefficient of friction between the two materials. Figure 9.27 shows the relationship between friction and wear for various conditions.

We can define several types of wear based on the mechanisms behind them (see Suh, 1986, and Stachowiak and Batchelor, 2001). Each of these wear mechanisms can appear alone or in combination with the others, but we can consider the following classifications.

Abrasive Wear The sliding of particles against a surface results in abrasive wear. Although most illustrations of abrasive wear incorporate a hard material as the abrader, even relatively soft materials can contribute to abrasion of an otherwise hard material. Even the erosive wear of a moving fluid and associated cavitation of bubbles can result in material removal. Mechanisms behind abrasive wear include cutting, fracture, cyclic deformation, and grain separation (grain pullout). The function of many surface finishing processes is based on abrasive wear.

Adhesive Wear When chemical bonding between mated materials is sufficiently high, adhesive wear can make important contributions. With the exception of gold and platinum, most metals exposed to ambient conditions have a thin oxide film at least several nanometers thick. Silver loses its protective oxide at temperatures above 200 to 300°C in air, but when cooled to room temperature, the surface film returns. Adhesive bonding between a metal surface and any other surface is limited unless this oxide is removed. For

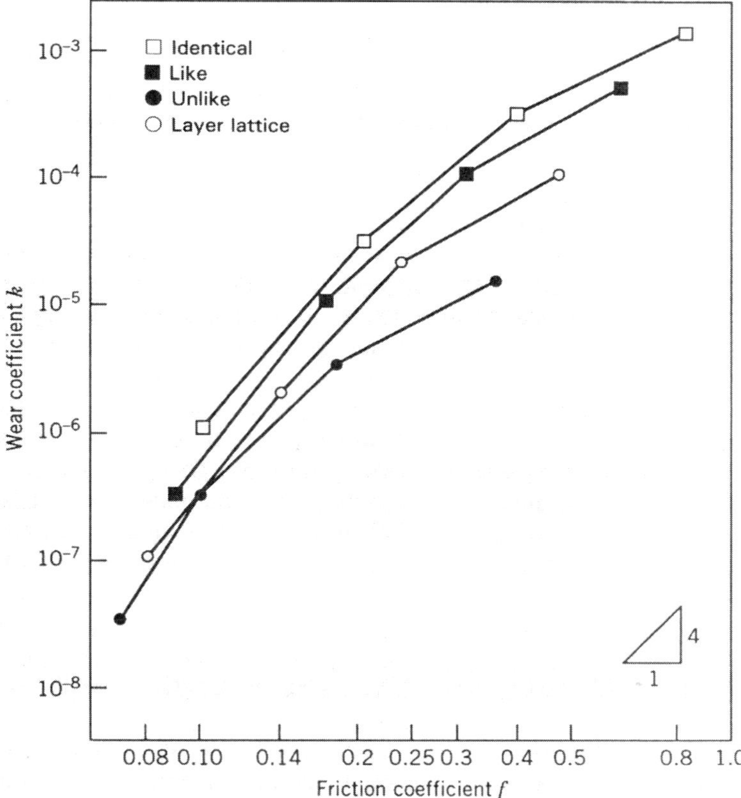

FIG. 9.27 Wear coefficient versus friction coefficient for metals in several conditions (Figure 6.2, p. 167, Rabinowicz, 1995, Wiley, used with permission of John Wiley & Sons, Inc.).

ceramics, the strong bonding and a tendency for surface impurity segregation prevent adhesive bonding unless high temperatures are employed. For polymers, the occurrence of adhesive wear is entirely dependent on the presence of favorable bonding sites.

Metal–polymer adhesion has been demonstrated, but transfer of the polymer does not occur unless the polymer has reactive nonmetals (e.g., fluorine in PTFE) as one of its constituents. Otherwise, van der Waals forces do not result in significant adhesion, as observed in crystals of mica. Metal–ceramic combinations can also result in adhesion, but normally the metal is transferred to the ceramic. Coherent transfer films are obtained with soft metals and alloys, but hard materials often form discontinuous transfer films and debris. Silicon-containing ceramics used for machining of steel or cast iron can undergo wear from the strong chemical affinity between silicon and iron at elevated temperatures.

Adhesion among polymer, ceramic, and polymer–ceramic combinations is usually based on weak or secondary bonding. Therefore, many favorable low-friction combinations are possible. In orthopedic applications, ceramic–polymer wear systems have demonstrated low friction and high wear resistance. In this application, thin polymer inserts serve as interfacial materials enabling solid lubrication. When used in artificial knees and hips, these combinations can be quite effective.

Fatigue Wear Fatigue-based wear is wear that forms through the propagation of fatigue cracks initiated and propagated by the repeated loading processes discussed at the beginning of this chapter. Unlike plastic plowing, the process of damage occurs through crack initiation followed by propagation and failure. The initial surface of contact in repeated or cyclic loading is smoothed of asperities in the initial wear cycles. Much of the damage can occur below these contact surfaces. Two possibilities that lead to most fatigue wear prevail. One is that a crack forms within highly deformed material at or near the surface and propagates to form wear debris. The second is that a subsurface crack forms and leads to debris formation by delamination. These voids are nucleated by preexisting flaws or inclusions, because heterogeneous elastic stresses occur.

9.4 REFERENCES

Fatigue References

ASM Handbook, Vol. 12, *Fractography,* 1987.

K. J. Bowman, Ph.D. Thesis, University of Michigan, 1987.

K. J. Bowman and R. Gibala, *Scripta Metallurgica,* **20**, 1451–1454, 1986.

K. J. Bowman and R. Gibala, *Acta Met.,* **40**, 193–200, 1992.

N. E. Dowling, *Mechanical Behavior of Materials,* Prentice Hall, 1993.

R. W. Hertzberg, *Deformation and Fracture Mechanics of Engineering Materials,* 4th Ed., Wiley, 1996.

M. Hollmann, J. Bretschneider, and C. Holste, *Cryst. Res. Technol.,* **35 (4)**, 479–492, 2000.

F. M. Howell and J. L. Miller, *Proc. ASTM,* **55**, 955–968, 1955.

D. Hull, *Fractography,* Cambridge University Press, 1999.

E. J. Imhof and J. M. Barsoum, *Proceedings of the 6th ASTM National Symposium on Fracture Mechanics,* ASTM STP 536, ASTM, 1973.

D. S. Jacobs and I.-W. Chen, *J. Am. Ceram. Soc.,* **78 (13)**, 513–520, 1995.

C. W. MacGregor and N. Grossman, NACA TN 2812, National Advisory Committee for Aeronautics, Washington, DC, 1952.

F. McClintock and A. Argon, *Mechanical Behavior of Materials,* Addison-Wesley, Reading, MA, 1966.

J. D. Morrow, *Internal Friction, Damping, and Cyclic Plasticity,* ASTM STP 378, ASTM, 1965.

M. Murayama, K. Hono, H. Hirukawa, T. Ohmura, and S. Matsuoka, "The combined effect of molybdenum and nitrogen on the fatigued microstructure of 316 type austenitic/stainless steel," *Scripta Materialia,* 41, 467–473, 1999.

P. Paris, *Fatigue–An Interdisciplinary Approach,* Proceedings of 10th Sagamore Conference, Syracuse University Press, 1964.

P. C. Paris, *Fat. Fract. Eng. Mat. Struct.,* **21 (5)**, 1998.

S. Rabinowitz and P. Beardmore, *J. Mater. Sci.,* **9**, 81, 1974.

D.V. Ramsamooj and T.A. Shugar, *International Journal of Fatigue,* **23** (Supplement) S287–S300, 2001.

M. N. Riddell, G. P. Koo, and J. L. O'Toole, *Poly. Eng. and Sci.,* **6**, 363–368, 1967.

G. M. SINCLAIR AND W. J. CRAIG, *Trans. ASM,* **44**, 929–948, 1952.
A. STEFFEN, R. DAUSKARDT, AND R. O. RITCHIE, *J. Am. Ceram. Soc.,* **74 (6)**, 1259–1268, 1991.
R. I. STEPHENS, A. FATEMI, R. R. STEPHENS, AND H. O. FUCHS, *Metal Fatigue in Engineering,* 2nd Ed., Wiley, 2000.
S. SURESH, *Fatigue of Materials,* 2nd Ed., Cambridge Solid State Series, Cambridge Press, 1999.

Friction and Wear References

I. M. HUTCHINGS, *Tribology: Friction and Wear of Engineering Materials,* Arnold, 1992.
A. KHURSHUDOV AND R. J. WALTMAN, *Wear,* **251**, 1124–1132, 2001.
E. RABINOWICZ, *Friction and Wear of Materials,* 2nd Ed., Wiley, New York, 1995.
G. W. STACHOWIAK AND A. W. BATCHELOR, *Engineering Tribology,* Butterworth-Heinemann, 2001.
N. P. SUH, *Tribophysics,* Prentice-Hall, Englewood Cliffs, NJ, 1986.

9.5 PROBLEMS

A.9.1 The hysteresis loops in Fig. 9.2 for copper and the hysteresis loops for tungsten in Fig. 9.8 have several important differences. Please summarize them.

A.9.2 The points A to E in Fig. 9.9 show a trend that depends on the stress level for loading. Please explain how this is related to adiabatic heating of the material.

A.9.3 How does hydrostatic stress (Chapter 2) play a role in the behavior shown in Fig. 9.10?

A.9.4 Why does surface hardening or compressive surface residual stress enhance the fatigue resistance of materials tested in bending?

A.9.5 Describe the effect of strengthening mechanisms on the S–N curves in Fig. 9.13.

A.9.6 Describe how load level may affect the contact of asperities shown in Fig. 9.24. Show a sketch that describes the progression of contact for the asperities.

A.9.7 What four factors control the magnitude of the friction coefficient between two surfaces?

B.9.1 Calculate the values of the accumulated or total plastic strain for each of the four cycle numbers given in Fig. 9.1, assuming that $\varepsilon_{pl,max} = |\varepsilon_{pl,min}| = 0.001$.

B.9.2 Plot the expected three cyclic hardening or softening behaviors depicted in Fig. 9.2. Your plot should consist of plastic strain amplitude versus number of cycles for these total strain controlled tests. For these same conditions, sketch your expectation for cyclic behavior of these materials if the testing was conducted in plastic strain control. For these plots, the maximum stress should be plotted versus the number of cycles.

B.9.3 For the test data in Fig. 9.7, calculate the number of cycles to reach the last data point in each curve for the given plastic strain amplitudes.

B.9.4 Using the stresses and cycles to failure given in Fig. 9.9, plot an S–N curve for PTFE.

B.9.5 Additional testing of the 40% cold drawn brass in Fig. 9.12 has indicated a yield stress of 400 MPa and a tensile strength of 590 MPa. If plans are to use the alloy in alternating stresses of 300 MPa and –200 MPa, would you expect this material to fail in fatigue? Plot a diagram to demonstrate your answer.

B.9.6 Find the values of C and m in the Paris law for 4340 steel from Fig. 9.20.

B.9.7 Find the values of C and m in the Paris law for the following materials shown in the figures of this chapter:

(a) Epoxy

(b) PMMA

(c) Nylon 66

(d) α-SiC

B.9.8 Using the data in Fig. 9.10, sketch the expected hysteresis loops for the first three cycles to scale.

B.9.9 One reason why Teflon (PTFE) and graphite offer very low sliding friction coefficients in contact with other materials is transfer of these materials to the other surface. What implications does this have for wear? Could this be a reason that Teflon is not a very good choice for wear surfaces in medical implants?

C.9.1 The discussion of test specimen orientation and symmetry suggests that the cyclic hardening behavior is different at microstrain and macrostrain levels for BCC metals. Discuss this relationship, and indicate which data in the provided figures support the claimed affect of test specimen symmetry.

C.9.2 Given the plastic hysteresis loop for the $[\bar{1}23]$ oriented W crystal in Fig. 9.8 and assuming a mobile dislocation density of 10^{11} m/m^3, estimate the average distance a dislocation must move to produce the strain shown.

∫ The author forgot to plot the Ni data from Fig. 9.5 in Fig. 9.6 to see if they agree with the normalized data. Convert from shear stress and shear strain (assuming a favorable Schmid factor) and perform the normalization to see if these data are consistent with the prior investigations plotted in Fig. 9.6.

C.9.4 Using the data given in Fig. 9.14, plot data for the method used for the three mean stress relations. Of the three models given in the text, which one do you believe is likely to fit the data best?

C.9.5 For the PVDF and the 2219 Al alloy shown in Fig. 9.21, calculate the Paris law parameters. For each material, calculate the number of cycles to failure for $\Delta\sigma = 10$ MPa, $\beta = 1$, and an initial crack length of 0.2 cm if the material has the following properties:

PVDF	$K_{Ic} = 1.2\ \text{MPa}\sqrt{\text{m}}$
2219 Al	$K_{Ic} = 19\ \text{MPa}\sqrt{\text{m}}$

C.9.6 Plot Eq. 9.11 for shear strength ratios from 0 to 1. Describe the implications of this relationship on the expected coefficient of friction. Are all values reasonable?

CHAPTER 10

DEFORMATION PROCESSING

THE DEFORMATION PROCESSES of forging, extrusion, drawing, and rolling are classic examples of forming operations critical to producing materials in final shapes or as precursors to subsequent forming processes. In the original development of these processes, the technical study of how they work and how they could be improved followed well after useful parts were being produced. Currently, commercial software is often available that applies many of the concepts explored in this book to forming and shaping operations. In addition, many additional processes, including some with exotic names such as stretch blow molding and hydroforming, have been developed in recent years for shaping of materials, with particular emphasis on polymers and other materials that can have strong temperature and rate dependencies. The presentation here is designed to offer some relatively simple tools for understanding forming processes and background information on common forming practices. For much more extensive coverage, the reader is encouraged to consult the presentations given by Hosford and Caddell (1993) and Wagoner and Chenot (1997). Although the foundations for this chapter are established in the descriptions of deformation in Section 2.4 and the discussions of yield criteria and plastic anisotropy in Chapter 3, nearly all of the topics discussed in the preceding chapters can impact success or failure in forming operations. (See also Johnson and Mellor, 1962.)

The number of forming operations available for shaping of materials is so extensive that a listing could go on for several pages. Squeezing, stretching, and bending operations take many forms and are strongly dependent on the type of material. As suggested above, the study of the science behind the forming operations lags quite far behind successful forming practice. In metal forming, traditional forming operations such as rolling, extrusion, and drawing are often preliminary to secondary forming operations such as bending, stamping, hydroforming, deep drawing, and ironing.

Rolling The process of flat rolling of materials shown in Fig. 2.9 is one of the most important ways to produce thin sections of materials and in particular structural metals. From the material bent to form the quarter-panels and deck lids of automobiles to the shaped structural sections that provide the underlying structural supports, coils of rolled products are continuously fed into the receiving gates of automobile manufacturing plants. Rolled products are also used in a wide array of other consumer products, including aluminum foil, but also serve as the precursor materials for food and drink cans. Rolling conducted at elevated temperatures is called *hot rolling*, and rolling conducted on metals at temperatures at which cold work takes place is called *cold rolling*.

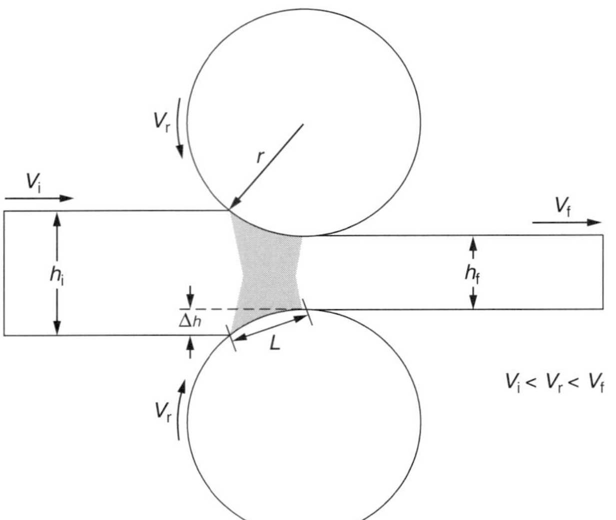

FIG. 10.1 Detailed rolling geometry showing two rolls moving in opposite directions to deform a slab of material to a height reduction of Δh. The approximate contact length of the rolls is L. As shown in this diagram, the plastic deformation takes place principally in the *roll bite*, which is shaded gray. The increase in velocity from the entrance to the exit is shown along with the intermediate velocity.

The contact length for the near plane strain compression conditions in the *roll bite* is given by the approximation $L \cong \sqrt{r\Delta h}$, as shown in Fig. 10.1. About midway in this length is a single position, called the neutral point, that indicates the line of contact wherein the roll surface velocity v_r is equal to the velocity of the material. The transition in relative velocities is shown in Fig. 10.2*a*. Because of the relative motion of the material with respect to the rolls, the frictional forces develop as illustrated in Fig. 10.2*b*, which shows a pressure maximum at the neutral point along what is called the *friction hill*. The friction hill can be described by the pressure function

$$P_{\text{roll}} = \frac{h}{\mu_F L}\left(\exp\frac{\mu_F L}{h} - 1\right)\sigma_o \tag{10.1}$$

where h is the average height of the material in the roll bite, $(h_i - h_f)/2$, μ_F is the frictional coefficient (discussed in Chapter 9), and σ_o is the average plane strain flow strength, $2k$, or the shear yield strength as defined by the Tresca relationship discussed in Chapter 3. If work hardening is considered, the value of the average flow strength is just $(2k_{\text{initial}} + 2k_{\text{final}})/2$, where $2k_{\text{initial}}$ is the initial flow strength and $2k_{\text{final}}$ is the strength following the rolling process. The deformation in the roll bite is nearly plane strain owing to the constraint provided by the material before and after the material within the roll bite, as shown in Fig. 10.3 and demonstrated in Example 10.1.

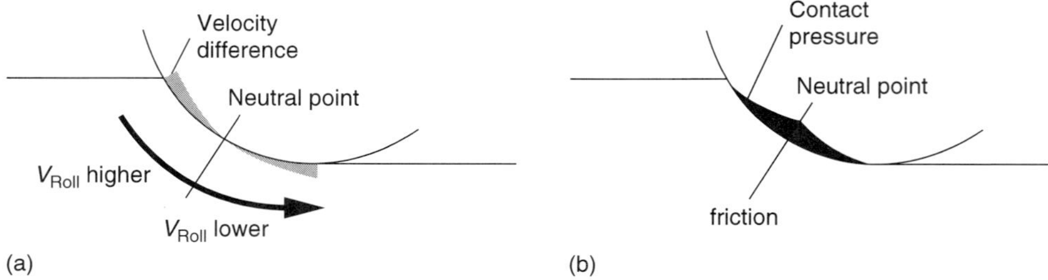

FIG. 10.2 (a) Schematic velocity differentials and neutral points in the roll bite. (b) Schematic pressure differentials along the roll–material contact length.

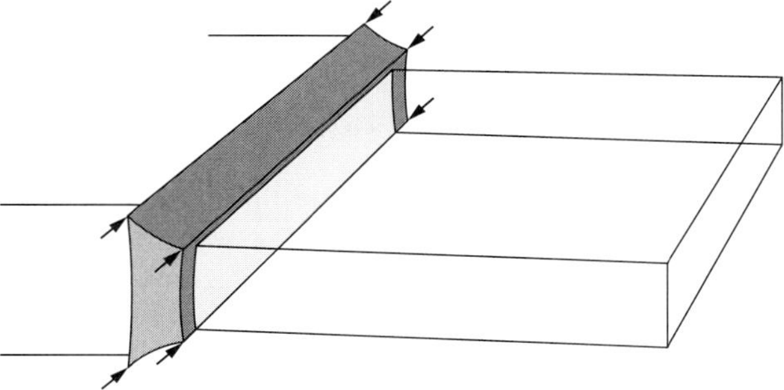

FIG. 10.3 The shaded material in dark gray is in the roll bite. If this material were deformed in simple compression, the strains expected in the rolling direction and the transverse direction would be expected to be equal tensile strains. Except at the very edges, the deformation of the material in the roll bite is constrained by the material entering and exiting the rolls. Similar to the case of plane strain fracture, the localized plastic deformation of the material in the roll bite is held back by the elastic rigidity of adjacent material that is not undergoing plastic deformation. This is why multiple rolling reductions of flat sheets result in very little change in the sheet width.

EXAMPLE 10.1 *Rolling of Copper Rod*

To demonstrate the constraint that suppresses changes in sheet width, we took some 5-mm-diameter commercial purity copper rod and ran it through a series of reductions using a laboratory rolling mill and no lubrication. The sequential changes in the height (or normal direction, ND) and width (or transverse direction, TD) are shown overlaid in Fig. 10.4*a*, and the actual dimensions and corresponding true strains for each pass are given in Table 10.1. We can calculate the length strain

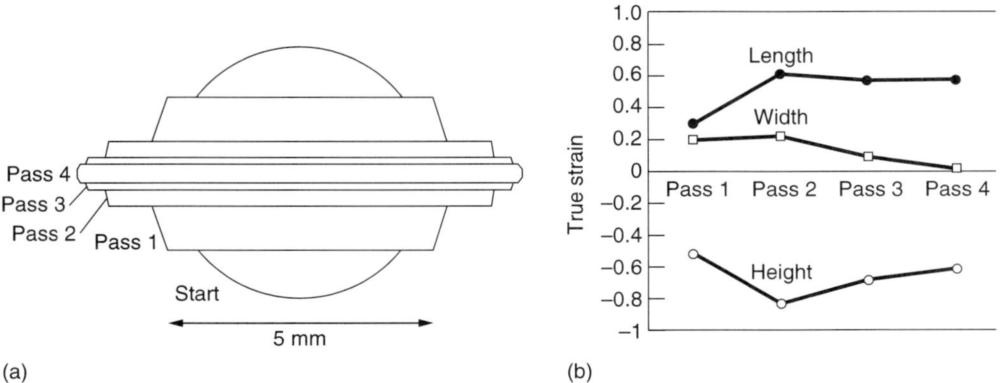

FIG. 10.4 (a) Sequential cross-sectional changes during rolling of copper rod. (b) The calculated true strains for each rolling pass.

TABLE 10.1 Dimensional Changes and Strains for Rolling of a Copper Rod

	Height (mm)	ε_{ND}	Width (mm)	ε_{TD}	ε_{RD}
Start	5	–	5	–	–
Pass 1	3	−0.51	6.1	0.2	0.31
Pass 2	1.3	−0.84	7.6	0.22	0.62
Pass 3	0.66	−0.68	8.4	0.1	0.58
Pass 4	0.36	−0.61	8.6	0.03	0.58

assuming that volume is conserved to give the strain in the rolling direction, ε_{RD}. By the fourth pass, it is almost equal to the strain in the normal direction whereas the width undergoes almost no change. Therefore, it is pretty safe to say that the conditions are plane strain for further reductions. The corresponding values of the three rolling strains are shown for each pass in Fig. 10.4*b*. ■

The friction hill is larger if h is small and r is large, resulting in a large force pushing back on the rolls. The larger values of P_{roll} result in a flattening of the roll contact and also can lead to bending of the rolls, as shown in Fig. 10.5. If the deflection of the rolls is sufficient, it can make rolling reductions of very thin stock almost impossible. The sheet thickness can be reduced by supporting the roll in contact with the material by placing other rolls against it, and the rolls can also be cambered, as shown in Fig. 10.5. The flattening of the roll in contact with the sheet is accounted for by the minimum sheet thickness

$$h_{min} \approx \frac{7\mu_F r}{E/(1-\nu^2)}(\sigma_o - \sigma_t) \tag{10.2}$$

where E is the Young's modulus of the rolls, ν is the Poisson's ratio of the rolls, and σ_t is a term representing the application of forward or backward tension to the material. Forward tension σ_{ft} and backward tension σ_{bt} can be designed into the rolling process and applied to exiting and entering material, respectively. The average of these two stresses is given in Eq. 10.2 as σ_t. The application of these tensile stresses to the band of material results in a reduction in the frictional stresses by similarly reducing the stress term in Eq. 10.1 and also can result in a shift in the neutral point.

Design of the rolling geometry can have a critical impact on the product exiting the rolls. If the rolls are not properly cambered, one of the effects shown in Fig. 10.6 is possible. Warping and wrinkling are likely with insufficient camber, and edge cracking can be a result of overly cambered rolls. Edge cracking can also occur if there is bulging of the material (similar to the example of bulging shown for axisymmetric compressive forging in Fig. 5.3). Rolls without camber can cause center cracking across the width. This is usually a result of the buildup of large tensile stresses in the rolling direction at the center of the sheet width and compressive stresses near the edges. The residual stresses can be reversed if the rolls are overcambered, resulting in edge cracking.

Other problems arising from inadequate roll geometry include deformation that may be very heterogeneous or *redundant,* as shown in Fig. 10.7. In addition to possibly result-

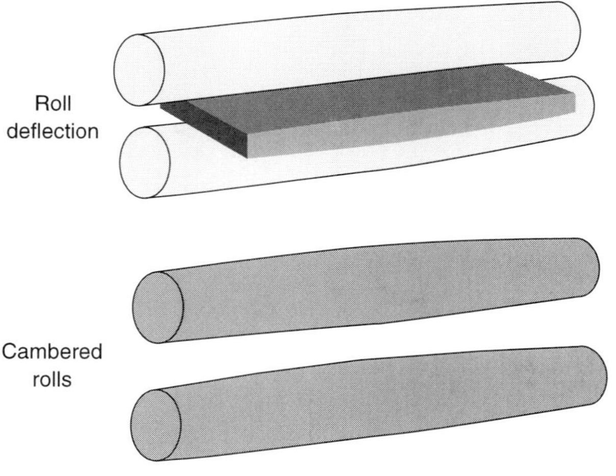

FIG. 10.5 High rolling pressures can result in deflection of the rolls, as shown in the upper set of rolls. One strategy is to use cambered rolls, as shown in the lower set.

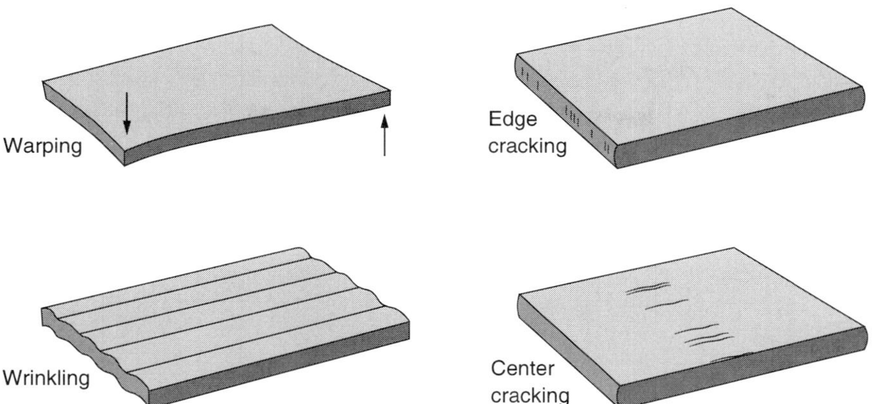

FIG. 10.6 Possible effects of improper roll geometry (after Hosford and Caddell, 1993).

ing in severe residual stresses, the work required to produce this deformation is part of the work that must be accomplished by the forming operation in addition to the work required to produce the desired shape change and overcome friction. When the deformation that occurs is heterogeneous, it can result in significant property variations through the thickness that can affect subsequent forming or heat treatment operations. For the same size of starting material and the same roll diameter, the amount of reduction taking place in a single pass can result in substantial hardness gradients, as shown in Fig. 10.8*a*. The gradient in hardness results from a difference in strain hardening with position across the thickness indicating a gradient in the amount of plastic deformation. A similar degree of heterogeneity in hardness takes place for light drawing passes, as shown in Fig. 10.8*b*. These gradients in plastic deformation history can also lead to gradients in grain size or preferred orientation (texture) across the material's thickness.

The deformation geometry, Δ, is commonly used to define the geometry of the deformation zone and the likelihood of heterogeneous deformation (see Backofen, 1972). It is defined as the ratio of the average thickness or diameter of the material being deformed, h, to the length of contact between the tool and material in the forming operation, L. Then

$$\Delta = \frac{h}{L} \tag{10.3}$$

For rolling, the definition of this parameter depends on the reduction in the normal direction, ε_n,[1] the roll radius r, and the initial thickness h, in the form

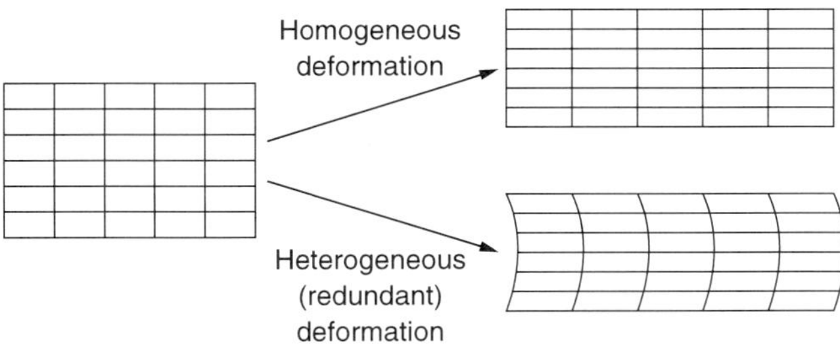

FIG. 10.7 Most forming operations result in heterogeneous or redundant deformation. Shown here is a comparison of homogeneous and heterogeneous deformation that may result from plane strain deformation caused by rolling. (After Hosford and Caddell, 1993)

[1] ε_n is negative in sign.

$$\Delta = \frac{(2+\varepsilon_n)}{2}\sqrt{\frac{h_i}{|\varepsilon_n| r}} \tag{10.4}$$

The magnitude of Δ increases with decreasing reduction or with an increase in the ratio of the sheet thickness to the roll radius, and increasing values of Δ generally result in greater heterogeneity and higher redundant strains near the surface. This heterogeneity also results in residual stresses that can build up over multiple forming passes. The buildup of residual stresses can reach such an extent that materials fracture along the center of the thickness on exiting the rolls, as shown in Fig. 10.9. Similar approaches can be used to evaluate heterogeneity in extrusion and drawing, as will be discussed later.

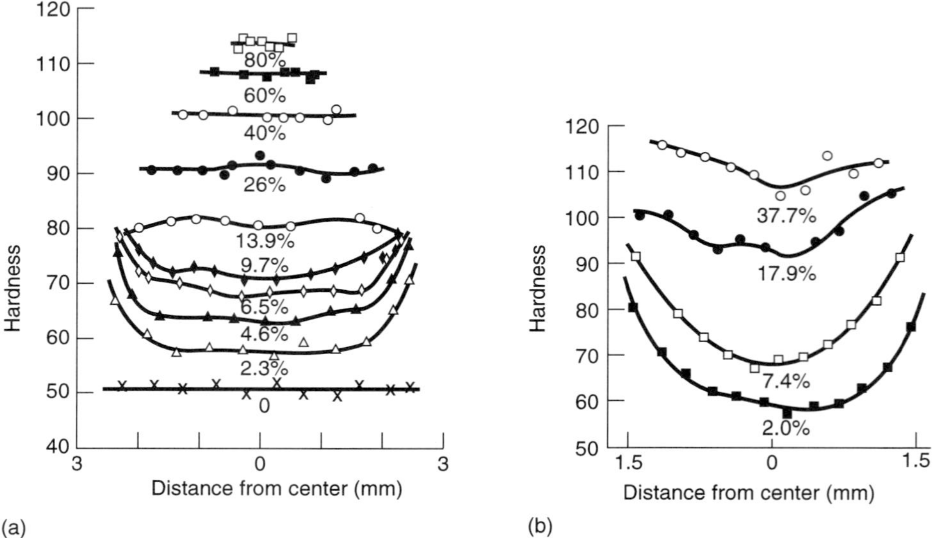

FIG. 10.8 (a) Hardness gradients in copper after a single rolling pass at the height reductions shown. The initial strip thickness was 5 mm and the roll radius was 127 mm. (b) Hardness gradients across the diameter for copper wire drawn through 30° dies. (Adapted from Hundy and Singer, 1954–1955, *J. Inst. Metals*, used with permission)

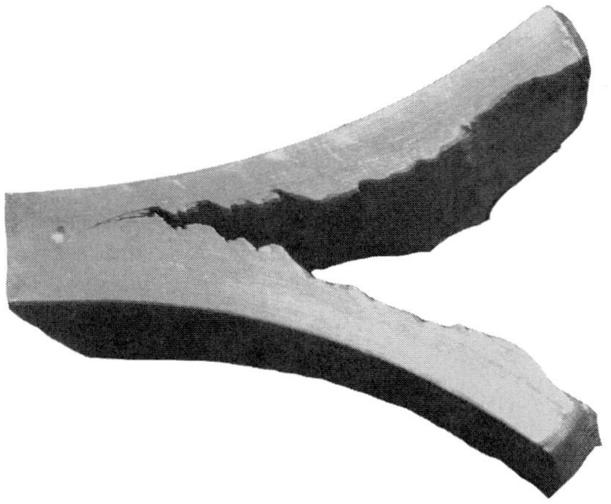

FIG. 10.9 "Alligatoring" fracture of aluminum strip from differential strains resulting from repeated rolling passes of less than 5 percent. The intact thickness of the strip is approximately 6 mm.

10.1 IDEAL ENERGY APPROACH FOR MODELING OF A FORMING PROCESS

The ideal energy approach relies on a very basic energy balance wherein the applied work is set equal to the deformation energy (see Wagoner and Chenot, 1997). No friction effects or heterogeneous deformation are considered in this model of a forming operation, so we can consider the process a lower bound. No die geometry is considered in this model of deformation. Simple tensile deformation of a reduced section that is well away from the enlarged heads wherein the material is gripped can be considered an ideal process. The work per unit volume conducted in a tensile test can be described as

$$w_{\text{ideal}} = \int_0^{\varepsilon_{\text{eff}}} \sigma_{\text{eff}} d\varepsilon_{\text{eff}} \tag{10.5}$$

where the effective strain $\varepsilon_{\text{eff}} = \ln(A_{\text{final}}/A_{\text{initial}})$. If we use the power law hardening expression introduced in Eq. 3.19, the ideal work is then

$$w_{\text{ideal}} = \frac{K\varepsilon_{\text{eff}}^{\,n+1}}{n+1} \tag{10.6}$$

We can apply this ideal concept to forward extrusion, as shown in Fig. 10.10. Then the amount of work applied per unit volume can be related to the extrusion pressure by $P_e > w_{\text{ideal}}$. Similarly, the minimum stress required to pull the material through the dies in drawing is σ_d, leading to $\sigma_d > w_{\text{ideal}}$. Some simple corrections can be made to evaluate the deformation efficiency of an actual process, but these mostly consist of factors that are better evaluated by the techniques given in Sections 10.2 and 10.3.

EXAMPLE 10.2 *Calculation of the Ideal Drawing Stress*

A rod of an aluminum alloy has been reported to have a strain hardening behavior given by $\sigma_{\text{eff}} = 350\varepsilon_{\text{eff}}^{0.3}$ (MPa). We would like to calculate the expected drawing stress if the aluminum alloy is to be reduced from an original diameter of 15 mm to 13 mm. First we would like to find the effective strain, which is given by

$$\varepsilon_{\text{eff}} = 2\ln\frac{15}{13} = 0.143$$

If we apply this strain in Eq. 10.6 we can write

$$w_{\text{ideal}} = \frac{350(0.143)^{1.3}}{1.3} = 21.5 \text{ MPa} = 21.5\times 10^6 \frac{J}{m^3}$$

■

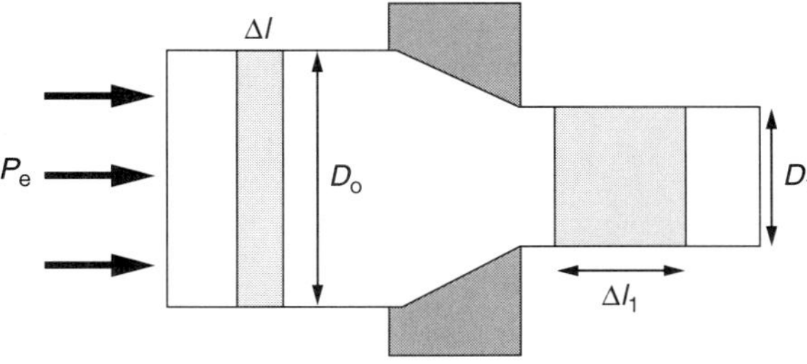

FIG. 10.10 Schematic figure showing extrusion of a round cross section and the changes in shape for a selected element.

10.2 INCLUSION OF FRICTION AND DIE GEOMETRY IN DEFORMATION PROCESSES: SLAB ANALYSIS

To develop a description of rolling, we can use two separate steps to include friction and roll geometry in what is called slab analysis (see Hosford and Caddell, 1993, and Wagoner and Chenot, 1997). The final goal is to be able to explain how the geometry of the rolls and the motion of the rolls relative to the material surface affect the forming operation. If we are able to use this technique to describe rolling, then it can be applied to any forming operation. Slab analysis employs a force balance that is resolved in only one direction. The main assumptions for slab analysis are as follows.

1. Principal stresses are defined by the direction of applied load.
2. Surface friction affects the force balance but does not cause internal gradients that would alter the orientation of the principal stresses.
3. All deformation is homogeneous.

Strip Drawing The first step toward evaluating rolling employs a model for plane strain deformation that consists of pulling a sheet of material through a set of wedge-shaped dies, called strip drawing, which is depicted in Fig. 10.11. The applied force is defined by $\sigma_d(wt_e)$. Figure 10.12 shows an enlarged view of the deformation geometry and defines a number of the steps in calculating the sum of forces in the rolling or x-direction.

The equilibrium of forces in the x-direction includes the directly applied forces defined in Fig. 10.12*a*, the component of the force P acting through the die walls in Fig. 10.12*b*, and the effect of the frictional force in Fig. 10.12*c*. After summing each of these contributions to the force balance, neglecting the smaller terms, and approximating P as the principal stress acting in the y-direction, we can write

$$\frac{d\sigma_x}{\mu_F \cot\alpha\,\sigma_x - 2k(1+\mu_F \cot\alpha)} = \frac{dt}{t} \tag{10.7}$$

If we allow friction to be represented by a single value, ignore work hardening, and use a constant die angle, we can integrate Eq. 10.7 to derive an expression for the drawing stress that incorporates an approximation of friction and die geometry effects as

$$\frac{\sigma_d}{2k} = \frac{(1+\mu_F \cot\alpha)}{\mu_F \cot\alpha}\left(1 - \exp(-\varepsilon_n \mu_F \cot\alpha\right) \tag{10.8}$$

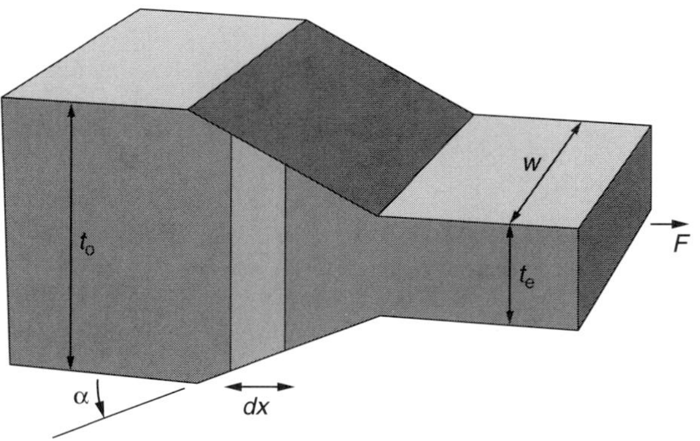

FIG. 10.11 Schematic figure showing strip drawing. The "slab" that will be undergoing loading is highlighted.

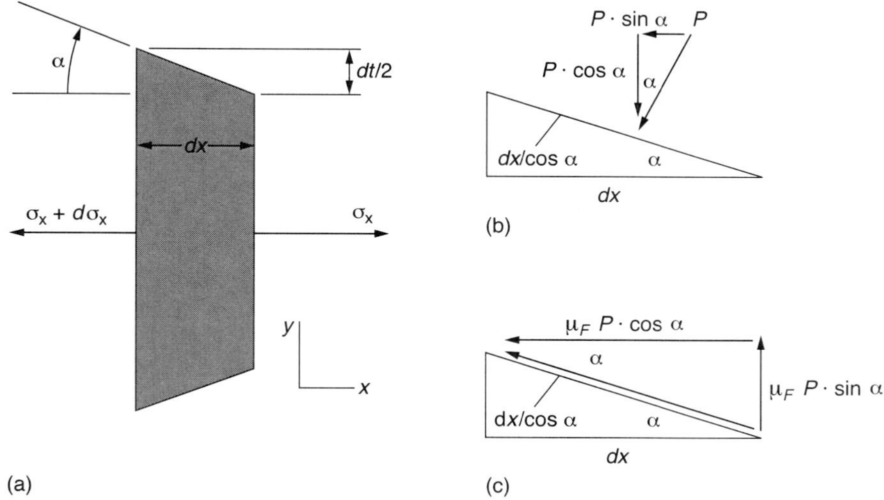

FIG. 10.12 Geometry of strip drawing. (a) Direct stresses in the x-direction. (b) Resolved components of the pressure P applied by the dies on the material. (c) Resolved components of the frictional force. (After Hosford and Caddell, 1993)

where ε_n is the absolute value of the strain in the y-direction. The value of $2k$ can be taken as the mean of the initial and final values. For the drawing of a round wire, $2k$ is replaced by the average of the initial and final yield stress, Y.

EXAMPLE 10.3 *Slab Method Compared with Ideal Energy Method*

Strip drawing is to occur on a material that is initially 1 cm thick and 20 cm wide. The dies through which it will be drawn have an angle of 20°. For an average shear yield stress of $k = 300$ MPa and an average coefficient of friction of 0.1, calculate the drawing force for a reduction of 15 percent. Then we will compare this example of slab analysis with the ideal energy method (without friction).

$$0.1 \cot 20^\circ = .027 \qquad \varepsilon_n = \ln\left(\frac{1}{1-0.15}\right) = 0.16$$

$$\sigma_d = \frac{2(300)(1+0.27)}{0.27}\left(1-\exp(-0.16(0.27)\right) = 119 \text{ MPa}$$

For the dimensions given, the force is then 240 kN.

For the ideal energy, the effective stress will be $2k = 600$ MPa, and we can calculate the effective strain as

$$\varepsilon_{\text{eff}} = \left[\frac{2}{3}\left(0.16^2 + (-0.16)^2 + 0^2\right)\right]^{\frac{1}{2}} = 0.18$$

Then the ideal energy is simply $600 \times 0.18 = 108$ MPa. ■

Plane Strain Compression The second consideration for rolling is how to approach the friction hill that arises in rolling. If we consider just the contact of the rolls across the contact length we saw in Fig. 10.1 and 10.2, we can flatten out the deformation to consider a two-dimensional deformation in which the relative translation of the material emanates from a neutral point, as shown in Fig. 10.13a. A friction hill is

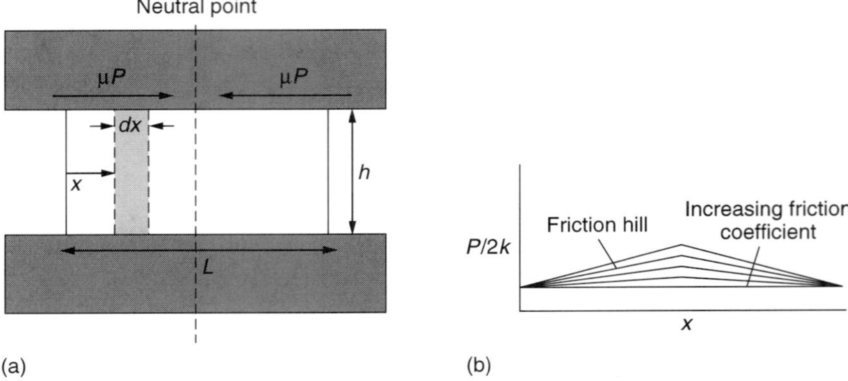

FIG. 10.13 (a) Plane strain compression geometry and definition of neutral point. (b) Friction hill resulting from plane strain compression.

expected as a result of the resistance of the material flow, as shown in Fig. 10.13*b*. This same type of deformation can also be considered as a model for a cross section of a round material that is under simple compression (see Hosford and Caddell, 1993, or Wagoner and Chenot, 1997).

The entire force balance of the small dx section in Fig. 10.13*a* is

$$\sigma_X h + 2\mu_F P dx - (\sigma_X + d\sigma_X)h = 0$$

If we take $\sigma_X + P = 2k$, then $d\sigma_X = -dP$. Rearranging the terms, we can write

$$\frac{dP}{P} = \frac{2\mu_F}{h} dx \tag{10.9}$$

Integration with $\sigma_X = 0$ at $x = 0$ and $P = 2k$ leads to

$$\frac{P}{2k} = \exp\frac{2\mu_F x}{h} \tag{10.10}$$

which gives the friction hill behavior shown in Fig. 10.13. This analysis requires the shear stress at the interface, $\mu_F P$, to be equal to or less than the shear yield strength of the material. If sticking friction is reached, deformation is very heterogeneous. For sticking friction, the derivation given above leads to

$$\frac{P}{2k} = 1 + \frac{x}{h} \tag{10.11}$$

When this approach is applied to direct compression (see Fig. 5.3) of a cylinder of height h, the pressure depends on the position ρ across the radius r as

$$P = Y \exp\left[\frac{2\mu_F}{h}(r - \rho)\right] \tag{10.12}$$

For sticking friction, the average pressure required to produce deformation can be substantial and can depend strongly on the radius-to-height ratio as

$$P_{\text{average}} = Y + \left(\frac{2kr}{3h}\right) \tag{10.13}$$

Problem C.10.3 considers the average pressure for loading at various levels of friction.

10.3 UPPER BOUND ANALYSIS

If it is your job to design or purchase the equipment for a mechanical forming operation, you want to be certain that the equipment has the capacity to complete the operation readily. Therefore, it is essential to know the required upper limit. Upper bound analysis uses the yield criterion with shape changes consistent with volume conservation (see Avitzur, 1977). The computation assumes that the external work must be equal to the internal energy required to produce changes in the material's shape. The tool we will use to calculate the upper bound is a velocity vector diagram known as a *hodograph.*

EXAMPLE 10.4 *Demonstration of Hodographs as Applied to Wind Shear*

The hodograph was initially developed to describe Newtonian attraction by William Hamilton in 1847. Since then it has found many uses as a descriptor of vector transitions with perhaps the most accessible being the application to changes in wind velocity with height called wind shear. Because wind shear has been implicated in flying accidents, it also has become a consideration in flying safety. The application of hodographs to wind velocity enables a depiction of the direction and the velocity changes as a function of height from the surface. Consider the diagram shown in Fig. 10.14. In this diagram, each of the numbers represents the distance from the earth's surface in kilometers, the axes define the compass directions, and each of the dotted circles represents an increasing level of velocity. The difference in wind velocity at each level is shown by the line that connects the tip of each of the arrowheads representing each layer of velocity. When it is used to evaluate threatening weather formations, the shape of this gray line, which is the hodograph of wind velocity, enables prediction of storm velocity and the likely severity of storms. ■

When hodographs are applied to material deformation, we need to make quite a few assumptions. We consider the material as being ideal, isotropic, and homogeneous with no strain hardening or rate effects considered. The initial deformation conditions assume that deformation is either frictionless or at a constant rate of friction and there is no adiabatic heating from the deformation. We will use the somewhat impractical forming operation of Fig. 10.15 as our first example. The material deformed in Fig. 10.15 undergoes an abrupt shearing deformation at the y-y' plane that introduces a simple shear across the material thickness. We will make the assumption that all of the deformation occurs at this position.

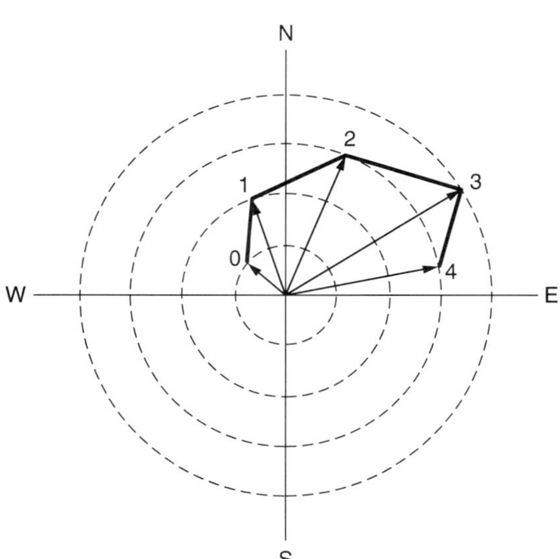

FIG. 10.14 A depiction of wind shear using a hodograph. The compass directions are defined on each of the axes, and the dotted circles represent increasing magnitude of velocity. The numbers 0 to 4 represent different layers from the earth's surface in kilometers. The hodograph is the line that connects the transitions in wind magnitude and direction with height.

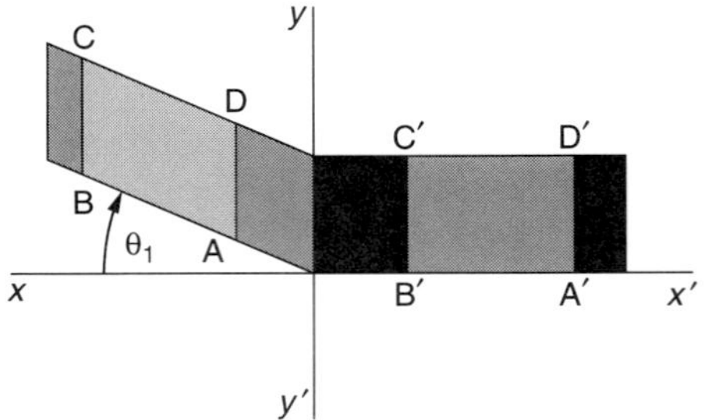

FIG. 10.15 Discrete shear at the y-y' plane of material entering a plane of discrete shear.

To construct the deformation hodograph for this deformation, we will employ the rules listed in Fig. 10.16. Next to these rules are shown the velocity vectors that represent the material before reaching the y-y' plane, v_1, and the material after passing through the y-y' plane, v_2. Before reaching the y-y' plane, an element of the material has the shape ABCD, and after passing through the y-y' plane, the shape is transformed by shear to the shape A′B′C′D′. The abrupt shearing that occurs at the y-y' plane to produce this shape change is the difference in the velocity vectors $v^*_{1\rightarrow 2}$ defining the deformation.

The rate of energy dissipated at y-y' should equal the work done on a volume of material crossing y-y'. For shear deformation with a shear yield strength, k, we can write this work as

$$\frac{\text{Work}}{\text{Volume}} = k\frac{dy}{dx} \tag{10.14}$$

The volume of material crossing y-y' per unit time can be defined as the distance from A to D, h, times the width w of the sheet being sheared times the velocity in the x-direction, which in this case is $v_{\text{x}} = v_2$ or

$$\frac{\text{Volume}}{\text{Time}} = h \cdot w \cdot v_x \tag{10.15}$$

Even if the exit velocity was not in the x-direction, the velocity used is the one normal to the plane of intense shear.

By combining Eq. 10.14 and Eq. 10.15, we can get the work dissipated per unit time as

$$\frac{\text{Work}}{\text{Time}} = k\frac{dy}{dx}h\,w\,v_{\text{x}} \tag{10.16}$$

Because the shear strain $dy/dx = v^*_{1\text{-}>2}/v_{\text{x}}$, we can write

$$\frac{\text{Work}}{\text{Time}} = k\,h\,w\,v^*_{1\rightarrow 2} \tag{10.17}$$

If multiple deformation fields are involved in modeling the forming operation, we can sum the contributions from each as

$$\frac{\text{Work}}{\text{Time}} = \sum_{1}^{i} k\,h_{\text{i}}\,w_{\text{i}}\,v_{\text{i}}^* \tag{10.18}$$

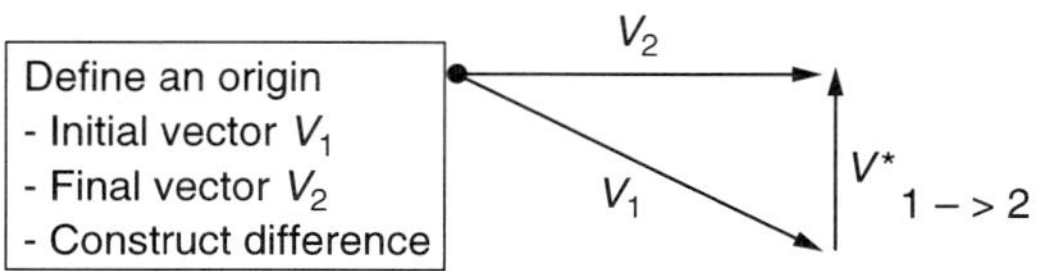

FIG. 10.16 Hodograph of the deformation in Fig. 10.15.

The summing of multiple abrupt shear steps is essential for constructing the deformation hodograph for the deformation process shown in Fig. 10.17, which is a demonstration of a plane strain extrusion process (we could consider this as strip extrusion). Just the top half of the symmetric deformation is shown for simplicity.

The material undergoes two abrupt shears in Fig 10.17, one at the plane AB and one at the plane BC. The magnitude of v_{final} is 50 percent higher than v_{initial} because there is a reduction of one-third. The two horizontal arrows in the deformation hodograph show this. The magnitude of the intermediate velocity is controlled by the angle β, but its trajectory is determined by the die angle. All of the material within the triangle ABC is considered to be moving at this same velocity, v_{ABC}. The two abrupt velocity changes, v^*_{AB} and v^*_{BC}, must be coupled to the dimensions $\overline{AB}$ and $\overline{BC}$, respectively, to calculate the total work consumed. The first abrupt shear v^*_{AB} changes the trajectory from v_{initial} to v_{ABC}. The second abrupt shear v^*_{BC} changes the trajectory from v_{ABC} to v_{final}. After completing the hodograph, it is important to be certain that it has dimensional consistency with the deformation process. A similar hodograph (but upside-down) can be drawn for the bottom half of the material if the deformation is symmetric about the thickness of the strip.

Now the next task is to calculate the work rate. To do so is straightforward and is derived directly from the geometry, beginning with

$$\frac{\text{Work}}{\text{Time}} = P_e h_{\text{initial}} w \, v_{\text{initial}} \tag{10.19}$$

We can equate this to the work expended in the abrupt shears at AB and BC by writing

$$P_e h_{\text{initial}} w \, v_{\text{initial}} = kw\left(v^*_{AB}\,\overline{AB} + v^*_{BC}\,\overline{BC}\right) \tag{10.20}$$

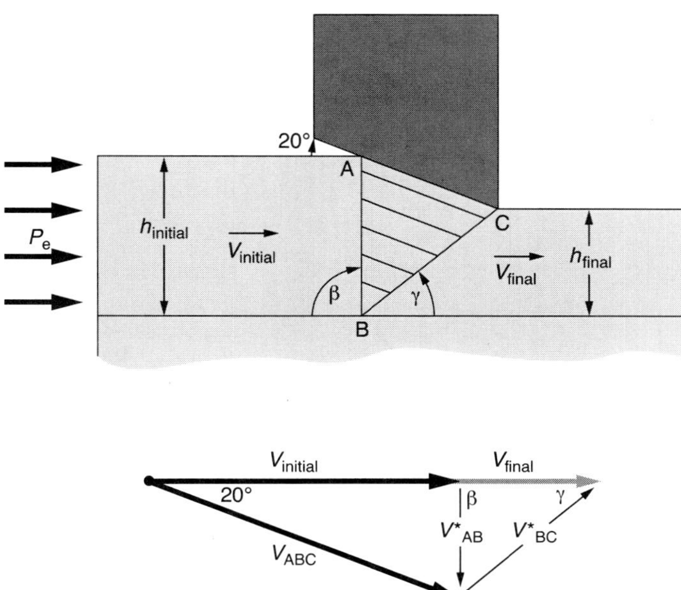

FIG. 10.17 Plane strain extrusion, showing the dimensions and changes in velocity for the top half of the billet and a deformation hodograph.

We can rearrange this to relate the pressure to the shear yield stress and cancel out the widths to yield

$$\frac{P_e}{2k} = \frac{1}{2h_{\text{initial}} w\, v_{\text{initial}}}\left(v^*_{AB}\overline{AB} + v^*_{BC}\overline{BC}\right) \tag{10.21}$$

The calculation can be done graphically if the drawings are to scale, but a general expression can be written using trigonometry. For the particular solution we have chosen to consider in Fig. 10.17, $\beta = 90^\circ$, which from geometry sets $\gamma = 36^\circ$. Then the solution becomes

$$v^*_{AB} = v_{\text{initial}} \sin(20^\circ) \qquad v^*_{BC} = \frac{0.5 v_{\text{initial}}}{\cos(\gamma)}$$

$$\overline{AB} = h_{\text{initial}} \qquad \overline{BC} = \frac{0.67 h_{\text{initial}}}{\sin(\gamma)}$$

If we enter each of these values in Eq. 10.21 and divide through by all of the equal terms, we get

$$\frac{P_e}{2k} = 0.52$$

which suggests that the extrusion pressure for this geometry is about half of the shear yield stress. This is about 20 percent greater than the lower bound solution $P_e/2k = \ln(1.5) = 0.40$. Consideration of the same geometry in Fig. 10.17 but with a die angle of 30°, a reduction of 50 percent, and $\beta = 90^\circ$ gives $P/2k = 0.87$. The effects of varying β shown in Fig. 10.18 demonstrate that changing the hodograph angles has only a small impact on the final solution. It is also possible to vary the solution to solve for the lowest extrusion pressure as a function of the die angle, reduction, and the chosen geometry of the hodograph.

Hodographs can also be constructed for the flow geometry in a hardness test. Using the two-dimensional geometry defined in Fig. 5.5, we can construct a hodograph that provides a solution to the constraint occurring in such a test. Figure 5.5 is reproduced in Fig. 10.19*a*. To construct a hodograph representing this deformation, we first define an origin, as shown in Fig. 10.19*b*, and then assign a velocity to the indenter, v_p. Immediately on crossing from the indenter into the material in triangle 0AB, the velocity must be v_1, thus suggesting that there is a shear across 0B, which we will define by its velocity change, v^*_{0B}.

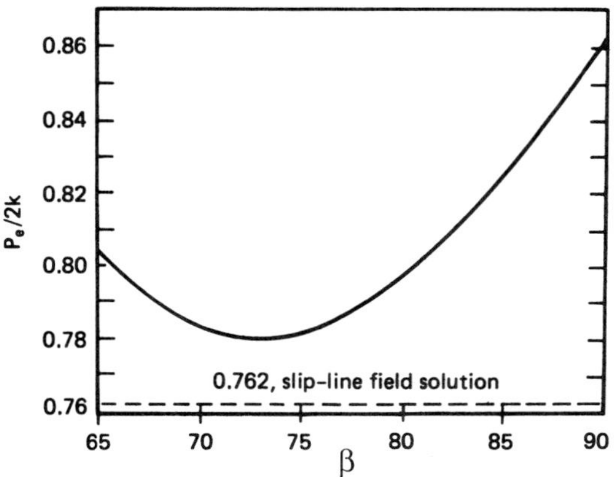

FIG. 10.18 Variation in extrusion pressure with β for a die angle of 30° using upper bound analysis. (Hosford and Caddell, 1993 (METAL FORMING, by Hosford/Caddell, © Reprinted by permission of Pearson Education, Inc., Upper Saddle River, NJ))

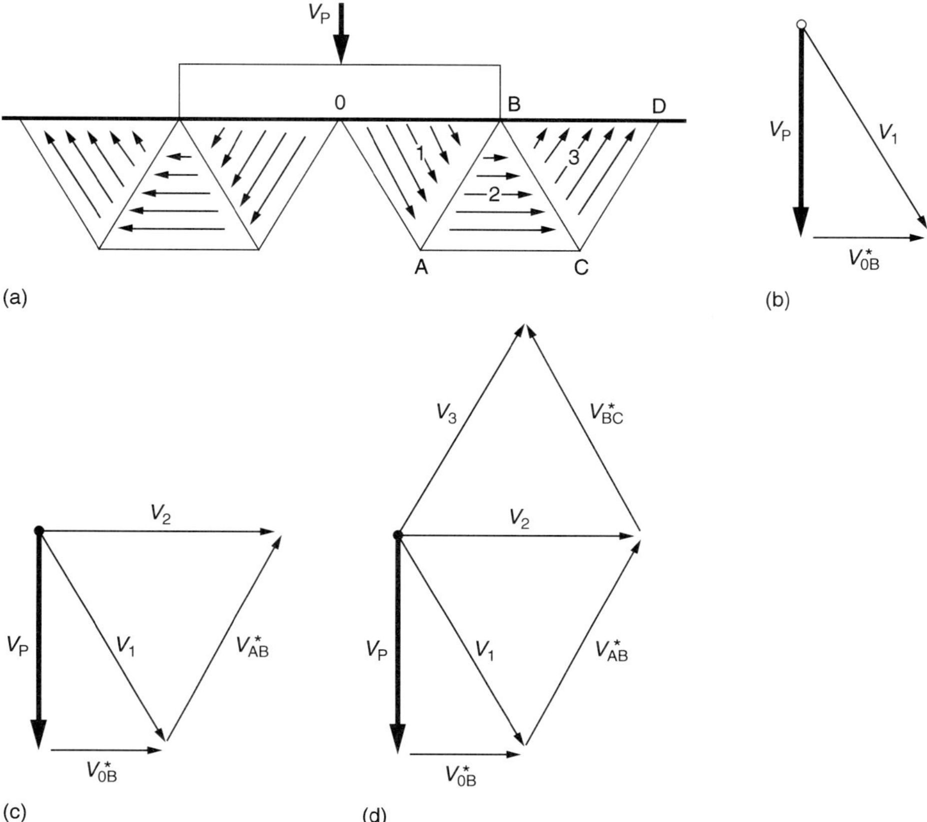

FIG. 10.19 Plane strain indentation geometry with three zones of deformation on each half of the indenter. The deformation zones are all equilateral triangles. (b) Partial hodograph for the right side of the indentation including the first deformation zone. (c) Partial hodograph for the right side of the indentation including the second deformation zone. (d) Partial hodograph for the right side of the indentation showing the third deformation zone. (After Hosford and Caddell, 1993)

Then, further on in Fig. 10.19*c*, we can consider the flow of the material in the second triangle, ABC, which will have a horizontal velocity, v_2, and the transition from v_1 to v_2 is defined as v^*_{AB}. Then, the last velocity we need to consider is the uplift of material outside the hardness indent taking place via the triangle BCD. This material will have a velocity v_3 and the transition to this velocity occurs by v^*_{BC}.

For the deformation in Fig. 10.19, begin by assuming frictionless conditions at the interface and no shear deformation along 0B. We will also not consider v^*_{0B}, since no material is flowing across this boundary at the instant the hardness test is conducted. The other abrupt shear that is occurring takes place at the interface between the hodograph and the material that is stationary. This interface acts as a condition of sticking friction, because the shear stress must be reached for deformation to occur. These abrupt shears are each parallel to the velocities within the three triangles of deformation and can be written as $v^*_{0A} = v_1$, $v^*_{AC} = v_2$, and $v^*_{CD} = v_3$. All five velocity transitions are equal to $2v_p/\sqrt{3}$, and each of the five lengths of shear discontinuity is half the width of the indenter. If we can assign the indenter width as w, we can write

$$\frac{\text{Work}}{\text{Time}} = 5k\left(\frac{w}{2} \cdot \frac{2v_p}{\sqrt{3}}\right)$$

If we set the applied external energy for the right half of the die, $P_p(w/2)v_p$, equal to this work, we can get the final result of $P/2k = 2.89$. Better designed hodographs that offer better solutions and inclusion of friction or sticking friction at the tool contact result in a solution approaching three times the shear yield strength of the material.

10.4 SLIP LINE FIELD ANALYSIS

Up to now, we have not used all of the information available to us to determine the deformation occurring within a material. If we look carefully at point D in Fig. 10.19*a*, it is clear that the hydrostatic stress (if we ignore atmospheric pressure) must be zero, and then if the deformation is really plane strain we know that the other two principal stresses must be equal to k and $-k$. If shear were really occurring along the line CD, we would expect it to meet the free surface at 45°. This kind of information is handy for developing a solution using what is called slip line field analysis. This slip line is different from those for single crystals and represents lines of maximum shear stress in what is otherwise considered to be a homogeneous body. We will make the same assumptions as we made for hodographs with the exception that now we will allow progressive rotation to take place. It is changes in the stress state that cause this rotation of the principal stresses to take place. As with the hodographs given above, the goal here is to present an introduction to reading an existing slip line field, and more detailed treatments are left to publications that provide more complete coverage (e.g., Hill, 1950, and Hosford and Caddell, 1993). To that end, the derivations of the governing relations can also be found in these sources.

In defining the plane strain principal stress state, we need to define the principal stresses. Two lines, defined as α- and β-lines, serve this purpose in slip line field analysis. First let us consider an element undergoing plane strain deformation, as shown in Fig. 10.20. The stress state with applied shears superimposed on a hydrostatic stress state is shown in Fig. 10.20*a* and then is rotated to show only principal stresses in Fig. 10.20*b*.

This allows us to identify a relatively compressive stress as $\sigma_m - k$ and a relatively tensile stress as $\sigma_m + k$. We can identify the α-line for the slip line field as a line of maximum shear that lies clockwise from the *maximum* principal stress, $\sigma_m + k$. Then the β-line is a line of maximum shear that is orthogonal to the α-line (see Fig. 10.21) or also clockwise from the *minimum* principal stress, $\sigma_m - k$.

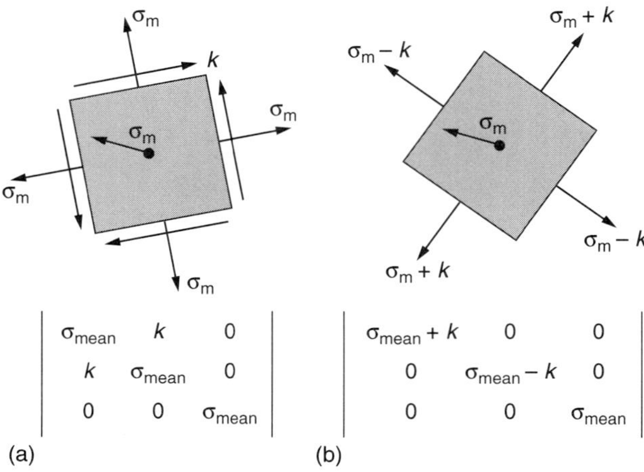

FIG. 10.20 Definition of the stresses and the relationship to the stress tensor for slip line field analysis as (a) a general tensor and (b) the principal stresses after rotation.

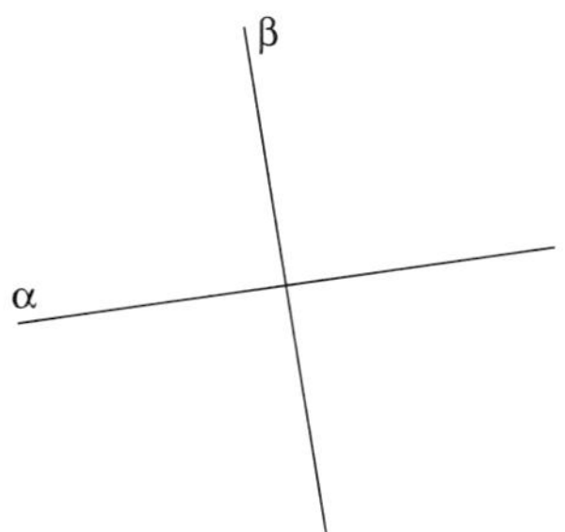

FIG. 10.21 Definition of α- and β-lines corresponding to Fig. 10.20.

We will consider the following boundary conditions for specific circumstances governing slip line fields.

Free surface (outside of tool contact): The α- and β-lines must meet the surface at 45°.

Frictionless interface: The α- and β-lines must meet the interface at 45°.

Sticking friction: Because the maximum shear must lie parallel to the surface, one of two lines, α or β, must be tangential to the interface.

If we take the case of a tilted free surface shown in Fig. 10.22, we can see a direct application of the definitions given for α- and β-lines. The stress normal to the free surface must be equal to zero, $-k + k = 0$, and the two other principal stresses must be $-2k$ and $-k$, as shown.

Curved α- and β-lines would indicate a change in the stresses within the tensor, and specifically the mean stress. The equations governing this change are called the Hencky equations and can be written as a change in pressure, $P_{final} - P_{initial} = \Delta P$, with a change in orientation, $\phi_{final} - \phi_{final} = \Delta\phi$, of the line where a counterclockwise rotation of ϕ (given in radians) is considered positive. The Hencky equations can be written as

$$\Delta P = -2k\Delta\phi_\alpha \text{ for an } \alpha\text{-line}$$

and (10.22)

$$\Delta P = 2k\Delta\phi_\beta \text{ for a } \beta\text{-line}$$

We will consider the effect of curvature by once again evaluating indentation. As shown in Fig. 10.23, we have solved part of the dead zone in earlier examples by having the zone immediately under the indenter move straight down along with the indenter. We already know the solution at point E, because it was given above for a free surface, so $P_E = -k$. Because there is no rotation as we proceed from E to D, there can be no change in the pressure since the lines do not rotate, thus $P_D = -k$. As we proceed past D to B, there is a rotation by $\pi/2$. We know this must be true if we consider that the rotation that must occur

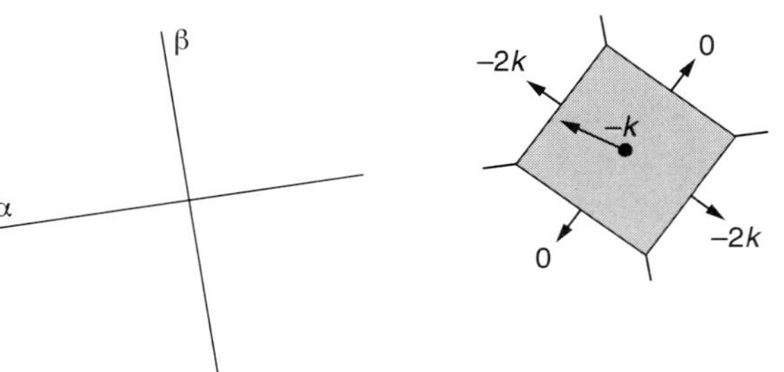

FIG. 10.22 Free surface and corresponding α- and β-lines and principal stresses.

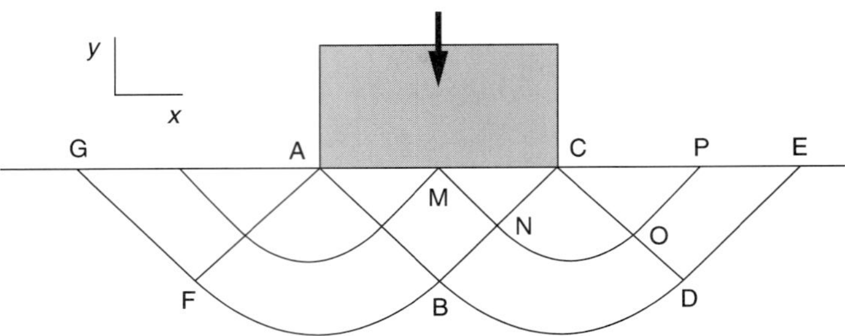

FIG. 10.23 Plane strain indentation slip line field.

for the stress tensor is 90°. The rotation is clockwise, which means that the sign of the rotation is negative.

Using the Hencky equations, we can consider the specific case

$$P_B - P_D = 2 \cdot k(\phi_B - \phi_D)$$

Then we insert the values and move everything to the same side of the equation

$$P_B - (-k) + 2 \cdot k((-\pi/2) - 0) = 0$$

leading to the final result

$$P_B = -k(\pi + 1)$$

From this result, we know the value of the mean stress at point B, $\sigma_{mean,B}$. All of the other points connected by straight lines to point B will have the same stress levels (e.g., M and N), with the exception that points A and C are "special" owing to the singularity of contact at these points. The downward pressure must represent the highest compressive stress, which will be an additional $-k$ in stress level from $\sigma_{mean,B}$, thus indicating that the minimum (most compressive) stress or the stress applied by the indenter will be

$$P_{indent} = \sigma_{minimum,B} = -k(\pi + 1) - k = 2k(1 + \pi/2)$$

With this we can solve for the pressure relative to the shear yield stress to yield

$$P_{indent}/2k = 2.57 \qquad (10.23)$$

If we use the von Mises criterion, the shear yield stress is 15 percent greater, resulting in a value for P_{indent} of 2.97.

Hodographs can also be constructed for the slip line fields to make certain that the deformation is reasonable. Because the rotation is gradual when in a particular section of a slip line field, the hodograph must then have curved sections. An example of a deformation geometry and the corresponding deformation hodograph adapted for curved sections is shown in Fig. 10.24 for the the plane strain indentation described in Fig. 10.19. Note that, as mentioned earlier, the angles BDC, AB0, etc., must be 45° for the proper shear deformation to reach the surface for a slip line field.

The application of slip line fields becomes very complex for extrusion, drawing, and rolling. Despite this complexity, they allow the evaluation of homogeneity in deformation in a way that is not readily accessible by other graphical techniques. When the thickness of the material being deformed is high relative to the length of tool contact, the deformation does not readily penetrate the material, leading to deformation mostly at the surface. To achieve homogeneous deformation, the reductions must be large enough to promote defor-

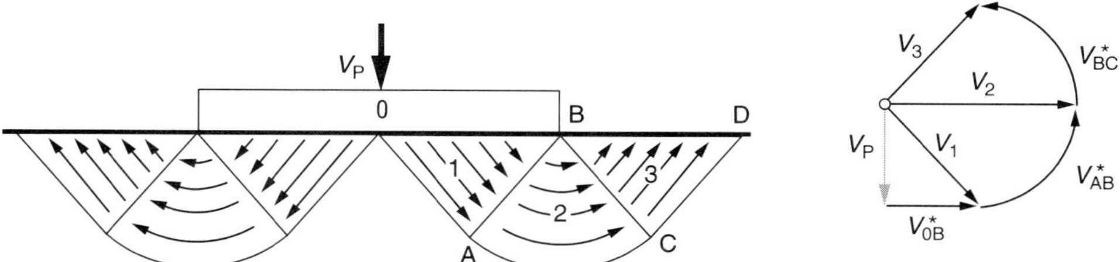

FIG. 10.24 Hodograph development for a slip line field of the indentation geometry shown in Fig. 10.19. The abrupt shears are converted to a sweeping curve using a slip line field approach. The angles were changed to correspond to a slip line field such that the angles BDC and AB0 are 45°.

mation through the thickness. Figure 10.25 shows a slip line field for a shallow die angle and a small reduction in strip drawing. The numbers on the slip line field represent the pressure, so that negative numbers here represent tensile hydrostatic stresses. As can be seen, the center line can have a negative pressure, which can clearly lead to void formation or even promote the type of failure shown in Fig. 10.9.

Guidance in the tool angle to reduce the centerline tensile stresses in Fig. 10.25 can be gleaned from Fig. 10.26. In Fig. 10.26, the amount of hydrostatic tension increases with increasing die angle and also increases with decreasing reduction for a given die angle.

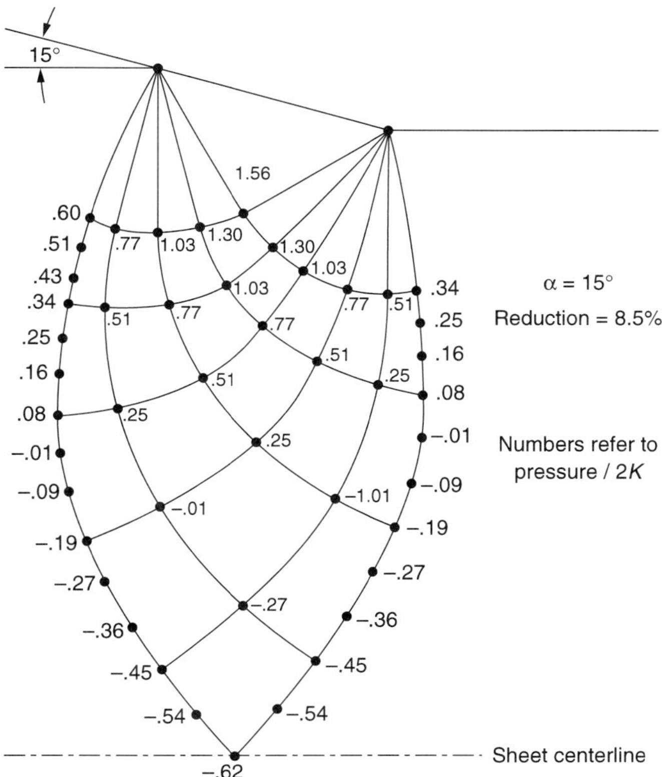

FIG. 10.25 Hydrostatic pressure in the deformation zone for strip drawing. The negative values are tensile in sign, because pressure is defined here as positive. (Coffin and Rogers, 1967, used with permission of ASM INTL.)

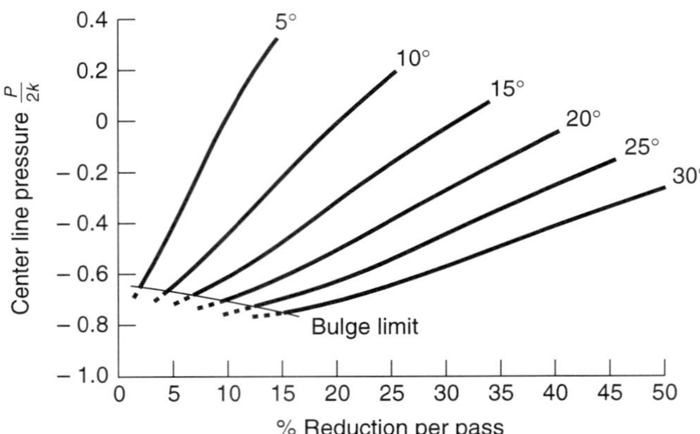

FIG. 10.26 Pressure at the center of the thickness for different reductions and die angles during strip drawing. Negative values represent hydrostatic tension. (Coffin and Rogers, 1967, used with permission of ASM INTL.).

10.5 FORMATION OF ALUMINUM BEVERAGE CANS: DEEP DRAWING, IRONING, AND SHAPING

The development behind the processes producing aluminum beverage cans around the world are part of a continuous effort to compete against other materials for soft drink and beer containers. In 2000, more than 100 billion aluminum beverage cans were produced in the United States. With such a high volume of production, improvements resulting in any degree of savings can have a very strong impact. The principal process used to produce the bottom sections of two-piece cans is deep drawing, as shown in Fig. 10.27*a*. The deep drawing process requires a thin sheet of aluminum, which is formed into a die using a punch that pulls the material into the die cavity. The holddowns provide a designated degree of friction to conduct this process. Because the diameter of the cupped material is reduced as the material is pulled into the die cavity, the deformation of the cup walls is essentially plane strain. There is some thickness increase as we proceed from top to bottom of the finished cupped material. If the plastic anisotropy for directions within the plane of the aluminum sheet used is significant, we can also form ears at the top of the cup (see Fig. 3.20), which suggests that the wall thickness will vary correspondingly. After the initial cupping process, the cup is then ironed to extend the sides of the cup walls and form the height of the can (see Fig. 10.27*b*). Repeated ironing processes can increase the height of the cup to form the can body by three or more times. At the end of this process, any nonuniformities introduced by the initial ear formation must be trimmed away as waste. Obviously, suppressing the plastic anisotropy that leads to ear formation is important in reducing the amount of waste that must be designed into the can forming operation.

At this point, a can could be made by simply cleaning and applying the desired coatings to the interior and exterior and attaching a top. But if one looks carefully at beverage cans, it will be noticed that an added step is used to form a neck at the top of the can. This neck is not added for cosmetic reasons, but rather for added weight savings. Because the top of the can must have better mechanical integrity than the sides to maintain an openable top, one of the easiest ways to reduce weight is to reduce the size of the top. Since the introduction of the aluminum can in the United States, the size of the top has been decreased several times. This has necessitated the added *necking-in* of the top to form a ridge and reduce the size required for the top. This also is a fairly delicate process, because the circumference undergoes a further compressive strain as a neck or ridge is formed at the top of the can.

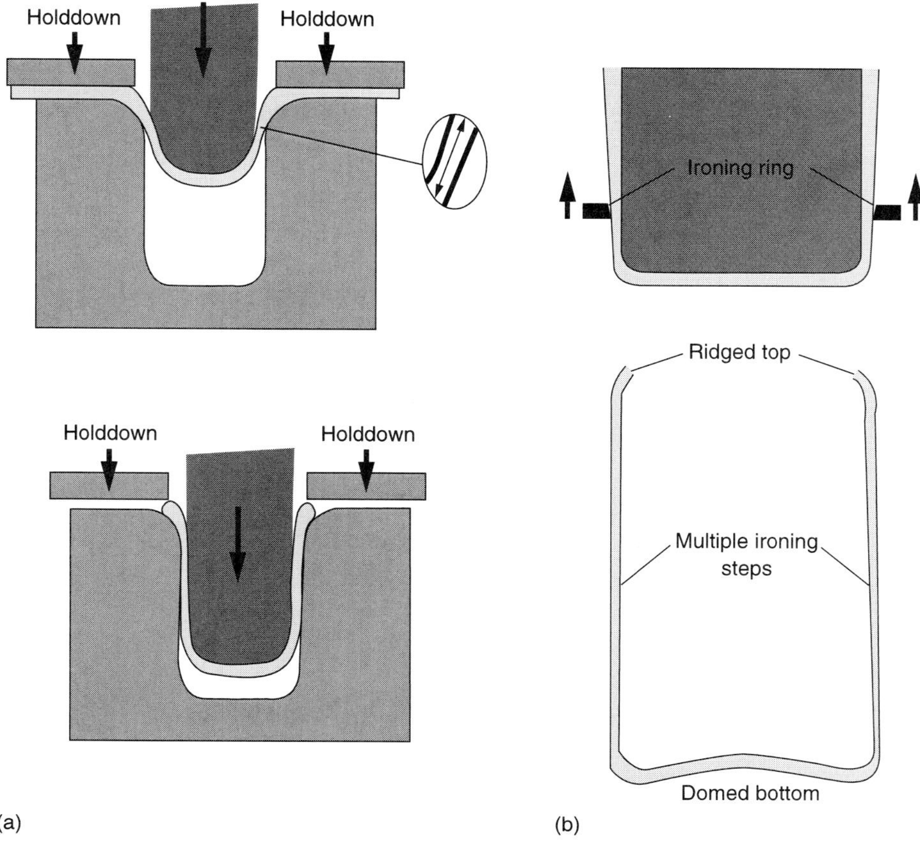

FIG. 10.27 The steps involved in converting flat rolled sheet into a beverage can include (a) deep drawing and (b) ironing. The holddown pressures shown in the deep drawing operations (a) allow slippage of the material as it enters the die cavity. As shown in the small inset of the die wall, the deformation is nearly plane strain with compressive plastic strains occurring in the circumferential or hoop direction. The ironing ring (b) is passed along the exterior of the can while it is supported internally by the internal die.

The *formability* of the material for the deep drawing process is strongly dependent on the yield behavior, strain hardening rate, and plastic anisotropy of the can stock. To express the formability for the cupping operation in deep drawing, a term called the drawing ratio is employed. The drawing ratio is the ratio of the diameter of the initial flat circular blank to the diameter of the cup being formed. The *limiting drawing ratio* (LDR) is the largest such ratio that can be drawn for a given stock. A simple approximation of this limit can be written as

$$\mathrm{LDR} \approx \exp\left(\frac{3}{4}\sqrt{\frac{(R+1)}{2}}\right) \tag{10.24}$$

where R is the plastic anisotropy factor from Eq. 3.34. Remember that when R is greater than 1, the material resists thinning during deformation. By resisting thinning, necking or fracture of the material can be suppressed. If there is significant ear formation, the LDR, which is typically slightly greater than 2, will be reduced.

10.6 FORMING AND RHEOLOGY OF GLASSES AND POLYMERS

The high-temperature forming of glasses and polymers is strongly dependent on the rate dependence and yield behavior of these materials. Soon after the development of polyethylenes in the 1930s, it was clear some type of simple grading process was necessary to predict flow behavior. Because most flowing polymers are very definitely non-Newtonian in their fluid flow behavior, an understanding of what was documented as strain rate sensitivity in Chapter 6 would be helpful, but testing of individual batches of polymer in the methods described in Chapter 6 is impractical. When approached from the direction of fluid flow, the field that investigates the flow of fluids and fluidlike materials is called rheology (see Carreau, DeKee, and Chabra, 1997). Most rheology investigations also involve complex evaluations of properties that are impractical for routine quality control. A simple test is called the melt flow index (MFI) (Fig. 10.28). Polymer is placed in a heated cylinder and then a known mass is placed atop the softened polymer. The softened polymer flows through a die of known length and diameter. Once the flow has reached steady-state conditions, the mass flowing through the die orifice in a fixed time is measured. Then the MFI is reported as mass per unit time—e.g., grams per 10 minutes.

Although the MFI is simple and provides one number for comparison, it has poor repeatability and a very narrow range of viscosities are applicable for a given geometry and mass. Otherwise, low-viscosity materials extrude very quickly and the measurement may require extrapolation to longer times. With high-viscosity materials, the low throughput results in a possibility of measurement error. The MFI does not describe details of the flow behavior that might be important for a given forming operation, which may require a certain rate of flow for the material or variable flow rates during the operation. Our earlier discussions of viscosity versus strain rate raised the possibility of a nonlinear relationship but did not discuss the implications for forming operations.

If we consider Eq. 3.43, we can write an expression for non-Newtonian shear viscosity as a power law, yielding

$$\eta = \frac{\tau}{\dot{\gamma}^m} \tag{10.25}$$

FIG. 10.28 Melt flow index is measured by placing a weight on a heated polymer and measuring the mass flow rate through a die of specific diameter and length.

material. Polymer melts exiting thin capillaries often expand significantly at the exit point, showing the dilation response following a reduction in hydrostatic pressure. The film blowing processes that are used to produce the thin plastic membranes used for garbage and food storage containers are an extreme example of extensional flow. The following example shows a more moderate strain level used for producing soft drink containers.

EXAMPLE 10.5 *Stretch Blow Molding*

The effect of stretch blow molding (SBM) shown in Fig. 10.30 on the production of polyethylene terephthalate (PET) containers was introduced at the end of Chapter 5. Figure 5.24*a* showed that the biaxial stretching of the polymer enables a strengthening derived from the orientation of the polymer chains. Then, Fig. 5.24*b* showed that even in uniaxial deformation the orientation strengthening in this material is accompanied by in increase in density. Clearly, SBM of PET is an example of a yield phenomenon that involves a change in volume, although in a different sense than for the examples at the end of Chapter 3. This unusual behavior has also led to the incredible success of carbonated beverage containers made from PET since it imparts a very low permeability of CO_2 through the container walls. The PET material used in the storage of carbonated beverages must be attractive and able to retain 85 percent of the CO_2 content for 4 months at room temperature, survive hot automobiles undergoing solar heating, and survive a drop of 2 meters.

The SBM process consists of heating a preform called a parison, which looks like a clear plastic test tube, to a temperature of about 100°C. The parison is expanded internally by gas pressure to produce a biaxial stretching of the parison walls. When the expanding part reaches the walls of a vented mold to form the desired shape, it is cooled until it is rigid enough to be ejected, and the process can begin again. ■

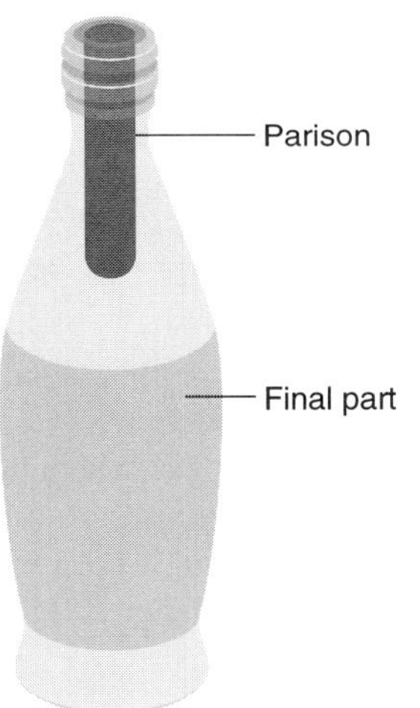

FIG. 10.30 The parison is a test-tube-shaped preform that is expanded by internal gas pressure to fill a vented mold, producing the final part shape.

BIOGRAPHY

NATHANIEL WYETH (1911–1990)

Wyeth and his company, Dupont, filed the patent for the PET soda bottle in 1973. Wyeth reported that they made thousands of attempts before they got the "truly beautiful" PET bottles he was seeking. By the end of that decade, 2-liter PET bottles were rapidly replacing the returnable glass bottles that previously filled store shelves. Although many other polymers could hold carbonated beverages, retention of CO_2 was key to the success of PET.

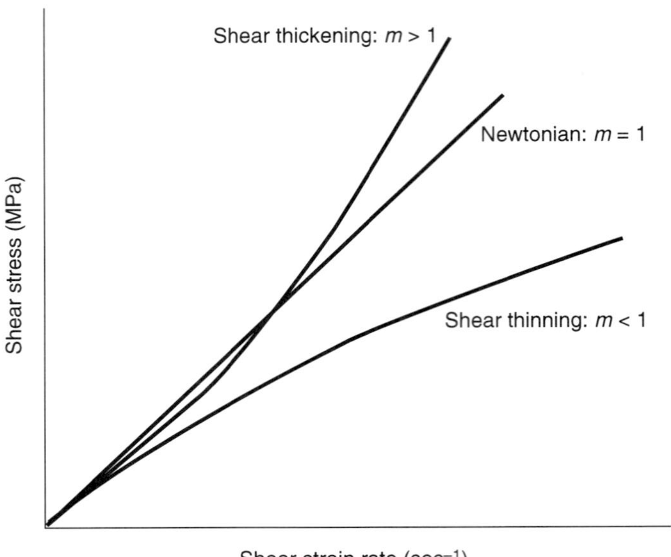

FIG. 10.29 Shear viscosity versus shear rate for various values of strain rate sensitivity.

where η is dependent on the shear rate. This relation enables us to consider directly the shear viscosities of polymers for different levels of strain rate sensitivity.

For $m < 1$, the material undergoes shear thinning behavior; for $m = 1$, we have a Newtonian response; and for $m > 1$, we have shear thickening. A diagram showing these three possibilities is presented in Fig. 10.29.

For most polymers undergoing forming operations at elevated temperatures, shear thinning occurs such that the more rapid the deformation the easier the polymers deform. This is the same behavior designed into the slurries used for wall paints. For paints, the high rate of spreading the slurry allows for easy coverage, but because low flow rates have a much more rigid response, the paint will not easily drip off the wall. Toothpastes, chocolate sauces, and facial creams have values of $m = 0.3$ to 0.5.

For filled polymer melts (polymer matrix particulate composites) with high composite reinforcement additions and liquid-particulate suspensions (colloids or slurries) with high particle loadings, the behavior can be shear thickening rather than shear thinning. In this type of material, attempts to deform it at higher rates are met with higher resistance.

If we measure the MFI value at two different loads, $m_1 < m_2$, we can see whether or not a material is shear thinning. We can take a ratio called the melt flow ratio (MFR) as

$$\text{MFR} = \frac{\text{MFI}_{m_1}}{\text{MFI}_{m_2}} \tag{10.26}$$

If the MFR is less than 1, the material clearly must be shear thinning. The separate MFI values do not give insight into the flow behavior at the higher flow rates typically employed in a forming process. Two materials can have the same viscosity at a given value of shear rate but very different behaviors for lower and higher rates. The MFI does not provide enough information to predict forming response.

Although a great deal of focus is typically placed on the shear flow of polymers, many of the processes used for shaping of polymers involve significant extensional (tensile) deformation. If the material is affected by dilatant yield behavior—that is, if the mean stress is an important component of the yield criteria—then the pressures involved in passing materials through various dies at high rates will be affected by the compressibility of the

EXPERIMENT: *Shrinking PET Bottles*

If you want to see how the stretch can be removed from a PET bottle, place the bottle on a piece of aluminum foil covering a cookie sheet and heat it in an oven to 300°F (150°C) after removing any paper label or glue that may be attached. Be sure it is not possible for the bottle to tip over and fall to the bottom of the oven (where it could catch fire). In a few minutes, you should see a bottle that has lost more than half of its volume.

10.7 TAPE CASTING OF CERAMIC SLURRIES

Just as rolling is a critical technology for producing large volumes of thin stock for the metallurgy industry, tape casting is one of the only viable ways to produce stand-alone ceramic materials as thin sheets. The production of multilayer capacitor stacks and any other process that requires a source of thin ceramic sheets rely on the slurry forming operation used in tape casting. Figure 10.31 depicts the process used in tape casting. A wide opening of controlled dimensions serves as an exit for the slurry from the reservoir. Although laboratory processes may use just a filled container, large-scale industrial processes use pumped supplies to maintain the slurry level in the reservoir.

The height and shape of the *doctor blade* help control the size of the slurry film that is formed on the typically polymeric carrying medium. However, the rate of motion of this carrying medium and its interfacial properties (interfacial energy with the slurry) also affect the thickness found soon after the doctor blade. An additional effect on the final thickness is any creep of the tape resulting from viscous flow and the amount of ceramic powder within the slurry system. As the powder dries, the loss of fluid content results in thinning of the tape.

Producing an appropriate size tape with a uniform thickness can be very complex and often requires significant development for any new combination of fluid, chemicals to enhance the colloidal or suspension characteristics, and ceramic powder. For the powder, the amount of solid volume per unit volume of the slurry (solids loading) has a strong influence on production.

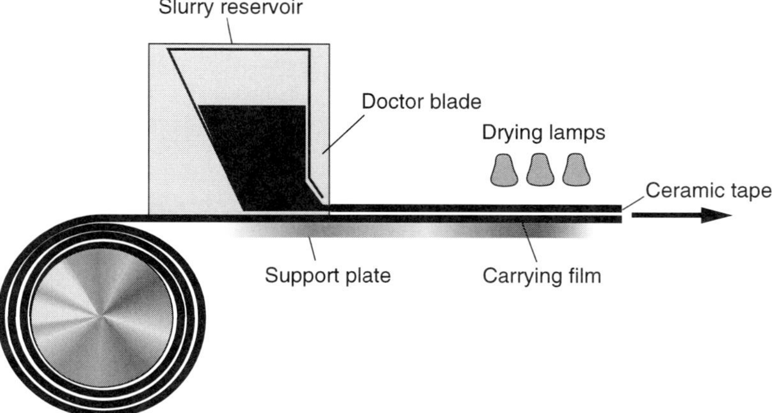

FIG. 10.31 Tape casting of ceramic slurries. For most industrial tape casting operations, the exit point of tape from the doctor blade is nearly simple shear with sticking friction between the tape and the carrying film.

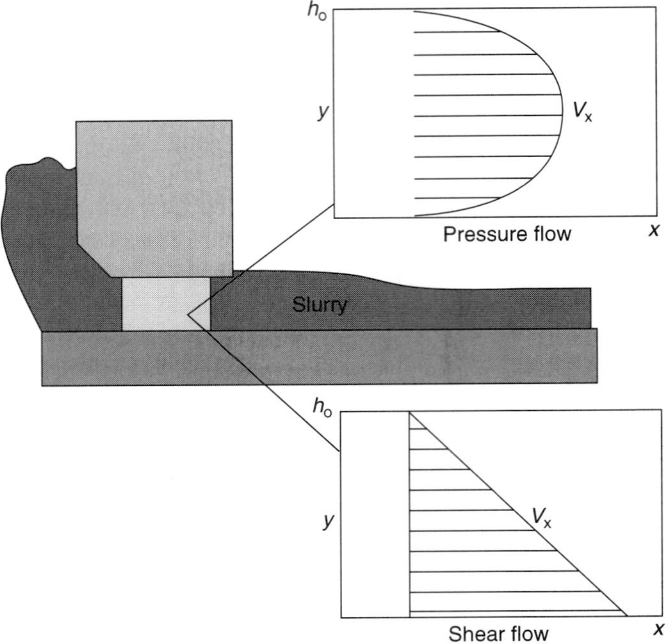

FIG. 10.32 The combined flow that could be expected between the doctor blade and the polymer carrying film for a tape cast slurry. The insets for pressure flow and shear flow show the profile of velocities expected for each type of flow.

The flow process for tape casting is a combination of pressure-driven flow and drag if we have the simple doctor blade geometry shown in Fig. 10.32.

Chou, Ko, and Yan (1987) have used this combined approach to model the expected flow of a tape cast slurry under these conditions. For the pressure flow, the velocity profile can be described as a function of position y

$$v_x = \frac{h_o P y}{2\eta L}\left(1 - \frac{y}{h_o}\right) \tag{10.27}$$

where h_o and L are given in Fig. 10.32, P is the pressure at $x = 0$ (which drops to zero at L), and η is the presumed Newtonian viscosity. For the shear flow (also called Couette flow), the velocity profile is simply

$$v_x = V\left(1 - \frac{y}{h_o}\right) \tag{10.28}$$

where V is the velocity of the moving carrying film. The respective volumetric flow rates can also be written so that we have for the pressure flow

$$Q_P = \frac{h_o^3 W P}{12\eta L} \tag{10.29}$$

and for shear flow

$$Q_s = \frac{1}{2} h_o W V \tag{10.30}$$

where W is the width of the tape.

If we assume that the Newtonian behavior allows the two velocities to be additive, we can sum them and convert from volumetric to mass flow with

$$Q_{\text{mass}} = \rho Q_{\text{volume}} = \frac{1}{2} h_o \rho W \left(V - \frac{h_o^2 P}{6\eta L} \right) \tag{10.31}$$

where ρ is the slurry density in the slurry reservoir. This enables us to predict the thickness of the tape before any spreading or drying takes place. To do so, we must assume that

$$Q_{\text{mass}} = \rho Q_{\text{volume o}} WV \tag{10.32}$$

where ρ_{exit} is the density after the pressure is relieved. At the exit, L, the in and out flow rates must be equal, so we can write for the thickness

$$\delta = \frac{1}{2} \frac{\rho_r}{\rho_{\text{exit}}} h_o \left(1 - \frac{h_o^2 P}{6\eta VL} \right) \tag{10.33}$$

Using Eq. 10.33, we can predict the tape thickness. Chou et al. (1987) have also shown that with corrections for sideways flow and allowing for drying, reasonable predictions of final tape dimensions can be made.

10.8 REFERENCES

B. AVITZUR, *Ann. Rev. Mat. Sci.*, **7**, 261–300, 1977.

W. A. BACKOFEN, *Deformation Processing,* Addison-Wesley, 1972.

P. J. CARREAU, D. C. R. DEKEE, AND R. P. CHABRA, *Rheology of Polymeric Systems: Principles and Applications,* Hanser-Gardner, 1997.

YE T. CHOU, YA T. KO, AND MAN F. YAN, *J. Am. Ceram. Soc.*, C-280–282, 1987.

L. COFFIN AND H. C. ROGERS, *Trans. ASM*, **60**, 672–686, 1967.

R. HILL, *The Mathematical Theory of Plasticity,* Oxford, 1950.

W. HOSFORD AND R, CADDELL, *Metal Forming: Mechanics and Metallurgy,* 2nd Ed., Prentice-Hall, 1993.

B. B. HUNDY AND A. R. E. SINGER, *J. Inst. Metals,* **83**, 401–407, 1954–1955.

W. JOHNSON AND P. B. MELLOR, *Plasticity for Mechanical Engineers,* Van Nostrand, 1962.

CH. W. MACOSKO, *Rheology: Principles, Measurements, and Applications,* Wiley, 1994.

R. WAGONER AND J. CHENOT, *Fundamentals of Metal Forming,* Wiley, 1997.

10.9 PROBLEMS

A.10.1 Describe all of the possible strategies for decreasing the thickness of the material that can be rolled in the presence of roll flattening. In each case, discuss the practicality of making the suggested changes.

A.10.2 Your supplier has stated that they have a new version of your alloy with a lower strain hardening coefficient. Because you later anneal the material anyway, the final hardness and yield strength after rolling are not very important. How would a reduced strain hardening rate change the roll pressure?

A.10.3 Calculate the ideal stress given the following values for drawing.

(a) $K = 500$ MPa, $n = 0.25$, $d_i = 2$ mm, $d_o = 1.5$ mm

(b) $K = 500$ MPa, $n = 0.2$, $d_i = 2$ mm, $d_o = 1.5$ mm

(c) $K = 500$ MPa, $n = 0.25$, $d_i = 2$ mm, $d_o = 1.2$ mm

(d) $K = 500$ MPa, $n = 0.2$, $d_i = 2$ mm, $d_o = 1.2$ mm

B.10.1 Calculate the value of L for a roll of radius 25 cm, an initial sheet thickness of 14 cm, and a final sheet thickness of 10 cm. Write a strain tensor that describes the plastic strains resulting from rolling of this material and then calculate the effective strain corresponding to this rolling reduction (see Chapter 3).

B.10.2 An existing specification at your company states that the maximum predicted roll pressure can be just 20 percent of the flow strength of the rolled product. For $h = 10$ mm and $L = 5$ mm, what is the maximum friction coefficient that can be allowed? If the friction coefficient could not be changed to produce the desired reduction, what other approach could you use to reduce the maximum pressure?

B.10.3 How would you expect forward tension to change the neutral point in rolling?

B.10.4 How would you expect backward tension to change the neutral point in rolling?

B.10.5 The deflection of uncambered rolls can result in residual stresses. Describe these residual stresses and why the variable deformation geometry with width could lead to their occurrence.

B.10.6 Overcambered rolls can lead to edge cracking that results from residual stresses. Why would overcamber result in this type of residual stresses?

B.10.7 Plot $P/2k$ versus position for $\mu = 0.1$, a contact length of 5 cm, and a height of 2 cm for a friction hill in plane strain compression.

B.10.8 Find the average contact pressure that applies to sticking friction geometrically for plane strain compression.

C.10.1 **(a)** Plot Eq. 10.1 for $L = 10$ cm and $h = 20$ cm for a material with $K = 500$ MPa and a strain hardening exponent of $n = 0.35$ in using Eq. 3.19 at four levels of friction, $\mu_F = 0.2$, 0.3, 0.4, and 0.5.

(b) Describe how the friction hill would change with increased reduction using the same rolls.

(c) If you did not use the average strength, but instead took into account the continuous hardening of the material as it proceeds through the rolls, how would you expect the friction hill to change? (Sketch your answer.)

C.10.2 Plot $P/2k$ in Eq. 10.9 for $\mu_F = 0.1$, $h = 3$ cm, and $L = 4$ cm. Find out how well the approximation

$$\frac{P_{\text{average}}}{2k} = 1 + \frac{\mu_F L}{2h}$$

applies to the numerical average given in your plot.

C.10.3 Plot P in Eq. 10.11 for a yield stress of 200 MPa, $\mu = 0.1$, $h = 3$ cm, and $r_c = 1$ cm. Find out how well the approximation

$$P_{\text{average}} = Y\left(1 + \frac{1}{3}\left(\frac{2\mu_F r_c}{h}\right) + \frac{1}{12}\left(\frac{2\mu_F r_c}{h}\right)^2 \cdots\right)$$

applies to the numerical average given in your plot.

C.10.4 If sticking friction prevails for direct compression of a cylinder, the friction term is $2k\rho d\rho$ and the applied stress produces the term $(-h\rho dP)$. Equate these two terms and integrate to find the expression equivalent to Eq. 10.11.

C.10.5 Calculate the solution $P_e/2k$ for the extrusion geometry shown in the figure and compare the solution with that shown in Fig. 10.17.

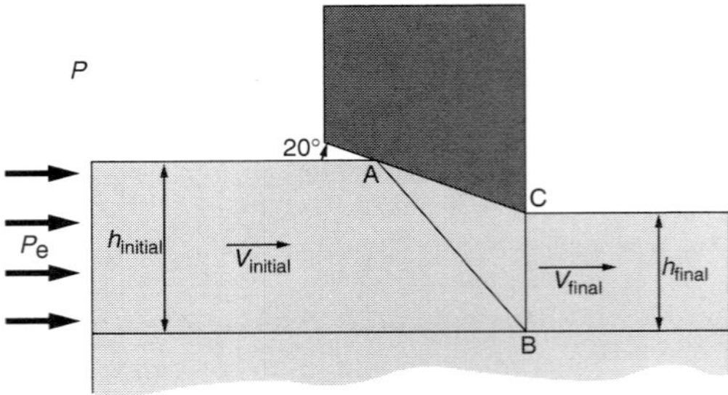

C.10.6 Calculate the solution $P_p/2k$ using a hodograph for the indentation geometry shown in the figure and compare the solution with that shown in Fig. 10.19.

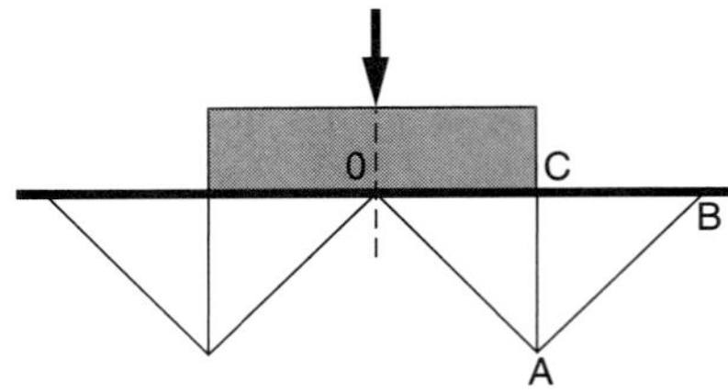

C.10.7 Construct a hodograph with curved sections for Fig. 10.23.

C.10.8 Calculate the slip line field solution for the indentation pressure for the diagram in Fig. 10.24 and compare the result with that found in Fig. 10.19.

C.10.9 The velocity profiles for pressure and shear flow are stated as being additive. Add the two velocity profiles together and show how the shape would change with viscosity. What effect would both types of non-Newtonian flow have on this flow behavior, and how would it affect tape casting?

C.10.10 Using a reservoir density of 4, an exit density of 3.8 gm/cm^3, a height of 250 μm, a pressure of 0.15 MPa, and a length of 2 cm, calculate the tape thickness versus velocity using a carrier velocity range of 2 to 10 cm per second for viscosities of 10, 1.0, and 0.1 Pa/sec. Describe how this process could go awry at very low velocities.

C.10.11 Calculate the Δ values for each of the reductions in Fig. 10.8*a*. Plot the difference in hardness (maximum-minimum) given for each reduction versus the Δ values. Explain the observed relation.

INDEX

H

I

J

K

L

M

N